WARRIOR'S WORDS

A QUOTATION BOOK

WARRIOR'S WORDS

A QUOTATION BOOK

From Sesostris III to Schwarzkopf
1871 BC to AD 1991

PETER G. TSOURAS

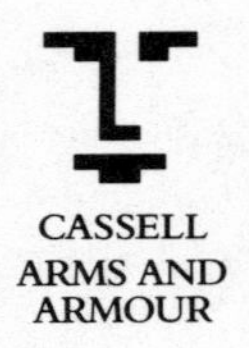

CASSELL
ARMS AND
ARMOUR

For my wonderful grandfather
Sergeant Petros Giorgos Tsouras
Royal Hellenic Army
who fought for the *Megali Idea*
1912–1913

Arms and Armour Press
A Cassell Imprint
Villiers House, 41–47 Strand, London WC2N 5JE.

Distributed in the USA by Sterling Publishing Co. Inc.,
387 Park Avenue South, New York, NY 10016-8810.

Distributed in Australia by Capricorn Link (Australia) Pty. Ltd.,
P.O. Box 665, Lane Cove, New South Wales 2066.

British Library Cataloguing in Publication Data
Tsouras, Peter G.
Warriors' words: a quotation book.
I. Title
828.02
ISBN 1-85409-088-7

Designed and edited by DAG Publications Ltd.
Designed by David Gibbons; edited by Michael Boxall;
typeset by Ronset Typesetters, Darwen, Lancashire;
printed and bound in Great Britain by Hartnolls Ltd, Bodmin.

Acknowledgements
The Publishers are grateful for the permissions granted
to use quotations in this book.

Contents

S

T

U–V

W–Y

Acknowledgments

Without the cheerful and efficient help of six librarians, whose profession is the unsung guardian of our heritage, this book could never have been prepared. My special thanks to Debbie Brannan, Meg Nicholas, Carol Norton, Anita Parrin, Holly Wilson, and Carol Wong for their diligence, good will and friendship. However, this invaluable assistance would have been for nought if my wife, Patricia, had not had the patience to help me with the mysteries of word processing, much of which will bewilders this computer coward more at home with reed pens and papyrus. Not the least proof of her forbearance and affection included sharing the family computer when she had equally pressing deadlines. My thanks also to Dr. John Turner for his help in suggesting sources for the ancient Middle East and to Bruce W. Watson, my friend and mentor.

For my colleagues and friends during Operations 'Desert Shield' and 'Desert Storm', I offer my salute and treasure my memories – to James Tate III, for his wisdom, friendship, and integrity; for 'Mo' – Elmo C. Wright, Jr. – comrade and companion who trudged the endless corridors of the Pentagon with me to deliver countless 'special' briefings; to Colonel James Solomon (Commander, US Army Intelligence and Threat Analysis Center), Colonel James Pardew (Chief, Foreign Intelligence, The Army Staff) and Brigadier General John Stewart (Commander, US Army Intelligence Agency and later J2, Third US Army) – the right men at the right time in the right place – whose revolution in intelligence support for the US Army prepared the way for victory. Another thank-you is in order for Colonel Pardew, whose good-natured appreciation for the right quote inspired me to finish this book.

Not least, I wish to thank my publisher for his calm and encouragement when the preparation of this book was interrupted by Saddam Hussein's recent thrashing in the Gulf.

P.G.T.

Preface

IN JUNE 1758, the British Army was laying siege for the second time in that century to the great French fortress of Louisburg, the iron gate of Canada. Then Brigadier General James Wolfe hurried the French defenders back into their works with his skilful use of his light infantry to turn one flank then another. The rest of the British force was no less impressed with the maneuver than the dazed French now shut up in Louisburg. A brother officer, more perceptive than the rest, remarked that Wolfe's tactics reminded him of an account by Xenophon in the *Anabasis*, the March of the Ten Thousand Greeks. It is not recorded if Wolfe smiled. If he did it was probably out of exasperation as he replied, 'I had it from Xenophon, but our friends here are astonished at what I have done because they have read nothing.'

It is my hope that this book of quotations will help the junior officer, the military professional, and the military enthusiast in general also to learn 'from Xenophon', and the more than 250 other soldiers and others quoted here in the almost four thousand years from Pharaoh Sesostris (Senruset) III *c.* 1871BC to General H. Norman Schwarzkopf in AD 1991. Here is much of the distilled experience of almost four thousand years of the profession of arms. This book, however, is not the easy one-stop answer to the art of war. It is a place to start.

I have chosen almost 350 subject areas for quotation. Basic terms such as helmet and rifle I have left to elementary experience and a general dictionary. Subject headings address concepts that are the essence of the profession of arms, the realm of ideas rather than things, e.g, Change, The Indirect Approach, and War itself, subdivided into twelve sub-categories.

The sources are a fraternity of soldiers, first and foremost. Most of them have long since turned to dust, but their deeds and experiences remain as vital and as insightful as when they trod the earth with thunder and acclaim. Follow them to their original sources, the histories, memoirs, textbooks, and works of theory. Open an old book that smells, feels and looks of another time. Be your own Odysseus when he summoned the ghosts of Achilles and Agamemnon from Hades to speak to him. If you let them, they will speak across time to you of problems of leadership and command, of logistics, of tactics and strategy, the sinews of their art. They will also speak to you of the human heart, comradeship, example, sacrifice, glory and death. They are not dead as long as they are remembered. And they are exhilarating company. If you are a serving soldier, the reward is not merely the professional satisfaction of a good read. It can also be as Wolfe so aptly stated, the road to success and at the price of a small butcher's bill.

This exercise of the profession of arms is the basic requirement for inclusion in this book of military quotations, unlike other such collections. Of necessity this fraternity is further limited to those military men who either had the talent, time and inclination to write or were lucky enough to have friend or military secretary who was inspired to record events and words. The hazards of time have also restricted this company. While Alexander the Great is quoted several times, it is nothing that could have been included if his *Royal Journal* and Ptolemy's history or any other first-hand source had survived. A student of Aristotle undoubtedly had much worth recording. Unfortunately, accident or scholars more interested in literary merit and moral lessons and not students of war preserved what little survived the destruction of the classical world, the destruction of the Library of Alexandria and later the burning of the libraries of Constantinople by French barbarians in 1204. Shaka Zulu is denied the place his military genius deserves simply by his membership of a pre-literate society. Others such as Tamerlane are omitted for lack of an English translation. A Dari language *Maxims of Tamerlane* may be in the process of translation at this time in the Washington area and would find an honored place in a later edition of this book. Recent modern scholarship has brought to English significant works by Byzantine soldiers, chief among whom is George Dennis' translation of a military classic, *The Strategikon* (1984) by the Emperor Maurice as well as his *Three Byzantine Military Treatises* (1985). Similarly, Thomas Cleary has brought to English-speaking readers the works of the great Chinese captains, Zhuge Liang and Liu Ji in *Mastering the Art of War* (1989) and William Wilson those of medieval Japanese soldiers in *Ideals of the Samurai* (1982). Although three English language biographies of the great Russian captain Aleksandr Suvorov have been written, there is a need for a military life of Suvorov. Nevertheless, he is well represented here if only from gleaning his sayings from the Soviet publications I peruse as a Soviet area analyst.

While often insightful and useful, the words of priest, scholar, diplomat and poet are no more appropriate than a lawyer's comments in a textbook on surgery even if the lawyer represents the surgeon for malpractice. There is no substitute for wielding either scalpel or sword in medicine and war. Of course, I have exercised the editor's prerogative of being inconsistent here and there to please myself and, it is to be hoped, the reader. Some opportunities are too good to miss. For example, Ralph Waldo Emerson – 'I hate quotations; tell me what you know.'

Homer finds an especially honored place in this Valhalla of paper, ink and cloth. In *The Iliad* I am convinced you hear many of the lines composed by the Achaean bards who accompanied their royal masters to Troy to record their glories, much as the Norse bards went aviking with their royal patrons. The purpose was both informative and entertaining however much they composed to royal taste. The modern military correspondent does much the same, though most soldiers consider his an odious profession, somewhere between pest and spy. The exceptions are like Homer, they put an unerring finger on the pulse of man in war, men like Kipling, the young Churchill, Konstantin Simonov and Ernie Pyle who were ever welcome in the mess.

I admit one more category to the short list of exceptions, mostly from antiquity – the military secretary and assistant such as Polybius, friend and advisor to Scipio Africanus the Younger; and Procopius, military secretary to Count Belisarius. To their first-hand observations we owe the only accounts of critical moments of history. On the other side of the world are the numerous scholarly commentators to Sun Tzu's *The Art of War*, whose words indicate that they did more than wet a writing brush.

Most soldiers quoted herein wrote at the end of their careers. Others were more prolific throughout military service and attempting to determine rank at the moment of literary conceptions presents an endless number of Gordian Knots. I have adopted an Alexandrine solution by cutting through to only the highest rank held. A few in their lifetimes wore the uniform only briefly, however gallantly, and are cited, for example, as

Justice Oliver Wendell Holmes, President Theodore Roosevelt, Sir Winston Churchill, and President John F. Kennedy. By common usage, soldiers in antiquity are not referred to by rank, and some great commanders are referred to by their titles alone such as the great English dukes, Marlborough and Wellington, whose reputations outshone any insignia of military rank. Of course, Napoleon, although officially Napoleon I, Emperor of the French, is simply Napoleon. That name alone kept the whole of Europe in terror for almost twenty years. Titles of nobility are also included. Sometimes they were a delight – in those cases where they were earned – such as Marshal of France Michel Ney, Duc d'Elchingen and Prince de la Moskova! I have also attempted wherever possible to give the source and date of the quotation. Unfortunately this has not always been possible; in such cases I have cited the birth and death dates of the originator.

Where there are a number of translations of works, mostly those from antiquity, I have freely chosen among them for those that best express the particular theme. The numerous versions of Napoleon's military maxims are a more recent example. I have also relied extensively on Samuel Griffith's translation of Sun Tzu's *The Art of War* for its vigor of expression despite all its recent competitors. For Homer, I find Robert Fagle's new translation of *The Iliad* consistently to be the best and with a few exceptions have relied on him alone. The delight of Fagle's translation is its fidelity to Homer's medium – the spoken not the written word. I cannot recommend it too highly.

> Rage – Goddess, sing the rage of Peleus' son Achilles,
> murderous, doomed, that cost the Achaeans countless losses,
> hurling down to the House of Death so many sturdy souls,
> great fighters' souls, but made their bodies carrion,
> feasts for the dogs and birds,
> and the will of Zeus was moving toward its end.
> Begin, muse, when the two first broke and clashed,
> Agamemnon lord of men and brilliant Achilles.

They are all there listening to him still in the great feast hall of Valhalla. At one table fire-eaters like Suvorov and Patton. At another, are the scholar-soldiers – Sun Tzu, Clausewitz, Fuller, and Liddell-Hart. Over there the mighty men of war, flinty commanders of great hosts – Scipio, Marlborough, and Montgomery. By them, the knightly warriors '*sans peur et sans reproche*' – Belisarius, Saladin, Bayard, and Lee. In the corner huddled together are the artists of the subtle strategem – Phillip II, Hannibal, and Machiavelli. Over there are those in love with glory – Ramses II, young again and boasting again of his courage at Kadesh to a smiling Alexander, shining in the midst of everlasting renown.

Peter G. Tsouras
Lieutenant Colonel, USAR

Alexandria, Virginia, 1992

Quotations

ABILITY

Win with ability, not with numbers.

Field Marshal Prince Aleksandr V. Suvorov, quoted in Danchenko and Vydrin, *Military Pedagogy*, 1973

Let no man be so rash as to suppose that, in donning a general's uniform, he is forthwith competent to perform a general's functions; as reasonably might he assume that in putting on the robes of a judge he was ready to decide any point of law . . .

Dennis Hart Mahan, 1864, in Dupuy, *The Military Heritage of America*, 1984

Unquestionably, victory or defeat in war is determined mainly by the military, political, economic and natural conditions on both sides. But not by these alone. It is also determined by each side's subjective ability in directing the war. In his endeavour to win a war, a military strategist cannot overstep the limitations imposed by the material conditions; within these limitations, however, he can and must strive for victory. The state of action for a military strategist is built upon objective material conditions, but on that stage he can direct the performance of many a drama, full of sound and colour, power and grandeur.

Mao Tse-tung, 'Problems of Directing China's Revolutionary War', December 1936, *Selected Works*

This is a long tough road we have to travel. The men that can do things are going to be sought out just as surely as the sun rises in the morning. Fake reputations, habits of glib and clever speech, and glittering surface performance are going to be discovered . . .

General of the Army Dwight D. Eisenhower, *At Ease: Stories I Tell My Friends*, 1967

ACTION

When people are entering upon a war they do things the wrong way round. Action comes first, and it is only when they have already suffered that they begin to think.

Thucydides, *History of the Peloponnesian War*, c. 404 BC

It is even better to act quickly and err than to hesitate until the time of action is past.

Major General Carl von Clausewitz, *On War*, 1832, tr. Howard and Paret

---

Action in war is like movement in a resistant element. Just as the simplest and most natural of movements, walking, cannot easily be performed in water, so in war it is difficult for normal efforts to achieve even moderate results. A genuine theorist is like a swimming teacher, who makes his pupils practise motions on land that are meant to be performed in water. To those who are not thinking of swimming the motions will appear grotesque and exaggerated. By the same token, theorists who have never swum, or who have not learned to generalize from experience, are impractical and even ridiculous: They teach only what is already common knowledge: how to walk.

Major General Carl von Clausewitz, *On War*, 1832, tr. Howard and Paret

---

There is always hazard in military movements, but we must decide between the possible loss of inaction and the risk of action.

General Robert E. Lee, quoted in Fuller, *Grant and Lee*, 1933

And perhaps the most influential reflection of all: 'In war it is often less important what one does than how one does it. Strong determination and perseverance in carrying through a simple idea are the surest routes to one's objective.'

Field Marshal Helmuth Graf von Moltke, quoted in Kessel, *Moltke*, 1957

In tactics, *action* is the *governing rule* of war.

Marshal of France Ferdinand Foch, *Precepts and Judgements*, 1919

The essential thing is action. Action has three stages: the decision born of thought, the order or preparation for execution, and the execution itself. All three stages are governed by the will. The will is rooted in character, and for the man of action character is of more critical importance than intellect. Intellect without will is worthless, will without intellect is dangerous.

Colonel General Hans von Seekt, *Thoughts of a Soldier*, 1930

When faced with the challenge of events, the man of character has recourse to himself. His instinctive response is to leave his mark on action, to take responsibility for it, to make it *his own business*. Far from seeking shelter behind his professional superiors, taking refuge in textbooks, or making the regulations bear the responsibility for any decision he may make, he sets his shoulders, takes a firm stand, and looks the problem straight in the face. It is not that he wishes to turn a blind eye to orders, or to sweep aside advice, but only that he is passionately anxious to exert his own will, to make up his own mind. It is not that he is unaware of the risks involved, or careless of consequences, but that he takes their measure honestly, and frankly accepts them. Better still, he embraces action with the pride of a master; for if he takes a hand in it, it will become his, and he is ready to enjoy success on condition that it is really *his own*, and that he derives no profit from it. He is equally prepared to bear the weight of failure, though not without a bitter sense of satisfaction. In short, a fighter who finds within himself all the zest and support he needs, a gambler more intent on success than profits, a man who pays his debts with his own money lends nobility to action. Without him there is but the dreary task of the slave; thanks to him, it becomes the divine sport of the hero.

Charles de Gaulle, *The Edge of the Sword*, 1932

As soon as a soldier does something, he becomes master of the situation . . . When men have been on the defensive for a long time, [they] send out patrols even if there be no special reason for patrols. The patrols instill a sense of self-confidence and superiority. Inactivity and waiting undermine morale and rub nerves raw.

Captain Adolf von Schell, lecture at the US Army Infantry School, in Marshal, *Infantry in Battle*, 1939

The relief which normally follows upon action after a long period of tension – that vast human sigh of relief – is one of the most recurrent phenomena of history, marking the onset of every great conflict. The urge to gain release from tension by action is a precipitating cause of war.

Captain Sir Basil Liddell Hart, *Thoughts on War*, 1944

Action is necessary to maintain the morale of a military force and no commander can afford to miss an opportunity for his troops to win glory by successful action. The effect is two-headed, it invigorates his forces and discourages the enemy. Final victory cannot be won by hit-and-run tactics. Real victories must be won in main engagements, and he who recognizes an opportunity for such an action should proceed, if necessary, without orders from above or even in the face of contrary orders from a higher command out of touch with the situation.

Lin Piao, 1946, in Ebon, *Lin Piao*

ACTIVITY

When the enemy is at ease, be able to weary him; when well fed, to starve him; when at rest, to make him move.

Appear at places at which he must hasten; move swiftly where he does not expect you.

Sun Tzu, *The Art of War*, c. 500 BC, tr. Griffith

The soldiers must always be doing something, even if no enemy is bothering us. Habitual idleness spells trouble for an army.

The Emperor Maurice, *The Strategikon*, c. AD 600

Once you engage in battle it is inexcusable to display any sloth or hesitation; you must breakfast on the enemy before he dines on you.

Kai Ka'us ibn Iskander, Prince of Gurgan (10th century Persian prince), in Reuben Levy, trans., *A Mirror for Princes, the Qabus nama*, 1082

. . . When you set out to war, be not inactive, depend not upon your captains, nor waste time in drinking, eating, or sleeping. Set the sentries yourselves, and take your rest only after you have posted them at night at every important point about your troops; then take your rest, but arise early. Do not put off your accoutrements without a quick glance about you, for a man may thus perish suddenly through his own carelessness.

Grand Prince Vladimir II Monomakh of Kiev (1053–1125), *The Testament of Vladimir Monomakh*, in Dmytryshyn, *Medieval Russia, A Source Book*, 1967

I am up and about when I am ill, and in the most appalling weather. I am on horseback when other men would be flat out on their beds, complaining. We are made for action, and activity is the sovereign remedy for all physical ills.

Frederick the Great (1712–1786), quoted in Duffy, *Military Life of Frederick the Great*, 1986

. . . what would I do if I were the enemy? What project could I form? Make as many as possible of these projects, examine them all, and above all reflect on the means to avert them. If you find yourself unable put it right at once! Often, through an hour's neglect, an unfortunate delay loses a reputation that has been acquired with a great deal of labour. Always presume that the enemy has dangerous designs and always be forehanded with the remedy.

Frederick the Great, *Instructions to His Generals*, 1747

You will not be in a Worse Situation, nor your arms in less Credit if you should meet with Misfortune than if you were to Remain Inactive.

Major General 'Mad' Anthony Wayne, November 1777, letter to George Washington

---

Accustom yourself to tireless activity . . .

Field Marshal Prince Aleksandr V. Suvorov (1729–1800), quoted in Blease, *Suvorof*, 1920

Activeness is the most important of all attributes of the military . . . Hurry, your excellency! Money is dear; human life still dearer, but time is dearest of all.

Field Marshal Prince Aleksandr V. Suvorov, 1799, during the Italian Campaign, quoted in Savkin, *Basic Principles of Operational Art and Tactics*, 1972

Activity, Activity, Speed! (Activité, Activité, Vitesse!)

Napoleon, 17 April 1809, order to Massena before Eckmühl

The history of warfare so often shows us the very opposite of unceasing progress toward the goal, that it becomes apparent that *immobility* and *inactivity* are the normal *state* of armies in war, and *action is the exception*. This might almost make us doubt the accuracy of our argument. But if this is the burden of much of military history, the most recent series of wars does substantiate the argument. Its validity was demonstrated and its necessity was proved only too plainly by the revolutionary wars. In these wars, and even more in the campaign of Bonaparte, warfare attained the unlimited degree of energy that we consider to be its elementary law. We see it is possible to reach this degree of energy; and if it is possible, it is necessary.

Major General Carl von Clausewitz, *On War*, iii, 1832, tr. Howard and Paret

---

Activity in war is movement in a resistant medium. Just as a man immersed in water is unable to perform with ease and regularity the most natural and simplest movement, that of walking, so in war, with ordinary powers, one cannot even keep the line of mediocrity.

Major General Carl von Clausewitz, *On War*, 1832, tr. Howard and Paret

---

A fundamental principle is never to remain completely passive, but to attack the enemy frontally and from the flanks, even when he is attacking us.

Major General Carl von Clausewitz, *On War*, 1832, tr. Howard and Paret

---

We must make this campaign an exceedingly active one. Only thus can a weaker country cope with a stronger; it must make up in activity what it lacks in strength. A defensive campaign can only be made successful by taking the aggressive at the proper time. Napoleon never waited for his adversary to become fully prepared, but struck him the first blow.

Lieutenant General Thomas J. 'Stonewall' Jackson, letter, April 1863

The most advantageous thing of all is an offensive operation against an opponent who stays where he is. However, an active enemy will not sit with his arms folded but will himself try to mount an attack.

Marshal of the Soviet Union Mikhail N. Tukhachevskiy, 1924, quoted in Simpkin, *Deep Battle*, 1987

It is during peace-time that we prepare for war, and, unless our preparation is systematic, unless it is based on some science of war, whether the one I have outlined or some other, for the ignorant – and all are ignorant who do not co-ordinate knowledge – there is one great maxim which throughout the history of war has more often than not proved successful, and this maxim is: 'When in doubt, hit out.' When the soldier, whether private or general, does not know what to do, he must strike; he must not stand still, for normally it is better to strike and fail than it is to sit still and be thrashed.

Major General J. F. C. Fuller, *Foundations of the Art of War*, 1926

October 31, 1942, in Stalingrad, the decision to counterattack: 'Was it permissible to show the enemy that we were capable only of defence and twisted about on the same spot, like bound rabbits before the hunter charging his rifle?'

Marshal of the Soviet Union Vassili I. Chuikov, 1962, quoted in Leites, *The Soviet Style in War*, 1982

ADAPTABILITY/ADAPTATION

As soon as they made these discoveries the Romans began to copy Greek arms, for this is one of their strong points: no people are more willing to adopt new customs and to emulate what they see is better done by others.

Polybius, *The Rise of The Roman Empire*, c. 125 BC

At the beginning of the war the instruction of the troops was conducted quite on German principles – they had simply translated the German Service Regulations into Japanese, and in the same way they had endeavoured to model the General Staff on German principles.

In this way our German principles for the command and the instruction of the army were tested in the war, and we can be satisfied with the results. By their success the Japanese were 'justified in the trust they had placed in our military system'.

After the war, when I presented my order of recall, I met General Fuji, the Chief of the General Staff of the 1st Army, and told him I was anxious to know what changes in the

Japanese Regulations would be introduced owing to their experiences in the war; he answered: 'So am I. We will wait to see what new Regulations for the Service Germany will issue on the basis of the reports that the officers who have been sent here will make, and we will translate these Regulations as we did the former ones.'

General Max Hoffmann, *War Diaries and Other Papers, Vol. II*, 1929; observation while attached as an observer to the Japanese Army during the Russo–Japanese War, 1904–5.

Victory smiles upon those who anticipate the changes in the character of war, not upon those who wait to adapt themselves after they occur.

General Giulio Douhet, *Command of the Air*, 1921

I have only one merit: I have forgotten what I taught and what I learned.

Marshal of France Ferdinand Foch, quoted in Monteilhet, *Les Institutions Militaires de la France*, 1932

It is also greatly in the commander's own interest to have a personal picture of the front and a clear idea of the problems his subordinates are having to face. It is the only way in which he can keep his ideas permanently up to date and adapted to changing conditions. If he fights his battles as a game of chess, he will become rigidly fixed in academic theory and admiration of his own ideas. Success comes most readily to the commander whose ideas have not been canalised into any one fixed channel, but can develop freely from the conditions around him.

Field Marshal Erwin Rommel, *The Rommel Papers*, 1953

In any problem where an opposing force exists, and cannot be regulated, one must foresee and provide for alternative courses. Adaptability is the law which governs survival in war as in life – war being but a concentrated form of the human struggle against environment.

Captain Sir Basil Liddell Hart, *Strategy*, 1954

ADMINISTRATION

The good condition of my armies comes from the fact that I devote an hour or two every day to them, and when I am sent the returns of my troops and my ships each month, which fills twenty large volumes, I set every other occupation aside to read them in detail in order to discern the difference that exists from one month to another. I take greater pleasure in this reading than a young lady would get from reading a novel.

Napoleon to Joseph, 20 August 1806, *Correspondance*, No. 10672, Vol. XIII, 1858–1870

My Lord,

If I attempted to answer the mass of futile correspondence that surrounds me, I should be debarred from all serious business of campaigning. I must remind your Lordship – for the last time – that so long as I retain an independent position, I shall see that no officer under my Command is debarred, by attending to the futile drivelling of mere quill-driving in your Lordship's office, from attending to his first duty – which is, and always has been, so to train the private men under his command that they may, without question, beat any force opposed to them in the field.

Attributed to the Duke of Wellington, 1810

I never knew what to do with a paper except to put it in a side pocket or pass it to a clerk who understood it better than I did.

General of the Army Ulysses S. Grant (1822–1885), quoted in S. L. A. Marshall, *The Armed Forces Officer*, 1950

There has been a constant struggle on the part of the military element to keep the end – fighting, or readiness to fight – superior to mere administrative considerations . . . The military man, having to do the fighting, considers that the chief necessity; the administrator equally naturally tends to think the smooth running of the machine the most admirable quality.

Rear Admiral Alfred Thayer Mahan, *Naval Administration and Warfare*, 1908

Whatever the system adopted, it must aim above all at perfect efficiency in military action; and the nearer it approaches to this ideal the better it is. It would seem that this is too obvious for mention. It may be for mention; but not for reiteration. The long record of naval history on the side of administration shows a constant predominance of other considerations, and the abiding necessity for insisting in season and out of season, that the one test of naval administration is not the satisfactory or economical working of the office, as such, but the readiness of the navy in all points for war. The one does not exclude the other; but there is between them the relation of greater and lesser.

Rear Admiral Alfred Thayer Mahan, *Naval Administration and Warfare*, 1908

The more I see of war, the more I realize how it all depends on administration and transportation . . . It takes little skill or imagination to see where you would like your forces to be and when; it takes much knowledge and hard work to know where you can place your forces and whether you can maintain them there.

Field Marshal Earl Wavell (1883–1950)

. . . there must be a clear-out, long-term relationship established between operational intentions and administrative resources. Successful administrative planning is dependent on anticipation of requirements.

Field Marshal Viscount Montgomery of Alamein, *Memoirs of Field Marshal Montgomery*, 1958

Success cannot be administered.

Admiral Arleigh Burke, speech, 1962

Paper-work will ruin any military force.

Lieutenant General Lewis B. 'Chesty' Puller, quoted in Davis, *Marine*, 1962

ADVANCEMENT

Hope encourages men to endure and attempt everything; in depriving them of it, or in making it too distant, you deprive them of their very soul. It is essential that the captain should be better paid than the lieutenant, and so for all grades. The poor gentleman should have the moral surety of being able to succeed by his actions and his services. When all these things are taken care of you can maintain the most austere discipline among the troops. Truly, the only good officers are the poor gentlemen who have nothing but their sword and their cape, but it is essential that they should be able to live on their pay.

Field Marshal Maurice Comte de Saxe, *My Reveries*, 1732

My own feeling now, after having been through two world wars, is that an extensive use of weedkiller is needed in the *senior* ranks after a war; this will enable the first class younger officers who have emerged during the war to be moved up.

Field Marshal Viscount Montgomery of Alamein, *Memoirs of Field Marshal Montgomery*, 1958

See also: PROMOTION, COMMAND SELECTION

ADVICE

I believe that a general who receives good advice from a subordinate officer should profit by it. Any patriotic servant of the state should forget himself when in that service, and look only to the interests of the state. In particular, he must not let the source of an idea influence him. Ideas of others can be as valuable as his own and should be judged only by the results they are likely to produce.

Frederick the Great, *General Principles of War*, 1748

There are generals who need no advice; who judge and decide for themselves, and their staffs merely execute orders. Such generals are stars of the first magnitude, appearing only once in century. In most cases, commanders of armies feel the need of advice. This can be given in conferences attended by large or small groups of men whose training and experience qualifies them to give expert judgment. But at the conclusion of such a conference, a single opinion should prevail to facilitate the infinitely more difficult task of deciding to execute the proposed plan. This responsibility rests solely upon the commander.

Field Marshal Helmuth Graf von Moltke, *Italian Campaigns of 1859*, 1904

AGGRESSION

When Assur, the great lord, in order to show to the peoples the immensity of my mighty deeds, extended [*lit.*, made powerful] my kingship over the kings of the four regions [of the world], and made great my name; [when] he caused my hands to bear a stern sceptre, for the annihilation of my foes, the land sinned against Assur, they treated [him] with scorn, they rebelled. To rob, to plunder, to extend the border of Assyria, they [the gods] empowered me [*lit.*, filled my hands]. After Assur and the great gods, my lords, commanded me to march over distant roads, wearying mountains and mighty [desert] sands, thirsty regions, – with trusting heart, I marched in safety.

Esarhaddon (680–669 BC), King of Assyria, quoted in Luckenbill, *Ancient Records of Assyria and Babylonia*, Vol. II, 1926

Men rise from one ambition to another: First, they seek to secure themselves against attack, and then they attack others.

Niccolo Machiavelli, *Discourses*, 1517

There is never a convenient place to fight a war when the other man starts it.

Admiral Arleigh A. Burke (1901 –)

AGGRESSIVENESS

He who brings danger upon another has more spirit than he who repels it. Add to this, that the terror excited by the unexpected is increased thereby. When you have entered the territory of an enemy you obtain a near view of his strong and weak points.

Provided no impediment is caused here, you will hear at once that I have landed, and that Africa is blazing with war; that Hannibal is preparing to depart from this country . . . Many things which are not now apparent at this distance will develop; and it is the part of a general not to be wanting when opportunity arises, and to bend its events to his designs. I shall, Quintus Fabius, have the opponent you assign me, Hannibal, but I shall rather draw him after me than be kept here by him.

Scipio Africanus, in Polybius, cited in Hart, *A Greater Than Napoleon*, 1926

Our Country will, I believe, sooner forgive an officer for attacking an enemy than for letting it [sic] alone.

Admiral Viscount Nelson, letter during the Attack on Bastin, 3 May 1794

. . . fire seldom but accurately. Thrust the bayonet with force. The bullet misses, the bayonet doesn't. The bullet's an idiot, the bayonet's a fine chap. Stab once and throw the Turk off the bayonet. Bayonet another, bayonet a third; a real warrior will bayonet half a dozen and more. Keep a bullet in the barrel. If three should run at you, bayonet the first, shoot the second and lay out the third with your bayonet. This isn't common but you haven't time to reload . . .

Field Marshal Prince Aleksandr V. Suvorov, *The Science of Victory*, 1796

I do not advise rashness, but I do desire resolute and actual fighting, with necessary casualties.

General of the Army Philip H. Sheridan, 23 September 1864, message to Major General Averell

My rule is: if you meet the weakest vessel, attack; if it is a vessel equal to yours, attack; and if it is stronger than yours, also attack . . .

Admiral Stephan O. Makarov (1849– 1904)

'The best method of gaining a goal is to act aggressively.' Nothing is said here to the effect that he who attacks *first* allegedly 'reveals a much stronger will'. The task of war is to annihilate the enemy. Annihilation is impossible without the offensive. The stronger will is revealed by him who creates the most favorable conditions for the offensive and utilizes them to the very end. But this does not at all mean that in order to reveal his will one must attack first. This is nonsense. If the material conditions of mobilization militate against it, then I would be a hopeless formalist and a dunderhead to build my plan on the notion that I must be the first to take the offensive. No, I shall reveal the superiority of my will by creating favourable conditions for my offensive – as the second one; by wresting the initiative at a certain limit fixed in advance; and by gaining victory even though I am the second to attack.

Leon Trotsky, 1922, *Military Writings*, 1969

How is it possible to vanquish the enemy except by a blow over the head? And for this it is necessary to attack him, to spring upon him. This was known to army leaders in biblical times.

Leon Trotsky, 1922, *Military Writings*, 1969

Hold what you've got and hit them where you can.

Admiral of the Fleet Ernest J. King, December 1941

Children of a free and sheltered people who have lived a generous life, we have not the pugnacious disposition of those oppressed beasts our enemies who must fight or starve. Our bravery is too negative. We talk too much of sacrifice, of the glory of dying that freedom may live. Of course we are willing to die but that is not enough. We must be eager to kill, to inflict on the enemy – the hated enemy – wounds, death, and destruction. If we die killing, well and good, but if we fight hard enough, viciously enough, we will kill and live. Live to return to our family and our girl as conquering heroes – men of Mars.

General George S. Patton, Jr., March 1942, *The Patton Papers*, Vol. II, 1974

Put your heart and soul into being expert killers with your weapons. The only good enemy is a dead enemy. Misses do not kill, but a bullet in the heart or a bayonet in the guts do. Let every bullet finds its billet – in the body of your foes.

General George S. Patton, Jr., December 1941, *The Patton Papers*, Vol. II, 1974

What counts is not necessarily the size of the dog in the fight – it's the size of the fight in the dog.

General of the Armies Dwight D. Eisenhower, 31 January 1958, Republican National Convention

---

An army which thinks *only* in defensive terms is doomed. It yields initiative and advantage in time and space to the enemy – even an enemy inferior in numbers. It loses the sense of the hunter, the opportunist.

General Sir David Fraser, *And We Shall Shock Them*, 1983

AIDE-DE-CAMP

Besides the confidence of the general officers, of which aides-de-camp render themselves worthy by indefatigable zeal, it is necessary that they should be extremely active, well acquainted with the different corps of the brigade or division to which they belong, the names of the several officers in command, and those of the commissaries, that they may be able to transmit orders with precision, and superintend their execution.

Marshal of France Michel Ney, Duc d' Elchingen, Prince de la Moskova, *Memoirs*, 1834

AIR POWER

Find the enemy and shoot him down, anything else is nonsense.

Captain Manfred Baron von Richthofen, 1917, opinion of air combat, quoted in Toliver and Constable, *Fighter Aces of the Luftwaffe*, 1977

The military mind always imagines that the next war will be on the same lines as the last. That has never been the case and never will be. One of the great factors in the next war will be aircraft obviously. The potentialities of aircraft attack on a large scale are almost incalculable.

Marshal of France Ferdinand Foch (1851–1929)

I have a mathematical certainty that the future will confirm my assertion that aerial warfare will be the most important element in future wars, and that in consequence not only will the importance of the Independent Air Force rapidly increase, but the importance of the army and navy will decrease in proportion.

General Giulio Douhet, *Command of the Air*, 1921

It is probable that future war will be conducted by a special class, the air force, as it was by the armored Knights of the Middle Ages.

Brigadier General William 'Billy' Mitchell, *Winged Defense*, 1924

The advent of air power, which can go straight to the vital centers and either neutralize or destroy them, has put a completely new complexion on the old system of making war. It is now realized that the hostile main army in the field is a false objective, and the real objectives are the vital centers.

Brigadier General William 'Billy' Mitchell, *Skyways: A Book on Modern Aeronautics*, 1930

In our victory over Japan, airpower was unquestionably decisive. That the planned invasion of the Japanese Home islands was unnecessary is clear evidence that airpower has evolved into a force in war co-equal with land and sea power, decisive in its own right and worthy of the faith of its prophets.

General Carl A. Spaatz, 'Evolution of Air Power', *Military Review*, 6/1947

The function of the Army and Navy in any future war will be to support the dominant air arm.

General James Doolittle, 1949, speech at Georgetown University

---

One of the outstanding characteristics of airpower proved to be its flexibility and the terrific concentration made possible by a unified air command – a unity only achieved by a faith born of mutual understanding between all branches and ranks of the air forces.

Chief Air Marshal Lord Tedder, *With Prejudice*, 1948

Anyone who has to fight, even with the most modern weapons, against an enemy in complete command of the air, fights like a savage against modern European troops, under the same handicaps and with the same chances of success.

Field Marshal Erwin Rommel, *Rommel Papers*, 1953

AMATEURS

Now, there are three ways in which a ruler can bring misfortune upon his army:

When ignorant that the army should not advance, to order an advance or ignorant that it should not retire, to order a retirement. This is described as 'hobbling the army.'

When ignorant of military affairs, to participate in their administration. This causes the officers to be perplexed.

When ignorant of command problems to share in the exercise of responsibilities. This engenders doubts in the minds of the officers.

If the army is confused and suspicious, neighbouring rulers will cause trouble. This is what is meant by the saying: 'A confused army leads to another's victory.'

Sun Tzu, *The Art of War*, c. 500 BC, tr. Griffith

I well know the character of that senseless monster the people, unable either to support the present or to foresee the future, always desirous of attempting the impossible, and of rushing headlong to its ruin. Yet your unthinking folly shall not induce me to permit your own destruction, nor to betray the trust committed to me by my sovereign and yours. Success in war depends less on intrepidity than on prudence to await, to distinguish, and to seize the decisive moment of fortune. You appear to regard the present contest as a game of hazard, which you might determine by a single throw of the dice; but I, at least, have learnt from experience to prefer security to speed. But it seems that you offer to reinforce my troops, and to march with them against the enemy. Where then have you acquired your knowledge of war? And what true soldier is not aware that the result of a battle must chiefly rest on the skill and discipline of the combatants? Ours is a real enemy in the field; we march to a battle, and not to a review.

Count Belisarius (c. AD 505–565), when the population of Rome during siege urged a battle out of desperation, in Mahon, *Life of Belisarius*, 1829

The commanders of armies are more to be pitied than one would think. Without listening to them, all the world denounces them, the newspapers ridicule them, and yet, of the thousands who condemn them, there is not one that could command even the smallest unit.

Frederick the Great (1712–1786)

. . . the kind of person who could not lead a patrol of nine men is happy to arrange armies in his imagination, criticise the conduct of a general, and say to his misguided self: 'My God, I know I could do better if I was in his place!'

Frederick the Great, 14 July 1745

This is a very suggestive age. Some people seem to think that an army can be whipped by waiting for rivers to freeze over, exploding powder at a distance, drowning out troops, or setting them to sneezing; but it will always be found in the end that the only way to whip an army is to go out and fight it.

General of the Army Ulysses S. Grant, January 1865, quoted in Porter, *Campaigning with Grant*, 1906

When, in complete security, after dinner in full physical and moral contentment, men consider war and battle, they are animated by a noble ardour which has nothing in common with reality. How many of them, however, at that moment would be ready to risk their lives? But oblige them to march for days and weeks to arrive at the battle front, and on the day of battle oblige them to wait a few minutes or hours to deliver it. If they were honest, they would testify how much the physical fatigue and the mental anguish that preceded action have lowered their morale, how much less eager to fight they are than a month before, when they arose from the table in a generous mood.

Colonel Charles Ardnant du Picq, *Battle Studies*, 1880

All those who criticize the dispositions of a general ought first to study military history, unless they have themselves taken part in a war in a position of command. I should like to see such people compelled to conduct a battle themselves. They would be overwhelmed by the greatness of their task, and when they realized the obscurity of the position, the exacting nature of enormous demands made on them, they would doubtless be more modest.

General Erich Ludendorff, *My War Memories, 1914–1918*, 1919

All the more honour to those gifted professional soldiers who have shown creative brains, but also all the more reason why we should preserve an open and receptive mind to the far larger stream of ideas which have come from 'amateurs', from Roger Bacon and Leonardo da Vinci down to the Volunteers who simplified our cumbruous drill and developed our musketry and tactical instructions. Nor can we forget that the decisive new weapon of the World War, the tank, owned its introduction principally to 'amateurs' in the face of stubborn opposition from the supreme professional opinion. It may also be significant that the one soldier who contributed notably to its causation was an engineer and an imaginative writer. (March 1923.)

Captain Sir Basil Liddell Hart, *Thoughts on War*, 1944

The atmosphere in Washington in that period [1964] is difficult, indeed impossible, to recreate. There were conferences almost weekly, at government expense, to gather in judgments from all possible quarters on what was to be done to carry the war to a successful conclusion. These huddles would bring together professors and self-appointed experts on guerrilla warfare from all part of the country. The collective contribution to the national cause may have been two degrees above zero. Government was proceeding according to the notion that, if enough persons are collected, some inspired thoughts of value will inevitably flow outward. It reminds me of the Army thesis that if two half-wits are assigned to a task, you get a whole wit, whereas the mathematical prospect is, more accurately, that what comes out will be a quarter-wit.

Brigadier General S. L. A. Marshall, 'Thoughts on Vietnam', in Thompson, *Lessons of Vietnam*, 1977

AMBITION

Campaign against Nubia

I sailed the King Okheperkere [Thutmose I], triumphant, when he ascended the river to Khenthennofer, in order to cast out violence in the highlands, in order to suppress the raiding of the hill region. I showed bravery in his presence in the bad water, in the [passage] of the ship by the bend. One appointed me chief of the sailors. His majesty was . . .

His majesty was furious thereat, like a panther; his majesty cast his first lance, which remained in the body of that fallen one. This was . . . powerless before his flaming uraeus, made [so] in an instant of destruction; their people were brought off as living prisoners. His majesty sailed down-river, with all countries in his grasp, that wretched Nubian Troglodyte being hanged head downward at the [prow] of the [barge] of his majesty, and landed at Karnak.

Asiatic Campaign

After these things one journeyed to Retenu to wash his heart among the foreign countries.

His majesty arrived at Naharin his majesty found that foe when he was [planning] destruction; his majesty made a great slaughter among them. Numberless were the living prisoners, which his majesty brought off from his victories. Meanwhile I was at the head of our troops, and his majesty beheld my bravery. I brought off a chariot, its horses, and him who was upon it as a living prisoner, and took them to his majesty. One presented me with gold in double measure.

Ahmose Son of Ebana (Egyptian naval officer, 18th Dynasty), in Breasted, *Ancient Records of Egypt*, 1906

A man's worth is no greater than the worth of his ambitions.

Emperor Marcus Aurelius Antonius, *Meditations*, c. AD 170

Ambition is the main driving power of men. A man expends his abilities as long as he hopes to rise; but when he has reached the highest round, he only asks for rest. I have created senatorial appointments and princely titles, in order to promote ambition, and, in this way, to make the senators and marshals dependent on me.

Napoleon, ed., Kircheisen, *The Memoirs of Napoleon I*, n.d.

But all that . . . he will learn will be of little use to him if he does not have the sacred fire in the depths of his heart, this driving ambition which alone can enable one to perform great deeds.

Napoleon (1769–1821)

A different habit, with worse effect, was the way that ambitious officers, when they came in sight of promotion to the general's list, would decide that they would bottle up their thoughts and ideas, as a safety precaution, until they reached the top and could put these ideas into practice. Unfortunately, the usual result, after years of such self-repression for the sake of ambition, was that when the bottle was eventually uncorked the contents had evaporated.

Captain Sir Basil Liddell Hart, quoted by Bunting, 'The Conscience of a Soldier', *Worldview*, 10/1973

ANALYSIS

Napoleon once said: 'I have always liked analysis; if I were to be seriously in love, I should analyse my love bit by bit. *Why?* and *How?* are questions so useful that they cannot be too often asked. I conquered rather than studied history; that is to say, I did not care to retain and did not retain anything that could not give me a new idea; I disdained all that was useless, but took possession of certain results which pleased me.' In other words, he taught himself 'How to think' which is the harvest of all true education, civil or military.

Napoleon and Major General J. F. C. Fuller, *Lectures on F. S. R. II*, 1931

The analysts write about the war as if it's a ballet . . . like it's choreographed ahead of time, and when the orchestra strikes up and starts playing, everyone goes out and plays a set piece.

What I always say to those folks is, 'Yes, it's choreographed, and what happens is the orchestra starts playing and some son of a bitch climbs out of the orchestra pit with a bayonet and starts chasing you around the stage.' And the choreography goes right out the window.

General H. Norman Schwarzkopf, 5 February 1991, interview in *The Washington Post*

ARMY

There is no need for me to enlarge upon the battles which the Romans fought and lost against Hannibal, for the defeats they suffered had nothing to do with weapons or formations, but were brought about by Hannibal's cleverness and military genius. This point I made sufficiently clear in my descriptions of the battles in question, and there are two pieces of evidence which support my conclusion. The first is the manner in which the war ended, for as soon as a general of ability comparable to Hannibal's appeared on the Roman side it was only a short while before victory was theirs. The second is provided by Hannibal himself, who as soon as he had won the first battle discarded the equipment with which he had started out, armed his troops with Roman weapons, and continued to use these till the end of the war.

Polybius, *The Rise of the Roman Empire, c.* 125 BC

To aid in understanding our treatise on tactical matters, let it be noted that the entire force on campaign is comprised of the following. First are the combat troops, the force armed for offensive action, which is called an armed force or an army. Then come the technical groups such as stonecutters, metalworkers, carpenters, and the like, who need to be on hand for siege operations and for other special work. Third, there are the supply services which provide bread, wine, meat, and other necessities.

Anonymous Byzantine general, *Strategy*, c. AD 534, in Dennis, *Three Byzantine Military Treatises*, 1985

Let us have a respectable army, and such as will be competent to every contingency.

General George Washington (1732–1799)

The country must have a large, efficient army, one capable of meeting the enemy abroad, or they must expect to meet him at home.

The Duke of Wellington, 10 December 1811, letter

The ideal army would be the one in which every officer would know what he ought to do in every contingency; the best possible army is the one that comes closest to this. I give myself only half the credit for the battles I have won, and a general gets enough credit when he is named at all, for the fact is that a battle is won by the army.

Napoleon, 1817, conversation, in Herold, ed. *The Mind of Napoleon*, 1955

War is a special activity, different and separate from any other pursued by man. This would still be true no matter how wide its scope, and though every able-bodied man in the nation were under arms. An army's military qualities are based on the individual who is steeped in the spirit and essence of this activity; who trains the capacities it demands, rouses them, and makes them his own; who applies his intelligence to every detail; who gains ease and confidence through practice, and who completely immerses his personality in the appointed task.

Major General Carl von Clausewitz, *On War*, 1832, tr. Howard and Paret

---

When an army starts upon a campaign, it resolves itself speedily into two parts, one that means to keep out of harm's way if possible, and the other that always keeps with the colors.

Major General George B. McClellan, quoted in Time-Life Books, *Tenting Tonight*, 1984

The Army has its common law as well as its statute law; each officer is weighed in the balance by his fellows, and these rarely err. In the barrack, in the mess, on the scout, and especially in battle, a man cannot – successfully – enact the part of a hypocrite or flatterer, and his fellows will measure him pretty fairly for what he is.

General of the Army William T. Sherman (1820–91)

An army is still a crowd, though a highly organized one. It is governed by the same laws . . . and under the stress of war is ever tending to revert to its crowd form. Our object in peace is so to train it that the reversion will become extremely slow.

Major General J. F. C. Fuller, *Training Soldiers for War*, 1914

One of the main arguments against armies is their futility; but, if this be true, this argument can with equal force be directed against peaceful organizations; for surely it is just as futile to keep vast numbers of a nation on the brink of starvation and prostitution, as happens in nearly all civilized countries today, as it is to keep an insignificant minority of this same nation on the brink of war.

Major General J. F. C. Fuller, *The Reformation of War*, 1923

All through history there have been recurring cycles of two schools of war; the one tending towards the adoption of small professional forces relatively well trained. The other toward the utilization of masses and mob psychology. Both schools have usually foundered on the rock of compromise, when the first sought to increase its numbers and the second its discipline . . .

General George S. Patton, Jr., January 1928, *The Patton Papers*, Vol. I, 1972

An army is an institution not merely conservative but retrogressive by nature. It has such natural resistance to progress that it is always insured against the danger of being pushed ahead too fast. Far worse and more certain, as history abundantly testifies, is the danger of it slipping backward. Like a man pushing a barrow up a hill, if the soldier ceases to push, the military machine will run back and crush him. To be deemed a revolutionary in the army is merely an indication of vitality, the pulse-beat which shows that the mind is still alive. When a soldier ceases to be a revolutionary it is a sure sign that he has become a mummy. (January 1931)

Captain Sir Basil Liddell Hart, *Thoughts on War*, 1944

An armed force differs from an armed mob, or crowd of men, not so much in its armament but in that it is disciplined, is organised, and is under control. A crowd may at times be courageous beyond belief, but it possesses little staying power, and is as quickly depressed as elated by outward circumstances. Courage in itself is insufficient, and to it must be added what is called morale, a complex quality depending upon honour to the cause, loyalty to the leader, confidence in one's skill and the skill of one's comrades, and in an innate feeling that everyone is doing his utmost to win and that no one will leave a comrade in the lurch. Morale endows a force of men with a feeling of superiority and invincibility, it is cultivated by commonsense, sound organisation, efficient arms and equipment, and the realisation that those in command are skilful, brave and just.

Major General J. F. C. Fuller, *Lectures on F. S. R. II*, 1931

Their task is to destroy. The balance sheet of their activities shows a hideous total of broken lives, of wealth destroyed, of nations ground to powder, of work brought to nothing, of efforts frustrated, and happiness maimed and killed. The wreckage of war is beyond counting. Land left fallow, fire, and famine, such are its material consequences. But, in brooding on the evil, do we not tend to forget the children born into safety, the lives lived out in security as the result of the toil and sacrifice of our soldiers? But for the strength of armies what tribe, what city, and what state could ever have been established? Protected by that human shield, fields have been sown and reaped and men have been able to work on undisturbed. The clash of arms has made possible all material progress. History cannot measure the debt owed by the wealth of nations, their network of communications, their ships upon the seas, to the lust for conquest.

Charles de Gaulle, *The Edge of the Sword*, 1932

The nature of armies is determined by the nature of the civilization in which they exist.

Captain Sir Basil Liddell Hart, *The Ghost of Napoleon*, 1933

Armies are created to serve the policy of states. And no one knows better than you yourself that strategy should include not only the given circumstances of military technique, but also the moral elements.

Charles de Gaulle, 5 January 1945, letter to Dwight D. Eisenhower, *Salvation: 1944–1946*, 1959

An army is an organized mob and the cement which bonds it together is discipline and mutual confidence. More therefore than all plans and schemes based on material factors, the art of battle consists in maintaining and strengthening the psychological cohesion of one's own troops while at the same time disrupting that of the enemy's.

General André Beaufre, *Introduction to Strategy*, 1965

ARTILLERY

Blessed by those happy ages that were strangers to the dreadful fury of these devilish instruments of artillery, whose inventor I am satisfied is now in hell, receiving the reward of his cursed invention, which is the cause that very often a cowardly base hand takes away the life of the bravest gentleman.

Miguel de Cervantes, *Don Quixote*, 1615

The object of artillery should not consist of killing men on the whole of the enemy's front, but to overthrow it, to destroy parts of this front, . . . then they obtain decisive effects; they make a gap.

Field Marshal François Comte de Guibert, *Essai général de tactique,* 1773

It used to be our custom to form regiments from the largest men possible. This was done for a reason, for in the early wars it was men and not cannon that decided victory, and battalions of tall men advancing with the bayonet scattered the poorly assembled enemy troops – with the first attack. Now artillery has changed everything. A cannon ball knocks down a man six feet tall just as easily as one who is only five feet seven. Artillery decides everything, and infantry no longer do battle with naked steel.

Frederick the Great, *A History of My Own Times*, 1789, tr. Holcroft

---

In siege warfare, as in the open field, it is the gun which plays the chief part; it has effected a complete revolution . . . It is with artillery that war is made.

Napoleon, 1809, after the Battle of Loebau

You can't describe the moral lift,
When in the fight your spirit weary
Hears above the hostile fire
Your *own* artillery.
Shells score the air like wavy hair
From a forward battery,
As regimental cannon crack
While, from positions further back,
In bitter sweet song overhead
Crashing discordantly
Division's pounding joins the attack;
Mother like she belches shell;
Gloriously it flies, and well,
As, with a hissing, screaming squall,
A roaring furnace, giving all,
She sears a path for the infantry . . .

Aleksandr Tvardovskiy's 1943 narrative poem, *Vasily Tyorkin,* tr. Bellemy, *Red God of War*, 1987

ART OF WAR

The art of war is, in the last result, the art of keeping one's freedom of action.

Xenophon (*c.* 430–355 BC)

The art of war is divided between art and strategem. What cannot be done by force, must be done by strategem.

Frederick the Great, *Instructions to His Generals*, 1747

The three military arts. First – Apprehension, how to arrange things in camp, how to march, how to attack, pursue, and strike; for taking up position, final judgement of the enemy's strength, for estimating his intentions.

Second – Quickness . . .

This quickness doesn't weary the men. The enemy doesn't expect us, reckons us 100 versts away, and if a long way off to begin with – 200, 300 or more – suddenly we're on him, like snow on the head; his head spins. Attack with what comes up, with what God sends; the cavalry to begin, smash, strike, cut off, don't let slip, hurra!

Brothers do miracles!

Third – Attack. Leg supports leg, arm strengthens arm; many men will die in the volley; the enemy has the same weapons, but he doesn't know the Russian bayonet. Extend the line – attack at once with cold steel; extend the line without stopping . . . the Cossacks to get through everywhere . . . In two lines is strength; in three, half as much again; the first breaks, the second drives into heaps, the third overthrows.

Field Marshal Prince Aleksandr V. Suvorov, *The Science of Victory*, 1796

One may teach tactics, military engineering, artillery work about as one teaches geometry. But knowledge of the higher branches of war is only acquired by experience and by a study of the history of the wars of great generals. It is not in a grammar that one learns to compose a great poem, to write a tragedy.

Napoleon, quoted in Foch, *Principles of War*, 1913

The art of war is an immense study, which comprises all others.

Napoleon, ed., Herold, *The Mind of Napoleon*, 1955

The art of war is a simple art; everything is in the performance. There is nothing vague about it; everything in it is common sense; ideology does not enter into it.

Napoleon, ed., Herold, *The Mind of Napoleon*, 1955

The whole art of war consists in a well-reasoned and extremely circumspect defensive, followed by rapid and audacious attack.

Napoleon (1769–1821)

The art of war is no more than the art of augmenting the chances which are in our favour.

Napoleon (1769–1821)

The art of war consists in bringing to bear with an inferior army a superiority of force at the point at which one attacks or is attacked.

Napoleon (1769–1821)

It simply consists in seeing to it that you have two soldiers on the battlefield to every one of the enemy.

Stendahl, quoted in Blanch, *Soldier of Paradise*, 1960

The art of war deals with living and with moral forces. Consequently, it cannot attain the absolute, or certainty; it must always leave a margin for uncertainty, in the greatest things as much as in the smallest. With uncertainty in one scale, courage and self-confidence must be thrown into the other to correct the balance. The greater they are, the greater the margin that can be left for accidents. Thus courage and self-confidence are essential in war, and theory should propose only rules that give ample scope to these finest and least dispensable of military virtues, in all their degrees and variations. Even in daring there can be method and caution; but here they are measured by a different standard.

Major General Carl von Clausewitz, *On War*, 1832, tr. Howard and Paret

---

Essentially, then, the art of war is the art of using the given means in combat; there is no better term for it than the *conduct of war*. To be sure in its widest sense the art of war includes all activities that exist for the sake of war, such as the creation of the fighting forces, their raising, armament, equipment, and training.

Major General Carl von Clausewitz, *On War*, 1832, tr. Howard and Paret

---

It must not be concluded that the art of war has arrived at that point that it cannot make another step towards perfection. There is nothing perfect under the sun! And if a committee were assembled under the presidency of the Archduke Charles or Wellington, composed of all the strategic and tactical notabilities of the age, together with the most skillful generals of engineers and artillery, this committee could not yet succeed in making a perfect, absolute and immutable theory on all branches of war, especially on tactics!

Lieutenant General Antoine-Henri Baron de Jomini, *Summary of the Art of War*, 1838

Of all theories on the art of war, the only reasonable one is that which, founded upon the study of military history, admits a certain number of regulating principles, but leaves to natural genius the greatest part in the general conduct of a war without trammelling it with exclusive rules.

On the contrary, nothing is better calculated to kill natural genius and to cause error to triumph, than those pedantic theories, based upon the false idea that war is a positive science, all the operations of which can be reduced to infallible calculations.

Lieutenant General Antoine-Henri Baron de Jomini, *Summary of the Art of War,* 1838

The art of war is simply enough. Find out where your enemy is. Get at him as soon as you can. Strike at him as hard as you can and as often as you can, and keep moving on.

General of the Army Ulysses S. Grant (1822–1885)

Science and *theory* are two entirely different things, for every form of art can and should have its theory, but it cannot be made into a science . . . Nobody would think today of claiming that there can be a *science of war*. That would be as absurd as a science of poetry, of painting, of music. But it does not follow that there is no theory of war, just as there is a theory of the arts of peace. Such theory alone does not create the Raphaels, the Beethovens, the Shakespeares, but it endows them with a technique without which they could not attain the heights they reach.

The *theory of the art of war* does not claim to produce Napoleons, but it teaches the properties of troops and ground. It points out the examples, the masterpieces achieved in the art of war, and in such manner it smooths the way for those who have natural military ability.

It does not give to any man the satisfaction of thinking that he knows all there is to know when as a matter of fact he only knows a part. Recipes for creating masterpieces such as Austerlitz, Friendland, Wagram, for conducting such campaigns as that of 1799 in Switzerland or for winning battles such as that of Koeniggrätz, we cannot obtain from theory. But it does explain these models as types for study, not to be blindly imitated but rather that the pupil may absorb their spirit and obtain inspiration from them.

If theory went wrong, it is due to the fact that very few theorists had seen war . . .

General Mikhail I. Dragomirov (1830–1905), quoted in Foch, *Principles of War,* 1913

They forget that the whole art of war is to gain your objective with as little loss as possible.

Field Marshal Viscount Montgomery of Alamein, 8 November 1917, letter to his mother, quoted in Hamilton, *Monty*, 1981

War rests on many sciences, but war itself is not a science – it is a practical art, a skill. The Prussian strategist, King Frederick II, was fond of saying that war is a trade for an ignoramus, an art for a man of talent and a science for a genius. But he told a lie. This is false. For an ignoramus war is not a trade because ignorant soldiers are the cannon fodder of war and not at all its 'tradesmen'. As is well known, each trade requires a certain schooling; and for those who are correctly schooled in military affairs war is therefore a 'trade'. It is a cruel, sanguinary trade, but a trade nonetheless, that is, a skill with certain habits which are elaborated by experience and correctly assimilated. For gifted people and those of genius, this skill become transformed into a high art.

Leon Trotsky, 1 April 1922, remarks at the 11th Congress of the Communist Party, *Military Writings*, 1969

There isn't and can't be a science of war, in the precise meaning of the word. There is an art of war. On the other hand, even a trade presupposes a schooling, and whoever has schooling is no ignoramus. It would be more correct to say that war is a skilled trade for for the average individual and an art for an outstanding one. As regards an ignoramus, he is only the raw material of war; its cannon fodder, and not at all a skilled man.

Leon Trotsky, 8 May 1922, remarks to the Military Scientific Society, *Military Writings*, 1969

War is a science which depends on art for its application.

Captain Sir Basil Liddell Hart, 'Strategy', *Encyclopaedia Britannica*, 1929

Whereas the other arts are, at their height, individual, the art of war is essentially orchestrated.

Captain Sir Basil Liddell Hart, *Thoughts on War*, 1944

I've been studying the art of war for forty-odd years. When a surgeon decides in the course of an operation to change its objective . . . he is not making a snap decision but one based on knowledge, experience and training. So am I.

General George S. Patton, Jr., on being accused of making snap decisions

---

The conduct of war, like the practice of medicine, is an art, and because the aim of the physician and surgeon is to prevent, cure, or alleviate the diseases of the human body, so should the aim of the statesman and soldier be to prevent, cure, or alleviate the wars which afflict the international body.

Major General J. F. C. Fuller, *The Conduct of War*, 1961

ATROCITY

The city was exceeding strong and and was surrounded by three walls. The men trusted in their mighty walls and in their hosts, and did not come down, and did not embrace my feet. With battle and slaughter I stormed the city and captured it. 3,000 of their warriors I put to the sword; their spoil and their possessions, their cattle and their sheep I carried off. Many captives from among them I burned with fire, and many I took as living captives. From some I cut off their hands and their fingers, and from others I cut off their noses, their ears, and their fingers [?], of many I put out the eyes. I made one pillar of the living, and another of heads, and I bound their heads to posts [tree trunks] round about the city. Their young men and maidens I burned in the fire, the city I destroyed, I devastated, I burned it with fire and consumed it. At that time the cities of the land of Nirbi and their strong walls I destroyed, I devastated, I burned with fire.

Assur-Nâsir-Pal II (883–859 BC), King of Assyria, quoted in Luckenbill, *Ancient Records of Assyria and Babylonia,* Vol I, 1926

They make a desert and call it peace.

Cornelius Tacitus (*c.* AD 56– *c.* 120), *Agricola*

No nation can safely trust its martial honor to leaders who do not maintain the universal code which distinguishes between those things that are right and those things that are wrong. The testimony shows a complete failure to comply with this simple but vital standard. The savageries which resulted have shocked the world. They have become synonyms of horror and mark the lowest ebb of depravity of modern times. There are few parallels in infamy and tragedy with the brutalization of troops who in good faith had laid down their arms. It is of peculiar aversion that the victims were a garrison whose heroism and valor has never been surpassed. Of all fighting men of all time none deserved more the honors of war in their hour of final agony. The callousness of denial has never been exceeded. This violation of a fundamental code of chivalry, which has ruled all honorable military men throughout the ages in treatment of defeated opponents, will forever shame the memory of the victorious troops.

General of the Armies Douglas MacArthur, February 1946, describing General Homma's conduct in The Philippines, 1942, in *Reminiscences*, 1962

ATTACK

The King also boasts of his wisdom in military matters. He knows when to strike and when to avoid combat . . . 'attacking him who attacks . . . since, if one is silent after attack, it strengthens the heart of the enemy. Valiance is eagerness, cowardice is to slink back', he tells us, adding 'he is truly a coward who is repelled upon his border'.

Sesostris III (1887–1849 BC), Pharaoh of Egypt, quoted in Breasted, *Ancient Records of Egypt*, Vol. I, 1906

Battle on the Orontes

First month of the third season [ninth month], day 26; his majesty crossed over the ford of the Orontes on this day, caused to cross . . . like the might of Montu of Thebes. His majesty raised his arm, in order to see the end of the earth; his majesty described a few Asiatics coming on horses . . . coming at a [gallop]. Behold, his majesty was equipped with his weapons of battle, his majesty conquered with the might [of Set] in his hour. They retreated when his majesty looked at one of them. Then his majesty himself overthrew their . . ., with his spear. . . . Behold, he carried away this Asiatic . . ., his horses, his chariot, and all his weapons of battle. His majesty returned with joy of heart [to] his father, Amon; he [his majesty] gave to him a feast . . .

Amenhotep II (1147–1420 BC), Pharaoh of Egypt, quoted in Breasted, *Ancient Records of Egypt*, Vol. II, 1906

Decline the attack unless you can make it with advantage.

Field Marshal Maurice Comte de Saxe, *My Reveries*, 1732

Gentlemen, the enemy stands behind his entrenchments, armed to the teeth. We must attack him and win, or else perish. Nobody must think of getting through any other way. If you don't like this, you may resign and go home.

Frederick the Great, 5 December 1757, to his officers before the battle of Leuthen

---

When we advance, the troops must move quietly, not speak a word, and not shoot.

Getting up to the wall, they are to rush forward quickly, and at the word of command shout 'Hurra!'

The ditch reached, without losing a second, throw your fascine into it, leap on to it, and put the ladder up against the wall; the marksmen to shoot the enemy down one by one – smartly, quickly, get up two by two! the ladder short? Bayonet into the wall – climb on to it, after him another and a third. Comrade help comrade! Up on the wall, thrust the enemy off with the bayonet, and in a twinkling form up beyond the wall.

Don't bother about firing; don't fire without need; beat the enemy and push him with the bayonet; work quickly, sharply, bravely, Russianly! Support your own men, in a body; don't leave your officers. Keep the front.

Don't run about into houses; give quarter to the enemy who ask for it; don't strike the unarmed; don't fight with women; don't touch boys and girls.

To those of us who die the Kingdom of Heaven; to the living glory, glory, glory!

Field Marshal Prince Aleksandr V. Suvorov, instructions before the storming of the Praga Fortress, 1794, quoted in Blease, *Suvorof*, 1920

The ditch isn't deep, the wall isn't high; fling yourself into the ditch; leap over the wall, charge with the bayonet, strike, chase, take prisoner.

Storm.

The enemy runs into the town, turn his guns against him, fire hard down the streets, keep up a lively bombardment; go after him at once . . . the enemy surrenders, spare him; the walls occupied, after the plunder.

Field Marshal Prince Aleksandr V. Suvorov, *The Science of Victory*, 1796

Inasmuch as the nature of our soldiers lends itself better to offense than to defence, the chiefs shall not forget that in case of emergency they should take the offensive.

General José de San Martin, April 1818, instructions before the battle of Maipú, in Rojas, *San Martin*, 1957

Attack inspires a soldier, it adds to his power, rouses his self-reliance, and confuses the enemy. The side attacked always overestimates the strength of the attacker.

Field Marshal August Graf von Gneisenau (1760–1831), quoted in Dupuy, *The Military Heritage of America*, 1984

The attack should be like a soap bubble, which distends itself until it bursts.

Major General Carl von Clausewitz, *On War*, 1832

I was too weak to defend, so I attacked.

General Robert E. Lee (1807–70) (attributed)

When we are fighting there is no need to think of defence. A positive attack is the best form of defence . . . the vital point in actual warfare is to apply to the enemy what we do not wish to be applied to ourselves and at the same time not to let the enemy apply it to us . . . we must always forestall them.

Admiral Marquis Togo Heihachiro, 15 May 1905, quoted in Warner, *The Tide at Sunrise*, 1974

Hard pressed on my right. My centre is yielding. Impossible to manoeuvre. Situation excellent. I am attacking.

Marshal of France Ferdinand Foch, 8 September 1914, message to Marshal Joffre during the Battle of the Marne

---

Ought I in these circumstances to assume the offensive myself or stand on the defensive? According to the invariable rule which I followed throughout the War, whenever there was the slightest possibility, I determined to attack vigorously.

General Aleksei A. Brusilov, September 1914, *A Soldier's Notebook*, 1931

Every attack, once undertaken, must be fought to the finish; every defence, once begun, must be carried on with the utmost energy.

Marshal of France Ferdinand Foch, *Precepts and Judgments*, 1919

One must create an irresistible battering ram.

Marshal of the Soviet Union Mikhail N. Tukhachevskiy, 1924, in Simpkin, *Deep Battle*, 1987

The normal purpose of an attack is the infliction of death wounds and destruction on the enemy troops with a view to establish both physical and moral ascendancy over them. The gaining of ground in such a combat is simply an incident; not an object.

General George S. Patton, Jr., 1 November 1926, letter to his wife, *The Patton Papers*, Vol. I, 1972

Napoleon once said, 'I attacked to be attacked'. What he meant was that he threw forward a small fraction of his forces for the enemy to bite on, and when his adversary's jaws were fixed he moved up his large reserves – the capital of his tactical bank – and struck his real blow . . .

Major General J. F. C. Fuller, 'Co-ordination of the Attack', *The Infantry Journal*, 1/1931

Attack is the chief means of destroying the enemy, but defence cannot be dispensed with. In attack the immediate object is to destroy the enemy, but at the same time it is self-preservation, because if the enemy is not destroyed, you will be destroyed. In defence the immediate object is to preserve yourself, but at the same time defence is a means of supplementing attack or preparing to go over to the attack. Retreat is in the category of defence and is a continuation of defence, while pursuit is a continuation of attack. It should be pointed out that destruction of the enemy is the primary object of war and self-preservation the secondary, because only by destroying the enemy in large numbers can one effectively preserve oneself. Therefore attack, the chief means of destroying the enemy, is primary, while defence, a supplementary means of destroying the enemy and a means of self-preservation, is secondary. In actual warfare the chief role is played by defence much of the time and by attack for the rest of the time, but if war is taken as a whole, attack remains primary.

Mao Tse-tung, *On Protracted War*, May 1938

We are so outnumbered there's only one thing to do. We must attack.

Admiral Andrew Browne Cunningham, 11 November 1940, before attacking the Italian fleet at Taranto

---

When the situation is obscure, attack.

Colonel General Heinz Guderian (1888–1954) (attributed)

Manton Eddy who took over Doc's corps asked me when I told him his job: 'How much shall I have to worry about my flank?'

I told him that depended on how nervous he was.

He has been thinking [that] a mile a day [was] good going. I told him to go fifty and he turned pale . . .

General George S. Patton, Jr., August 1944, *The Patton Papers*, Vol. II, 1974

ATTRITION

The enemy must fight his battles far from his home base for a long time . . . We must further weaken him by drawing him into protracted campaigns. Once his initial dash is broken, it will be easier to destroy him.

Marshal Tran Hung Dao (Vietnam), c. AD 1284

Then I had the opportunity of seeing the Dutch Army, and their famous general Prince Maurice. It is true that the men behaved themselves well enough in action, when they were put to it, but the prince's way of beating his enemies without fighting, was so unlike the gallantry of my royal instructor that it had no manner of relish with me. Our way in Germany was always to seek out the enemy and fight him, and, give the imperialists their due, they were seldom hard to be found, but were as free of their flesh as we were.

Whereas Prince Maurice would lie in a camp till he had starved half his men, if by lying there he could but starve two-thirds of his enemies; so that, indeed, the war in Holland had more of fatigue and hardships in it, and ours had more of fighting and blows. Hasty marches, long and unwholesome encampments, winter parties, countermarching, dodging, and intrenching, were the exercises of his men, and often time killed more men with hunger, cold and diseases than he could do with fighting; not that it required less courage, but rather more, for a soldier had at any time rather die in the field a la coup de mousquet, than be starved with hunger, or frozen to death in the trenches.

Nor do I think to lessen the reputation of that great general, for it is most certain he ruined the Spaniards more by spinning the war thus out in length, than he could possibly have done by swift conquest; for had he, Gustavus like, with a torrent of victory, dislodged the Spaniard from all the twelve provinces in five years (whereas he was forty years in beating them out of seven) he had left them rich and strong at home, and able to keep the Dutch in constant apprehension of the return of his power; whereas, by the long continuance of the war, he so broke the very heart of the Spanish monarchy, so absolutely and irrecoverably impoverished them that they have ever since languished of the disease, till they are fallen from the most powerful to be the most despicable nation in the world.

Daniel Defoe (1659–1731), *Memoirs of a Cavalier*, 1889

My object in war was to exhaust Lee's army. I was obliged to sacrifice men to do it. I have been called a butcher. Well, I never spared lives to gain an object; but then I gained it, and I knew it was the only way.

General of the Army Ulysses S. Grant, quoted in Chancellor, *An Englishman in the American Civil War*, 1971

In the stage of the wearing out struggle losses will necessarily be heavy on both sides, for in it the price of victory is paid.

Field Marshal Earl Haig, *Dispatches*, 1919

An enemy may be worn out by physical and moral action; this, though the usual method of defeating him, is also, frequently, the most uneconomical method, for the process of disintegration is mutually destructive.

Major General J. F. C. Fuller, *The Reformation of War*, 1923

From the delusion that the armed forces themselves were the real objective in war, it was the natural sequence of ideas that the combatant troops who composed the armies should be regarded as the target to strike at.

Thus progressive butchery, politely called 'attrition', becomes the essence of war. To kill, if possible, more of the enemy troops than your own side loses, is the sum total of this military creed, which attained its tragicomic climax on the Western front in the Great War.

The absurdity and wrong-headedness of this doctrine should have been apparent to any mind which attempted to think logically instead of blindly accepting inherited traditions. War is but a duel between two nations instead of two individuals. A moment's unprejudiced reflection on the analogy of a boxing match would be sufficient to reveal the objective dictated by common sense. Only the most stupid boxer would attempt to beat his opponent by merely battering and bruising the latter until at last he weakens and yields. Even if this method of attrition finally succeeds, it is probably that the victor himself will be exhausted and injured. The victorious boxer, however, has won his stake, and can afford not to worry over the period of convalescence, whereas the recovery of a nation is a slow and painful process.

Captain Sir Basil Liddell Hart, *Thoughts on War*, 1944

But attrition is a two-edged weapon and, even when skilfully wielded, puts a strain on the users. It is especially trying to the mass of the people, eager to see a quick finish – and always inclined to assume that this can only mean the enemy's finish.

Captain Sir Basil Liddell Hart, *Strategy*, 1954

Attrition is not a strategy. It is, in fact, irrefutable proof of the absence of any strategy. Any commander who resorts to attrition admits his failure to conceive of an alternative. He rejects warfare as an art . . . He uses blood in lieu of brains.

General David R. Palmer, *Summons of the Trumpet*, 1978

AUDACITY

Impetuosity and audacity often achieve what ordinary means fail to achieve.

Niccolo Machiavelli, *Discourses*, 1517

In audacity and obstinacy will be found safety.

Napoleon, *Maxims of War*, 1831

With audacity one can undertake anything, but not do everything.

Napoleon, *Maxims of War*, 1831

If the theory of war does advise anything, it is the nature of war to advise the most decisive, that is the most audacious.

Major General Carl von Clausewitz, *Principles of War*, 1812

Never forget that no military leader has ever become great without audacity.

Major General Carl von Clausewitz, *Principles of War*, 1812

For great aims we must dare great things.

Major General Carl von Clausewitz, *Principles of War*, 1812

They want war too methodical, too measured; I would make it brisk, bold, impetuous, perhaps sometimes even audacious.

Lieutenant General Antoine-Henri Baron de Jomini, *Summary of the Art of War*, 1838

I objected as forcefully as I knew how. I argued and railed. I told them in the bluntest terms that they should not interfere. They did not know the target or the terrain as I did. They did not know our units and commanders as I did. They did not know the enemy and what could be expected of him as I did. It was their business to tell me *what* to do. The goal, the target – their job was to define that as carefully as they were able. If they did not want this particular target, fine. They should choose another, or none. But they should not take it on themselves to tell me *how* to do it. When it came to how, there should be as little interference as possible with the commander in the field. My position was then (as it is now) that the upper echelon should intervene only if they are actually on the battlefield, if they know everything intimately, if they are forward where they can see and understand all the elements that affect the conduct of the battle.

General Ariel Sharon, discussing the raid on the Jordanian headquarters at Kalkilya in 1956, *Warrior*, 1989

AUFTRAGSTAKTIK

. . . obedience is the principle, but man stands above the principle . . . who is right in battle is decided in most cases by success . . .

Field Marshal Helmuth Graf von Moltke (1880–1891)

Auftragstaktik is what I would like to call the leadership action which we saw for the first time in full action in our Exercise Rule 88 and also emphasized for lower leadership in Exercise Rule 06 in the same sense, by which the higher leader does not give his subordinate a binding order, but more an excerpt from his own thought process, through which he demands from [the subordinate] the intellectual cooperation for the accomplishment of the mission.

General Otto von Moser, 1912, quoted in *Infantry*, January–February 1991

Command and control of armed forces is an art, a creative activity based on character, ability, and mental power . . . Mission-oriented command and control is the first and foremost command and control principle in the army, of relevance in war even more than in peace. It affords the subordinate leader freedom of action in the execution of his mission, the extent depending on the type of mission to be accomplished.

HDv 100/100, Leadership Manual of the Bundeswehr

. . . when a detachment is made, the commander thereof should be informed of the object to be accomplished, and left as free as possible to execute it in his own way . . .

General of the Army William T. Sherman, *Memoirs of General W. T. Sherman*, 1875

I do not propose to lay down for you a plan of campaign . . . but simply lay down the work it is desirable to have done and leave you free to execute it in your own way. Submit to me, however, as early as you can, your plan of operations.

General of the Army Ulysses S. Grant, 4 April 1864, instructions to General William Sherman, *Personal Memoirs of U.S. Grant,* 1885

AUTHORITY

But we can go one step beyond General Harbord's suggestion that multiplied individual acceptance of a command alone gives that command authority. It is not less true that the multiplied rejection of a command nullifies it. In other words authority is the creature rather than the creator of discipline and obedience. In the more recent experiences of our arms, under the stresses of battle, there are many instances of troops being given orders, and refusing to obey. In every case, the root cause was lack of confidence in the wisdom and ability of those who led. When a determining number of men in ranks have lost the will to obey, their erstwhile leader has *ipso facto* lost the capacity to command. *In the final analysis, authority is contingent upon respect far more truly than respect is founded upon authority.*

Brigadier General S. L. A. Marshall, *The Armed Forces Officer*, 1950

If you take chance, it usually succeeds, presupposing good judgement.

Lieutenant General Sir Gifford Martel, quoted in Marshall, *The Armed Forces Officer*, 1950

Audacity is nearly always right, gambling nearly always wrong.

Captain Sir Basil Liddell Hart, quoted in Simpkin, *Race to the Swift*, 1985

Audace, audace, toujours audace.

Motto on plaque outside U.S. Army Command and General Staff College, Fort Leavenworth, Kansas

---

When soldiers forge ahead and do not dare to fall back, this means they fear their own leaders more than they fear the enemy. If they dare to fall back and dare not forge ahead, this means they fear the enemy more than they fear their own leaders. When he can get his troops to plunge right into the thick of raging combat, it is his authority and sternness that brings this about.

The rule is 'To be awesome and yet caring makes a good balance'.

Liu Ji (AD 1310–1375), *Lessons of War*

Discipline and morale influence the inarticulate vote that is constantly taken by masses of men when the order comes to move forward – a variant of the crowd psychology that inclines it to follow a leader. But the Army does not move forward until the motion has carried. 'Unanimous consent' only follows cooperation between the individual men in the ranks.

Lieutenant General James G. Harbord, *The American Army in France*, 1936

Military authority, directing the armed forces, is a matter of the authoritative power of the leading general.

If the general can hold the authority of the military and operate its power, he oversees his subordinates like a fierce tiger with wings, flying over the four seas, going into action whenever there is an encounter.

If the general loses his authority and cannot control the power, he is like a dragon cast into a lake; he may seek the freedom of the high seas, but how can he get there?

Zhuge Liang (AD 180–234), *The Way of the General*

God favors the bold and strong of heart.

General Alexander A. Vandergrift, quoted in Marshall, *The Armed Forces Officer*, 1950

The commander must establish personal and comradely contact with his men but without giving away an inch of his authority.

Field Marshal Erwin Rommel, *The Rommel Papers*, 1953

In war nothing is impossible, provided you use audacity.

General George S. Patton, Jr., *War As I Knew It*, 1947

BATTLE/BATTLES

At the word of Assur, the great lord, my lord, on flank and front I pressed upon the enemy like the onset of raging storm. With the weapons of Assur, my lord, and the terrible onset of my attack, I stopped their advance, I succeeded in surrounding them, I decimated the enemy host with arrow and spear. All of their bodies I bored through like a sieve . . . Like the many waters of a storm I made the contents of their gullets and entrails run down upon the wide earth. My prancing steeds harnessed my riding, plunged into the streams of their blood as into a river. The wheels of my chariot, which brings down the wicked and evil, were bespattered with blood and filth. With the bodies of their warriors I filled the plain, like grass. Their testicles I cut off, and tore out their privates like the seeds of cucumbers of June. Their hands I cut off . . . The chariots and their horses, whose riders had been slain at the beginning of the terrible onslaught, and who had been left to themselves, kept running back and forth . . ., – I put an end to their riders' fighting . . . They abandoned their tents and to save their lives they trampled the bodies of their fallen soldiers, they fled like young pigeons that are pursued. They were besides themselves; they held back their urine, but let their dung go into their chariots. In pursuit of them I dispatched my chariots and horses after them. Those among who had escaped, who had fled for their lives, wherever my charioteers met them, they cut them down with the sword.

Sennacherib (705–681 BC), King of Assyria, quoted in Luckenbill, *Ancient Records of Assyria and Babylonia*, 1926

Good officers never engage in general actions unless induced by opportunity or obliged by necessity.

Flavius Vegetius Renatus, *Military Institutions of the Romans*, c. AD 378

In half an hour, you will see how we shall lose one.

Prince Louis II, 'The Great Condé', 14 June 1658, comment to ally the Duke of Gloucester who told him he had never seen a battle before, as they were about to engage Marshal Turenne in the Battle of the Dunes, quoted in Weygand, *Turenne, Marshal of France*, 1930

It is a paradox to hope for victory without fighting. The goal of the man who makes war is to fight in the open field to win a victory.

Field Marshal Prince Raimond Montecuccoli, *Memoirie della guerra*, 1703

Battles concerning which one cannot say *why* and to what purpose they have been delivered are commonly the resource of ignorant men.

Field Marshal Maurice Comte de Saxe, *My Reveries*, 1732

The man who does things without motive or in spite of himself is either insane or a fool. War is decided by battles, and it is not finished except by them. They have to be fought, but is should be done opportunely and with all the advantages on your side . . .

Advantages are procured in battles every time that you determine to fight, or when a battle that you have meditated upon for a long time is a consequence of the manoeuvres that you have made to bring it on.

Frederick the Great, *Instruction for his Generals*, 1747

Battles decide the fate of a nation. In war it is absolutely necessary to come to decisive actions either to get out of the distress of war or to place the enemy in that position, or even to settle a quarrel which otherwise perhaps would never be finished. A wise man will make no movement without good reason, and a general of an army will never give battle if it does not serve some important purpose. When he is forced by his enemy into battle it is surely because he will have committed mistakes which force him to dance to the tune of his enemy.

Frederick the Great, *Instruction militaire*, 1761

A battle sometimes decides everything; and sometimes the most trifling thing decides the fate of a battle.

Napoleon, 9 November 1816, letter at Saint Helena

There is a moment in engagements when the least manoeuvre is decisive and gives the victory; it is the one drop of water that makes the vessel run over.

Napoleon, 1820, dictation at Saint Helena, ed. Herold, *The Mind of Napoleon*, 1955

The business of the English Commander-in-Chief being first to bring an Enemy's fleet to Battle on the most advantageous terms to himself (I mean that of laying his Ships close on board the Enemy, as expeditiously as possible); and secondly to continue them there until the Business is decided.

Admiral Lord Nelson, 1805, from the Order to the Fleet

Between a battle lost and a battle won, the distance is immense and there stand empires.

Napoleon, 15 October 1813, on the eve of the Battle of Leipzig

---

The issue of a battle is the result of a single instant, a single thought. The adversaries come into each other's presence with various combinations; they mingle; they fight for a length of time; the decisive moment appears; a psychological spark makes the decision; and a few reserved troops are enough to carry it out.

Napoleon, 1815, conversation on Saint Helena, ed., Herold, *The Mind of Napoleon*, 1955

A battle is a dramatic action which has its beginning, its middle, and its end. The battle order of the opposing armies and their preliminary manoeuvres until they come to grips form the exposition. The countermanoeuvres of the army which has been attacked constitute the dramatic complication. They lead in turn to new measures and bring about the crisis, and from this results the outcome or dénouement.

Napoleon (1769–1821), dictation at Saint Helena, ed. Herold, *The Mind of Napoleon*, 1955

MAXIM 15. In giving battle a general should regard it as his first duty to maintain the honour and glory of his arms. To spare his troops should be put a secondary consideration. But the same determination and perseverance which promote the former object are the best means of securing the latter. In a retreat you lose, in addition to the honour of your arms, more men than in two battles.

Napoleon, *The Military Maxims of Napoleon*, 1827, ed. Burnod

---

. . . the concept of the engagement lies at the root of all strategic action, since strategy is the use of force, the heart of which, in turn, is the engagement. So in the field of strategy we can reduce all military activity to the unitary concept of the single engagement, and concern ourselves exclusively with its purposes . . . every engagement, large or small, has its own particular purpose which is subordinate to the general one. That being so, the destruction and subjugation of the enemy must be regarded simply as a means toward the general end, which it obviously is.

Major General Carl von Clausewitz, *On War*, iv, 1832, tr. Howard and Paret

---

The battle may therefore be regarded as War concentrated, as the centre of effort of the whole war or campaign. As the sun's rays unite in the focus of a concave mirror in perfect image, and in the fulness of their heat; so the forces and circumstances of war unite in a focus in the great battle for one concentrated utmost effort.

Major General Carl von Clausewitz, *On War*, 1832

Combats may be quite independent of scientific combinations; they may become essentially dramatic; personal qualities and inspirations and a thousand other things frequently are the controlling elements.

Lieutenant General Antoine-Henri Baron de Jomini, *Summary of the Art of War*, 1838

Battles have been stated by some writers to be the chief and deciding features of war. This assertion is not strictly true, as armies have been destroyed by strategic operations without the occurrence of pitched battles, merely by a succession of inconsiderable affairs. It is also true that a complete and decided victory may bring similar results even though there may have been no grand strategic combinations. But it is the morale of armies, as well as of nations, more than anything else, which makes victories and their results decisive.

Lieutenant General Antoine-Henri Baron de Jomini, *Summary of the Art of War*, 1838

Battle is the ultimate to which the whole life's labor of an officer should be directed. He may live to the age of retirement without seeing a battle; still, he must always be getting ready for it as if he knew the hour and the day it is to break upon him. And then, whether it come late or early, he must be willing to fight – he must fight.

Brigadier General C. F. Smith to Colonel Lew Wallace, September 1861

It would be a service to humanity and to one's people to dispel this illusion and show what battles are. They are buffooneries, the spilling of blood. The actors, heroes, in the eyes of the crowd, are only poor folk torn between fear, discipline and pride. They play some hours at a game of advance and retreat, without ever meeting, closing with, even seeing closely, the other poor folk, the enemy, who are as fearful as they but who are caught in the same web of circumstance.

Colonel Charles Ardant du Picq, *Battle Studies*, 1880

It is always necessary in battle to do something which would be impossible for men in cold blood.

Colonel de Grandmaison, February 1911, lecture, Ecole de Guerre

---

War as we study it, positive in its nature, permits only of positive answers: there is no result without cause; if you seek the result, develop the cause, employ force.

If you wish your opponent to withdraw, beat him; otherwise nothing is accomplished, and there is only one means to that end: the battle.

Marshal of France Ferdinand Foch, *Principles of War*, 1913

Every engagement is a bloody and destructive test of physical and moral strength. Whoever has the greater sum of both left at the end is the victor.

In the engagement, the loss of morale has proved the major decisive factor. Once the outcome has been determined, the loss continues to increase, and reaches its peak only at the end of the action. This becomes the means of achieving the margin of profit in the destruction of the enemy's physical forces which is the real purpose of the engagement.

Major General Carl von Clausewitz, *On War*, iv, 1832, tr. Howard and Paret

---

Battle is the bloodiest solution. While it should not simply be considered as mutual murder – its effect . . . is rather a killing of the enemy's spirit than of his men – it is always true that the character of a battle, like its name, is slaughter (*schlact*), and its price is blood.

Major General Carl von Clausewitz, *On War*, iv, 1832, tr. Howard and Paret

---

Mountains, rivers, grass and trees: utter desolation. For ten miles the wind smells of blood from this new battlefield. The steed advances not, men speak not.

General Nogi Maruseki, after the Battle of Nanshan, 1904, quoted in Warner, *The Tide at Sunrise*, 1974

. . . Battles are decided in favor of the troops whose bravery, fortitude, and especially, whose endurance, surpasses that of the enemy's; the army with the higher breaking point wins the decision.

General of the Army George C. Marshall, *Memoirs of My Service in the World War* (written 1919–1923), published 1976

Battles are won by slaughter and manoeuvre. The greater the general, the more he contributes to the manoeuvre, the less he demands in slaughter.

Sir Winston S. Churchill, *The World Crisis*, Vol. II, 1923

In modern battle, which is delivered with combatants so far apart, man has come to have a horror of man. He comes to hand-to-hand fighting only to defend his body, or if forced to it by some fortuitous encounter . . . It may be said that seeks to catch the fugitive only for fear that he will turn and fight.

Colonel Charles Ardant du Picq, *Battle Studies*, 1880

Battle is an orgy of disorder. No level lawns or marker flags exist to aid us strut ourselves in vain display, but rather groups of weary wandering men seek gropingly for means to kill their foe. The sudden change from accustomed order to utter disorder – to chaos, but emphasize the folly of schooling to precision and obedience where only fierceness and habituated disorder are useful.

General George S. Patton, Jr., 27 October 1927, lecture 'Why Men Fight', *The Patton Papers*, Vol. I, 1972

Irrespective of plans and organization, all battles quickly resolve themselves into a series of more or less isolated head-on conflicts between small groups of combatants.

General George S. Patton, Jr., 30 January 1928, letter, *The Patton Papers*, Vol. I, 1972

But battle must not be seen as some kind of smooth-running conveyor belt on which the various technical combat resources are merged. Battle is a complex and fickle thing. So command and control must be ready to deal with abrupt changes in the situation, and sometimes to reshape an earlier plan radically.

Marshal of the Soviet Union Mikhail N. Tukhachevskiy, 1931–32, quoted in Simkin, *Deep Battle*, 1987

Battles are the principal milestones in secular history. Modern opinion resents this uninspiring truth, and historians often treat the decisions of the field as incidents in the dramas of politics and diplomacy. But great battles, won or lost, change the entire course of events, create new standards of values, new moods, new atmospheres, in armies and nations, to which all must conform.

Sir Winston S. Churchill, *Marlborough*, 1933

A 'decisive victory' is apt in military language to have a mystical sense which is by no means synonymous with its actual effect. The latest edition, in general much improved, of our own *Field Service Regulations* states that 'Battle is usually the decisive act in land warfare'. The statement hardly accords with the last two or three thousand years of experience. A survey of history, indeed, here gives a new meaning to the saying that the exception proves the rule; this military rule seems to be founded on a large preponderance of exceptions. It is extraordinary how many victories have been gained in wars without the effect that in normal language is meant by the word 'decisive'. (March 1936)

Captain Sir Basil Liddell Hart, *Thoughts on War*, 1944

The acid test of battle brings out the pure metal.

General George S. Patton, Jr., *War As I Knew It*, 1947

Battles are won by fire and by movement. The purpose of the movement is to get the fire in a more advantageous place to play on the enemy flank. This is from the rear of flank.

General George S. Patton, Jr., *War As I Knew It*, 1947

While the battles the British fight may differ in the widest possible ways, they have invariably two common characteristics – they are always fought uphill and always at the junction of two or more map sheets.

Field Marshal Viscount Slim, *Unofficial History*, 1959

Battle is more than a combination of fire and movement. It is the integration of fire, movement, and consciousness. The commander, therefore, cannot rest content with guiding the fire and directing the movement; he must guide the soldier's mental reactions to battle. Hence the commander is responsible for the mental preparation of his men no less than the for their physical and technical training and their being brought to battle. Paternal concern for the soldier and his welfare does not mean pampering him; far from it. Soldiers wrapped in cotton-wool will fall helpless victims to the terrors of war, being unprepared to meet the most terrible of all dangers. Sincere concern, on the other hand, will win the soldier's confidence and will prepare him to face the most trying circumstances.

General Yigal Allon, *The Making of Israel's Army*, 1960

The art of battle consists in maintaining and strengthening the psychological cohesion of one's troops while at the same time disrupting that of the enemy.

General André Beaufre, *An Introduction to Strategy*, 1965

Battles in which no one believes should not be fought.

Field Marshal Earl Wavell, 'Recollections' (unpublished), 1946

Battle is the most magnificent competition in which a human being can indulge. It brings out all that is best; it removes all that is base.

General George S. Patton, Jr., 27 June 1943, to the officers of the 45th Infantry Division before the invasion of Sicily

---

Battle is not a terrifying ordeal to be endured. It is a magnificent experience wherein all the elements that have made man superior to the beasts are present: courage, self-sacrifice, loyalty, help to others, devotion to duty. As you go in, you will perhaps be a little short of breath, and your knees may tremble . . . This breathlessness, this tremor, are not fear. It is simply the excitement every athlete feels just before the whistle blows – no, you will not fear for you will be borne up and exalted by the proud instinct of our conquering race. You will be inspired by a magnificent hate.

General George S. Patton, Jr., 20 December 1941, address to the 2nd Armored Division, *The Patton Papers*, Vol. II, 1974

Battle should no longer resemble a bludgeon fight, but should be a test of skill, a manoeuvre combat, in which is fulfilled the great principle of surprise by striking 'from an unexpected direction against an unguarded spot'.

Captain Sir Basil Liddell Hart, *Thoughts on War*, 1944

When things are going badly in battle the best tonic is to take one's mind off one's own troubles by considering what a rotten time one's opponent must be having.

Field Marshal Earl Wavell, *Other Men's Flowers*, 1944

BATTLEFIELD

It was a beautiful, calm, moonlight night. Suddenly a dog, which had been hiding under the clothes of a dead man, came up to us with a mournful howl, and then disappeared again immediately into his hiding place. He would lick his master's face, then run up to us again, only to return once more to his master. Whether it was the mood of the moment, whether it was the place, the time, the weather, or the action itself, or whatever it was, it is certainly true that nothing on any battlefield ever made such an impression on me. I involuntarily remained still, to observe the spectacle. This dead man, I said to myself, has perhaps friends, and he is lying there abandoned by all but his dog! What a lesson nature teaches us by means of an animal.

Napoleon (1769–1821), ed. Kircheisen, *The Memoirs of Napoleon I*, n.d.

There is in every battlefield a decisive point, the possession of which, more than of any other, helps secure the victory, by enabling its holder to make proper application of the principles of war . . .

The decisive point of a battlefield is determined . . . by the character of the position, the bearing of different localities upon the strategic object in view, and, finally, by the arrangement of the contending forces.

Lieutenant General Antoine-Henri Baron de Jomini, *Summary of the Art of War*, 1838

The dominant feeling of the battlefield is loneliness.

Field Marshal Viscount Slim, June 1941, to the officers of the 10th Indian Infantry Division

---

The village of Fuentes de Onoro, having been the field of battle, has not been much improved by the circumstance.

The Duke of Wellington, dispatch, 3–5 May 1811

The battlefield is the epitome of war. All else in war, when war is perfectly conducted, exists but to serve the forces of the battlefield and to assure success on the field.

Brigadier General S. L. A. Marshall, *Men Against Fire*, 1947

The battlefield is cold. It is the loneliest place which men may share together.

To the infantry soldier new to combat, its most unnerving characteristic is not that it invites him to a death he does not seek. To the extent necessary, a normal man may steel himself against the chance of death.

The harshest thing about the field is that it is empty. No people stir about. There is little or no signs of action. Overall there is a great quiet which seems more ominous than the occasional tempest of fire.

It is the emptiness which chill's a man's blood and makes the apple harden in his throat. It is the emptiness which grips him with paralysis. The small dangers which he had faced in his earlier life had always paid their dividend of excitement. Now there is great danger, but there is not excitement about it.

Brigadier General S. L. A. Marshall, *Men Against Fire*, 1947

I have always regarded the forward edge of the battlefield as the most exclusive club in the world.

General Sir Brian Horrocks, *A Full Life*, 1960

THE BAYONET

By push of bayonets, no firing till you see the whites of their eyes.

Frederick the Great, at the Battle of Prague, 6 May 1757

The bullet is a mad thing; only the bayonet knows what it is about.

Field Marshal Prince Aleksandr V. Suvorov, *The Science of Victory*, 1796

Strike once – throw the pagan from your bayonet; dead on your bayonet, one strikes at your neck with his sword. Sword at your neck – jump back a pace, hit again, strike another, strike a third; a champion will kill half-a-dozen, and I have seen more. Keep the bullet in your musket; three leap at you – knock down the first, shoot the second, do in the third with the bayonet.

Field Marshal Prince Aleksandr V. Suvorov, *The Science of Victory*, 1796

The bayonet has always been the weapon of the brave and the chief tool of victory.

Napoleon (1769–1821)

The bristling points and the glitter of the bayonets were fearful to look upon as they were levelled in front of a charging line; but they were rarely reddened with blood. The day of the bayonet is passed.

General John B. Gordon, *Reminiscences of the Civil War*, 1904

Rangers of Connaught! It is not my intention to expend any powder this evening. We'll do this business with cold steel.

General Sir Thomas Picton, 6 April 1812, to the 88th Foot before the assault on Badajoz

---

I beg leave to remind the [Cavalry] Board that very few people have ever been killed with the bayonet or the sabre, but the fear of having their guts explored with cold steel in the hands of battle maddened men has won many a fight.

General George S. Patton, Jr., January 1941, reply to the Cavalry Board's solicitation of his views, *The Patton Papers*, Vol. II, 1974

The attack will be pressed with the ruthless vigor that has routed every enemy formation opposing the 3rd Div. All men will be brought to the highest possible state of offensive spirit prior to the jump-off. *Bayonets will be sharpened.*

General John W. O'Daniel, March 1945, order to the 3rd Infantry 'Rock of the Marne' Division, as it attacked into Germany, quoted in Weigley, *Eisenhower's Lieutenants*, 1981. (The 3rd Infantry Division was the only Allied infantry division mentioned by name in the German Army war diary and was described as the best among the Allies.)

If you're close enough to stick 'em, you're close enough to shoot 'em.

Bill Mauldin's 'Willie and Joe' to a recruit in the movie *Willie and Joe*

Few men are killed by the bayonet; many are scared by it. Bayonets should be fixed when the fire fight starts.

General George S. Patton, Jr., *War As I Knew It*, 1947

. . . That weapon ceased to have any major tactical value at about the time the inaccurate and short-range musket was displaced by the rifle. But we have stubbornly clung to it – partly because of tradition which makes it inevitable that all military habits die a slow death, but chiefly because of the superstition that the bayonet makes troops fierce and audacious, and therefore likely to close with the enemy.

Brigadier General S. L. A. Marshall, *The Soldier's Load and the Mobility of a Nation*, 1950

Many a man has lost his life at close quarters by trying to 'pig-stick' a man who was cool enough to use his rifle instead. So long as a bare couple of yards separates men, the bullet can outreach the bayonet. (March 1932)

Captain Sir Basil Liddell Hart, *Thoughts on War*, 1944

BLITZKRIEG

Therefore, when I have won a victory I do not repeat my tactics but respond to circumstances in an infinite variety of ways.

Now an army may be likened to water, for just as flowing water avoids the heights and hastens to the lowlands, so an army avoids strength and strikes weakness.

And so as water shapes its flow in accordance with the ground, so an army manages its victory in accordance with the situation of the enemy.

And as water has no constant form, there are in war no constant conditions.

Thus, one able to gain the victory by modifying his tactics in accordance with the enemy situation may be said to be divine.

Sun Tzu, *The Art of War*, c. 500 BC, tr. Griffith

The rule of offensive warfare is that those who go the easy way prevail over their opponents. If your enemies are garrisoned in several places, there will inevitably be some places stronger and better manned than others. In that case, you should keep your distance from their strong points and attack their weak points; avoid places where they have many troops and strike where they have few – then you will not fail to win.

Liu Ji (AD 1310–1375), *Lessons of War*

Mobility, Velocity, Indirect Approach . . .

Colonel General Heinz Guderian (1888–1954) as quoted by Liddell Hart

---

In the series of swift German conquests, the air force combined with the mechanized elements of the land forces in producing the paralysis and moral disintegration of the opposing forces and of the nations behind. Its effect was terrific, and must be reckoned fully as important as that of the panzer forces. The two are inseparable in any valuation of the elements that created the new style of lightning war – the *blitzkrieg*.

Captain Sir Basil Liddell Hart, *Strategy*, 1954

BLOODSHED

I do not want them to get used to shedding blood so young; at their age they do not know what it means to be a Moslem or an infidel, and they will grow accustomed to trifling with the lives of others.

Saladin to Beha-ed-Din, when Saladin's children asked if they might kill a prisoner, AD 1191

We must spare our enemies or it will be our loss, since they and all that belong to them must soon be ours.

Anonymous Inca prince

If you wish to be loved by your soldiers, husband their blood and do not lead them to slaughter.

Frederick the Great, *Instructions to His General,* 1747

Nothing honours a general more than keeping his serenity in danger and facing it when there are chances of winning; but nothing so much eclipses his name as the useless shedding of the blood of his men.

General José de San Martin, 1818, quoted in Rojas, *San Martin,* 1957

Kindhearted people might . . . think there was some ingenious way to disarm or defeat an enemy without too much bloodshed, but pleasant as it sounds, it is a fallacy that must be exposed . . . It would be futile – even wrong – to try and shut one's eyes to what war really is from sheer distress at its brutality.

Major General Carl von Clausewitz, *On War*, iii, 1832, tr. Howard and Paret

---

Let us not hear of generals who conquer without bloodshed. If bloody slaughter is a horrible sight, then that is a ground for paying more respect to war, but not for making the sword we wear blunter and blunter by degrees from feelings of humanity, until some one steps in with one that is sharp and lops off the arm from our body.

Major General Carl von Clausewitz, *On War*, 1832, tr. Graham

---

War is only a means to results . . . These being achieved, no man has the right to cause another drop of blood to be shed.

Marshal of France Ferdinand Foch, quoted in Liddell Hart, *Through the Fog of War*, 1938

What the American people want to do is fight a war without getting hurt. You cannot do that any more than you can go into a barroom brawl without getting hurt.

Lieutenant General Lewis 'Chesty' Puller, 1951, quoted in Davis, *Marine*, 1962

. . . in such dangerous things as war, the errors which proceed from a spirit of benevolence are the worst . . . He who uses force unsparingly, without reference to the bloodshed involved, must obtain a superiority if his adversary uses less vigour in its application. The former then dictates the law to the latter, and both proceed to extremities, to which the only limitations are imposed by the amount of counteracting force on each side.

Major General Carl von Clausewitz, *On War*, 1832, tr. Graham

Wounds which in 1861, would have sent a man to the hospital for months, in 1865 were regarded as mere scratches, rather the subject of a joke than of sorrow. To new soldiers the sight of blood and death always has a sickening effect, but soon men become accustomed to it, and I have heard them exclaim on seeing a dead comrade borne to the rear, 'Well, Bill has turned up *his* toes to the daisies.'

General of the Army William T. Sherman, *Memoirs of General W.T. Sherman*, 1875

BOLDNESS

A decent boldness ever meets with friends.

Homer, *The Odyssey*, vii, *c.* 800 BC, tr. Pope

My conclusion is, then, that as fortune is variable and men fixed in their ways, men will prosper as long as they are in tune with the times and will fail when they are not. However, I will say that in my opinion it is better to be bold than cautious, for fortune is a woman and whoever wishes to win her must importune and beat her, and we may observe that she is more frequently won by this sort than by those who proceed more deliberately.

Niccolo Machaivelli, *The Prince*, 1513

I wish to have no connection with any ship that does not sail *fast*; for I intend to go *in harm's way*.

Admiral John Paul Jones (1747–1792)

Desperate affairs require desperate measures.

Admiral Lord Nelson (1758–1805)

Boldness governed by superior intellect is the mark of a hero. This kind of boldness does not consist in defying the natural order of things and in crudely offending the laws of probability; it is rather a matter of energetically supporting that higher form of analysis by which genius arrives at a decision: rapid, only partly conscious weighing of the possibilities. Boldness can lend wings to intellect and insight; the stronger the wings then, the greater the heights, the wider the view, and the better results; though a greater prize, of course, involves greater risks.

Major General Carl von Clausewitz, *On War*, iii, 1832, tr. Howard and Paret

---

. . . this noble capacity to rise above the most menacing dangers should also be considered as a principle in itself, separate and active. Indeed, in what field of human activity is boldness more at home than in war?

A soldier, whether drummer boy or general, possess no nobler quality; it is the very metal that gives edge and lustre to the sword.

Major General Carl von Clausewitz, *On War*, iii, 1832, tr. Howard and Paret

---

'Safety first' is the road to ruin in war.

Sir Winston S. Churchill, 3 November 1940, telegram to Anthony Eden

---

A bold general may be lucky, but no general can be lucky unless he is bold. The general who allows himself to be bound and hampered by regulations is unlikely to win a battle.

Field Marshal Earl Wavell, *General and Generalship*, 1941

It is my experience that bold decisions give the best promise of success. But one must differentiate between strategical or tactical boldness and military gamble. A bold operation is one in which success is not a certainty but which in case of failure leaves one with sufficient forces in hand to cope with whatever situation may arise. A gamble, on the other hand, is an operation which can lead either to victory or to the complete destruction of one's forces. Situations arise where even a gamble may be justified – as, for instance, when in the normal course of events defeat is merely a matter of time, when the gaining of time is therefore pointless and the only chance lies in an operation of great risk.

Field Marshal Erwin Rommel, *The Rommel Papers*, 1953

. . . a distinguished commander without boldness is unthinkable.

Major General Carl von Clausewitz, *On War*, 1832

The power of the various emotions is sharply reduced by the intervention of lucid thought and more, by self-control. Consequently, boldness grows *less common in the higher ranks*. Even if the growth of an officer's perception and intelligence does not keep pace with his rise in rank, the realities of war will impose their conditions and concerns on him. Indeed their influence on him will be greater the less he really understands them. In war, this is the main basis for the experience expressed in the French proverb, 'The same man who shines at the second level is eclipsed at the top.' Nearly every general known to us from history as mediocre, even vacillating, was noted for dash and determination as a junior officer.

Major General Carl von Clausewitz, *On War*, iii, 1832, tr. Howard and Paret

---

The measure may be thought bold, but I am of the opinion the boldest are the safest.

Admiral Lord Nelson, 24 March 1801, to Sir Hyde Parker urging vigorous action against the Russians and Danes

---

Perhaps I should not insist on this bold manoeuvre, but it is my style, my way of doing things.

Napoleon, 1813, letter to Prince Eugene

A general should show boldness, strike a decided blow, and manoeuvre upon the flank of his enemy. The victory is in his hands.

Napoleon, *The Military Maxims of Napoleon*, 1831, tr. D'Aguilar

---

There are occasions when daring and risky operations, boldly executed, can pay great dividends.

General Mathew B. Ridgway, *Soldier*, 1956

BRAVERY

. . . Whoever wants to see his own people again must remember to be a brave soldier: that is the only way of doing it. Whoever wants to keep alive must aim at victory. It is the winners who do the killing and the losers who get killed. And those who want money must try to win battles. The winners can not only keep what they have themselves, but can take what belongs to the losers.

Xenophon (c. 432–c. 352 BC), *The Persian Expedition (Anabasis)*, tr. Warner

---

Few men are born brave; many become so through training and force of discipline.

Flavius Vegetius Renatus, *Military Institutions of the Romans*, c. AD 378

He supposes all men to be brave at all times and does not realize that the courage of the troops must be reborn daily, that nothing is so variable, and that the true skill of a general consist in knowing how to guarantee it by his dispositions, his positions, and those traits of genius that characterize great captains. . . . It is of all the elements of war the one that is most necessary to learn.

Field Marshal Maurice Comte de Saxe, *My Reveries*, 1732

One cannot think that blind bravery gives victory over the enemy.

Field Marshal Prince Aleksandr V. Suvorov (1729–1800)

No, shoot them all, I do not wish them to be brave.

Lieutenant General Thomas 'Stonewall' Jackson, in response to someone who deplored the necessity to kill so many brave enemy troops, quoted in Fuller, *Grant and Lee*, 1933

Each man must think not only of himself, but think of his buddy fighting beside him. We don't want yellow cowards in this army, to come back to the States after the war and breed more like them. The brave men will breed brave men. One of the bravest men I saw in the African campaign was one of the fellows I saw on top of a telegraph pole in the midst of furious fire while we were plowing towards Tunis. I stopped and asked him what in the hell he was doing there at a time like that. He answered, 'Fixing the wire, sir.' Isn't it a little unhealthy right now? I asked. 'Yes sir, but this God damn wire has got to be fixed.'

General George S. Patton, Jr., 1944, speech

If we take the generally accepted definition of bravery as a quality which knows not fear, I have never seen a brave man. All men are frightened. The more intelligent they are, the more they are frightened. The courageous man is the man who forces himself, in spite of his fear, to carry on. Discipline, pride, self-respect, self-confidence, and the love of glory are attributes which will make a man courageous even when he is afraid.

General George S. Patton, Jr., *War As I Knew It*, 1947

When I was in the navy I served in an office staffed by over a hundred temporary officers of the Royal Navy Voluntary Reserve. As civilians in uniform we found in many instances that the only way to get things done quickly was to short-circuit the system by getting verbal decisions and letting the paper work tag along later. This required senior officers of the regular navy to give verbal decisions involving the expenditure of thousands of pounds without any paper cover. These officers were as brave as lions and would have risked their lives in a destroyer torpedo attack without a second thought, but they balked at risking their jobs on a verbal decision. Now and again we would find some cheerful young commander or captain who did not hesitate, who was as brave in the office as he was at sea. Commenting on such a regular officer and on his way of doing business we would say, 'He's a good one. I bet he's got private means.' Invariably investigation proved we were right. The officers who were brave in the Admiralty were the officers who had an independent income, who could afford to resign from the navy if necessary without bringing financial disaster to their wives and children. It started as a joke with us to say that brave officer in the office probably had private means, and then it got beyond a joke and turned into an axiom.

Nevil Shute, *Slide Rule* and *Other Proceedings*

I do not believe that there is any man who would not rather be called brave than have any other virtue attributed to him.

Field Marshal Viscount Slim, *Courage* and Other Broadcasts, 1957

BUREAUCRACY

The ancients had a great advantage over us in that their armies were not trailed by a second army of pen-pushers.

Napoleon, ed. Herold, *The Mind of Napoleon*, 1955

Gentlemen:

Whilst marching to Portugal to a position which commands the approach to Madrid and the French forces, my officers have been diligently complying with your request which has been sent to H.M. ship from London to Lisbon and then by dispatch rider to our headquarters.

We have ennumerated our saddles, bridles, tents and tent poles, and all manner of sundry items for which His Majesty's Government holds me accountable. I have dispatched reports on the character, wit, spleen of every officer. Each item and every farthing has been accounted for, with two regrettable exceptions for which I beg you your indulgence.

Unfortunately, the sum of one shilling and ninepence remains unaccounted for in one infantry battalion's petty cash and there has been a hideous confusion as to the number of jars of raspberry jam issued to one cavalry regiment during a sandstorm in western Spain. This reprehensive carelessness may be related to the pressure of circumstances since we are at war with France, a fact which may come as a bit of surprise to you gentlemen in Whitehall.

This brings me to my present purpose, which is to request elucidation of my instructions from His Majesty's Government, so that I may better understand why I am dragging an army over these barren plains. I construe that perforce it must be one of two alternative duties, as given below. I shall pursue one with the best of my ability but I cannot do both.

1. To train an army of uniformed British clerks in Spain for the benefit of the accountants and copy-boys in London, or perchance,

2. To see to it that the forces of Napoleon are driven out of Spain.

Your most obedient servant,

Wellington

Although a revolutionary government, none was ever so much under the domination of red tape as the one at Richmond. The martinets who controlled it were a good deal like the hero of Molière's comedy, who complained that his antagonist had wounded him by thrusting in *carte*, when according to the rule, it should have been in *tierce*. I cared nothing for the form of a thrust if it brought blood. I did not play with foils.

Colonel John S. Mosby, *Mosby's War Reminiscences*, 1887

The business of even the most military department . . . is in itself more akin to civil than to military life: but it by no means follows that those departments would be better administered under men of civil habits of thought than by those of military training. The method exists for the result, and an efficient fighting body is not to be attained by weakening the appreciation of military necessities at the very fountainhead.

Rear Admiral Alfred Thayer Mahan, *Naval Administration and Warfare*, 1908

If [the bureaucracy] consequently tends to overvalue the orderly routine and observant of the system by which it receives information, transmits orders, checks expenditures, files returns, and in general, keeps with the Service the touch of paper; in short, the organization which has been created for facilitating its own labors.

Rear Admiral Alfred Thayer Mahan, *Naval Administration and Warfare*, 1908

This man, with all his charm, sincerity and self-sacrificing hard work, was one of the type who, through a profound sense of duty, coupled with a microscopic vision of the reality of things, spent hours every day blowing blue pencil dust into the military machine, and in consequence nearly brought our part of it to a standstill. To get a new establishment through his office was like playing golf on course with a hundred bunkers between each green. Every item was scrutinised, criticised and discussed, and then re-scrutinised, re-criticised and re-discussed *ad infinitum*, with the result that delays were never ending. It was bureaucracry in a nightmare.

Major General J. F. C. Fuller, *Memoirs of an Unconventional Soldier*, 1936

. . . While giving lip service to the humanitarian values and while making occasional spectacular and extravagant gestures of sentimentality, those whose task it was to shape personnel policy have tended to deal with man power as if it were motor lubricants or sacks of potatoes. They have destroyed the name and tradition of old and honored regiments with the stroke of a pen, for convenience's sake. They have uprooted names and numbers which had identity with a certain soil and moved them willy-nilly to another soil. They have moved men around as if they were pegs and nothing counted but a specialist classification number. They have become fillers-of-holes rather than architects of the human spirit.

Brigadier General S. L. A. Marshall, *Men Against Fire*, 1947

Well, you're apt to get so tied up in administrative processes, in peacetime that you forget exactly what you're in business for. The emphasis is on something else – not on a ruckus which doesn't exist. But when you get out there and start fighting, it is the people who are actually *fighting* who see these things and realize that they are necessary. Then they've still got to prod and slug and struggle, all the way up through the normal bureaucracy which every military organization owns.

General Curtis LeMay, *Mission With LeMay*, 1965

Several years in this field have led me to think that the mentality of our military leaders, indeed the mentality of the bulk of both military and civilian mid-level executives in the Pentagon, has become largely that of efficiency-worshipping functionaries. That in itself is a harmless enough state of affairs as long as business-as-usual and riding the bicycle of bureaucratic procedure continues to be the order of the day. But my experience – as a fighter pilot operating from aircraft carriers at sea, observing from first hand the unprecedented battle and intelligence scenes of the Vietnam War, and eight years of prison-camp revelations – impressed me how such mindsets breed disaster when the unexpected occurs, and then it becomes necessary to steer an institution into uncharted waters.

Admiral James Stockdale, 'Educating Leaders', *Washington Quarterly*, Winter 1983

It is a good thing to see inside of the War Office for a short time, as it prevents one from having any respect for an official letter, but it is a mistake to stay there too long.

Field Marshal Earl Haig, 1902, letter to his nephew

CALCULATION

The reasonable course of action in any use of arms starts with calculation. Before fighting, first assess the relative sagacity of the military leadership, the relative strength of the enemy, the size of the armies, the lie of the land, and the adequacy of provisions. If you send troops out only after making these calculations, you will never fail to win.

Liu Ji, 1310–1375, *Lessons of War*

War is essentially a calculation of probabilities.

Napoleon (1761–1821)

Military science consists in calculating all the chances accurately in the first place, and then in giving accident exactly, almost mathematically, its place in one's calculations. It is upon this point that one must not deceive oneself, and yet a decimal more or less may change all. Now this apportioning of accident and science cannot get into any head except that of a genius . . . Accident, hazard, chance, call it what you may, a mystery to ordinary minds, becomes a reality to superior men.

Napoleon, quoted in Chandler, *The Campaigns of Napoleon*, 1966

In short, absolute, so-called mathematical factors never find a firm basis in military calculations. From the very start there is an interplay of possibilities, probabilities, good luck and bad that weaves its way throughout the length and breadth of the tapestry. In the whole range of human activities, war most closely resembles a game of cards.

Major General Carl von Clausewitz, *On War*, i, 1832, tr. Howard and Paret

CANNAE

. . . For although the Romans had beyond any dispute been worsted in battle and their reputation annihilated, yet through the peculiar virtues of their constitution and their ability to keep their heads they not only won back their supremacy in Italy and later defeated the Carthaginians, but within a few years had made themselves masters of the whole world.

Polybius, *The Rise of the Roman Empire*, c. 125 BC

The analysis of Cannae is ended. Before passing to the recital of Pharsalus, we cannot resist the temptation . . . to say a few words about the battles of Hannibal.

These battles have a particular character of stubbornness explained by the necessity for overcoming the Roman tenacity. It may be said that to Hannibal victory was not sufficient. He must destroy. Consequently he always tried to cut off all retreat for the enemy. He knew that with Rome, destruction was the only way of finishing the struggle.

He did not believe in the courage of despair in the masses; he believed in terror and he knew the value of surprise in inspiring it.

Colonel Charles Ardnant du Picq, *Battle Studies*, 1880

A complete battle of Cannae is rarely met in history. For its achievement, a Hannibal is needed on the one side and a Terrentius Varro on the other, both cooperating for the attainment of the great objective.

A Hannibal must possess, if not a superiority in numbers, the knowledge of how to create one. It is desirable for this purpose that the general combine in himself something of a Scharnhorst . . . to weld together a strong army; something of a Moltke to assemble it solely against the principal enemy; of a Frederick the Great or a Napoleon to direct the principal attack against the flank or rear . . . Lastly there are needed to comprehend the scheme of manoeuvre of higher headquarters.

All the desirable qualifications will not be met with or combined in a single person on either side. A few of Hannibal's qualities and some of the means at his disposal have been possessed by other generals in history. Terrentius Varro, on the other hand, has existed during all periods of history.

Field Marshal Alfred Graf von Schlieffen, *Cannae*, 1903

. . . The basic laws of battle remained unchanged since Hannibal's victory over the Consul Terrentius Varro at Cannae. A battle of annihilation can still be fought according to the same plan which Hannibal devised many centuries ago. The enemy front should not be the objective of the main attack. Neither the main concentration of force nor the reserves should be used against the enemy front. Only the smashing of the enemy's flanks is essential. Annihilation is complete if the enemy is also attacked from the rear.

Field Marshal Alfred Graf von Schlieffen, *Cannae*, 1903

According to the principle of Cannae a broad battle line goes forth against a narrower, but generally deeper one. The overlapping wings turn against the flanks, the cavalry preceding them, against the rear. Should the flanks be separated from the centre, for some reason or other, it is not necessary to assemble them against the latter in order to continue jointly the march for a surrounding attack, as they can immediately advance, by the shorter road, against flank or rear. This was what Moltke called 'the junction of separated units on the field of battle' and declares it the highest achievement of a general. It is also the most effective and, of course, the most risky.

Field Marshal Alfred Graf von Schlieffen, *Cannae*, 1903

The battle of Cannae clearly shows that the decisive victory is not the result of a brilliant strategic idea, nor of its effective and skillful execution alone. Victory is also dependent upon the attitude of the enemy commander, who must be caught unawares and ignorant of the true intentions of his opponent until it is no longer possible for him to act on his own initiative. The blow must fall swiftly and unexpectedly; strength must be met by weakness if a battle of annihilation shall materialize . . . One belligerent must surprise, the other must be surprised. Only and when the two commanders play these respective roles will a battle lead to the annihilation of one army.

General Waldemar Erfurth, *Surprise*, 1943

THE CAPTAIN

A captain cannot be too careful of the company the state has committed to his charge. He must pay the greatest attention to the health of his men, their discipline, arms, accoutrements, ammunition, clothes and necessaries.

His first object should be, to gain the love of his men, by treating them with every possible kindness and humanity, enquiring into their complaints, and when well founded, feeling them redressed. He should know every man of his company by name and character. He should often visit those who are sick, speak tenderly to them, see that the public provision, whether of medicine or diet, is duly administered, and procure them besides such comforts and conveniences as are in his power. The attachment that arises from this kind of attention to the sick and wounded, is almost inconceivable; it will moreover be the means of preserving the lives of many valuable men.

Major General Friedrich Baron von Steuben, *Revolutionary War Drill Manual*, 1794

The company is the true unit of discipline, and the captain is the company. A good captain makes a good company, and he should have the power to reward as well as punish. The fact that soldiers would naturally like to have a good fellow for their captain is the best reason why he should be appointed by the colonel, or by some superior authority, instead of being elected by the men.

General of the Army William T. Sherman, *Memoirs of General W. T. Sherman*, 1875

A loud-mouthed, profane captain who is careless of his personal appearance will have a loud-mouthed, profane, dirty company. Remember what I tell you. Your company will be the reflection of yourself. If you have a rotten company it will be because you are a rotten captain.

Major C. A. Bach, US Army, 1917, farewell instructions to the graduating student officers, Fort Sheridan, Wyoming

CASUALTIES

. . . all great captains, chieftains, and men of charge have holden for a maxim to preserve by all means possible the lives of their soldiers, and not to employ and hazard them upon every light occasion, and therewithal to esteem the preservation of the lives of a very few of their soldiers before the killing of great numbers of the enemy . . .

John Smythe (*c.* 1580–1631), *Certain Discourses Military*

It is very difficult to do one's duty. I was considered a barbarian because at the storming of the Praga 7,000 people were killed. Europe says that I am a monster. I myself have read this in the papers, but I would have liked to talk to people about this and ask them: is it not better to finish a war with the death of 7,000 people rather than to drag it on and kill 100,000?

Field Marshal Prince Aleksandr V. Suvorov (1729–1800), quoted in Longworth, *The Art of Victory*, 1966

Well if there was economy of life, it is something for which I will be grateful to the end of my days. When I thought about the number who were killed, nothing could console me except the thought maybe by God's grace and hard effort we had saved some that might otherwise have been slain.

General of the Army Douglas MacArthur, 11 May 1949, quoted by Douglas Southall Freeman at the US Naval War College

---

. . . Far from being a handicap to command, compassion is the measure of it. For unless one values the lives of his soldiers and is tormented by their ordeals, he is unfit to command. He is unfit to appraise the cost of an objective in terms of human lives.

To spend lives, knowingly, deliberately – even cruelly – he must steel his mind with the knowledge that to do less would cost only more in the end. For if he becomes tormented by the casualties he must endure, he is in danger of losing sight of his strategic objectives. Where the objective is lost, the war is prolonged and the cost becomes infinitely worse.

General of the Army Omar Bradley, *A Soldier's Story*, 1951

It is a familiar experience that the winner's casualties in the course of an engagement show little difference from the loser's. Frequently there is no difference at all, and sometimes even an inverse one. The really crippling losses, those the vanquished does not share with the victor, only start with his retreat. The feeble remnants of badly shaken battalions are cut down by cavalry; exhausted men fall by the wayside; damaged guns and caissons are abandoned, while others are unable to get away quickly enough on poor roads and taken by the enemy's cavalry; small detachments get lost in the night and fall defenceless into the enemy's hands. Thus a victory usually only starts to gather weight after the issue has already been decided . . .

Major General Carl von Clausewitz, *On War*, iv, 1832, tr. Howard and Paret

---

The concerned officer taking advantage of the situation will always seek a means of economizing manpower and thus reducing the number of people exposed to danger.

Admiral P. S. Nakhimov, during the siege of Sevastopol, 1855, *Documenti i Materialy*, 1954

His majesty's millions conquer the strong foe.
Field battles and siege result in mountains of corpses.

How can I, in shame, face their fathers?
Songs of triumph today, but how many have returned?

General Nogi Maresuki, 1905, at the surrender of Port Arthur, quoted in Warner, *The Tide at Sunrise*, 1974

Let there be dismissed at once, as preposterous, the hope that war can be carried on without some one or something being hurt.

Rear Admiral Alfred Thayer Mahan, *Some Neglected Aspects of War*, 1907

After all, the most distressing and the most expensive thing in war is – to get men killed.

Major General J. F. C. Fuller, *Memoirs of an Unconventional Soldier*, 1936

A big butcher's bill is not necessarily evidence of good tactics.

Field Marshal Earl Wavell, 1940, reply to Churchill when reproached for evacuating British Somaliland with only 260 casualties

---

The successful soldier wins his battles cheaply so far as his own casualties are concerned, but he must remember that violent attacks, although costly at the time, save lives in the end.

General George S. Patton, Jr., *War As I Knew It*, 1947

In battle, casualties vary directly with the time you are exposed to effective fire. Your own fire reduces the effectiveness and volume of the enemy's fire, while rapidity of attack shortens the time of exposure. A pint of sweat will save a gallon of blood.

General George S. Patton, Jr., *War As I Knew It*, 1947

War consumes men; that is its nature . . . That the large masses which were led into battle would suffer heavy casualties, in spite of all tactical counter-measures, was unfortunately a matter of course.

General Erich von Ludendorff, *My War Memories, 1914–1918*, 1919

Every waking and sleeping moment, my nightmare is the fact that I will give an order that will cause countless numbers of human beings to lose their lives. I don't want my troops to die. I don't want my troops to be maimed.

It is an intensely personal, emotional thing for me . . . Any decision you have to make that involves the loss of human life is nothing you do lightly. I agonize over it.

It's not purely a question of accomplishing the mission . . . But it's a question of accomplishing the mission with a minimum loss of human life and within an effective time period.

General H. Norman Schwarzkopf, 5 February 1991, interview in the *Washington Post*

I personally have experienced the job of writing letters of condolence by the hundreds, by the thousands, to bereaved mothers and wives. That is a very sobering experience.

General of the Army Dwight D. Eisenhower (1890–1969)

You find yourself encountering considerable tragedy . . . in overcoming a natural grief concerning casualties among the people whom you've led. You possess a natural inborn repugnance against killing people; yet you know you're going to have to do it. It's rough. It used to be particularly vile when I realized that I'd lost someone, and felt that I shouldn't have lost them because I made a mistake or somebody else made a mistake.

You must wring the greatest possible benefit out of every lesson.

And you must train yourself grimly to adopt a philosophical attitude with regard to those losses. If you're going to fight you're going to have some people killed. But if you have done everything humanly possible to prepare for that mission and plan it properly, and you have observed that it was properly executed, and you have attained the results which you wished to attain – Then you can think, and feel in your heart, 'The losses were paid for'.

I think every man who has had to nerve himself into the grim responsibility of leading other men in combat will agree to this:

You have to pay a price in warfare, and part of the price is human life.

So, while you don't like to see people killed, if you're going to fight you're going to have those losses. And, if you keep flying under simulated combat conditions and in every sort of weather, you're going to have losses. But if you've done the best you can in the way of preparation, planning, and execution, then you feel that if those people who were sacrificed came up in front of your desk and looked you in the eye, you could look *them* in the eye and say, 'I think it was a good operation.'

If you can't imagine yourself doing that, then you ought to start worrying.

While it tortured me to lose people in the ETO and in the Pacific, I think that in most cases I would be willing to meet them, and I would say, 'Well, you were properly expended, Gus. It was part of the price.'

Can't tell. I might meet them eventually.

General Curtis LeMay, *Mission With LeMay*, 1965

CATCHWORDS

There are three things against which the human mind struggles in vain: stupidity, bureaucracy, and catchwords. All three are perhaps alike inasmuch as they are necessary. I prefer to leave the hopeless fight against stupidity to my more sagacious contemporaries; admit unqualified defeat in my struggle with military bureaucracy; and propose . . . to join battle with a few catchwords in the military sphere at home.

If I admit the necessity for catchwords I mean that they are necessary for all those who are unable to think for themselves. This necessity or utility of catchwords is thus unimpugnably demonstrated, and the following observations have no other object than to stimulate some one or other of my aforesaid contemporaries to think for himself and, whenever a catchword is uttered, to confront him with the question, Is it true?

In my own professional military sphere I make war on catchwords and slogans for the very definite reason that here their effect can and indeed must be deadly, in the literal sense of the word, because human lives are sacrificed to military catchwords – assuredly not from any evil intention, but simply from lack of independent thought.

Colonel General Hans von Seekt, *Thoughts of a Soldier*, 1930

CAVALRY

Those who face the dizzying heights and cross the dangerous defiles, who can shoot at a gallop as if in flight, who are in the vanguard when advancing and in the rearguard when withdrawing, are called cavalry generals.

Zhuge Liang (AD 180–234), *The Way of the General*

Why then should it seem wonderful that a firm and compact body of infantry should be able to sustain a cavalry attack, especially since horses are prudent animals and when they are apprehensive of danger cannot easily be brought to rush into it? You should also compare the force that impels them to advance with that which makes them retreat; you will then find that the latter is much more powerful than the former. In the one case, they feel nothing but the prick of a spur, but in the other they see a rank of pikes and the sharp weapons presented to them. So, you can see from both ancient and modern examples that good infantry will always be able not only to make headway against cavalry but generally to get the better of them. But if you argue that the fury with which the horses are driven to charge an enemy makes them consider a pike no more than a spur, I answer that even though a horse has begun to charge, he will slow down when he draws near the pikes and, when he begins to feel their points, will either stand stock still or wheel off to the right or left. To convince yourself of this, see if you can ride a horse against a wall; I fancy you will find very few, if any – however spirited they may be – that can be made to do that.

Niccolo Machiavelli, *The Art of War*, 1521

. . . as for the cavalry attack, I have considered it necessary to make it so fast and in such close formation for more than one reason: (1) so that this large movement will carry the coward along with the brave man; (2) so that the cavalryman will not have time to reflect; (3) so that the power of our big horses and their speed will certainly overthrow whatever tries to resist them; and (4) to deprive the simple cavalryman of any influence in the decision of such a big affair.

Frederick the Great, *Instructions to His Generals*, 1747

. . . the possession of a horse furnishes a man in the hour of his greatest danger with the means of saving himself and it cannot be expected of him that he should not avail of it.

Field Marshal Colmar Baron von der Goltz, *The Nation in Arms*, 1887

MAXIM 51. It is the function of cavalry to follow up the victory, and to prevent the beaten army from rallying.

MAXIM 86. A cavalry general should be a master of practical science, know the value of seconds, despise life and not trust to chance.

Napoleon, *The Military Maxims of Napoleon*, 1827, ed. Burnod

---

Useful as cavalry may be against simple bodies of broken, demoralized troops, still opposed to the whole, it becomes again only the auxiliary arm, because the troops in retreat can re-employ fresh reserves to cover their movement, and therefore, at the next trifling obstacle of ground, by combining all arms they can make a stand with success.

Major General Carl von Clausewitz, *On War*, iv, 1832, tr. Howard and Paret

---

The principal value of cavalry is derived from its rapidity and mobility. To these characteristics may be added its impetuosity, but we must be careful lest a false application be made of the last.

Lieutenant General Antoine Henri Baron de Jomini, *Summary of the Art of War*, 1838

. . . The cavalry is always the aristocratic arm which loses very lightly, even if it risks at all. At least is has the air of risking all, which is something at any rate. It has to have daring and daring is not so common. But the merest infantry engagements in equal numbers costs more than the most brilliant cavalry raid.

Colonel Charles Ardnant du Picq, *Battle Studies*, 1880

. . . 'No longer like Frederick at the end of the day do we hurl our jingling squadrons upon the tottering foe.' No indeed; better not: they might still have one of those machine-guns with them which refuses to totter.

General Sir Ian Hamilton, comment in his introduction to *Thoughts of a Soldier* by Hans von Seekt, 1930

In one respect a cavalry charge is very like ordinary life. So long as you are all right, firmly in your saddle, your horse in hand, and well armed, lots of enemies will give you a wide berth. But as soon as you have lost a stirrup, have a rein cut, have dropped your weapon, are wounded, or your horse is wounded, then is the moment when from all quarters enemies rush upon you.

Sir Winston Churchill, *The River War*, 1899

For my part I maintain it would be as reasonable to introduce the elephants of Porus on to a modern battle-field as regiments of lancers and dragoons who are too much imbued with the true cavalry spirit to use fire-arms and too sensible, when it comes to the pinch, to employ their boasted *armes blanches*.

General Sir Ian Hamilton, military attaché's report on the Russo–Japanese War, *Soul of an Army*, 1921

In war only what is simple can succeed. I visited the staff of the Cavalry Corps. What I saw there was not simple.

Field Marshal Paul von Hindenburg, at the 1932 German Army manoeuvres, quoted in Guderian, *Panzer Leader*, 1953

. . . As to the Army, heaven knows that we have suffered enough from what is known in the services and outside as the 'cavalry mind'. For the past century and more, even to the outbreak of war in 1939, nobody had much chance of promotion to high command in the army unless he was a cavalry soldier, and it is perfectly well known that in order to get cavalry soldiers as far as high command, it was often necessary to excuse them from the more difficult qualifications in the intervening stages of their career. Unfortunately, they were too apt to develop the mentality of the animals they were so enthusiastic about. If some such provision had not been made, the working infantry, sappers, and gunners would generally have reached the higher posts first, but, as it is, how many of the high commanders of the last 150 years, up to 1939, were cavalry officers? Because of the cavalry influence, and consequently, of this system of making appointments, the idea of using cavalry inevitably persisted for a quarter of a century after the presence of a horse on a battlefield could only be considered a symptom of certifiable lunacy.

Air Marshal Sir Arthur 'Bomber' Harris, *Bomber Offensive*, 1947

CENTRE OF GRAVITY

. . . one must keep the dominant characteristics of both belligerents in mind. Out of these characteristics a certain centre of gravity develops, the hub of all power and movement, on which everything depends. That is the point against which all our energies should be directed.

Major General Carl von Clausewitz, *On War*, viii, 1832, tr. Howard and Paret

---

For Alexander, Gustavus Adolphus, Charles XII, and Frederick the Great, the centre of gravity was their army. If the army had been destroyed, they would all have gone down in history as failures. In countries subject to domestic strife, the centre of gravity is generally the capital. In small countries that rely on large ones, it is usually the army of their protector. Among alliances, it lies in the community of interest, and in popular uprisings it is the personalities of the leaders and public opinion. It is against these that our energies should be directed. If the enemy is thrown off balance, he must not be given time to recover. Blow after blow must be aimed in the same direction: the victor, in other words, must strike with all his strength and not just against a fraction of the enemy's. Not by taking things the easy way – using superior strength to filch some province, preferring the security of this minor conquest to great great success – but by constantly seeking out the centre of his power, by daring to win all, will one really defeat the enemy.

Major General Carl von Clausewitz, *On War*, viii, 1832, tr. Howard and Paret

. . . They [the North Vietnamese Politburo] were . . . correct in the selection of a strategic center of gravity. They knew, especially after their battlefield defeat in the Ia Drang, that it could not be the US Army. Neither could it be the Army of the Republic of Vietnam (ARVN) as long as the United States was acting as its protector. They knew they did not have the military means to threaten our homeland or even to seize Saigon, the capital of South Vietnam. Based on their successful experiences against the French the center of gravity that they identified was the alliance between the United States and South Vietnam, and particularly the 'community of interest . . . personalities of the leaders and public opinion'.

Colonel Harry G. Summers, Jr., *On Strategy: The Vietnam War in Context*, 1981

CHANCE

So now, each of you, turn straight for the enemy, live or die – that is the lovely give and take of war.

Hector to the Trojan allies in Homer, *The Iliad*, xvii, *c.* 800 BC, tr. Fagles

---

If it really *was* Achilles who reared besides the ships,
all the worse for him – if he wants his fill of war.
I for one, I'll never run from his grim assault,
I'll stand up to the man – see if he drags off glory
or I drag it off myself! The god of war is impartial:
he hands out death to the man who hands out death.

Hector in Homer, *The Iliad*, *c.* 800 BC, xviii, tr. Fagles

In war we must always leave room for strokes of fortune, and accidents that cannot be foreseen.

Polybius, *The Rise of the Roman Empire*, *c.* 125 BC

. . . they did not reflect on the common chances of war where such trifling causes as groundless suspicion or sudden panic or superstitious scruples frequently produce great disasters, or when a general's mismanagement or a tribune's mistake is a stumbling block to an army.

Julius Caesar, *The Civil War*, *c.* 45 BC, tr. Moses Hadad

We should make war without leaving anything to chance, and in this especially consists the talent of a general.

Field Marshal Maurice Comte de Saxe, *My Reveries*, 1732

The French are what they were in Caesar's time, and as he described them, brave to excess but unstable . . . As it is impossible for me to make them what they ought to be, I get what I can out of them and try to leave nothing of importance to chance.

Field Marshal Maurice Comte de Saxe, September 1746, letter to Frederick the Great

---

When a general conducts himself with all prudence, he still can suffer ill fortune; for how many things oppose his labours! Weather, harvest, his officers, the health of his troops, blunders, the death of an officer on whom he counts, discouragement of the troops, exposure of his spies, negligence of the officers who should reconnoitre the enemy and, finally, betrayal. These are the things that should be kept continually before your eyes so as to be prepared for them and prevent good fortune from blinding us.

Frederick the Great, *Instructions to His Generals*, 1747

In war something must be allowed to chance and fortune seeing it is, in its nature, hazardous and an option of difficulties.

Major General Sir James Wolfe, 5 November 1757, letter

Indeed the fate of states and the reputation of generals rest on the most trifling incidents. A few seconds are enough to determine their fortune . . .

Frederick the Great, *Oeuvres de Frederic le Grand* (1846–57)

Something must be left to chance; nothing is sure in a sea fight above all.

Admiral Lord Nelson, 9 October 1805, in a memorandum to the fleet off Cadiz before the Battle of Trafalgar

---

MAXIM 95. War is composed of nothing but accidents, and, although holding to general principles, a general should never lose sight of everything to enable him to profit from these accidents; that is the mark of genius.

In war there is but one favourable moment; the great art is to seize it.

Napoleon, *Military Maxims of Napoleon*, 1827, ed. Burnod

It is now quite clear how greatly the objective nature of war makes it a matter of assessing probabilities. Only one more element is needed to make war a gamble – chance: the very first thing that war lacks. No other human activity is so continuously or universally bound up with chance. And through the element of chance, guesswork and luck come to play a great part in war.

Major General Carl von Clausewitz, *On War*, i, 1832, tr. Howard and Paret

---

War is the realm of chance. No other human activity gives it greater scope: no other has such incessant and varied dealings with this intruder. Chance makes everything more uncertain and interferes with the whole course of events.

Major General Carl von Clausewitz, *On War*, i, 1832, tr. Howard and Paret

---

One must consider two factors, the first of which is known: one's own will, the other unknown: the will of the opponent. To these must be added factors of another kind, impossible to foresee, such as the weather, sickness, railroad accidents, misunderstandings, mistakes, in short all factors of which man is neither the creator nor master, let them be called luck, fate or providence. War must not, however, be waged arbitrarily or blindly. The laws of chance show that these factors are bound to be as often favourable as unfavourable to one or the other opponent.

Field Marshal Helmuth Graf von Moltke, cited in Foch, *Principles of War*, 1913

. . . no operation plan extends with any certainty beyond the first encounter with the main body of the enemy.

Field Marshal Helmuth Graf von Moltke (1800–1891)

Always remember, however sure you are that you can easily win, there would not be a war if the other man did not think he also had a chance.

Sir Winston S. Churchill, *My Early Life*, 1930

In war, chance – *sa majesté le hazard* – plays a great role. It is through prudence and foresight that he deprives chance of everything that can be torn away from it.

General Friedrich von Boetticher, *The Art of War: Principles of the German General Staff in light of our time. A Military Testament*, 1951

CHANGE

The essence of the principles of warriors is responding to changes; expertise is a matter of knowing the military. In any action it is imperative to assess the enemy first. If opponents show no change or movement, then wait for them. Take advantage of change to respond accordingly, and you will benefit.

The rule is 'The ability to gain victory by changing and adapting according to opponents is called genius'.

Liu Ji (AD 1310–1375), *Lessons of War*

There is nothing more difficult to take in hand, more perilous to conduct, or more uncertain in its success, than to take the lead in the introduction of a new order of things.

Niccolo Machiavelli, *The Prince*, 1513

Gentlemen: I hear many say, 'What need so much ado and great charge in caliver, musket, pike, and corselet? Our ancestors won many battles with bows, black bills, and jacks.' But what think you of that?
Captain: Sir, then was then, and now is now. The wars are much altered since the fiery weapons first came up.

Robert Barnet, *The Theory and Practice of Modern Wars*, 1598

. . . To say that the enemy will adopt the same measures is to admit the goodness of them; nevertheless they will probably persist in their errors for some time, and submit to be repeatedly defeated, before they will be reconciled to such a change – so reluctant are all nations to relinquish old customs.

Field Marshal Maurice Comte de Saxe, *My Reveries*, 1732

The student will observe that changes in tactics have not only taken place *after* changes in weapons, which reasonably is the case, but that the interval between such changes has been unduly long.

An improvement of weapons is due to the energy of one or two men, while changes in tactics have to overcome the inertia of a conservative class.

Rear Admiral Alfred Thayer Mahan (1840–1914)

We must try, by correctly foreseeing what is coming, to anticipate developments, and thereby to gain an advantage which our opponents cannot overcome on the field of battle. That is what the future expects us to do.

General Friedrich von Bernhardi, quoted in Foertsch, *The Art of Modern War*, 1940

Victory always smiled on one who was able to renew traditional forms of warfare, and not the one who hopelessly tied himself to those forms.

General Giulio Douhet (1869–1930)

In most armies we see weapons evolving on no rational plan. New arms are invented and introduced without a definite tactical reason, and without a definite relationship to structure, maintenance, and control. Old weapons are maintained; the old and new are mixed irrespective of their elemental values. Proportions are not logically arrived at, but are the outcome of ignorant opposition on the one side and enthusiastic aggressiveness on the other. The whole process is alchemical, is slow and costly and inefficient; ultimately trial and error wins through.

. . . Thus for a hundred years we find the French knights charging English archers; for another hundred years or so, cavalry charging musketeers and riflemen; and I suppose we shall see for yet another years infantry charging tanks. What for, indeed what for? Not to win a battle, for the impossiblity of this is obvious to a rhinoceros. No; but to maintain the luxury of mental indolence in the head of some military alchemist. Thinking to some people is like washing to others. A tramp cannot tolerate a hot bath, and the average general cannot tolerate any change in preconceived ideas; prejudice sticks to his brain like tar to a blanket.

Major General J. F. C. Fuller, *The Foundations of the Science of War*, 1926

Philosophers and scientists have shown that adaptation is the secret of existence. History, however, is a catalogue of failures to change in time with the need. And armies, which because of their role should be the most adaptable of institutions, have been the most rigid – to the cost of the causes they upheld. Almost every great soldier of the past has borne witness to this truth. But it needs no such personal testimony, for the facts of history, unhappily, prove it in overwhelming array. No one can in honesty ignore them if he has once examined them. And to refrain from emphasizing them would be a crime against the country. For it amounts to complicity after the event, which is even more culpable when the life of a people, not merely of one person, is concerned. In the latter case there may be some excuse for discreet silence, as your testimony cannot restore the dead person to life. But in the former case there is no such excuse – because the life of a people will again be at stake in the future. (December 1932)

Captain Sir Basil Liddell Hart, *Thoughts on War*, 1944

The only thing harder than getting a new idea into the military mind is to get an old one out.

Captain Sir Basil Liddell Hart, *Thoughts on War*, 1944

In military affairs, especially in contemporary war, one cannot stand in one place; to remain, in military affairs, means to fall behind; and those who fall behind, as is well known, are killed.

Joseph Stalin (1879–1930), quoted in Garthoff, *Soviet Military Doctrine*, 1953

. . . the history of wars convincingly testifies . . . to the constant contradiction between the means of attack and defence. The appearance of new means of attack has always inevitably led to the creation of corresponding means of counteraction, and this in the final analysis has led to the development of new methods for conducting engagements, battles, and operations and the war in general.

Marshal of the Soviet Union Nikolai V. Ogarkov, quoted in *Kommunist*, 1978

CHARACTER

A man should be upright, not kept upright.

Emperor Marcus Aurelius Antonius, *Meditations*, c. AD 170

One should not associate with people whose conduct is poor. In the *Shih Chi* it says, 'If you don't know a man's character, investigate who his friends are.'

Takeda Nobushige, *Opinions in 99 Articles*, 1588, in Wilson, *Ideals of the Samurai*, 1982

One should not carry maliciousness in his heart. In the *Chun Ch'an* it says, 'If a malicious man is in a high position, his troops will fight among themselves to the last man.'

Takeda Nobushige, *Opinions in 99 Articles*, 1958, in Wilson, *Ideals of the Samurai*, 1982

The success of my whole project is founded on the firmness of the conduct of the officer who will command it.

Frederick the Great, *Instructions to His Generals*, 1747

War must be carried on systematically, and to do it you must have men of character activated by principles of honor.

General George Washington (1732–1799)

My character and good name are in my own keeping, Life with disgrace is dreadful. A glorious death is to be envied.

Admiral Lord Nelson, 10 March 1795, journal entry

A military leader must possess as much character as intellect. Men who have a great deal of intellect and little character are the least suited; they are like a ship whose masts are out of proportion to the ballast; it is preferable to have much character and little intellect. Those men whose intellect is mediocre and whose character is in proportion are likely to succeed in their profession. The base must equal the height.

Napoleon, dictation at Saint Helena, ed. Herold, *The Mind of Napoleon*, 1955

We repeat again: strength of character does not consist solely in having powerful feelings, but in maintaining one's balance in spite of them. Even with the violence of emotion, judgment and principle must still function like a ship's compass, which records the slightest variations however rough the sea.

Major General Carl von Clausewitz, *On War*, i, 1832, tr. Howard and Paret

---

A strong faith in the overriding truth of tested principles is needed; the *vividness* of transient impressions must not make us forget that such truth as they contain is of a lesser stamp. By giving precedence, in case of doubt, to our earlier convictions, by holding to them stubbornly, our actions acquire that quality of steadiness and consistency which is termed strength of character.

Major General Carl von Clausewitz, *On War*, i, 1832, tr. Howard and Paret

---

It is not the critic who counts, not the man who points out how the strong man stumbled, or where the doer of deeds could have done them better, the credit belongs to the man who is actually in the arena, whose face is marred by dust and sweat and blood; who strives valiantly; who errs and comes short again and again; who knows the great enthusiasms, the great devotions, and spends himself in a worthy cause; who, at the best, knows in the end the triumph of high achievement; and who, at the worst, if he fails, at least fails while daring greatly, so that his place shall never be with those cold and timid souls who know neither victory nor defeat.

President Theodore Roosevelt (1858–1919)

A sound body is good; a sound mind is better; but a strong and clean character is better than either.

President Theodore Roosevelt, 24 May 1904, address at Groveton College

---

. . . By greatness of character a general gains command over himself, and by goodness of character he gains command over his men, and those two moods of command express the moral side of generalship.

Major General J. F. C. Fuller, *The Foundations of the Science of War*, 1926

Some virtue is required which will provide the army with a new ideal, which, through the military élite, will unite the army's divergent tendencies and fructify its talent. This virtue is called character which will constitute the new ferment – character the virtue of hard times.

Charles de Gaulle, *The Edge of the Sword*, 1932

What is the test of human character? It is probably this: that man will know how to be patient in the midst of hard circumstance, and can continue to be personally effective while living through whatever discouragements beset him and his companions.

Brigadier General S. L. A. Marshall, *The Armed Forces Officer*, 1950

Of the many personal decisions that life puts upon the military officer, the main one is whether he chooses to swim upstream. If he says yes to that, and means it, all things then begin to fit into place. Then will develop gradually but surely that well-placed inner confidence that is the foundation of military character. From knowing of what to do comes the knowing of how to do, which is likewise important. The pre-eminent quality that all great commanders have owned in common is a positiveness of manner and of viewpoint, the power to concentrate on means to a given end, to the exclusion of exaggerated fears of the obstacles that lie athwart the course.

Brigadier General S. L. A. Marshall, *The Armed Forces Officer*, 1950

Character is the bedrock on which the whole edifice of leadership rests. It is the prime element for which every profession, every corporation, every industry searches in evaluating a member of its organization. With it, the full worth of an individual can be developed. Without it – particularly in the military profession – failure in peace, disaster in war, or, at best, mediocrity in both will result.

General Mathew B. Ridgway, 'Leadership', *Military Review*, 10/1966

He must have 'character', which simply means that he knows what he wants and has the courage and determination to get it. He should have a genuine interest in, and a real knowledge of, humanity, the raw material of his trade; and, most of all, he must know and recognize it in sport, the man who plays his best when things are going badly, who has the power to come back at you when apparently beaten, and who refuses to acknowledge defeat. There is one other moral quality I would stress as the mark of the really great commander as distinguished from the ordinary general. He must have a spirit of adventure, a touch of the gambler in him.

Field Marshal Earl Wavell, *Soldiers and Soldiering*, 1953

The average general envies the buck private; when things go wrong, the private can blame the general, but the general can blame only himself. The private carries the woes of one man; the general carries the woes of all. He is conscious always of the responsibility on his shoulders, of the relatives of the men entrusted to him, and of their feelings. He must act so that he can face those fathers and mothers without shame or remorse. How can he do this? By constant care, by meticulous thought and preparation, by worry, by insistence on high standards in everything, by reward and punishment, by impartiality, by an example of calm and confidence. It all adds up to character.

Q: If a man has enough character to be a good commander, does he ever doubt himself? He should not. In my case, I doubt myself. Therefore, I am probably not a good commander.

General Joseph Stilwell, quoted in preface to Myrer, *Once an Eagle*, 1968

A man of character in peace is a man of courage in war. Character is a habit. The daily choice of right and wrong. It is a moral quality which grows to maturity in peace and is not suddenly developed in war.

General Sir James Glover, 'A Soldier and His Conscience', *Parameters*, 9/1973

Glib, cerebral, detached people can get by in positions of authority until the pressure is on. But when the crunch comes, people cling to those they know can trust – those who are not detached, but involved – those who have consciences, those who can repent, those who do not dodge unpleasantness. Such people can mete out punishment and look their charges in the eye as they do it. In different situations, the leader with the heart, not the bleeding heart, not the soft heart, but the Old Testament heart, the hard heart, comes into his own.

Admiral James Stockdale, 'Educating Leaders', *Washington Quarterly*, Winter 1983

Of great important in shaping the character of a young man is both the school, the public organizations, the traditions of the heroic feat and labour which is cultivated in the nation, but still an inestimable role here is played by the family. I am speaking of myself. In the circle of parents and sisters the emotional principles and the core were formed and to which I have kept all my life.

Admiral of the Fleet of the Soviet Union Sergei S. Gorshkov, quoted in Ablamanov, *Admiral*, 1986

CHIEF OF STAFF

My lord again showed favour to me; the King of Upper and Lower Egypt, Nibmare [Amenhotep III], he put all the people subject to me, and the listing of their number under my control, as superior king's scribe over recruits. I levied the military classes of my lord, my pen reckoned the numbers of millions; I put them in classes in the place of their elders . . . I appointed all their troops; I placed troops at the heads of the ways to turn back the foreigners in their places. The two regions were surrounded with a watch scouting for the Sandrangers. I did likewise at the heads of the river-mouths, which were closed under my troops except to the troops of royal marines. I was the guide of their ways, they depended upon my command.

I was the chief at the head of the mighty men, to smite the Nubians and the Asiatics, the plans of my lord were a refuge behind me;

when I wandered his command surrounded me; his plans embraced all lands and all foreigners who were by his side. I reckoned up the captives of the victories of his majesty, being in charge of them. I did according to that which he said, I followed according to the things which he commanded me, I found them excellent things for the future.

Amenhotep, son of Hapi, minister to Pharaoh Amenhotep III, 1411–1375 BC, quoted in Breasted, *Ancient Records of Egypt*, Vol. I, 1906

What ought to be done, I know only too well; what is going to be done, only the gods know.

General Gerhard von Scharnhorst, 1806, while chief of staff (completely ignored) to the Duke of Brunswick, commander of the Prussian armies destroyed at the Battles of Jena and Auerstadt

---

These reports, as you know, Monsieur le Maréchal, are not for my personal benefit; for I am nothing in the army. I receive in the Emperor's name the reports of the marshals and I sign on his behalf, so personally I have no axe to grind. But His Majesty stipulates that detailed reports on everything which occurs are to be sent to me: for better or worse, nothing should be concealed from the Emperor. I require you therefore to be so kind as to keep me advised of all that occurs in your corps, in the same way as the other marshals.

Marshal of France Louis Berthier, Prince de Neuchatel, Prince de Wagram, March 1807, to Marshal Soult, quoted in Watson, *By Command of the Emperor, A Life of Marshal Berthier*, 1957

In my campaigns Berthier was always to be found in my carriage. During the journey I used to study the plans of the situation and the reports sent in, sketch out my plans for battle from them, and arrange the necessary moves. Berthier would watch me at work, and at the first stopping-place or rest, whether it was day or night, he made out the orders and arrangements with a method and an exactness that was truly admirable. For this work he was always ready and untiring. That was Berthier's special merit. It was very great and valuable, and no one else could have replaced Berthier.

Napoleon, ed. Herold, *Memoirs of Napoleon*, 1955

To know the country thoroughly; to be able to conduct a *reconnaissance* with skill; to superintend the transmission of orders promptly; to lay down the most complicated movements intelligibly, but in a few words and with simplicity; these are the leading qualifications which should distinguish an officer selected for the head of the staff.

Napoleon, *Military Maxims of Napoleon*, ed., tr. D'Aguilar

Well! If I am to become a doctor, you must at least make Gneisenau an apothecary, for we two belong together always . . . Gneisenau makes the pills which I administer.

Field Marshal Prince Gehardt von Blücher (1742–1819), describing his chief of staff, August von Gneisenau, upon receiving an honorary degree at Oxford after the Napoleonic Wars

---

The best means of organizing the command of an army, in default of a general approved by experience, is to: (1) Give the command to a man of tried bravery, bold in the fight and of unshaken firmness in danger. (2) Assign as his chief of staff a man of high ability, of open and faithful character, between whom and the commander there may be perfect harmony. The victor will gain so much glory that he can spare some to the friend who has contributed to his success.

Lieutenant General Antoine-Henri Baron de Jomini, *Summary of the Art of War*, 1838

. . . I don't believe in a chief of staff at all, and any general commanding an army, corps, or division, that has a staff-officer who professes to know more than his chief, is to be pitied.

General of the Army William T. Sherman, *Memoirs of General W. T. Sherman*, 1875

It has sometimes been thought that the defects of an army commander, such an inexperience, inadequate training, weakness of character, may be made good by assigning an exceptionally qualified man as chief of staff. Experience, however, has shown that these attempts invariably fail. Why the experiment has been tried so often through history calls for study.

All men have some weak points and the more vigorous and brilliant a person may be, the more strongly these weak points stand out. It is highly desirable, even essential, therefore, for the more influential members of a general's staff not to be too much like the general, or the shadows will be even more distinct. Ideally, the general and his principal assistants should in a sense supplement one another. Such a mutually supplementary relationship, however, produces desirable results only when the general himself is fully qualified for his office in all professional respects. Mutual supplementation is often misinterpreted, however, to mean that a lack of some indispensable quality in the general may be compensated for by some special talents in the chief of staff. This is a dangerous error.

Archduke Albert of Austria (1817–1895), quoted in von Freytag-Loringhoven, *The Power of Personality in War*, 1911

. . . As I knew from my own experience, the relations between the Chief of Staff and his General, who has the responsibility, are not theoretically laid down in the German Army. The way in which they work together and the degree to which their powers are complementary are much more a matter of personality. The boundaries of their respective powers are therefore not clearly demarcated. If the relations between the General and his Chief of Staff are what they ought to be, these boundaries are easily adjusted by soldierly and personal tact and the qualities of mind on both sides.

I myself have often characterized my relations with General Ludendorff as those of a happy marriage. In such a relationship how can a third party clearly distinguish the merits of the individuals? They are one in thought and action, and often what the one says is only the expression of the wishes and feelings of the other.

Field Marshal Paul von Hindenburg, *Out of My Life*, 1920

. . . the commander himself takes the responsibility for giving orders and must listen to the advice of one man alone: the Chief of Staff assigned to him. The decision is taken by these two together and when they emerge from their privacy there is one decision and one only. They have reached it together; they two are one. If there was any difference of opinion, then by evening the two parties to this happy matrimonial council no longer know which give way. The outside world and the history of war learn nothing of any tiff. In this unification of two personalities lies the security of command. It is all the same whether the order is signed by the General Officer Commanding or, according to our old custom, by the Chief of Staff on his behalf . . . Thus the relation between the two is built up entirely on confidence; if there is no confidence they should part without delay. A relation of this kind cannot be reduced to rule; it will and must always vary according to situation and personality. Thus it is obvious that the correct combination of personalities is of decisive importance for success.

Colonel General Hans von Seekt, *Thoughts of a Soldier*, 1930

. . . It is only by the appointment of the right Chief of Staff that a Commander's brilliant talents for leadership can be fully displayed, or that compensation can be found for his weaknesses.

Colonel General Hans von Seekt, *Thoughts of a Soldier*, 1930

. . . The chief of staff of 48th Panzer Corps was Colonel von Mellenthin, whom I had known from the Chir River days. An extraordinarily happy partnership began with this outstanding staff officer. On the Chir we had already got to understand one another quickly and with few words. Now a brief discussion every morning – often simply a short indication of something on the map – was enough to put the two of us in tune for the whole day.

During years of working together not the slightest ill-feeling ever arose between us, and we also agreed that one should get things down with as little staff work as possible. We never loaded units with unnecessary bits of paper. Whatever could be disposed of verbally – and that meant most things, including operations orders – was done verbally.

We often changed places on visits to the front line, as a staff can also only work properly when he maintains the closest contact with the men on the spot and knows the terrain himself.

General Hermann Balck, November 1943, quoted in von Mellenthin, *German Generals of World War II*, 1977

Have a good chief of staff.

Field Marshal Viscount Montgomery of Alamein, advice on the requirements for a general

---

Between the commander and his chief of staff in a division or larger unit there should be thorough mutual respect, understanding, and confidence with no official secrets between them. Together they form a single dual personality, and the instructions issuing from the chief of staff must have the same weight and authority as those of the commander himself.

But this does not mean that a commander who delegates such authority to his chief of staff can allow his chief to isolate him from the rest of the staff. If that happens, the commander will soon find himself out of touch, and the chief of staff will be running the unit.

General Matthew B. Ridgway, 'Leadership', *Military Review*, 10/1966

CHIVALRY

His helmet flashed as Hector nodded: 'Yes, Ajax,
Since god has given you power, build and sense
and you are the strongest spearman of Achaea,
let us break off this duelling to the death,
at least for today. We'll fight again tommorow,
until some fatal power decides between our armies,
handing victory down to one side or another . . .
Come,
let us give each other gifts, unforgettable gifts,
so any man may say, Trojan soldier or Argive,
"First they fought with heart-devouring hatred,
then they parted, bound by pacts of friendship."'

With that he gave him his silver-studded sword,
slung in its sheath on a supple, well-cut sword-strap,
and Ajax gave his war-belt, glistening purple.
So both men parted, Ajax back to Achaea's armies,
Hector back to his thronging Trojans – overjoyed
to see him still alive, unharmed, striding back,
free of the rage and hands of Ajax still unconquered.
They escorted him home to Troy – saved, past all their hopes –
While far across the field the Achaean men-at-arms
escorted Ajax, thrilled with victory, back to Agamemnon.

Homer, *The Iliad*, vii, tr. Fagles

When he fights with his foes in battle, let him not strike with weapons concealed, nor with such as are barbed, poisoned, or the points of which are blazing with fire.

Let him not strike one who in flight has climbed on an eminence, nor a eunuch, nor one who joins the palms in supplication, nor one who flees with flying hair, nor one who sits down, nor one who says 'I am thine';

Nor one who sleeps, nor one who has lost his coat of mail, nor one who is naked, nor one who is disarmed, nor one who looks on without taking part in the fight, nor one who is fighting with another foe.

Nor one whose weapons are broken, nor one who has been grievously wounded, nor one who is in fear, nor one who has turned to flight; but in all cases let him remember the duty of honourable warriors.

The Laws of Manu, India, *c.* AD 200

One who is a samurai must before all things keep constantly in mind, by day and by night, from the morning when he takes up his chopsticks to eat his New Year's breakfast to Old Year's night when he pays his yearly bills, the fact that he has to die. That is his chief business.

Daidoji Yuznan (early 17th century), ed., Sadler, *The Code of the Samurai*, 1988

Strike at the enemy with humane treatment as effectively as with weapons.

Field Marshal Prince Aleksandr V. Suvorov (1729–1800), quoted in *Soviet Military Review*, 11/1979

As a soldier preferring loyal and chivalrous warfare to organized assassination if it be necessary to make a choice, I acknowledge that my prejudices are in favour of the good old times when the French and English Guards courteously invited each other to fire first – as at Fontenoy – preferring them to the frightful epoch when priests, women, and children through Spain plotted the murder of isolated soldiers.

Lieutenant General Antoine-Henri Baron de Jomini, *Summary of the Art of War*, 1838

Alexander, informed of his (Porus') approach, rode out to meet him, accompanied by a small party of his Companions. When they met, he reined in his horse, and looked at his adversary with admiration: he was a magnificent figure of a man, over seven feet high and of great personal beauty; his bearing had lost none of its pride; his air was of one brave man meeting another, of a king in the presence of a king, with whom he had fought honourably for his kingdom.

Alexander was the first to speak, 'What', he said, 'do you wish that I should do with you?'

'Treat me as a king ought,' Porus is said to have replied.

'For my part,' said Alexander, pleased by his answer, 'your request shall be granted. But is there not something you would wish for yourself? Ask it.'

'Everything', said Porus, 'is contained in this one request.'

The dignity of these words gave Alexander even more pleasure, and he restored to Porus his sovereignity over his subjects, adding to his realm other territory of even greater extent. Thus he did indeed use a brave man as a king ought, and from that time forward found him in every way a loyal friend. Such was the result of the battle with Porus and the Indians beyond the Hydaspes.

Flavius Arrianus Xenophon (Arrian), 326 BC, The meeting of Alexander and Porus after the Battle of the Hydaspes, *The Campaigns of Alexander the Great*, *c.* AD 150, tr. de Selincourt

---

GENERAL ORDERS, No. 73

Headquarters, Army of Northern Virginia
Chambersburg, Pennsylvania
June 27, 1863

The commanding general has observed with marked satisfaction the conduct of the troops on the march, and confidently anticipates results commensurate with the high spirit they have manifested.

No troops could have displayed greater fortitude or better performed the arduous marches of the past ten days.

Their conduct in other respects has with few exceptions been in keeping with their character as soldiers, and entitles them to approbation and praise.

There have however been instances of forgetfulness on the part of some, that they have in keeping the yet unsullied reputation of the army, and that the duties exacted of us by civilization and Christianity are not less obligatory in the country of the enemy than in our own.

The commanding general considers that no greater disgrace could befall the army, and through it our whole people, than the perpetration of the barbarous outrages upon the unarmed, and defenceless and the wanton destruction of private property that have marked the course of the enemy in our own country.

Such proceedings not only degrade the perpetrators and all connected with them, but are subversive of the discipline and efficiency of the army, and destructive of the ends of our present movement.

It must be remembered that we make war only upon armed men, and that we cannot take vengeance for the wrongs our people have suffered without lowering ourselves in the eyes of all whose abhorrence has been excited by the atrocities of our enemies, and offending against Him to whom vengeance belongeth, without whose favor and support our efforts must all prove in vain.

The commanding general therefore earnestly exhorts the troops to abstain with most scrupulous care from unnecessary or wanton injury to private property, and he enjoins upon all officers to arrest and bring to summary punishment all who shall in any way offend against the orders on this subject.

R. E. Lee
General

General Robert E. Lee (1807–1870), *Wartime Papers of Robert E. Lee*, 1987

We want to make our nation entirely one of gentlemen, men who have a strong sense of honour, of chivalry towards others, of playing the game bravely and unselfishly for their side, and to play it with a sense of fair play and happiness for all.

Lieutenant General Sir Robert Baden Powell, *Headquarters Gazette*, 1914

. . . Chivalry governed by reason is an asset both in war and in view of its sequel – peace. Sensible chivalry should not be confounded with the quixotism of declining to use a strategical or tactical advantage, of discarding the supreme moral weapon of surprise, of treating war as if it were a match on the tennis court – such quixotism as is typified by the burlesque of Fontenoy, 'Gentlemen of France, fire first.' This is merely stupid. So also is the traditional tendency to regard the use of a new weapon as 'hitting below the belt', regardless of whether it is inhuman or not in comparison with existing weapons.

But chivalry . . . is both rational and far-sighted, for it endows the side which shows it with a sense of superiority, and the side which falls short with a sense of inferiority. The advantage in the morale sphere reacts on the physical.

Captain Sir Basil Liddell Hart, *A Greater Than Napoleon: Scipio Africanus*, 1926

The military code which he perpetuates has come down to us from even before the age of knighthood and chivalry. It embraces the highest moral laws and will stand the test of any ethics or philosophies ever promulgated for the uplift of mankind. Its requirements are for the things that are right, and its restraints are from the things that are wrong. Its observance will uplift everyone who comes under its influence.

General of the Army Douglas MacArthur, 11 July 1935, address to veterans of the Rainbow Division (42nd Infantry) cited in MacArthur, *A Soldier Speaks*, 1965

Rommel gave me and those who served under my command in the Desert many anxious moments. There could never be any question of relaxing our efforts to destroy him, for if ever there was a general whose sole preoccupation was the destruction of the enemy, it was he. He showed no mercy and expected none. Yet I could never translate my deep detestation of the regime for which he fought into personal hatred of him as an opponent. If I say, now that he is gone, that I salute him as a soldier and a man and deplore the shameful manner of his death, I may be accused of belonging to what Mr. Bevin has called the 'trade union of generals'. So far as I know, should such a fellowship exist, membership in it implies no more than recognition in an enemy of the qualities one would wish to possess oneself, respect for a brave, able and scrupulous opponent and a desire to see him treated, when beaten, in the way one would have wished to be treated had he been the winner and oneself the loser. This used to be called chivalry: many will now call it nonsense and say that the days when such sentiments could survive a war are past. If they are, then I, for one, am sorry.

Field Marshal Sir Claude Auchinleck, foreword to Young, *The Desert Fox*, 1950

From CO
Royal Dragoons

Dear Major von Luck,

We have had other tasks and so were unable to keep in touch with you. The war in Africa has been decided, I'm glad to say, not in your favour.

I should like, therefore, to thank you and all your people, in the name of my officers and men, for the fair play with which we have fought against each other on both sides.

I and my battalion hope that all of you will come out of the war safe and sound and that we may find the opportunity to meet again sometime, in more favourable circumstances.

With greatest respect.

Message to Major Hans von Luck, commander of the Afrika Corps' reconnaissance group, November 1942, von Luck, *Panzer Commander*, 1990

The patience of a saint in hardship.
The tenacity of a bulldog in adversity.
The courage of a lion when aroused.
The chivalry of a Knight in all his dealings.

Field Marshal Viscount Alanbrooke, 'The Ideal Soldier', quoted in Fraser, *Alanbrooke*, 1982

. . . The soldier, be he friend or foe, is charged with the protection of the weak and unarmed. It is the very essence and reason for his being. When he violates this sacred trust, he not only profanes his entire cult but threatens the very fabric of international society. The traditions of fighting men are long and honourable. They are based upon the noblest of human traits – sacrifice.

General of the Army Douglas MacArthur, *Reminiscences*, 1964

CIRCUMSTANCES

MAXIM 105. Conditions on the ground should not alone decide the organization for combat, which should be determined from consideration of all attending circumstances.

Napoleon, *The Military Maxims of Napoleon*, 1827, ed. Burnod

---

This plasticity of mind cannot be cultivated during war, except by an occasional genius; the generality of soldiers simply cannot change if they are dogma-ridden. The only way to prevent this ossification of mind is to accept nothing as fixed, to realize that the circumstances of war are ever-changing, and that, consequently, organization, administration, strategy and tactics must change also; and if during peace time we cannot change them in fact, we can nevertheless change them in theory, and so be mentally prepared when circumstances require that changes should be made.

Major General J. F. C. Fuller, *Lectures on F. S. R. III*, 1932

The art of war has no traffic with rules, for the infinitely varied circumstances and conditions of combat never produce exactly the same situation twice . . .

It follows, then, that the leader who would become a competent tactician must first close his mind to the alluring formulae that well-meaning people offer in the name of victory. To master his difficult art he must learn to cut to the heart of a situation, recognize its decisive elements and base his course, of action on these. The ability to do this is not God-given, nor can it be acquired overnight; it is a process of years. He must realize that training in solving problems of all types, long practice in making clear, unequivocal decisions, the habit of concentrating on the question at hand, and an elasticity of mind, are indispensable requisites for the successful practice of the art of war.

The leader who frantically strives to remember what someone else did in some slightly similar situation has already set his feet on a well-travelled road to ruin.

. . . Every situation encountered in war is likely to be exceptional. The schematic solution will seldom fit. Leaders who think that familiarity with blind rules of thumb will win battles are doomed to disappointment. Those who seek to fight by rote, who memorize an assortment of standard solutions with the idea of applying the most appropriate when confronted by actual combat, walk with disaster. Rather, it is essential that all leaders – from subaltern to commanding general – familiarize themselves with the art of clear, logical thinking. It is more valuable to be able to analyse one battle situation correctly, recognize its decisive elements and devise a simple, workable solution for it, than to memorize all erudition ever written of war.

General of the Army George C. Marshall, ed., *Infantry in Battle*, 1939

In economy as in politics or strategy, there exists, I believe, no absolute truth. There are only circumstances.

Charles de Gaulle, *Salvation 1944–1946*, 1960

. . . one must not think in absolutes either in military theory or in practical military affairs.

Marshal of the Soviet Union Nikolai V. Ogarkov, *History Teaches Vigilance*, 1985

CITIZEN SOLDIERS

Cease to hire your armies. Go yourself, every man of you, and stand in the ranks.

Demosthenes, Third Philippic, 341 BC

After they have elected the consuls, they proceed to appoint military tribunes; fourteen are drawn from those who have seen five years' service and ten from those who have seen ten. As for the rest, a cavalryman is required to complete ten years' service and an infantryman sixteen before he reaches the age of forty-six, except for those rated at less than 400 *drachmae* worth of property who are assigned to naval service. In periods of national emergency the infantry are called upon to serve for twenty years and no one is permitted to hold any political office until he has completed ten years' service.

Polybius, *The Rise of the Roman Empire*, vi, c. 125 BC

When we assumed the soldier, we did not lay aside the citizen.

General George Washington, address to the provincial Congress of New York, 26 June 1775

When men are irritated and their passions inflamed, they fly hastily and cheerfully to arms: but, after the first emotions are over . . . a soldier reasoned with upon the goodness of the cause he is engaged in and the inestimable rights he is contending for, hears you with patience and acknowledges the truth of our observations, but adds that it is of no more importance to him than others. The officer makes you the same reply, with this further remark, that his pay will not support him, and he cannot ruin himself and his family to serve his country, when every member of the community is equally interested and benefited by his labors.

General George Washington, *The Writings of George Washington*, Vol. VI, 1931–1944

Remember, soldiers, that first and foremost you are citizens. Let us not become a greater scourge to our country than the enemy themselves.

General Lazare Carnot, 1792, address to the Armée du Nord

The individual who refuses to defend his rights when called by his government, deserves to be a slave, and must be punished as an enemy of his country and friend to his foe.

Major General Andrew Jackson, Proclamation to the people of Louisiana upon the British invasion, 21 September 1814

The armies of Europe are machines: the men are brave and the officers capable; but the majority of the soldiers in most of the nations of Europe are taken from a class of people who are not very intelligent and who have very little interest in the contest in which they are called upon to take part. Our armies were composed of men who were able to read, men who knew what they were fighting for, and could not be induced to serve as soldiers, except in an emergency when the safety of the nation was involved, and so necessarily must have been more than equal to men who fought merely because they were brave and because they were thoroughly drilled and inured to hardships.

General of the Army Ulysses S. Grant, *Personal Memoirs of U.S. Grant,* 1885

Montgomery understood this 'civilian army' as few before him. The rigid old type discipline was not enforced. Human weaknesses were fully appreciated, and the man's lot made as easy for him as possible. This is why he was so lenient as regards dress, and why a certain amount of 'personal commandeering' – technically I suppose it might be called 'looting' – was winked at. Even when on occasions a unit did not behave in battle as well as it might have done, Montgomery gave them a chance to put things right. All in all he realized that the Prussian type of discipline was not suited to the 'civilian' soldier of the Empire.

Major General Sir Francis de Guingand, *Operation Victory*, 1947

CIVIL AFFAIRS

But of all the methods that can be taken to gain the hearts of a people, none contribute so much as remarkable example of continence and justice; such was the example of Scipio in Spain when he returned a most beautiful young lady, safe and untouched, to her father and husband; this was a circumstance which was more conducive to the reduction of Spain than any force of arms could have ever been. Caesar acquired such reputation for his justice in paying for the wood which he cut down to make palisades for his camps in Gaul, that it greatly facilitated the conquest of that province.

Niccolo Machiavelli, *The Art of War,* 1521

Set off with all diligence. Give orders for this affair. You shall be answerable to me for it. What! If they ruin my people, who will feed me? Who will support the expenses of the state? Who will pay your pensions, gentlemen? As God's alive, to take from my people is taking from myself.

Henry IV (Henry of Navarre) (1553–1610), King of France, upon ordering the investigation of pillage by his troops

---

It is evident that a taste for plunder has up until now been one of the principal attractions that has caused many to choose the soldier's trade and that in the present time debauchery and plunder have on occasion taken the place of pay for very many of the troops. A wise prince may tolerate it, although for the apparent good that it seems to accomplish, it causes an infinite number of great evils. All writers on political subjects are agreed that if there is one area where severity is absolutely necessary for a sovereign, it is with regard to his soldiers. The oppression that people suffer from them is so evident that there is no need to embark on a long discourse to convince you of it, or to show how the sovereign suffers indirect losses from it through a reduction in his revenues . . . Being by honour bound to secure their states from the violence of foreigners, they should be even more eager to protect them from pillage by their own troops. And you must not delude yourself, in weighing the interests of the peasant and the soldier, as many others do, that those men who have enlisted in your troops should mean that much more to you than the rest of the subjects, or that it is to your interests to favour them more than the others.

Louis XIV, *Mémoires de Louis XIV,* 1860, quoted in Ralston, *Soldiers and States*, 1966

When entering the territory of an offender do no violence to the shrines of the deities. Do not hunt over the rice fields or damage the earth-works. Do not burn houses or cut down trees. Do not seize domestic animals or grain or agricultural implements. Where you find old people or children allow them to go home unharmed, and do not antagonize even able-bodied men if they do not challenge you. And see that the enemy wounded have medical treatment.

Ssu Ma Jang Chu (near contemporary of Sun Tzu), *The Precepts of Ssu Ma Jang Chu, c.* 500 BC, quoted in Sadler, *Three Military Classics of China*, 1944

If you can win over the whole country so much the better. At least organize your partisans. The friendship of the neutral country is gained by requiring the soldiers to observe good discipline and by picturing your enemies as barbarous and badly intentioned; if they are Catholic, do not speak about religion; if they are Protestant, make the people believe that a false ardour for religion attaches you to them . . .

Frederick the Great, *Instructions to His Generals*, 1747

The conduct of a general in a conquered country is full of difficulties. If severe, he irritates and increases the number of his enemies. If lenient, he gives birth to expectations which only render the abuses and vexations inseparable from war the more intolerable. A victorious general must know how to employ severity, justice, and mildness by turns, if he would allay sedition, or prevent it.

Napoleon, Military Maxims of Napoleon, 1831, tr. D'Aguilar

We have already arrived at the place of our destiny, and now courage must finish the labour of constancy. Remember that your great duty is to console America and that you are not here to make conquests but to liberate peoples. The Peruvians are your brothers. Embrace them and respect their rights, as you respected those of the Chileans after Chacabuco.

General José de San Martin, 1820, address to his army, quoted in Rojas, *San Martin*, 1957

Although I have served in my profession in several countries, and among foreigners, some of whom professed various forms of the Christian religion, while others did not profess it at all; I never was in one in which it was not the bounden duty of the soldier to pay proper deference and respect to whatever happened to be the religious institutions or ceremonies of the place.

The Duke of Wellington, 8 April 1829, speech to the House of Lords

---

Friends, I promise you this conquest; but there is one condition you must swear to fulfill – to respect the people whom you liberate, to repress the horrible pillaging committed by soundrels incited by our enemies. Otherwise you would not be the liberators of the people; you be their scourge . . . Plunderers will be shot without mercy; already, several have been . . .

Napoleon, 27 March 1796

We come to give you liberty and equality, but don't lose your heads about it – the first person who stirs without permission will be shot.

Marshal of France Pierre-François Lefebre (1755–1820), upon occupying a Franconian town

---

. . . The Vietnamese fighter has always taken care to observe point 9 of its Oath of Honour:

'In contacts with the people, to follow these three recommendations:

– To respect the people

– To help the people

– To defend the people . . . in order to win their confidence and affection and achieve a perfect understanding between the people and the army.'

General Vo Nguyen Giap, *People's War, People's Army*, 1961

As commander-in-chief of the Peloponnesus I have led many thousands of men, I have conducted many campaigns and have fought many battles; but I confess in all conscience that during the whole duration of the struggle I did not spend a single obol of my own; for I had none. The people fed me, and supplied my troops with all they required; my sword supplied the rest, for all the horses and arms we needed we took from the Turks.

Theodoros Kolokotronis, *Theodoros Kolokotronis: The Old Man of the Morea*, 1852

CIVIL-MILITARY RELATIONS

The advance and retirement of the army can be controlled by the general in accordance with prevailing circumstances. No evil is greater than commands of the sovereign from the court.

Chia Lin, Tang Dynasty, AD 618–905

. . . that a general during the campaign need not follow court orders is sound military law.

Emperor Kublai Khan (1214–1294)

. . . A general to whom the sovereign has entrusted his troops should act on his own initiative. The confidence which the sovereign reposes in the general's ability is authorization enough to conduct affairs in his own way.

Frederick the Great, *General Principles of War*, 1748

. . . it may be proper constantly and strongly to impress upon the Army that they are mere agents of Civil power: that out of Camp, they have no other authority than other citizens, that offences against the laws are to be examined, not by a military officer, but by a Magistrate; that they are not exempt from arrests and indictments for violations of the law.

General George Washington, letter to Daniel Morgan, 27 March 1795, *The Writings of George Washington*, Vol. 34, 1934–44

To leave a great military enterprise or the plan for one to purely military judgment and decision is a distinction which cannot be allowed, and is even prejudicial; indeed, it is an irrational proceeding to consult professional soldiers on the plan of a war, that they give a purely military opinion upon what the cabinet ought to do . . .

Major General Carl von Clausewitz, *On War*, 1832, tr. Graham

If in ordinary times, under the restrictions of constitutional forms, governments subjected to all the changes of an elective legislature are less suitable than others for the creation or preparation of a formidable military power, nevertheless in great crises these deliberative bodies have sometimes attained very different results and have concurred in developing the national strength to the fullest.

Lieutenant General Antoine-Henri Baron de Jomini, *Summary of the Art of War*, 1838

The action of a cabinet in reference to the control of armies influences the boldness of their operations. A general whose genius and hands are tied by an Aulic council five hundred miles distant cannot be a match for one who has liberty of action, other things being equal.

Lieutenant General Antoine-Henri Baron de Jomini, *Summary of the Art of War*, 1838

I do not desire to place myself in the most perilous of all positions:- *a fire upon my rear, from Washington, and the fire, in front, from the Mexicans.*

Lieutenant General Winfield Scott, 21 May 1846, letter to the Secretary of War

I am a soldier. It is my duty to obey orders. It is enough to turn one's hair grey to spend one day in the Congress. The members are patriotic and earnest, but they will neither take the responsibility of action nor will they clothe me with the authority to act for them.

General Robert E. Lee, cited in John B. Gordon, *Reminiscences of the Civil War*, 1903

Keep your hands off the regiment, ye iconoclastic civilian officials who meddle and muddle in Army matters. Clever politicians you may be, but you are not soldiers and you do not understand them; they are not pawns on a chessboard. Leave the management of our fighting men to soldiers of experience in our British Army of old renown, and do not parody us by appearing in public decked for the nonce in a soldier's khaki coat. You might as well put your arm in a sling, or tie your head up in the bandage of some poor mained soldier, to whom, when wounded and unable to earn a livelihood, your regulations allow a pension of sixpence a day!

Field Marshal Viscount Wolseley, *The Story of a Soldier's Life,* 1903

In seeking to trace all the great mistakes and blunders committed during the Civil War to defects in our military system, it is important to bear in mind the respective duties and responsibilites of soldiers and statesmen. The latter are responsible for the creation and organization of our resources, and, as in the case of the President, may further be responsible for their management or mismanagement. Soldiers, while they should suggest and be consulted on all details of organization under our system, can alone be held responsible for the control and direction of our armies in the field.

So long as historians insist upon making our commanders alone responsible for disasters in time of war, so long will the people and their representatives fail to recognize the importance of improving our system.

Colonel Emory Upton, *The Military policy of the United States Since 1775*, 1904

. . . a popular outcry will drown the voice of military experience.

Rear Admiral Alfred Thayer Mahan, *Naval Strategy,* 1911

. . . disastrous results . . . must follow a failure [in civil military relations] in Washington . . . Who is to blame? Surely not the civilians in the government who have long since learned to regard professional advice with suspicion. It is we ourselves who are at fault and we can fairly blame neither congress, our form of government, the un-military characteristics of the people nor any civilian official. [What we] imperatively need first of all is a conception of war. To reach the ultimate goal of war efficiency, we must begin with principles, conceptions, and major doctrine . . . We must build from the foundation upwards not from the roof downwards.

Admiral Dudley W. Knox, *Naval Institute Proceedings,* March–April 1915

The well-being of the people equally with the well-being of the Army requires a common sympathy and a common interest between them.

Major General John Pope (1822–1892), to the veterans of the Army of the Tennessee

---

I have heard in such a way as to believe it, of your recently saying that both the army and the government needed a dictator . . . Only those generals who gain successes can set up dictators. What I ask you now is military success, and I will risk the dictatorship.

President Abraham Lincoln, letter to Major General Joseph Hooker, 26 January 1863

It has always been an article of my creed that Army and people have but one body and one soul, and that the Army cannot remain sound for ever if the nation is diseased.

General Erich Ludendorff, *My War Memories, 1914–1918*, 1919

There is a certain book, *Vom Kriege*, which never grows old. Its author is Clausewitz. He knew war, and he knew men. We had to listen to him, and whenever we followed him it was to victory. To do otherwise meant disaster. He gave a warning about the encroachment of politics on the conduct of military operations. In saying this, I am far from passing a judgment upon the orders we now received. I may have criticized in thought and word in 1914, but to-day I have completed my education in the rough school of reality, the conduct of operations in a coalition war. Experience tempers criticism, indeed frequently reveals how unfounded it has been. During the war we have times without number attempted to think: 'He is a lucky man who has an easier soldier's conscience than ours, and who has won the battle between his military convictions and the demands of politics as easily as we have.' The political tune is a ghastly tune! I myself during the war seldom heard in that tune those harmonies which would have struck an echo in a soldier's heart. Let us hope that if ever our Fatherland's dire necessity involves a summons to arms again, others will be more fortunate in this respect than we were!

Field Marshal Paul von Hindenburg, *Out of My Life*, 1920

A matter of more immediate importance is that the nation should realize the necessity for having educated leaders – trained statesmen – to conduct its war business, if and when war should again come along. This is a direction in which much-needed preparation can be made without the expenditure of cash, and it may be the means of saving tens of thousands of lives and hundreds of millions of money. In all trades and professions the man who aims at taking the lead knows that he must first learn the business he purposes to follow: that he must be systematically trained in it. Only in the business of war – the most difficult of all – is no special training or study demanded from those charged with, and paid for, its management.

Field Marshal Sir William Robertson, *From Private to Field Marshal*, 1921

The real headquarters of Armies in these days are not to be found in the field abroad, but at the seat of government at home; and plans of campaigns are, and must be, now analysed and criticised by civilian Ministers in a way quite unknown a few decades ago. The military chief must be prepared to expound and justify, lucidly and patiently, the plans for which he seeks ministerial sanction, and he must also be prepared to explain and substantiate his objections to such alternative plans as Ministers themselves may suggest.

Field Marshal Sir William Robertson, *From Private to Field Marshal*, 1921

What is the true bastion of iron? It is the masses, the millions upon millions of people who genuinely and sincerely support our revolution. That is the bastion which it is impossible for any force on earth to smash.

Mao Tse-tung, 'Be Concerned with the Well-Being of the Masses', 27 January 1934, *Selected Works*

Civilians provide the manpower for our huge armies. Parents provides the sons who fight. They make sacrifices, enormous sacrifices for the cause. Wives lose their husbands, children lose their fathers, families lose their breadwinners. They go short of food, clothing and the comforts they are used to. They pay the heavy taxes required to finance our war effort . . . [hence] when they know that serious mistakes have been made they want to know why. After all they pay the cost of those mistakes.

Lieutenant General H. Gordon Bennett, *Why Singapore Fell*, 1944

The people are like water and the army is like fish.

Mao Tse-tung, 'Aspects of China's Anti-Japanese Struggle', 1948

Once war had come, however, the die was cast. I was quite clear in my own mind that for me, as a member of the Armed Forces, there was only one possible course – to fight against our external enemies. The stronger the moral determination of the forces under my command, the greater would be their fighting value. It is the moral obligation of every fighting man unreservedly to support the government of his country when it goes to war. To expect a commander to accept the heavy responsibilities which devolve upon him in war and at the same time to bother his head about internal conditions, let alone to enter upon a struggle against the political leadership of his own country, is asking altogether too much of him.

Admiral Karl Doenitz, *Memoirs, Ten Years and Twenty Days*, 1959

While political goals shape war and influence it from one moment to the next, it is the soldier's duty to insure that he will not be confronted with insoluble tasks by his political leaders (as happened in June 1941).

General Reinhard Gehlen, *The Service*, 1972

This is not to say that the civilian strategists were wrong. The political scientists provided a valuable service in tying war to its political ends. They provided answers to 'why' the United States ought to wage war. In like manner the systems analysts provided answers on 'what' means we would use. What was missing was the link that should have been provided by the military strategists – 'how' to take the systems analyst's *means* and use them to achieve the political scientist's *ends*.

Colonel Harry G. Summers, Jr., *On Strategy: The Vietnam War in Context*, 1981

What a society gets in its armed services is exactly what it asks for, no more or less. What it asks for tends to be a reflection of what it is. When a country looks at its fighting forces it is looking in a mirror; the mirror is a true one and the face that it sees will be its own.

General Sir John Hackett, *The Profession of Arms*, 1983

All the reproaches that have been levelled against the leaders of the armed forces by their countrymen and by the international courts have failed to take into consideration one very simple fact: that policy is not laid down by soldiers, but by politicians. This has always been the case and is so today. When war starts, the soldiers can only act according to the political and military situation as it then exists. Unfortunately it is not the habit of politicians to appear in conspicuous places when the bullets begin to fly. They prefer to remain in some safe retreat and to let the soldiers carry out 'the continuation of policy by other means'.

Colonel General Heinz Guderian, *Panzer Leader*, 1953

CLOSE WITH THE ENEMY

In battle most men expose themselves enough to satisfy the needs of honour; few wish to do more than this, or more than enough to carry to success the action in which they are engaged.

François de la Rochefoucauld, *Réflexions ou sentences et maximes morales*, 1665

By push of bayonets, no firing till you see the whites of their eyes.

Frederick the Great, at Prague, 6 May 1757

Rascals, would you live forever?

Frederick the Great, when the Guards hesitated at Kolin, 18 June 1757

If your bayonet breaks, strike with the stock; if the stock gives way, hit with your fists; if your fists are hurt, bite with your teeth.

General Mikhail I. Dragomirov, *Notes for Soldiers*, 1890

If your sword is too short, take one step forward.

Admiral Marquis Togo Heihachiro, 15 May 1905

The laurels of victory are at the point of the enemy bayonets. They must be plucked there; they must be carried by a hand-to-hand fight if one really means to conquer.

Marshal of France Ferdinand Foch, *Precepts and Judgments*, 1919

But now from the direction of the enemy there came a succession of grisly apparations; horses spouting blood, struggling on three legs, men staggering on foot, men bleeding from terrible wounds, fish-hook spears stuck right through them, arms and faces cut to pieces, bowels protruding, men gasping, crying, collapsing, expiring . . . The blood of our leaders cooled . . . They remembered for the first time that we had carbines.

Sir Winston S. Churchill, *My Early Life*, 1930

Infantry must move forward to close with the enemy. It must shoot in order to move . . . To halt under fire is folly. To halt under fire and not fire back is suicide. Officers must set the example.

General George S. Patton, Jr., *War As I Knew It*, 1947

COALITION WARFARE

It is better to have a known enemy than a forced ally.

Napoleon, *Political Aphorisms*, 1848

The allies we gain by victory will turn against us upon the bare whisper of our defeat.

Napoleon, *Political Aphorisms*, 1848

Granting the same aggregate of force, it is never as great in two hands as in one, because it is not perfectly concentrated.

Rear Admiral Alfred Thayer Mahan, *Naval Strategy*, 1911

I was no more than conductor of an orchestra . . . A vast orchestra, of course . . . Say, if you like, that I beat time well! Has the music stopped? Are we tired of the tune? We must start a new one. Never stop . . . The true meaning of the unified command is not to give orders, but to make suggestions . . . One talks, one discusses, one persuades . . . One says, 'That is what should be done; it is simple; it is only necessary to will it.'

Marshal of France Ferdinand Foch, commander of allied forces in World War I, quoted in Liddell Hart, *Foch: The Man of Orleans*, 1931

. . . Negative goals frequently are more widespread when waging a coalition war because the egotistic interests of each ally thrust him toward offering the others the honour of striking the enemy while he himself vigilantly guards and protects his own forces for the final hour to force consideration of his interests when concluding the peace. Therefore, coalition wars rather assume a positional nature than the single combat of two states . . .

General A. A. Svechin, *Strategy*, 1927

My life is a mixture of politics and war. The latter is bad enough – but I've been trained for it! The former is straight and unadulterated venom! But I have to devote lots of my time, and much more of my good disposition, to it.

General of the Army Dwight Eisenhower, 27 September 1943, letter to his wife

---

In war I would deal with the Devil and his grandmother.

Joseph Stalin (1879–1953)

. . . History testifies to the ineptitude of coalitions in waging war. Allied failures have been so numerous and their inexcusable blunders so common that professional soldiers had long discounted the possibility of effective allied action unless available resources were so great as to assure victory by inundation. Even Napoleon's reputation as a brilliant military leader suffered when students in staff college came to realize that he always fought against coalitions – and therefore against divided counsels and diverse political, economic, and military interests.

General of the Army Dwight D. Eisenhower, *Crusade in Europe*, 1948

Allied effectiveness in World War II established for all time the feasibility of developing and employing joint control machinery that can meet the sternest tests of war. The key to the matter is a readiness, on highest levels, to adjust all nationalistic differences that affect the strategic employment of combined resources, and, in the war theater, to designate a single commander who is supported to the limit. With these two things done, success rests in the vision, the leadership, the skill, and the judgment of the professionals making up command and staff groups; if these two things are not done, only failure can result.

General of the Army Dwight D. Eisenhower, *Crusade in Europe*, 1948

In war it is not always possible to have everything go exactly as one likes. In working with allies it sometimes happens that they develop opinions of their own.

Sir Winston Churchill, *The Hinge of Fate*, 1950

. . . a too complete victory inevitably complicates the problem of making a just and wise peace settlement. Where there is no longer the counter-balance of an opposing force to control the appetites of the victors, there is no check on the conflict of views and interests between the parties to the alliance. The divergence is then apt to become so acute as to turn the comradeship of common danger into the hostility of mutual dissatisfaction – so that the ally of one war becomes the enemy in the next.

Captain Sir Basil Liddell Hart, *Strategy*, 1954

A clue to why coalition warfare failed in Vietnam, was given by a North Vietnamese Army (NVA) major in Hanoi a week before Saigon's fall. 'You should not feel badly,' he said. 'You have done more than enough . . . more than enough.' Even worse than the mortification of being in the enemy's capital while he is celebrating your defeat is the humiliation of his trying to lessen your discomfort with reassuring words. Yet the NVA major was more correct than he knew. We failed in Vietnam because we attempted to do too much.

Colonel Harry G. Summers, Jr., *On Strategy: The Vietnam War in Context*, 1981

COHESION

Every man reported and one mustered an army of Pharaoh which was under command of the king's son. He made troops, commanded by commanders, each man with his village . . .

Pharaoh Amenhotep III (1411–1375 BC), quoted in Breasted, *Ancient Records of Egypt*, Vol. II, 1906

And the two Aentes ranged all points of the rampart, calling
out commands to spur their comrades' fury.
Now cheering a soldier on, tongue-lashing the next
if they marked a straggler hanging back from battle:
'Friends – you in the highest ranks of Argives,
you in the midst and you in rank and file,
we cannot all be equal in battle, ever,
but now the battle lies before us all –
come, see for yourselves, look straight down.
Now let no fighter be turned back to the ships,
not with his captain's orders ringing in his ears –
keep pressing forward, shouting each other on!'

Homer, *The Iliad*, xii, *c.* 800 BC, tr. Fagles

The worst cowards, banded together, have their power.

Homer, *The Iliad*, xiii, *c.* 800 BC, tr. Fagles

I do not regard it, soldiers, as of small account that there is not a man among you before whose eyes I have not often achieved some military exploit; and to whom, in like manner, I, the spectator and witness of his valour, could not recount his own gallant deeds, particularized by time and place. With soldiers who have a thousand times received my praises and gifts, I, who was the pupil of you all before I became your commander, will march out in battle-array against those who are unknown and ignorant of each other.

Hannibal, 218 BC, address to his troops upon entering Italy after crossing the Alps

It is the part of a wise general to station brothers in rank besides brothers, friends besides friends, and lovers besides their favourites. For whenever that which is in danger near by is more than ordinarily dear the lover necessarily fights more recklessly for the man beside him. And of course one is ashamed not to return a favour that he has received, and is dishonoured if he abandons a benefactor and is the first to flee.

Onasander, *The General*, AD 58

Having formed his company . . . he (the captain) will then arrange comrades. Every Corporal, Private and Bugler will select a comrade of the rank differing from his own, i.e., front and rear rank, and is never to change him without permission of his captain. Comrades are always to have the same berth in quarters; and that they may be as little separated as possible, in either barracks or the field, will join the same file on parade and go on the same duties with arms when it is with baggage also.

British Army *Regulations of the Rifle Corps*, 1800, prepared under the aegis of Sir John Moore

---

I looked alongst the line; it was enough to assure me. The steady determined scowl of my companions assured my heart and gave me determination.

Anonymous soldier of the 71st Regiment at the Battle of Vimeiro, 1808

Victory and disaster establish indestructible bonds between armies and their commanders.

Napoleon, *Political Aphorisms*, 1848

Solidarity and confidence cannot be improvised . . . It is time we should understand the lack of power in mob armies.

Colonel Charles Ardnant du Picq, *Battle Studies*, 1880

Today the soldier is often unknown to his comrades. He is lost in the smoke, the dispersion, the confusion of battle. He seems to fight alone. Unity is no longer insured by mutual surveillance. A man falls, and disappears. Who knows whether it was a bullet or the fear of advancing farther that struck him down?

Colonel Charles Ardnant du Picq, *Battle Studies*, 1880

A wise organization ensures that the personnel of combat groups changes as little as possible, so that comrades in peacetime shall be comrades in war.

Colonel Charles Ardnant du Picq, *Battle Studies*, 1880

It is therefore noted as a principle that, all other things being equal, the tactical unity of men working together in combat will be in the ratio of their knowledge and sympathetic understanding of each other. Lacking these things, though they be well-trained soldiers, they are not likely to adhere unless danger has so surrounded them that they must do so in order to survive, and even then, quick surrender is the most probable result.

Brigadier General S. L. A. Marshall, *Men Against Fire*, 1947

. . . the best tactical results obtain from those dispositions and methods which link the power of one man to that of another. Men who feel strange with their unit, having been carelessly received by it, and indifferently handled, will rarely, if ever, fight strongly and courageously. But if treated with common decency and respect, they will perform like men.

Brigadier General S. L. A. Marshall, *The Armed Forces Officer*, 1951

. . . During the period of four to six weeks we had been enabled to complete the training and equipment of the troops, to make them familiar with the country, to continue to build field fortifications, to prepare work of destruction, and to lay mines and minefields. The homogeneity of the units thus became very different from what is usually the case with hastily mobilized troops immediately thrown into battle. These circumstances were calculated to produce calm and confidence before the trials in front of us.

Field Marshal Carl Gustav Baron von Mannerheim, citing the preparation of the Finnish Army before the Soviet attack in 1939, *The Memoirs of Marshal Mannerheim*, 1953

In such a unit as the 101st, it was a constant task to impress on the officers the folly, indeed the unfairness, of unnecessarily exposing themselves to enemy fire. The strength of a fighting outfit is the mutual respect of all its members of whatever rank. Shared danger breeds admiration for the hard, utter intolerance for the weak, and a fierce loyalty to comrades. Students of military history have often tried to determine why some men fight well and others run away. It never seemed to me that ideological motives or political or moral concepts had much to do with it. If I could get any of my men to discuss a matter so personal as their honest reaction to combat, they would tell me that they fought, though admittedly scared, because 'I couldn't let the other boys down' or 'I couldn't look chicken before "Dog Company."' These are simple reasons for simple virtues in simple men whom it is an ennobling privilege to command. For their officers, that privilege carries with it the responsibility to stay alive and look after them. This was a sound precept but one hard to impress on the officers of the 101st who, like their men, didn't want to 'look chicken'.

General Maxwell Taylor, *Swords and Ploughshares*, 1972

. . . the inherent unwillingness of the soldier to risk danger on behalf of men with whom he has no social identity. When a soldier is unknown to the men who are around him he has relatively little reason to fear losing the one thing that he is likely to value more highly than life – his reputation as a man among other men.

However much we may honor the 'Unknown Soldier' as the symbol of sacrifice in war, let us not mistake the fact that it is the 'Known Soldier' who wins battles. Sentiment aside, it is the man whose identity is well known to his fellows who has the main chance as a battle effective.

Brigadier General S. L. A. Marshall, *Men Against Fire*, 1947

My first wish would be that my Military family, and the whole Army, should consider themselves as a band of brothers, willing and ready to die for each other.

General George Washington, 21 October 1798, letter to Henry Knox

THE COLOURS

The soldiers should make it an article of faith never to abandon their standard. It should be sacred to them; it should be respected; and every type of ceremony should be used to make it respected and precious.

Field Marshal Maurice Comte de Saxe, *My Reveries*, 1732

Landing was difficult for the reasons following: The size of the ships made it impossible for them to run ashore except in deep water; the soldiers did not know the ground, and with their hands burdened and themselves weighed down by their cumbrous arms, they had to jump down from their vessels, keep a foothold in the surf, and fight the enemy all at once; while the enemy had their hands free, knew the ground perfectly, and could stand on dry land or advance a little into the water, throw their missiles boldly, and put their horses into the sea, to which they were trained. Our men were unnerved by these handicaps and being inexperienced in this kind of warfare, they did not show the dash and energy they generally did in land battles.

. . . Then, while our soldiers were hesitating, chiefly because of the depth of the water, the standard-bearer of the Tenth, with a prayer that his act might redound to the success of the legion, cried, 'Leap down, men, unless you want to betray your eagle to the enemy; I, at least shall have done my duty to my country and my general.' As he said this, in a loud voice, he threw himself overboard and began to advance against the enemy with the eagle. Then our men called upon one another not to suffer such a disgrace, and with one accord leaped down from the ships. Seeing this their comrades from the nearest ships followed them and advanced close to the enemy.

Julius Caesar, *The Gallic War*, c. 51 BC, tr. Hadas, 1957

And hence it was that when one man in every two, or even two in every three, had fallen in Hoghton's Brigade the survivors were still in line by their colours, closing in towards the tattered silk which represented the ark of their convenant – the one thing supremely important in the World.

Sir John Fortescue, in the Battle of Albuera, 1811, quoted in Richardson, *Fighting Spirit*, 1978

A man is not a soldier until he is no longer homesick, until he considers his regiment's colours as he would his village steeple; until he loves his colours, and is ready to put hand to sword every time the honour of the regiment is attacked.

Marshal of France Thomas R. Bugeaud (1778–1846), quoted in Thomas, *Les Transformations de la'Armée Française*, 1887

COMBAT

Combat is the only effective force in the world; its aim is to destroy the enemy's forces as a means to a further end. That holds good even if no actual fighting occurs, because the outcome rests on the assumption that if it came to fighting, the enemy would be destroyed.

Major General Carl von Clausewitz, *On War*, 1832

Combats may be quite independent of scientific combinations; they may become essentially dramatic; personal qualities and inspirations and a thousand other things frequently are the controlling elements.

Lieutenant General Antoine-Henri Baron de Jomini, *Summary of the Art of War*, 1838

Nothing is more exhilarating than to be shot at without result.

Sir Winston Churchill, *The Malakand Field Force*, 1898

I define combat as a violent, planned form of physical interaction (fighting) between two hostile opponents, where at least one party is an organized force, recognized by governmental or *de facto* authority, and one or both opposing parties hold one or more of the following objectives: to seize control of territory or people; to prevent the opponent from seizing and controlling territory or people; to protect one's own territory or people; to dominate, destroy, or incapacitate an opponent . . .

There are two key points in this definition that I wish to emphasize. Though there may be much in common between military combat and a brawl in a barroom, there are important differences. The opponents in military combat are to some degree organized, and both represent a governmental or quasi-governmental authority. There is one other essential difference: the all-pervasive influence of *fear* in a *lethal* environment. People have been killed in barroom brawls, but that is exceptional. In military combat there is the constant danger of death from lethal weapons employed by opponents with deadly intent. Fear is without question the most important characteristic of combat.

Colonel N. T. Dupuy, *Understanding War*, 1987

Combat is organized armed conflict between the sub-units, units, and formations of belligerents and is the main form of tactical strikes coordinated in terms of objective, place, and time, fire, and manoeuvre aimed at routing (destroying) the enemy or taking him prisoner and of taking or holding important areas (lines, objectives). Combat is the only means of achieving victory.

Colonel General V. G. Reznichenko, *Taktika*, 1984

COMBINED ARMS

You hear that Philip goes where he pleases not by marching his phalanx of infantry, but by bringing in his train light infantry, cavalry, archers, mercenaries, and other such troops.

Demosthenes, cited in Duncan Head, *Armies of the Macedonian and Punic Wars*, 1982

Infantry, cavalry, and artillery are nothing without each other. They should always be so disposed in cantonments as to assist each other in case of surprise.

Napoleon, *The Military Memoirs of Napoleon*, 1831, tr. D'Aguilar

---

We know well what happens when a single arm is opposed to two others.

Major General Carl von Clausewitz (1780–1831)

It is not so much the mode of formation as the proper combined use of the different arms which will ensure victory.

Lieutenant General Antoine-Henri Baron de Jomini, *Summary of the Art of War*, 1838

There is still a tendency in each separate unit . . . to be a one-handed puncher. By that I mean that the rifleman wants to shoot, the tanker to charge, the artilleryman to fire . . . That is not the way to win battles. If the band played a piece first with the piccolo, then with the brass horn, then with the clarinet, and then with the trumpet, there would be a hell of a lot of noise but no music. To get the harmony in music each instrument must support the others. To get harmony in battle, each weapon must support the other. Team play wins. You musicians of Mars must not wait for the band leader to signal you . . . You must each of your own volition see to it that you come into this concert at the proper place and at the proper time . . .

General George S. Patton, Jr., 8 July 1941, address to the men of the 2nd Armored Division, *The Patton Papers*, Vol. II, 1974

To achieve victory on any scale it is important to secure the cooperation between all arms of the service both in operational elements and tactical formations on the terrain (or at least on the sand table) . . .

Marshal of the Soviet Union Georgi K. Zhukov, 1941, *Reminiscences and Reflections*, 1974

COMMAND

A prince should therefore have no other aim or thought, nor take up any other thing for his study, but war and its organization and discipline, for that is the only art that is necessary to one who commands, and it is of such virtue that it not only maintains those who are born princes, but often enables men of private fortune to attain to that rank.

Niccolo Machiavelli, *The Prince*, 1513

He who wishes to be obeyed must know how to command.

Niccolo Machiavelli, *Discourses*, 1517

. . . in all well-ordered militias the commendation and sufficiency of all generals, colonels, captains, and other officers hath consisted in knowing how to command, govern, and order their armies, regiments, bands, and companies, and to win the love of their soldiers by taking great care of their healths and safties, as also by all examples of virtue and worthiness not only by instruction but also by action in their own persons, venturing their lives in all actions against the enemy amongst them, and therewithal accounting of them in sickness and health or wounds received as of their own children; and whereas, again, all colonels and captains of horsemen according to all discipline have used to serve amongst their horsemen on horseback, and all colonels and captains of footmen, yea, even the very lieutenants general and kings themselves, if their armies and forces of the field have consisted more of footmen than of horsemen, have always used by all discipline military upon the occasion of any battle to put their horses from them and to serve on foot, and to venture their lives in the former ranks . . .

John Smythe (c. 1580–1631), *Certain Discourses Military*

There are no bad regiments; there are only bad colonels.

Napoleon (1769–1821)

If officers desire to have control over their commands, they must remain habitually with them, industriously attend to their instruction and comfort, and in battle lead them well.

Lieutenant General Thomas 'Stonewall' Jackson, *Letter of Instruction to Commanding Officers*, November 1861

I cannot trust a man to control others who cannot control himself.

General Robert E. Lee (1807–1870)

The only prize much cared for by the powerful is power. The prize of the general is not a bigger tent, but command.

Justice Oliver Wendell Holmes, 15 February 1913, speech at Harvard Law School, quoted in Howe, ed., *The Occasional Speeches of Justice Oliver Wendell Holmes*, 1962

Only the Head of the Government, the Statesmen who has decided for war, and that with a clear conscience, shoulders the same or a bigger burden of responsibility than that of the Commander-in-Chief. In his case it is a question of one great decision only, but the Commander of an army is faced with decisions daily and hourly. He is continuously responsible for the welfare of many hundred-thousands of persons, even of nations. For a soldier there is nothing greater, but at the same time more overwhelming, than to find himself at the head of an army or the entire field army of his country.

General Erich Ludendorff, *My War Memories, 1914–1918*, 1919

It is often argued that the possession of original ideas is different from, or even incompatible with, the power of executive command. The gulf between these two faculties is narrower than is commonly supposed. The conception of ideas implies mental initiative, while the moral courage required to express them publically and with vigour in the face of tradition-bound seniors indicates a strong personality, for the role of pioneer or reformer is rarely a popular one. These two qualities of mental initiative and strong personality, or determination, go a long way towards the power of command in war – they are, indeed, the hallmark of the Great Captains. (March 1923)

Captain Sir Basil Liddell Hart, *Thoughts on War*, 1923

There were those who had been relieved from their commands who came in to tell me of the injustice which had been done them. It was hard to talk to men of this class, because in most instances I was convinced by what they told me that their relief was justified. To discuss with an old friend the smash-up of his career is tragic and depressing at best, and more particularly when he feels that he has been treated unfairly and an honorable record forever besmirched. In decided contrast were those who stopped by on their way to the front, selected to replace commanders of units enaged in battle. Enthusiastic in anticipation of the opportunity before them, but a little fearful of following in the footsteps of their deposed predecessors, they pressed for tips or advice on how to succeed under the conditions of battle. If a man had not laboriously prepared himself for his duties, there was little profit that could be said at that late hour, but I am inclined to think that I helped a few by briefly reciting the more-or-less specific errors which seemed to be prevalent. One could form a fair estimate of the chances for success of these men by their manner during discussion. While all asked questions, some rarely waited for a reply, occupying the time in dogmatically outlining their own views and opinions. Those who later were successful on the battlefield usually 'pumped' me with the relentless persistence of a lawyer cross-examining a reluctant witness.

General of the Army George C. Marshall, *Memoirs of My Service in the World War*, 1976 (written between 1923 and 1926)

The intelligent leadership of troops, and the ability to appreciate and predict the way operations develop call for firm and precise direction of forces. Any suggestion of the exercise of independent command by junior commanders is unacceptable. Not knowing the general situation, junior commanders are always liable to take decisions incompatible with it; and this may engender catastrophe. It may cause a boldly conceived and executed operation, requiring precise co-operation between its component parts, to start coming apart at the seams.

Marshal of the Soviet Union Mikhail N. Tukhachevskiy, 1924, quoted in Simpkin, *Deep Battle*, 1987

The art of command is not that of thinking and deciding for one's subordinates as though one stood in their shoes.

Marshal of France Ferdinand Foch, *Precepts and Judgments*, 1919

The power to command has never meant the power to remain mysterious.

Marshal of France Ferdinand Foch, *Precepts and Judgments*, 1919

Your greatness does not depend upon the size of your command, but on the manner in which you exercise it.

Marshal of France Ferdinand Foch, quoted in Aston, *Biography of Foch*, 1929

To command troops is certainly the greatest pleasure of military life, but above all is this case when one is a captain or a colonel. For the captain, by reason of his immediate contact with the rank and file, which in the French Army is intelligent, active and devoted; for the colonel, because of the influence he exerts on a body of officers imbued with the noblest sentiments, competent in their trade and of proved devotion. It is through his officers that a colonel moulds his regiment so that it becomes in time the very image of the commander.

Marshal of France Ferdinand Foch, *Memoirs of Marshal Foch*, 1931

'Firm' command of 'one's' troops by no means always signifies real control in battle. Frequently 'firm' command may even run counter to development of the tactical process. Commanders who seek to control their entire order of battle 'firmly', 'on a tight rein', are apt to hold back the offensive during a penetration or pursuit and thus damage their chances of success.

Marshal of the Soviet Union Mikhail N. Tukhachevskiy, 1931–2, quoted in Simpkin, *Deep Battle*, 1987

The Regular Army to-day has a high proportion of competent staff officers, but a much lower proportion of senior officers with the gift for training or the personality for command. And it may be remarked that the value of 'tact' can be over-emphasized in selecting officers for command: positive personality will evoke a greater response than negative pleasantness. (May 1939)

Captain Sir Basil Liddell Hart, *Thoughts on War*, 1944

Command is not a prerogative, but rather a responsibility to be shared with all who are capable of filling up the spaces in orders and of carrying out that which is not openly expressed though it may be understood. Admittedly, it is not easy for a young officer, who by reason of his youth is not infrequently lacking in self-assurance and in the confidence that he can command respect, to understand that as a commander he can grow in strength in the measure that he succeeds in developing the latent strength of his subordinates. But if he stubbornly resists this premise as he goes along in the service, his personal resources will never become equal to the strain which will be imposed upon him, come a war emergency. The power to command resides largely in the ability to see when a proper initiative is being exercised and in giving it moral encouragement. When an officer feels that way about his job and his men, he will not be ready to question any action by a junior which might be narrowly construed as an encroachment upon his own authority. Of this last evil come the restraints which reduce men to automatons, giving only that which is asked, or less, according to the pressing of a button.

Brigadier General S. L. A. Marshall, *The Armed Forces Officer*, 1950

Bob, I'm putting you in command at Buna. Relieve Harding . . . [all] officers who won't fight . . . If necessary put sergeants in charge of battalions and corporals in charge of companies – anyone who will fight. Time is of the essence . . . I want you to take Buna, or not come back alive.

General of the Army Douglas MacArthur, 29 November 1942, MacArthur's oral orders to General Robert L. Eichelberger, quoted in Luvaas, ed., *Dear Miss Em: General Eichelberger's War in the Pacific*, 1972

. . . From General Marshall I learned the rudiments of effective command. Throughout the war I deliberately avoided intervening in a subordinate's duties. When an officer performed as I expected him to, I gave him a free hand. When he hesitated, I tried to help him. And when he failed, I relieved him.

General of the Army Omar Bradley, *A Soldier's Story*, 1951

I believe in 'personal' command, i.e., that a commander should never attempt to control an operation or a battle by remaining at his H.Q. or be content to keep touch with his subordinates by cable, W/T or other means of communication. He must as far as possible see the ground for himself to confirm or correct his impressions of the map; his subordinate commanders to discuss their plans and ideas with them; and the troops to judge of their needs and their morale. All these as often as possible. The same of course applies to periods of preparation and periods between operations. In fact, generally the less time a commander spends in his office and the more he is with the troops the better.

Field Marshal Earl Wavell, *Soldiers and Soldiering*, 1953

. . . As to the morale factors in command, it is always worth while to bear in mind the following:

(a) Two-thirds of the reports which are received in war are inaccurate; never accept a single report of success or disaster as necessarily true without confirmation.
(b) Always try to devise means to deceive and outwit the enemy and throw him off balance; the British in war are usually very lacking in this low cunning.
(c) Attack is not only the most effective but easiest form of warfare and the morale difference between advance and retreat is incalculable. Even when inferior in numbers, it pays to be as aggressive as possible.
(d) *Finally, when things look bad and one's difficulties appear great, the best tonic is to consider those of the enemy.*

Field Marshal Earl Wavell, *Soldiers and Soldiering*, 1953

You are probably busier than I am. As a matter of fact, commanding an army is not such a very absorbing task except that one has to be ready at all hours of the day and night . . . to make some rather momentous decision, which frequently consists of telling somebody who thinks he is beaten that he is not beaten.

General George S. Patton, Jr., letter, 11 October 1944, *The Patton Papers*, Vol. II, 1974

At the top there are great simplifications. An accepted leader has only to be sure of what it is best to do, or at least have his mind made up about it. The loyalties which centre upon number one are enormous. If he trips, he must be sustained. If he makes mistakes, they must be covered. If he sleeps, he must not be wantonly disturbed. If he is no good, he must be pole-axed.

Sir Winston Churchill, *Their Finest Hour*, 1949

The leader who depends for success on the strength of his personality soon learns that in combat the discount rate is very high indeed. Manners and appearance continue to command respect for the individual only when he is capable of carrying his proportionate part of the burden. The criterion of command is the ability to think clearly and work hard rather than to strike attitudes or accept disproportionate risks.

Brigadier General S. L. A. Marshall, *Men Against Fire*, 1947

Command must be direct and personal.

Field Marshal Viscount Montgomery of Alamein, *The Memoirs of Field Marshal Montgomery*, 1958

Men are of no importance. What counts is who commands.

Charles de Gaulle, interview in the *New York Times*, 1968.

It is absolutely vital that a senior commander should keep himself from becoming immersed in details, and I always did so. I would spend many hours in quiet thought and reflection in thinking out the major problems. In battle a commander has got to think how he will defeat the enemy. If he gets involved in details he cannot do this since he will lose sight of the essentials which really matter; he will then be led off on side issues which will have little influence on the battle, and he will fail to be that solid rock on which his staff can lean. Details are their province. No commander whose daily life is spent in the consideration of details, and who has not time for quiet thought and reflection, can make a sound plan of battle on a high level or conduct large-scale operations efficiently.

Field Marshal Viscount Montgomery of Alamein, *The Memoirs of Field Marshal Montgomery*, 1958

The command of troops in war is an art, a free creative activity based on character, ability and powers of intellect.

HDV 100/1, *Truppenfuehrung*, 1962 (troop leadership in the Bundeswehr)

It was my first command. It was the small unit, the group unit. I knew the four squadrons. I could never know any other four squadrons as well as I knew those four, because, when I had more rank and more command and more responsibility, I would be bound to lose out on the proportionate intimacies. I went on to command air wings and air divisions, and then to command even larger assemblages in the field. And finally to command all of SAC; and eventually to my job as Chief of Staff of the United States Air Force.

But the 305th – It was my 305th, our 305th.

General Curtis Le May, *Mission with Le May*, 1965

It was good fun commanding a division in the Iraq desert. It is good fun commanding a division anywhere. It is one of the four best commands in the service – platoon, battalion, a division, and an army. A platoon, because it is your first command, because you are young, and because, if you are any good, you know the men in it better than their mothers do and love them as much. A battalion, because it is a unit with a life of its own; whether it is good or bad depends on you alone; you have at last a real command. A division, because it is the smallest formation that is a complete orchestra of war and the largest in which every man can know you. An army, because the creation of its spirit and its leadership in battle give you the greatest unit of emotional and intellectual experience that can befall a man.

Field Marshal Viscount Slim, *Defeat Into Victory*, 1962

COMMAND SELECTION

In choosing their centurions the Romans look not so much for the daring or fire-eating type, but rather for men who are natural leaders and possess a stable and imperturbable temperament, not men who will open the battle and launch attacks, but those who will stand their ground even when worsted or hard-pressed, and will die in defence of their posts.

Polybius, *The Rise of the Roman Empire*, vi, c. 125 BC

. . . the general should be manly in his attitudes, naturally suited for command, profound in this thinking, sound in his judgment, in good physical condition, hardworking, emotionally stable. He should instill fear in the disobedient, while he should be gracious and kind to others. His concern for the common good should be such that he will neglect nothing at all that may be to its advantage. The general must be judged by his actions, and it is preferable that he be chosen for command on the basis of his record.

Anonymous Byzantine general, c. AD 527–565, *On Strategy* (*Peri Strategias*) in Dennis, *Three Byzantine Military Treatises*, 1985

Daily in the Meuse-Argonne, officers were assigned to command of regiments, brigades, divisions, or other important positions to replace casualties, some of which were not due to the enemy. Frequently they took over their duties in a situation which might be characterized as a bad mess; their officers and troops were scattered about on the terrain, under the fire of the enemy; deficiencies and special difficulties had to be ascertained. Yet there was rarely time for a pause in which to size up the officers and the troops, check over the supplies and material, and make a calm survey of the situation. Action was demanded and often within the first hour. A renewed attack was usually required the following morning. It should be obvious that the successful handling of such situations required a very special type of man, a type that frequently is in difficulties in a peace regime. The more conservative individual who had had combat experience under less strenuous circumstances was indeed fortunate, for he was in an important measure prepared for the seemingly chaotic scramblings of the battlefield.

General of the Army George C. Marshall, *Memoirs of My Service in the World War* (written between 1923–26), finally published 1976

. . . once men who are capable of leadership have been absorbed into the army, it is necessary to choose the best among them, and so to arrange matters that they shall occupy the highest posts. No task could well be more difficult, for though, time and peace permitting, it is not impossible to judge of the intelligence, and even of the authority, of those in positions of command, there is no opportunity to find out to what extent they possess the true instincts of the fighter. No doubt manoeuvres and field exercises provide useful training in appreciation and in reaching quick decisions, but only in an academic and conventional sense. Such training is fed on theories rather than on facts, for the facts of war cannot be convincingly simulated in a field day's mimic combat. It lacks the great test – that of events – and, more often than not, makes it impossible to distinguish real aptitude from superficial cleverness. It follows that more value is attached to a candidate's ability to learn than to his possible possession of the creative instinct; the art of grasping the immediate features of a situation more than the power to see its essentials; flexibility more than genuine understanding. As Scharnhorst said: 'Minds which have been trained mechanically triumph in peacetime over those possessing true insight and genius.' Furthermore, powerful personalities equipped with the qualities needed in the fighting soldier, and capable of standing up to the tests of great events, frequently lack that surface charm which wins popularity in ordinary life. Strong characters are, as a rule, rough, disagreeable, and aggressive. The man in the street may, somewhat shyly, admit their superiority and pay them lip service, but they are not often liked and, therefore, seldom favoured. When it comes to choosing men for high positions, the lot usually falls on the pleasing and docile rather than on the meritorious.

Charles de Gaulle, *The Edge of the Sword*, 1932

THE COMMANDER

. . . If wise, a commander is able to recognize changing circumstances and to act expediently. If sincere, his men will have no doubt of the certainty of rewards and punishments. If humane, he loves mankind, sympathizes with others, and appreciates their industry and toil. If courageous, he gains victory by seizing opportunity without hesitation. If strict, his troops are disciplined because they are in awe of him and afraid of punishment.

Tu Mu, AD 803–852, Chinese commentator to Sun Tzu, *The Art of War*, c. 500 BC, tr. Griffith

. . . the details depend on the circumstances, the judgment, the skill and on the bravery of the man in command.

Field Marshal Prince Aleksandr V. Suvorov (1729–1800)

An army of lions commanded by a deer will never be an army of lions.

Napoleon, 'Précis des guerres de Frédéric II', *Correspondance*, Vol. XXX, 1958–1870

In the profoundest sense, battles are lost and won in the mind of the commander, and the results are merely registered in his men.

Captain Sir Basil Liddell Hart, *Colonel Lawrence: The Man Behind the Legend*, 1934

In seeking victory, those who direct a war cannot overstep the limitations imposed by the objectives conditions; within these limitations, however, they can and must play a dynamic role in striving for victory. The stage of action for commanders in a war must be built upon objective possibilities, but on that stage they can direct the performance of many a drama, full of sound and colour, power and grandeur.

Mao Tse-tung, *On Protracted War*, 1938

A commander should have a profound understanding of human nature, the knack of smoothing out troubles, the power of winning affection while communicating energy, and the capacity for ruthless determination where required by circumstances. He needs to generate an electrifying current, and to keep a cool head in applying it.

Captain Sir Basil Liddell Hart, *Thoughts on War*, 1944

Sorting out muddles is really the chief job of a commander.

Field Marshal Earl Wavell, 'Training of the Army for War', *Journal, Royal United Services Institution*, February 1933

Among the ever-pressing problems of the commander is the seeking of means to break down the natural timidity of the great majority of his men. This he can never do unless he is sufficient master of himself that he can come out of his own shell and give his men a chance to understand him as a human being rather than as an autocrat giving orders. Nothing more unfortunate can happen to the commander than to come to be regarded by his subordinates as unapproachable, for such a reputation isolates him from the main problems of command as well as its chief rewards.

Brigadier General S. L. A. Marshall, *Men Against Fire*, 1947

In seeking to upset the enemy's balance, a commander must not lose his own balance. He needs to have the quality which Voltaire described as the keystone of Marlborough's success – 'that calm courage in the midst of tumult, that serenity of soul in danger, which the English call a cool head'. But to it he must add the quality for which the French have found the most aptly descriptive phrase – '*le sens du practicable*'. The sense of what is possible, and what is not possible – tactically and administratively. The combination of both these two 'guarding' qualities might be epitomised as the power of cool calculation. The sands of history are littered with the wrecks of finely conceived plans that capsized for want of ballast.

Captain Sir Basil Liddell Hart, introduction to Erwin Rommel, *The Rommel Papers*, 1953

Intelligence, knowledge, and experience are telling prerequisites. Lack of these may, if necessary, be compensated for by good general staff officers. Strength of character and inner fortitude, however, are decisive factors. The confidence of the men in the ranks rests upon a man's strength of character.

Field Marshal Erich von Manstein, *Lost Victories*, 1957

The commander is the backbone of the military unit. The command personnel constitute the skeleton which holds the separate limbs of an army together and supports them as a single, closely-knit military organization. The capability, devotion and courage of the combatants – in other words the practical ability of the army – are related directly to the ability of their commanders, of each of their commanders individually within the limits of his particular responsibilities, of all their commanders together in so far as they are collectively responsible for the army as a whole and all its undertakings.

General Yigal Allon, *The Making of Israel's Army*, 1960

I have developed almost an obsession as to the certainty with which you can judge a division, or any other large unit, merely by knowing its commander intimately. Of course, we have had pounded into us all through our school courses that the exact level of a commander's personality and ability is always reflected in his unit – but I did not realize, until opportunity came for comparisons on a rather large scale, how infallibly the commander and unit are almost one and the same thing.

General of the Army Dwight D. Eisenhower, *At Ease: Stories I Tell My Friends*, 1967

For confidence, spirit, purposefulness, aggressiveness flow down from the top and permeate a whole command. And in the same way do anxiety and lack of resolution on the part of a commander put their indelible stamp upon his men.

General Mathew B. Ridgway, quoted in *Military Review*, June 1987

In my experience, all very successful commanders are prima donnas, and must be so treated. Some officers require urging, others require suggestions, very few have to be restrained.

General George S. Patton, Jr., *War As I Knew It*, 1947

There is required for the composition of a great commander not only massive common sense and reasoning power, not only imagination, but also an element of legerdemain, an original and sinister touch, which leaves the enemy puzzled as well as beaten.

Sir Winston S. Churchill, *The World Crisis*, 1923

To be a successful commander, one must combine qualities of leadership with a knowledge of his profession. Either without the other is not of much avail.

Admiral Raymond A. Spruance, 'Thesis on Command', US Naval War College, 11 September 1926

Initiative and the desire to crawl into any crack in the enemy combat formations must be the main qualities of every commander.

Marshal of the Soviet Union Mikhail N. Tuckhachevskiy, quoted in Savkin, *The Basic Principles of Operational Art and Tactics*, 1972

No victory is possible unless the commander be energetic, eager for responsibilities and bold undertakings; unless he possess and can impart to all the resolute will of seeing the thing through; unless he be capable of exerting a personal action composed of will, judgment, and freedom of mind in the midst of danger.

Marshal of France Ferdinand Foch, *Precepts and Judgments*, 1919

COMMANDING GENERAL/ COMMANDER-IN-CHIEF

The responsibility for a martial host of a million lies in one man. His is the trigger of its spirit.

Wu Ch'i (430–381 BC) quoted by Ho Yen-hsi in a commentary to Sun Tzu, *The Art of War*, c. 500 BC, tr. Griffith

---

Having by this service won a universally acknowledged reputation for bravery, he in subsequent times refrained from exposing his person without sufficient reason when his country reposed her hopes of success on him – conduct characteristic not of a commander who relies on luck, but on one gifted with intelligence.

Polybius, *c.* 125 BC, of Scipio Africanus, quoted in Liddell Hart, *A Greater Than Napoleon: Scipio Africanus*, 1926

MAXIM 73. The first qualification in a general-in-chief is a cool head – that is, a head which receives just impressions, and estimates things and objects at their real value. He must not allow himself to be elated by good news, or depressed by bad.

The impressions he receives, either successively or simultaneously in the course of the day, should be so classed as to take up only the exact place in his mind which they deserve to occupy; since it is upon a just comparison and consideration of the weight due to different impressions that the power of reasoning and of right judgment depends.

Some men are so physically and morally constituted as to see everything through a highly coloured medium. They raise up a picture in the mind on every slight occasion and give to every trivial occurrence a dramatic interest. But whatever knowledge, or talent, or courage, or other good qualities such men may possess, nature has not formed them for the command of armies, or the direction of great military operations.

Napoleon, *Military Maxims of Napoleon*, 1831, tr. D'Aguilar

MAXIM 77. Generals-in-chief must be guided by their own experience or their genius. Tactics, evolutions, the duties and knowledge of an engineer or an artillery officer may be learned in treatises, but the science of strategy is only to be acquired by experience, and by studying the campaigns of all the great captains.

Gustavus, Adolphus, Turenne, and Frederick the Great, as well as Alexander, Hannibal, and Caesar, have all acted upon the same principles. These have been; to keep their forces united; to leave no weak part unguarded; to seize with rapidity on important points.

Such are the principles which lead to victory, and which, by inspiring terror at the reputation of your arms, will at once maintain fidelity and secure subjection.

Napoleon, *Military Maxims of Napoleon*, 1831, tr. D'Aguilar

The character of the man is above all other requisites in a commander-in-chief.

Lieutenant General Antoine-Henri Baron de Jomini, *Summary of the Art of War*, 1838

An army is to a commander what a sword is to a soldier, it is only worth anything in so far as it receives from him a certain emulsion.

Marshal of France Ferdinand Foch (1851–1929), quoted in Liddell Hart, *Through the Fog of War*, 1938

Whenever a General was offered an inferior command and refused it, he did very wrong. You should always accept the command offered to you when it is an active one. I would have taken over any command, however small, without hesitation. And I would have commanded it in such a way that I should soon have been given a more important one.

Marshal of France Ferdinand Foch (1851–1929), quoted in Recouly, *Marshal Foch*, 1929

In my youthful days I used to read about commanders of armies and envied them what I supposed to be a great freedom in action and decision. What a notion! The demands upon me that must be met make me a slave rather than a master. Even my daily life is circumscribed with guards, aides, etc., etc., until sometimes I want nothing so much as complete seclusion.

General of the Army Dwight D. Eisenhower, 27 May 1943, letter to his wife, Mamie, quoted in *Letters to Mamie*, 1978

A commander-in-chief, therefore, whose power and dignity are so great and to whose fidelity and bravery the fortunes of his countrymen, the defence of their cities, the lives of the soldiers, and the glory of the state, are entrusted, should not only consult the good of the army in general, but extend his care to every private soldier in it. For when any misfortunes happen to those under his command, they are considered as public losses and imputed entirely to his misconduct.

Flavius Vegetius Renatus, *Military Institutions of the Romans*, c. AD 378

I have formed a picture of a general commanding which is not chimerical – I have seen such men.

The first of all qualities is COURAGE. Without this the others are of little value, since they cannot be used. The second is INTELLIGENCE, which must be strong and fertile in expedients. The third is HEALTH.

He should possess a talent for sudden and appropriate improvisation. He should be able to penetrate the minds of other men, while remaining impenetrable himself. He should be endowed with the capacity of being prepared for everything, with activity accompanied by judgment, with skill to make a proper decision on all occasions, and with exactness of discernment.

Field Marshal Maurice Comte de Saxe, *My Reveries*, 1732

The first quality of a general-in-chief is a great knowledge of the art of war. This is not intuitive, but the result of experience. A man is not born a commander. He must become one. Not to be anxious; to be always cool; to avoid confusion in his commands; never to change countenance; to give his orders in the midst of battle with as much composure as if here perfectly at ease. These are proofs of valour in a general.

Field Marshal Prince Raimondo Montecuccoli (1609–1680), *Commentarii Bellici*, 1740

Let what will arrive, it is the part of the general-in-chief to remain firm and constant in his purposes: he must be equally superior to elation in prosperity and depression in adversity, for in war good and bad fortune succeed each other by turns, and form the ebb and flow of military operations.

Field Marshal Prince Raimondo Montecuccoli (1609–1680), *Commentarii Bellici*, 1740

There are three types of commanders in the higher grades:

1. Those who have faith and inspiration, but lack the infinite capacity for taking pains and preparing for every foreseeable contingency – which is the foundation of all success in war. These fail.
2. Those who possess the last-named quality to a degree amounting to genius. Of this type I would cite Wellington as the perfect example.
3. Those who, possessing this quality, are inspired by a faith and conviction which enables them, when they have done everything possible in the way of preparation and when the situation favours boldness, to throw their bonnet over the moon. There are moments in war when, to win all, one has to do this. I believe such a moment occurred in August 1944, after the Battle of Normandy

had been won, and it was missed. Nelson was the perfect example of this – when he broke the line at St. Vincent, when he went straight in to attack at the Nile under the fire of the shore batteries and with night falling, and at the crucial moment at Trafalgar.

Field Marshal Viscount Montgomery of Alamein, *The Memoirs of Field Marshal Montgomery*, 1958

In my opinion, generals – or at least the Commanding General – should answer their own telephones in the daytime. This is not particularly wearisome because few people call a general, except in emergencies, and then they like to get him at once.

General George S. Patton, *War As I knew It*, 1947

COMMON SENSE

I attribute it entirely to the application of good sense to the circumstances of the moment.

The Duke of Wellington (1769–1852), when asked to what characteristic of mind he attributed his invariable success, quoted in Fraser, *Words of Wellington*, n.d.

I thought I detected in the management what I had never discovered before on the battlefield – a little common sense. Dash is handsome, genius glorious; but modest, old-fashioned, practical everyday sense is the trump after all.

A staff officer's observation of Ulysses S. Grant, quoted in Williams, *McClellan, Sherman, and Grant*, 1962

While exceptional ability is desirable in a general, common sense is essential. Yet there is a peculiar danger in the latter quality unless it is accompanied by the former faculty. For many men of strong common sense but limited education are apt to feel uneasy in dealing with those who are intellectually better equipped. And the too common result is that they become more than normally sensitive to the arguments, especially the negative arguments, of intellectual second-raters, clever but uncreative – the class which produces most of the mouthpieces of orthodoxy. And the purely practical man is too often led to yield, against his better judgment, because he feels incapable of combating their objections. (April 1929)

Captain Sir Basil Liddell Hart, *Thoughts on War*, 1944

COMMUNICATIONS

An army ought only to have one line of operation. This should be preserved with care, and never abandoned but in the last extremity.

Napoleon, *Military Maxims of Napoleon*, 1831, tr. D'Aguilar

The line that connects an army with its base of supplies is the Heel of Achilles – its most vital and vulnerable point.

Colonel John S. Mosby, *Mosby's War Reminiscences*, 1887

Communications dominate war; broadly considered, they are the most important single element in strategy, political or military.

Rear Admiral Alfred Thayer Mahan, *The Problem of Asia and its Effects on International Relations Naval and Civil*, 1900

Free supplies and open retreat are two essentials to the *safety* of an army or a fleet.

Rear Admiral Alfred Thayer Mahan, *Naval Strategy*, 1911

All military organizations, land or sea, are ultimately dependent upon open communications with the basis of national power.

Rear Admiral Alfred Thayer Mahan, *Naval Strategy*, 1911

If intercommunication between events in front and ideas behind are not maintained, then two battles will be fought – a mythical headquarters battle and an actual front-line one, in which case the real enemy is to be found in our own headquarters. Whatever doubt exists as regards the lessons of the first war, this is one which cannot be controverted.

Major General J. F. C. Fuller, quoted in George Marshall, ed., *Infantry in Battle*, 1939

Nine times out of ten an army has been destroyed because its supply lines have been severed.

General of the Army Douglas MacArthur, 23 August 1950, to the Joint Chiefs of Staff

---

In planning any stroke against the enemy's communications, either by manoeuvre round his flank or by rapid penetration of a breach in his front, the question will arise as to the most effective point of aim – whether it should be directed against the immediate rear of the opposing force, or further back.

In general, the nearer to the force that the cut is made, the *more immediate* the effect; the nearer to the base, the *greater* the effect . . .

A further consideration is that while a stroke close in rear of the enemy force may have more effect on the minds of the enemy troops, a stroke far back tends to have more effect on the mind of the enemy commander.

Captain Sir Basil Liddell Hart, *Strategy*, 1954

To exploit the principles of war for our purpose and base ourselves upon strategic indirect approach, so as to determine the issue of the fighting even before fighting has begun, it is necessary to achieve the three following things:

(a) to cut the enemy's line of communication, thus paralyzing his physical build-up.

(b) to seal him off from his lines of retreat, thus undermining the enemy's will and destroying his morale;

(c) to hit his centres of administration and disrupt his communications, thus severing the link between his brain and his limbs.

Reflection on these three aims proves the truth of Napoleon's saying: 'The whole secret of the art of war lies in the ability to become master of the lines of communications.'

General Yigael Yadin, *Bamachaneh* (The Israel Forces' Journal) September, 1949, condensed in Liddell Hart, *Strategy*, 1954)

COMPETENCE

A scorpion will sting because it has poison; a soldier can be brave when he can rely on his equipment. Therefore when their weapons are sharp and their armour is strong, people will readily do battle. If the armour is not strong, it is the same as baring one's shoulders. If a bow cannot shoot far, it is the same as close-range weapon. If a shot cannot hit the mark, it is the same as having no weapon. If a scout is not careful, it is the same as having no eyes. If a general is not brave in battle, it is the same as having no military leadership.

Zhuge Liang (AD 180–234), *The Way of the General*

The general should be ignorant of none of the situations likely to occur in war. Who can attempt to accomplish what he does not understand? Who is able to furnish assistance in situations whose dangers he does not understand?

The Emperor Maurice, *The Strategikon*, c. AD 600

Our object ought to be to have a good army rather than a large one.

General George Washington, 15 September 1780

Officers can never act with confidence until they are masters of their profession.

Major General Henry Knox (1750–1806)

One must understand the mechanism and power of the individual soldier, then that of a company, a battalion, a brigade and so on, before one can venture to group divisions and move an army. I believe I owe most of my success to the attention I always paid to the inferior part of tactics as a regimental officer. There are few men in the Army who knew these details better than I did; it is the foundation of all military knowledge.

The Duke of Wellington, quoted in Bryant, *The Great Duke*, 1971

The true way to be popular with the troops is not to be free and familiar with them, but to make them believe you know more than they do.

General of the Army William T. Sherman, 11 November 1864, letter to Rev. Henry Lay

---

A competent leader can get efficient service from poor troops, while on the contrary an incapable leader can demoralize the best of troops.

General of the Armies John J. Pershing, *My Experiences in the World War*, 1931

The officers know that I myself am not ashamed to work at this . . . Suvorov was Major, and Adjutant, and everything down to Corporal; I myself looked into everything and could teach everybody.

Field Marshal Prince Aleksandr V. Suvorov (1729–1800), quoted in Blease, *Suvorof*

COMRADESHIP

My first wish would be that my military family, and the whole Army, should consider themselves as a band of brothers, willing and ready to die for each other.

General George Washington, 21 October 1798, letter to Henry Knox, *The Writings of George Washington*, Vol. 36, 1931–44

These voices, these quiet words, these footsteps in the trench behind me recall me at a bound from the terrible loneliness and fear of death by which I had been almost destroyed. They are more to me than life, those voices, they are more than motherliness and more than fear, they are the strongest, most comforting thing there is anywhere, they are the voices of my comrades.

Erich Maria Remarque, *All Quiet On the Western Front*, 1929

The soldier's field of activity is man, who controls science, technics and material. The army is a combination of many men with the same serious aim. This gives the soldier's profession a quite peculiar bond of unity, a corporate sense which we call comradeship. The term is extremely comprehensive. If we start out from the notion of responsibility we find that 'comradeship' means 'one for all', for each man bears, in his own way and in his own place, a share of the responsibility for the welfare, the ability, the achievements, and the life of others. For the senior, the leader, the superior, this means the duty of correcting, of training, of supervising others; for the junior, the novice, the subordinate, it means the duty of conscious, voluntary subordination. Love and confidence are the two great components of comradeship.

Colonel General Hans von Seekt, *Thoughts of a Soldier*, 1930

Great achievements in war and peace can only result if officers and men form an indissoluble band of brothers.

Field Marshal Paul von Hindenburg, 1934, Proclamation to the German Army

---

. . . not because their presence was going to make a lot of difference to the big scheme of the war, and not to uphold the traditions of the umpteenth regiment. A lot of guys don't know the name of their regimental commanders. They went back because they knew their companies were shorthanded, and they were sure that if someone else in their squad or section were in their shoes, and the situation was reversed, those friends would come back to make the load lighter on *them*.

Bill Mauldin, commenting on 'how men sneaked back to their units rather than spend time convalescing after discharge from hospital', quoted in Keegan, *Soldiers*, 1983?

A mysterious fraternity born out of smoke and danger of death.

Stephen Crane, *The Red Badge of Courage*, 1893

I hold it to be one of the simplest truths of war that the thing which enables an infantry soldier to keep going with his weapons is the near presence or the presumed presence of a comrade.

Brigadier General S. L. A. Marshall, *Men Against Fire*, 1947

Men who have been in battle know from first-hand experience that when the chips are down, a man fights to help the man next to him, just as a company fights to keep pace with its flanks. Things have to be that simple. An ideal does not become tangible at the moment of firing a volley or charging a hill. When the hard and momentary choice is death, the words once heard at an orientation lecture are clean forgot, but the presence of a well-loved comrade is unforgettable. In battle the most valued thing at hand is that which becomes most stoutly defended. All values are interpreted in terms of the battlefield itself. Yet above and beyond any symbol – whether it be the individual life or a pillbox commanding a wadi in the Sahara – are all of the ideals which press upon men, causing them to accept a discipline and to hold to the line even though death may be at hand.

Brigadier General S. L. A. Marshall, *Men Against Fire*, 1947

The bald, red hills with the sandbag bunkers, the banter and frolic of dirt-covered grunts, the fearful intensity of contact . . . Down south his men were on patrol, or digging new perimeters, or dying, and he was nothing if he did not share that misery.

James Webb, *Fields of Fire*, 1978

Why does the soldier leave the protection of his trench or hole in the ground and go forward in the face of shot and shell? It is because of the leader who is in front of him and his comrades who are around him. Comradeship makes a man feel warm and courageous when all his instincts tend to make him cold and afraid.

Field Marshal Viscount Montgomery of Alamein, *A History of Warfare*, 1968

CONCENTRATION

If I am able to determine the enemy's dispositions while at the same time I conceal my own then I can concentrate and he must divide. And if I concentrate while he divides, I can use my entire strength to attack a fraction of his. There, I will be numerically superior. Then, if I am able to use many to strike few at the selected point, those I deal with will be in dire straits.

The enemy must not know where I intend to give battle. For if he does not know where I intend to give battle he must prepare in a great many places. And when he prepares in a great many places, those I have to fight in any one place will be small.

For if he prepares to the front his rear will be weak, and if to the rear, his front will be fragile. If he prepares to the left, his right will be vulnerable and if to the right, there will be few on his left. And when he prepares everywhere he will be weak everywhere.

Sun Tzu, *The Art of War*, c. 500 BC, tr. Griffith

There is an ancient rule of war that cannot be repeated often enough: hold your forces together, make no detachments, and, when you are ready to fight the enemy, assemble all your forces and seize every advantage to make sure of success. This rule is so certain that most of the generals who have neglected it have been punished promptly.

Frederick the Great, *Instructions to His Generals*, 1747

. . . Petty geniuses attempt to hold everything; wise men hold fast to the most important points. They parry great blows and scorn little accidents. There is an ancient apothegm: he who would preserve everything, preserves nothing. Therefore, always sacrifice the bagatelle and pursue the essential! The essential is to be found where big bodies of the enemy are. Stick to defeating them decisively, and the detachments will flee by themselves or you can hunt them down without difficulty.

Frederick the Great, *Instructions to His Generals*, 1747

One should never risk one's whole fortune unless supported by one's entire forces.

Niccolo Machiavelli, *Discourses*, 1517

. . . It is an ancient rule of war, and I am just repeating it – if you separate your forces you will be beaten in detail. When you give battle you must concentrate all the troops you can – you cannot find a better use for them.

Frederick the Great, *Principes Généraux de la Guerre*, 1748, quoted in Duffy, *The Military Life of Frederick the Great*, 1985

True military art consists of the ability to be stronger than the enemy at the given moment.

Military art is the art of separating for life and uniting for battle.

On the battlefield there is no such thing as surplus battalion or squadron.

Commanders who save fresh forces for operations after the battle will almost always be defeated.

Napoleon, quoted in Savkin, *Basic Principles of Operational Art and Tactics*, 1972

Never divide your forces where the outcome is to be decided.

. . . Major goals may be achieved only by decisive attacks. Therefore the most important art of a general consists of the following: *correctly determine the moment and points when and where such decisive attacks can be delivered* with the greatest probability of favourable results. Such a decisive attack is possible only with a superiority in forces at the point at which it is delivered . . . These principles, lying in the nature of war and being the sole ones leading to decisive results, permit a definition of military art as follows: *it consists of the art of concentrating and using numerically superior forces at the decisive point.*

Archduke Charles of Austria (1771–1847), quoted in Savkin, *Basic Principles of Operational Art and Tactics*, 1972

The best strategy is always *to be very strong*; first in general, and then at the decisive point. Apart from the effort needed to create military strength, which does not always emanate from the general, there is no higher and simpler law of strategy than of *keeping one's forces concentrated.*

Major General Carl von Clausewitz, *On War*, iii, 1832, tr. Howard and Paret

---

MAXIM 28. No force should be detached on the eve of a battle, because affairs may change during the night, either by the retreat of the enemy, or by the arrival of large reinforcements to enable him to resume the offensive, and counteract your previous dispositions.

Napoleon, *Military Maxims of Napoleon*, 1831, tr. D'Aguilar

MAXIM 29. When you have resolved to fight a battle, collect your whole force. Dispense with nothing. A single battalion sometimes decides the day.

Napoleon, *Military Maxims of Napoleon*, 1831, tr. D'Aguilar

In any military scheme that comes before you, let your first question be, Is this consistent with the requirement of concentration?

Rear Admiral Alfred Thayer Mahan, *Naval Strategy*, 1911

The fundamental object in all military combinations is to gain local superiority by concentration.

Rear Admiral Alfred Thayer Mahan, *Naval Strategy*, 1911

War is the impact of opposing forces. It follows that the stronger force not only destroys the weaker, but that its impetus carries the weaker force along with it. This would seem not to allow a protracted, consecutive employment of force: instead, the simultaneous use of all means intended for a given action appears as an elementary law of war.

Major General Carl von Clausewitz, *On War*, iii, 1832, tr. Howard and Paret

---

Concentration sums up in itself all the other factors, the entire alphabet of military efficiency in war.

Lieutenant General Antoine-Henri Baron de Jomini, quoted in Simpkin, *Race to the Swift*, 1985

. . . one must not rely on the heroism of the troops. *Strategy must furnish tactics with tasks easy to accomplish.* This is obtained in the first place by the concentration in the place of the main blow of forces many times superior to those of the enemy.

Marshal of the Soviet Union Mikhail N. Tuhkachevskiy, 1920, quoted in Leites, *The Soviet Style of War*, 1982

The Principle of Concentration. Concentration, or the bringing of things or ideas to a point of union, presupposes movements; movement of ideas, especially in an army, is a far more difficult operation than the movement of men. Nevertheless, unless ideas, strategical, tactical and adminstrative, be concentrated, cohesion of effort will not result; and in proportion as a unity of action is lacking, so will an army's strength, moral and physical, be squandered in detail until a period be arrived at in which the smallest result will be obtained from every effort.

Major General J. F. C. Fuller, *The Reformation of War*, 1923

Another rule – never fight against heavy odds, if by any possible maneuvering you can hurl your own forces on only a part, and then the weakest part, of your enemy and crush it. Such tactics will win every time, and a small army may thus destroy a large one in detail, and repeated victory will make it invincible.

Lieutenant General Thomas 'Stonewall' Jackson, 1862, quoted in Henderson, *Stonewall Jackson*, 1898

I always make it a rule to get there first with the most men.

Lieutenant General Nathan Bedford Forrest (1821–1877)

To remain separated as long as possible while operating and to be concentrated in good time for the decisive battle, that is the task of the leader of large masses of troops,

Field Marshal Helmuth Graf von Moltke, *Instructions for Superior Commanders of Troops*, 1869

. . . To have at the decisive moment, at the decisive point, an overwhelming superiority of force – this law of military success is also the law of political success, especially in that seething and bitter war of classes which is called revolution.

V. I. Lenin, quoted in Trotsky, *The History of the Russian Revolution*, 1932

One great principle underlies all the operations of war – a principle which must be followed in all good combinations. It is embraced in the following maxims:

1. To throw by strategic movements the mass of an army, successively, upon the decisive points of a theatre of war, and also upon the communications of the enemy as much as possible without compromising one's own.

2. To manoeuvre to engage fractions of the hostile army with the bulk of one's own forces.

3. On the battlefield, to throw the mass of the forces upon the decisive point, or upon that portion of the hostile line which it is of the first importance to overthrow.

4. To so arrange that these masses shall not only be thrown upon the decisive point, but

that they shall engage at the proper times and with ample energy.

This principle has too much simplicity to escape criticism . . .

Lieutenant General Antoine-Henri Baron de Jomini, *Summary of the Art of War*, 1838

All through history, from the days of the great phalanx of the Roman Legion, the master law of tactics remains unchanged; this law is that to achieve success you must be superior at the point where you intend to strike the decisive blow.

Field Marshal Viscount Montgomery, 'The Growth of Modern Infantry Tactics', *Antelope*, 1/1925

My strategy is one against ten, my tactic ten against one.

Mao Tse-tung, quoted in Bloodworth, *The Chinese Machiavelli*, 1976

I don't care how many tanks you British have so long as you keep splitting them up the way you do. I shall just continue to destroy them piecemeal.

Field Marshal Erwin Rommel, 1941, to a captured British brigadier, quoted in Moorehead, *The March to Tunis*, 1943

The principles of war could, for brevity, be condensed into a single word – 'Concentration'.

Captain Sir Basil H. Liddell Hart, *Thoughts on War*, 1944

Concentration is supposed to be exceptional only because people do not try and, in this, as in so many things, starve within an inch of plenty.

Brigadier General S. L. A. Marshall, *The Armed Forces Officer*, 1950

The principles of war not merely one principle, can be condensed into a single word – 'concentration'. But for truth this needs to be amplified as the 'concentration of strength against weakness'. And for any real value it needs to be explained that the concentration of strength against weakness depends on the dispersion of your opponent's strength, which in turn is produced by a distribution of your own that gives the appearance and partial effect of dispersion. Your dispersion, his dispersion, your concentration – such is the sequence, and each is a sequel. True concentration is the fruit of calculated dispersion.

Captain Sir Basil Liddell Hart, *Strategy*, 1954

The enemy found himself face to face with a contradiction: without scattering his forces it was impossible for him to occupy the invaded territory; in scattering his forces, he put himself in difficulties. His scattered units would fall easy prey to our troops, his mobile forces would be more and more reduced and the shortage of troops would be all the more acute. On the other hand if he concentrated his forces to move from the defensive position to cope with us with more initiative, the occupation forces would be weakened and it would be difficult for him to hold the invaded territory. Now, if the enemy gives up occupied territory, the very aim of the war of reconquest is defeated.

General Vo Ngyyen Giap, referring to General Navarre's dilemma in the First Indo–China War, *People's War, People's Army*, 1961

Oppose the strategy of striking with two 'fists' in two directions at the same time, and uphold the strategy of striking with one 'fist' in one direction at a time.

Mao Tse-tung, *Selected Military Writings*, 1963

CONFIDENCE

. . . Now the method of conducting the march of an army is this. The established order of advancing and of halting is not to be violated; the right times for eating and drinking should not be missed. Do not deplete the strength of men and animals. In these three matters the troops must have confidence in the orders of their seniors. The orders of their superiors is the source whence discipline is born. If advances and halts are not well regulated, if food and drink are not suitable, the horses will be exhausted and the men fatigued and not able to be relaxed and sheltered. Such are the situations in which troops do not trust the orders or their superiors. If orders of superiors are ineffective, when the army encamps it will be disorderly, and when it fights, it will be defeated.

Wu Ch'i (430–381 BC), commentary in Sun Tzu, *The Art of War*, c. 500 BC, tr. Griffith

---

Troops are not be led into battle unless confident of success.

Flavius Vegetius Renatus, *Military Institutions of the Romans*, c. AD 378

When the general leads his men out to battle, he should present a cheerful appearance, avoiding any gloomy look. Soldiers usually estimate their prospects by the appearance of the general.

The Emperor Maurice, *The Strategikon*, c. AD 600

The general who wants to keep his plans concealed from the enemy should never take the rank and file into his confidence.

The Emperor Maurice, *The Stategikon*, c. AD 600

In connection with military matters, one must never say that something can absolutely not be done. By this, the limitations of one's heart will be exposed.

Asakura Soteki (1477–1555) *Soteki Waki*, c. 1550, in Wilson, *Ideals of the Samurai*, 1982

Whenever despondency or fear has fallen on an army because the enemy has received reinforcements or gained an advantage, then especially the general should show himself to his soldiers gay, cheerful and undaunted. For the appearance of the leaders brings about a corresponding change in the minds of the subordinates, and if the general is cheerful and has a joyful look, the army also takes heart, believing that there is no danger; but should he have a frightened, worried appearance, the spirits of the soldiers fall with his, in the belief that disaster is impending. On this account, the general must inspire cheerfulness in the army, more by the strategy of his facial expressions than by his words; for many distrust speeches on the ground that they have been concocted especially for the occasion, but believing a confident appearance to be unfeigned they are convinced of his fearlessness; and it is an excellent thing to understand these two points, how to say the right word and how to show the right expression.

Onasander, *The General*, AD 58

One should never display a weak attitude, even though he may be with sympathetic relatives or retainers. In the *San Lueh* it says, 'If a man loses his courage, his servants and soldiers will lose their respect for him.'

Takeda Nobushige (1525–1561), *Opinions in 99 Articles*, 1558, in Wilson, *Ideals of the Samurai*, 1982

This is a maxim that the most experienced captains of this era hold to be true: that many more battles are won by well-ordered ranks and by a fine appearance than by gun shot or blows from a sword. Such well-ordered ranks make one appear confident, and it seems that it is sufficient merely to give the appearance of being brave, since most often our enemies will not await us long enough to let us show if we are in actual fact . . .

Louis XIV, *Mémoires de Louis XIV*, 1860, quoted in Ralston, *Soldiers and States*, 1966

A great and good general is . . . in himself a host; for his influence, insinuating itself into every member of the military body, connects and binds the whole together imperceptibly, but firmly and securely. Such confidence in a leader is the charm against a panic.

Robert Jackson, *A Systematic View of the Formation, Discipline, and Economy of Armies*, 1804

Every man passed through my hands, and he was told that nothing more remained for him to know, if only he did not forget what he had learned. Thus he was given confidence in himself, the foundation of bravery.

Field Marshal Prince Aleksandr V. Suvorov (1729–1800), quoted in Blease, *Suvorof*, 1920

Take nobody into your confidence, not even your chief of staff.

Napoleon, 26 May 1812, to Jerome, *Correspondance*, No. 18727, Vol. XXIII, 1858–1870

I think, in summing up my reflections, that it would be useless to advise colonels of infantry regiments to avoid, with the greatest care, a reverse at the opening of the campaign. The least check has more influence than is generally supposed upon remaining operations of the campaign: it diminishes confidence in the men, by raising mistrust of the commander's talents. The least success, on the contrary, impresses upon the troops, from the very beginning, that just military pride which doubles their strength, and serves as a presage of a series of brilliant feats.

Marshal of France Michel Ney, Duc d'Elchingen, Prince de la Moskova, *Memoirs of Marshal Ney*, 1834

No matter what may be the ability of the officer, if he loses the confidence of his troops, disaster must sooner or later ensue.

General Robert E. Lee, 8 August 1863, letter to Jefferson Davis

---

If confidence be indeed 'half the battle', then to undermine the enemy's confidence is more than the other half – because it gains the fruits without an all-out fight. (April 1929)

Captain Sir Basil H. Liddell Hart, *Thoughts on War*, 1944

One of the most interesting points to my mind about all this business of making war is the way that people try and shake your confidence in what you are doing, and suggest that your plan is not good, and that you ought to do this, or that. If I had done all that was suggested I would still be back in the Alamein area.

Field Marshal Viscount Montgomery of Alamein, quoted in Hamilton, *Master of the Battlefield*, 1983

The most vital quality a soldier can possess is self-confidence, utter, complete and bumptious.

General George S. Patton, Jr., 6 June 1944, letter to his son, *The Patton Papers*, Vol. II, 1974

For two weeks Puller had commanded the rear of the First Marine Division, cut off in the Chosin Reservoir region by hundreds of thousand of Chinese Communist troops. The colonel was visiting a hospital tent where a priest was administering last rites to Marine wounded when a messenger came:

'Sir, do you know they've cut us off? We're entirely surrounded.'

'Those poor bastards,' Puller said. 'They've got us right where we want 'em. We can shoot in every direction now.'

Lieutenant General Louis 'Chesty' Puller, 1950, as narrated by Burke Davis, *Marine*, 1962

Finally, that confidence is the *sine qua non* of all useful military power. The moral strength of an organic unit comes from the faith in the ranks that they are being wisely directed and from faith up top that orders will be obeyed. When forces are tempered by this spirit, there is no limit to their enterprise. They become invincible. Lacking it, however, any military body, even though it has been compelled to toe the mark in training, will deteriorate into a rabble under conditions of extraordinary stress in the field, as McDowell's Army did at Bull Run in the American Civil War, and as Hitler's Armies did in 1945 after the Rhine had been crossed at Remagen.

Brigadier General S. L. A. Marshall, *The Armed Forces Officer*, 1950

Personal bravery of a single individual alone is not decisive on the day of battle, but rather bravery of the corps, and the latter rests on the good opinion and the confidence that each individual soldier places in the corps to which he belongs.

Field Marshal Colmar von Der Goltz (1843–1916) *Rossbach und Jena*

In each battle there is a moment when even the bravest soldiers feel like fleeing. This panic is caused by the lack of confidence in their own courage. However, any event, any pretext can return confidence to them. The highest military art is to create these events or pretexts.

Mikhail V. Frunze (1885–1925), quoted in Gareyev, M. V. *Frunze, Military Theorist*, 1985

However desperate the situation, a senior commander must always exude confidence in the presence of his subordinates. For anxiety, topside, can spread like cancer down through the command.

General of the Army Omar Bradley, *A Soldier's Story*, 1951

The general who sees that the soldier is well fed and looked after, and who puts him into a good show and wins battles, will naturally have his confidence. Whether he will also have his affection is another story . . . But does it matter to a general whether he has his men's affection so long as he has their confidence? He must certainly never court popularity. If he has their appreciation and respect it is sufficient. Efficiency in a general his soldier have a right to expect; geniality they are usually right to suspect. Marlborough was perhaps the only great general to whom geniality was always natural.

Field Marshal Earl Wavell, *Soldiers and Soldiering*, 1953

. . . Probably one of the greatest assests a commander can have is the ability to radiate confidence in the plan and operations even (perhaps especially) when inwardly he is not too sure about the outcome.

Field Marshal Viscount Montgomery of Alamein, *The Memoirs of Field Marshal Montgomery*, 1958

Confidence is the cornerstone of success in battle: *each soldier's belief in his own competence, his trust in that of other members of his unit, and their collective pride, cohesion, and effectiveness.*

FM 100–5 *Operations*, Department of the Army, 1 July 1976

Senior officers decrying the low morale of their forces evidently do not realize that the esprit of the men is but a mirror of their confidence in their leadership. Confidence in leaders is an accepted ingredient of organizational spirit. However, dissentions among the top command, like a single drop of poison in wine, can destroy all partakers.

General of the Army Omar Bradley, *A General's Life*, 1983

CONTACT

In war as in love, we must achieve contact ere we triumph.

Napoleon, *Political Aphorisms*, 1848

Wherever the enemy goes, let our troops go also.

General of the Army Ulysses S. Grant, 1 August 1864, dispatch to General Halleck about Sheridan's operations in the Shenandoah Valley

---

Bear in mind, the object is to drive the enemy south; and to do this you want to keep him always in sight. Be guided in your course by the course he takes.

General of the Army Ulysses S. Grant, 5 August 1864, orders to General Sheridan about the upcoming operations in the Shenandoah Valley, quoted in Sheridan, *Personal Memoirs of P. H. Sheridan*, 1888

. . . *Contact* (a word which perhaps better than any other indicates the dividing line between tactics and strategy).

Rear Admiral Alfred Thayer Mahan, *The Influence of Sea Power Upon History*, 1890

Contact is information of the most tangible kind, an enemy met with is an enemy at grips, and, as in a wrestling match, contact is likely to be followed by much foot-play. Time still remains the decisive factor, time wherein to modify a plan according to the information contact gains.

Major General J. F. C. Fuller, *Armoured Warfare*, 1943

CONTINGENCY

MAXIM 2. In forming the plan of a campaign, it is requisite to foresee anything the enemy may do, and to be prepared with the necessary means to counteract it.

Plans of campaign may be modified *ad infinitum* according to circumstances, the genius of the general, the character of the troops, and the features of the country.

Napoleon, *The Military Maxims of Napoleon*, 1831, tr. D'Aguilar

---

MAXIM 8. A general-in-chief should ask himself frequently in the day, What should I do if the enemy's army appeared now in my front, or on my right, or on my left? If he have any difficulty in answering these questions he is ill posted, and should seek to rememdy it.

Napoleon, *The Military Maxims of Napoleon*, 1831, tr. D'Aguilar

---

The sharp general takes into account not only probable dangers, but also those which may be totally unexpected.

The Emperor Maurice, *The Strategikon*, c. AD 600

In battles and in every action against the enemy the wise general, even the most courageous, will keep in mind the possibility of failure and defeat and will plan for them as actually occurring.

The Emperor Maurice, *The Strategikon*, c. AD 600

CONTRACTORS

In the councils of government we must guard against the acquisition of unwarranted influence, whether sought or unsought, by the military-industrial complex. The potential for the diastrous rise of misplaced power exists and will persist.

President Dwight D. Eisenhower, 17 January 1961, Farewell Address as president

---

There is nothing performed by contractors which may not be much better executed by intelligent officers. They make immense fortunes at the expense of the state which ought to be saved. They destroyed the army, horse and foot and even hospitals, by furnishing the worst of everything.

Major General Henry Lloyd (1720–1783), *History of the Late War in Germany*, 1766–1782, quoted in Liddell Hart, *The Sword and the Pen*, 1976

. . . bureau, not to speak of the contractors, are the born enemies of all that tends to put the details of military administration into military hands.

Field Marshal François Comte de Guibert (1744–1790) quoted in Vagts, *A History of Militarism*, 1937

CONTROL

Generally, management of the many is the same as management of the few. It is a matter of organization.

And to control many is the same as to control few. This is a matter of formations and signals.

Sun Tzu, *The Art of War*, c. 500 BC, tr. Griffith

And therefore those skilled in war avoid the enemy when his spirit is keen and attack him when it is sluggish and his soldiers homesick. This is control of the morale factor.

In good order they await a disorderly enemy; in serenity, a clamorous one. This is control of the mental factor.

Close to the field of battle, they await an enemy coming from afar; at rest, an exhausted enemy; with well-fed troops, hungry ones. This is control of the physical factor.

Sun Tzu, *The Art of War, c.* 500 BC, tr. Griffith

After gaining a victory the general who pursues the enemy with a scattered and disorganized army gives away his victory to the foe.

The Emperor Maurice, *The Strategikon, c.* AD 600

I have known commanders who considered that once their plan was made and orders issued, they need take no further part in the proceedings, except to influence the battle by means of their reserves. Never was there a greater mistake. The modern battle can very quickly go off the rails. To succeed, a C-in-C must ensure from the beginning a very firm grip on his military machine; only in this way will his force maintain balance and cohesion and thus develop its full fighting potential. This firm grip does not mean interference, or cramping the initiative of subordinates; indeed, it is by the initiative of subordinates that the battle is won. The firm grip is essential in order that the master plan will not be undermined by the independent ideas of individual subordinate leaders at particular moments in the battle. Operations must develop within a predetermined pattern of action. If this is not done the result will be a compromise between the individual conceptions of subordinates about how operations should develop; alternatively, operations will develop as a result of situations created by subordinate action and in a way which does not suit the master plan. A third alternative is that the initiative might pass to the enemy. The master plan must never be so rigid that the C-in-C cannot vary it to suit the changing tactical situation; but nobody else may be allowed to change it at will – and, especially, not the enemy.

Field Marshal Viscount Montgomery of Alamein, *The Memoirs of Field Marshal Montgomery*, 1958

Half of control during battle comes from the commander's avoiding useless expenditure of the physical resources of his men while taking action to break the hold of fear.

Brigadier General S. L. A. Marshall, *The Soldier's Load and the Mobility of the Nation*, 1980

COOPERATION

When two work side-by-side, one or the other
spots the opening first if a kill's at hand.
When one looks out for himself, alert but alone,
his reach is shorter – his sly moves miss the mark.

King Nestor of Pylos, in Homer, *The Iliad*, x, *c.* 800 BC, tr. Fagles

---

True cooperation requires that a subordinate should not carry out his superior's ideas in a merely wooden manner, but should do everything in his power to contribute to the full execution of his superior's intentions, using his intelligence to foresee any obstacles which may arise to prevent his carrying out the orders he receives, and taking measures accordingly. On the other hand, it is for the superior to furnish his subordinates with full indication of his intentions, so that the latter may bring his intelligence, instead of mere mechanical obedience, to bear upon the problems which will confront him. (December 1919)

Captain Sir Basil Liddell Hart, *Thoughts on War,* 1944

The Principle of Co-operation. Co-operation is a cementing principle; it is closely related to economy of force, and therefore to concentration, but it differs from both of these principles, for while mass is the concentrated strength of the organism and economy of force the dispersed strength which renders the former stable, co-operation may be likened to the muscular tension which knits all the parts to the whole. Without co-operation an army falls to pieces. In national wars, the value of co-operation is enormously enhanced, fusing, as it does, the body and soul of a nation into one intricate self-supporting organism. All must pull together, for such wars are the wars of entire nations, and, whatever may be the size of the armies operating, these should be looked upon as national weapons, and not as fractions of nations whose duty is to fight while the civil population turns thumbs up or thumbs down. Gladiatorial wars are dead and gone.

Major General J. F. C. Fuller, *The Reformation of War,* 1923

My personal philosophy is that the best outfits are those wherein a procedure is developed whereby every man who has an idea on a particular subject may bring it forward at the time of the discussion, without the slightest criticism or hesitation. He argues for his point of view *when you're discussing* exactly how you're going to proceed. He shouldn't hang back because his idea may appear radical, or because the bulk of the crowd may not agree with it.

Everyone steps forward and expresses an idea.

Once the decision is *made*, however –

'*This* is the way we're going to do it.'

Bang. Everybody complies. If a man doesn't comply, his official head should roll. He may have thought originally that we should all zig instead of zag. After hearing every argument, after listening carefully to all testimony relating to the subject, the commander has decided to zag. Everybody had better zag.

General Curtis Le May, *Mission With Le May,* 1965

CORRUPTION

They say the gods themselves
Are moved by gifts and gold does more with men than words.

Euripides (480–406 BC), *Medea,* tr. Warner

There are five types of harm in decadence among national armed forces.

First is the formation of factions that band together for character assassination, criticizing and vilifying the wise and the good.

Second is luxury in uniforms.

Third is wild tales and confabulations about the supernatural.

Fourth is judgment based on private views, mobilizing groups for personal reasons.

Fifth is making secret alliances with enemies, watching for where the advantage may lie.

All people like this are treacherous and immoral. You should distance yourself from them and not associate with them.

Zhuge Liang (AD 180–234), *The Way of the General*

An avaricious general can be the ruin of his own people and an object of contempt to the enemy.

A general who loves luxury can destroy the whole army.

The Emperor Maurice, *The Strategicon,* c. AD 600

. . . such of our men of war as, neglecting and contemning all true honour and discipline military, have brought in amongst us a most shameful and detestable art and disicipline of carousing and drunkenness, turning all matters military to their own profit and gain, neglecting to love and to win the love of their soldiers under their governments and charges, making in a manner no accompt of them nor of their lives. In such sort as by their evil conduction, starving and consuming of great numbers and many thousands of our most brave English people, as also by their infinite other disorders, they have made a far greater war upon the crown and realm of England and English nation than anyways upon the enemies of our country.

John Smythe (*c.* 1580–1631), *Certain Discourses Military*

Such dearth of public spirit, and such want of virtue, such stock-jobbing, and fertility in all the low arts to take advantage of one kind or another . . . I never saw before and I pray God's mercy that I may never be witness to again!

General George Washington (1732–1799), quoted in Montross, *War Through the Ages*, 1944

There is something in corruption which, like a jaundiced eye, transfers the colour of itself to the object it looks upon, and sees everything stained and impure.

Thomas Paine, *The American Crisis*, 1776–83

COUNCIL OF WAR

. . . it is not a proper subject for deliberation; that courage and action, and not deliberation, were necessary in such a calamity. That those who desired the safety of the state would attend him in arms forthwith; that in no place was the camp of the enemy more truly than where such designs were mediated.

Scipio Africanus, referring to the panic council of war after the Roman disaster at Cannae, quoted in Liddell Hart, *A Greater Than Napoleon, Scipio Africanus*, 1926

It does not help to assemble the whole army in council, or to keep sending for men when they are off duty. These things only cause discord in the army.

The Emperor Maurice, *The Strategicon*, c. AD 600

My saddle is my council chamber.

Saladin, c. 1180, quoted by Ibn al-Athir in *Recueil des Historiens des Croisades*, 1872–1906

The general should either choose a staff to participate in all his councils and share in his decisions, men who will accompany the army especially for this purpose, or summon as members of his council a selected group of the most respected commanders, since it is not safe that the opinions of one single man, on his sole judgment, should be adopted. For the isolated decision of one man, unsupported by others, can see no farther than his own ingenuity, but that which has the additional testimony of councillors guarantees against mistake. However, the general must neither be so undecided that he entirely distrusts himself, nor so obstinate as not to think that anyone can have a better idea than his own; for such a man, either because he listens to every one else and never to himself, is sure to meet with frequent misfortune, or else, through never listening to others but always to himself, is bound to make many costly mistakes.

Onasander, *The General* AD 58

MAXIM 65. The same consequences which have uniformly attended long discussions and councils of war will follow at all times. They will terminate in the adoption of the worst course, which in war is always the most timid, or, if you will, the most prudent. The only true wisdom in a general is determined courage.

Napoleon, *The Military Maxims of Napoleon*, 1831, tr. D'Aguilar

---

Hold no council of war, but accept the views of each, one by one . . . the secret is to make each alike . . . believe he has your confidence.

Napoleon, 12 January 1806, letter to Joseph, *Correspondance*

Councils of war are a deplorable resource and can be useful only when concurring in opinion with the commander, in which case they may give him more confidence in his own judgment, and, in addition, may assure him that his lieutenants, being of his opinion will use every means to insure the success of the movement.

Lieutenant General Antoine-Henri Baron de Jomini, *Summary of the Art of War*, 1838

I never held a council of war in my life.

General of the Army Ulysses S. Grant, quoted in the *Washington Post*, 12 July 1985

Otherwise we should have wasted all our time in discussions, diplomatical, tactical, enigmatical; they would have smothered me, and the enemy would have settled our arguments by smashing up our tactics.

Field Marshal Prince Aleksandr V. Suvorov (1729–1800)

If a man consults whether he is to fight, when he has the power in his own hands, it is certain that his opinion is against fighting.

Admiral Lord Nelson, August 1801, letter

Whenever I hear of Councils of War being called, I always called them as 'cloaks for cowardice' – so said the brave Boscawen, and from him I imbibed this sentiment.

Admiral Lord St. Vincent, 1809, speech at the opening of parliament

---

If the commander . . . feels the need of asking others what he ought to do, the command is in weak hands.

Field Marshal Helmuth Graf von Moltke, 20 January 1890, letter

---

A council of war never fights, and in a crisis the duty of the leader is to lead and not to take refuge behind the generally timid wisdom of a multitude of councillors.

President Theodore Roosevelt, *Autobiography*, 1913

Not every act is favoured with such happy conception or such easy birth. Meetings, discussion, committees, councils of war, etc., are the enemies of vigorous and prompt decision and their danger increases with their size. They are mostly burdened with doubts and petty responsibilities, and the man who pleads for action ill endures the endless hours of discussion.

Colonel General Hans von Seekt, *Thoughts of a Soldier*, 1930

On principle, I am opposed to all councils of war, so I did not send for the Army commanders to ask their advice as to the possibilities of success or the general plan; but, since momentous events were impending, I thought it very important to convene my immediate collaborators so as to explain to them my decisions, and, in case of any misunderstanding, clear up any points they might find obscure or hard to grasp. If this is done, those leaders who are neighbours in the field can talk things over together and thus avoid friction and bickerings which otherwise would be inevitable.

General Aleksei A. Brusilov, *A Soldier's Notebook*, 1931

To hear some people talk, however, one would think that the way to win the war is to make sure that every Power contributing armed forces and branches of these armed forces is represented on all councils and organizations which have to be set up, and that everybody is fully consulted before anything is done. That is, in fact, the most sure way to lose a war.

Sir Winston S. Churchill, 27 January 1942, Speech in the House, *Maxims and Reflections*, 1949

You've told me all the reasons why the project is not feasible. This meeting is now adjourned and will reconvene in fifteen minutes. When you return I want to hear from you the action you propose to take in order to get the job done.

Admiral William F. 'Bull' Halsey, to his staff, 1943, quoted in Potter, *Bull Halsey*, 1985

It is essential to understand the place of the 'conference' when engaged on active operations in the field. By previous thought, by discussion with his staff, and by keeping in close touch with his subordinates by means of visits, a commander should know what he wants to do and whether it is possible to do it. If a conference of his subordinates is then necessary, it will be for the purpose of giving orders. He should never bring them back to him for such a conference; he must go forward to them. Then nobody looks over his shoulder. A conference of subordinates *to collect ideas* is the resort of a weak commander.

Field Marshal Viscount Montgomery of Alamein, *The Memoirs of Field Marshal Montgomery*, 1958

I could almost hear my father's voice telling me as he had so many years before, 'Doug, councils of war ever breed timidity and defeatism'.

General of the Army Douglas MacArthur, *Reminiscences*, 1964

COUNTER-ATTACK/SPOILING ATTACK

After a victory the general should not allow the men to break ranks right away. For it has happened often enough that the enemy, on noticing that our men have let down their guard in their rejoicing and have broken ranks, have regained their courage, come back to fight and turned victory into defeat.

The Emperor Maurice, *The Strategicon*, c. AD 600

MAXIM 23. When you are occupying a position which the enemy threatens to surround, collect all your force immediately, and menace *him* with an offensive movement. By this manoeuvre you will prevent him from detaching and annoying your flanks, in case you should judge it necessary to retire.

Napoleon, *The Military Maxims of Napoleon*, 1831, tr. D'Aguilar

---

Lo, while his majesty sat talking with the princes, the vanquished chief of Kheta came, and the numerous countries, which were with him. They crossed over the channel on the south of Kadesh, and charged into the army of his majesty while they were marching, and not expecting it. Then the infantry and chariotry of his majesty retreated before them, northward to the place where his majesty was. Lo, the foes of the vanquished chief of Kheta surrounded the bodyguard of his majesty, who were by his side.

When his majesty saw them, he was enraged against them, like his father, Montu, lord of Thebes. He seized the adornments of battle, and arrayed himself in his coat of mail. He was like Baal in his hour. Then he betook himself to his horses, and led quickly on, being alone by himself. He charged into the foes of the vanquished chief of Kheta, and the numerous countries that were with him. His majesty was like Sutekh, the great in strength, smiting and slaying among them; his majesty hurled them headlong, one upon another into the water of the Orontes.

I charged all countries, while I was alone, my infantry and my chariotry having forsaken me. Not one among them stood to turn about. I swear, as Re loves me, as my father, Atum, favours me, that, as for every matter which his majesty has stated, I did it in truth, in the presence of my infantry and my chariotry.

Ramses II, 1284 BC, his personal counter-attack at the Battle of Kadesh, quoted in Breasted, *Ancient Records of Egypt*, 1906

Yesterday we were on the defence, we retreated, but today we went over to the offensive.

. . . We all . . . had thought thus: first we would stop the enemy, then we would bring up forces, prepare, and, finally . . . throw ourselves on the enemy. Reality turned out to be different, harsher, and more exacting: . . . we did not find the time to . . . prepare . . . It became necessary, figuratively speaking, just to turn around one's left shoulder to strike the enemy under whose pressure we has still been retreating yesterday.

Marshal of the Soviet Union Kirill S. Moskalenko, early December 1941 on the approaches to Moscow, quoted in Leites, *The Soviet Style of War*, 1982

Counter-attack is the soul of defense. Defense is in a passive attitude, for that is the negation of war. Rightly conceived it is an attitude of alert expectation. We wait for the moment when the enemy shall expose himself to a counter-stroke, the success of which will so far cripple him as to render us relatively strong enough to pass to the offensive ourselves.

Julian Corbett, *Some Principles of Maritime Strategy*, 1911

I calculated that it is only by a counter-strike that one can disrupt the enemy's . . . preparation for a new offensive. To force the enemy to take the offensive earlier than at the time which he had set is more advantageous for us than to sit and wait until he is fully prepared.

Marshal of the Soviet Union Vassili I. Chuikov, the Soviet offensive in Stalingrad, 12 October 1942, quoted in Leites, *The Soviet Style of War*, 1982

COUP DE MAIN

The success of a *coup de main* depends absolutely upon luck rather than judgment.

Napoleon, *Political Aphorism*, 1848

COUP D'OEIL

The *coup d'oeil* is a gift of God and cannot be acquired; but if professional knowledge does not perfect it, one only sees things imperfectly and in a fog, which is not enough in these matters where it is so important to have a clear eye . . . To look over a battlefield, to take in at the first instance the advantages and disadvantages is the great quality of a general.

Chevalier Jean-Charles de Folard (1669–1752), *Nouvelles Découvertes sur la Guerre*, 1724

. . . the ability to assess a situation at a glance, to know how to select the site for a camp, when and how to march, and where to attack.

Field Marshal Prince Aleksandr V. Suvorov, *The Science of Victory*, 1796

Nine-tenths of tactics are certain, and taught in books: but the irrational tenth is like the kingfisher flashing across the pool and this is the test of generals. It can only be ensured by instinct, sharpened by thought practising the strokes so often that at the crisis it is as natural as a reflex.

Colonel T. E. Lawrence, 'The Science of Guerrilla Warfare', *Encyclopaedia Britannica*, 1929

There is a gift of being able to see at a glance the possibilities offered by the terrain . . . One can call it the *coup d'oeil* and it is inborn in great generals.

Napoleon (1769–1821)

A vital faculty of generalship is the power of grasping *instantly* the picture of the ground and the situation, of relating the one to the other, and the local to the general. It is that flair which makes the great executant. (October 1933.)

Captain Sir Basil Liddell Hart, *Thoughts on War*, 1944

A general thoroughly instructed in the theory of war but not possessed of military *coup d'oeil*, coolness, and skill, may make an excellent strategic plan and be entirely unable to apply the rules of tactics in presence of an enemy. His projects will not be successfully carried out, his defeat will be probable. If he is a man of character he will be able to diminish the evil results of his failure, but if he lose his wits he will lose his army.

Lieutenant General Antoine-Henri Baron de Jomini, *Summary of the Art of War*, 1838

The problem is to grasp, in innumerable special cases, the actual situation which is covered by the mist of uncertainty, to appraise the facts correctly and to guess the unknown elements, to reach a decision quickly and then to carry it out forcefully and relentlessly.

Field Marshal Helmuth Graf von Moltke, (1800–1891)

. . . The acid test of an officer who aspires to high command is his ability to be able to grasp quickly the essentials of a military problem, to decide rapidly what he will do, to make it quite clear to all concerned what he intends to achieve and how he will do it, and then to see that his subordinate commanders get on with the job. Above all, he has got to rid himself of all irelevant detail; he must concentrate on the essentials, and on those details and only those details which are necessary to the proper carrying out of his plan – trusting his staff to effect all necessary co-ordination.

Field Marshal Viscount Montgomery of Alamein, *The Memoirs of Field Marshal Montgomery*, 1958

On the field of battle the happiest inspiration (*coup d'oeil*) is often only a recollection.

Napoleon (1769–1821), quoted in Simpkin, *Race to the Swift*, 1985

My great talent, the one that distinguished me the most, is to see the entire picture distinctly.

Napoleon, quoted in Gourgand, *Journal inédit de 1815 à 1818*, n.d.

In war everything is perception – perception about the enemy, perception about one's own soldiers. After a battle is lost, the difference between victor and vanquished is very little; it is, however, incommensurable with perception, for two or three cavalry squadrons are enough to produce a great effect.

Napoleon, 22 September 1808, letter to Joseph, *Correspondance*, No. 14343, Vol. XVIII, 1858–1870

If the mind is to emerge unscathed from this relentless struggle with the unforeseen, two qualities are indispensible: *first, an intellect that, even in the darkest hour, retains some glimmerings of the inner light which leads to truth; and second, the courage to follow this faint light wherever it may lead.* The first of these qualities is described by the French term *coup d'oeil*; the second is determination.

Major General Carl von Clausewitz, *On War*, i, 1832, tr. Howard and Paret

. . . *Coup d'oeil* therefore refers not alone to the physical but, more commonly, to the inward eye. The expression, like the quality itself, has certainly always been more applicable to tactics, but it must also have its place in strategy, since here as well quick decisions are often needed. Stripped of metaphor and of the restrictions imposed by the phrase, the concept merely refers to the quick recognition of a truth that the mind would ordinarily miss or would perceive only after long study and reflection.

Major General Carl von Clausewitz, *On War*, i, 1832, tr. Howard and Paret

---

Clear insight is not enough to prevent irresolution. Clear insight can help us chart an excellent course of action, but adhering to this course is another matter. As soon as a decision is made, and throughout the life of any plan based on that decision, a steady stream of ideas and events tending to refute the basis of the decision impinges on the leader's consciousness and weakens his determination. Often, too, a decision is made which cannot be justified by any accepted principle, and the urge to modify such a decision, no matter how well-thought-out it may have been, becomes particularly hard to resist. Remaining steadfast here calls for a vast amount of confidence in oneself, as well as a certain degree of scepticism. One's main help . . . is faith in the time-tested principles used to frame the decision, backed up by this maxim: In case of doubt, hold to your first decision, and make no change until positive conviction forces it.

General der Infantrie Hugo Baron von Freytag-Loringhoven, *The Power of Personality in War*, 1911

The ability of a commander to comprehend a situation and act promptly is the talent which great men have of conceiving in a moment all the advantages of the terrain and the use that they can make of it with their army. When you are accustomed to the size of your army you soon form your *coup d'oeil* with reference to it, and habit teaches you the ground that you can occupy with a certain number of troops.

Use of this talent is of great importance on two occasions. First, when you encounter the enemy on your march and are obliged instantly to choose ground on which to fight. As I have remarked, within a single square mile a hundred different orders of battle can be formed. The clever general perceives the advantages of the terrain instantly; he gains advantage from the slightest hillock, from a tiny marsh; he advances or withdraws a wing to gain superiority; he strengthens either his right or his left, moves ahead or to the rear, and profits from the merest bagatelles.

Frederick the Great, *Instructions to His Generals,* 1747

COURAGE

I am well aware, soldiers, that words cannot inspire courage, and that a spiritless army cannot be rendered active, or a timid army valiant, by the speech of the commander. Whatever courage is in the heart of a man, whether from nature or from habit, so much will be shown by him in the field; and on him whom neither glory nor danger can move, exhortation is bestowed in vain; for the terror in his breast stops his ears.

Catiline (Lucius Sergius Catilina), 62 BC, speech delivered to his soldiers before the battle in which he suffered defeat and death

---

Now be men, my friends! Courage, come, take heart!
Dread what comrades say of you here in bloody combat!
When men dread that, more men come through alive –
when soldiers break and run, good-bye glory, good-bye all defences!

Menelaus, in Homer, *The Iliad,* v., *c.* 800 BC, tr. Fagles

This is courage in a man:
to bear unflinchingly what heaven sends.

Euripides, *Heracles, c.* 422 BC, tr. Arrowsmith

The principle on which to manage an army is to set up one standard of courage which all must reach.

Sun Tzu, *The Art of War, c.* 500 BC

A coward turns away but a brave man's choice
Is danger

Euripides, *Iphigenia in Tauris, c.* 414–412 BC, tr. Bynner

But the bravest are surely those who have the clearest vision of what is before them, glory and danger alike, and yet notwithstanding go out to meet it.

Thucydides, *The History of the Peloponnesian War,* ii, 404 BC

When people discuss a general they always pay attention to his courage. As far as a general is concerned, courage is but one quality. Now a valiant general will be certain to enter an engagement recklessly and if he does so he will not appreciate what is advantageous.

Wu Ch'i (*c.* 430–381 BC), quoted in commentary to Sun Tzu, *The Art of War, c.* 500 BC, tr. Griffith

---

The courage of the soldier is heightenend by the knowledge of his profession.

Flavius Vegetius Renatus, *Military Institutions of the Romans, c.* AD 378

. . . There is, of course, such a thing as individual courage, which has a value in war, but familiarity with danger, experience in war and its common attendants, and personal habit, are equally valuable traits, and these are the qualities with which we usually have to deal in war. All men naturally shrink from pain and danger, and only incur their risk from some higher motive, or from habit; so that I would define true courage to be a perfect sensibility of the measure of danger, and a mental willingness to incur it, rather than that insensibility to danger of which I have heard far more than I have seen. The most courageous men are generally unconscious of possessing the quality; therefore, when one professes it too openly, by words or bearing, there is reason to mistrust it. I would further illustrate my meaning by describing a man of true courage to be one who possesses all his faculties and senses perfectly when serious danger is actually present.

General of the Army William T. Sherman, *Memoirs of General W. T. Sherman*, 1875

Cowards may fear to die; but courage stout,
Rather than live in snuff, will be put out.

Sir Walter Raleigh (1554–1618), the night before his death

Perfect courage and utter cowardice are two extremes that rarely occur.

No one can answer for his courage when he has never been in danger.

Courage, in soldiers, is a dangerous profession they follow to earn their living.

François Duke de la Rochefoucauld (1613–1680), *Réflexions ou sentences et maximes morales*, 1665

The courage of the troops must be reborn daily . . . nothing is so variable . . . the true skill of the general consists in knowing how to guarantee it.

Field Marshal Maurice Comte de Saxe, *My Reveries*, 1732

As to moral courage, I have rarely met with two o'clock in the morning courage: I mean unprepared courage.

Napoleon, quoted in Las Cases, *Mémorial de Saint Hélène*, 1823

. . . A great many men, when they smell battle afar off, chafe to get into the fray. When they say so themselves they generally fail to convince their hearers that they are as anxious as they would like to make believe, and as they approach danger they become more subdued. This rule is not universal, for I have known a few men who were always aching for a fight when there was no enemy near, who were as good as their word when the battle did come. But the number of such men is small.

General of the Army Ulysses S. Grant, *Personal Memoirs of U.S. Grant*, 1885

At the grave of the hero we end, not with sorrow at the inevitable loss, but with the contagion of courage; and with a kind of desperate joy we go back to the fight.

Justice Oliver Wendell Holmes (1841–1935)

In sport, in courage, and in the sight of Heaven, all men meet on equal terms.

Sir Winston S. Churchill, *The Malakand Field Force*, 1898

The more comfort the less courage there is.

Field Marshal Prince Aleksandr V. Suvorov, quoted in *Soviet Military Review*, 11/1979

Pay not attention to those who would keep your far from fire: you want to prove yourself a man of courage. If there are opportunities, expose yourself conspicuously. As for real danger, it is everywhere in war.

Napoleon, 2 February 1806, to Joseph, *Correspondance*, No. 9738, Vol. XI, 1858–1870

When soldiers brave death, then drive him into the enemy's ranks.

Napoleon, 14 November 1806, to a regiment of chasseurs before the Battle of Jena.

Courage is like love; it must have hope for nourishment.

Napoleon, Maxims, quoted in *International Thesaurus of Quotations*, 1970

Courage . . . is that firmness of spirit, that moral backbone, which, while fully appreciating the danger involved, nevertheless goes on with undertaking. Bravery is physical; courage is mental and moral. You may be cold all over; your hands may tremble; your legs may quake; your knees may be ready to give way – that is fear. If, nevertheless, you go forward; if in spite of this physical defection you continue to lead your men against the enemy, you have courage. The physical manifestations of fear will pass away. You may never experience them but once.

Major C. A. Bach, 1917, address to graduating new officers, Fort Sheridan, Wyoming

---

. . . It is not so much a question of destroying the enemy troops as of destroying their courage. Victory is yours as soon as you convince your opponent his cause is lost . . . One defeats the enemy not by individual and complete annihilation, but by destroying his hopes of victory.

Field Marshal Colmar Baron von der Goltz, quoted in Foch, *Principles of War*, 1913

War is the realm of danger; therefore *courage* is the soldier's first requirement.

Courage is of two kinds: courage in the face of personal danger, and courage to accept responsibility, either before the tribunal of some outside power or before the court of one's own conscience.

Major General Carl von Clausewitz, *On War*, i, 1832, tr. Howard and Paret

---

The most essential qualities for a general will always be: *first*, a high moral courage, capable of great resolution; *second*, a physical courage which takes no account of danger. His scientific or military acquirements are secondary to these.

Lieutenant General Antoine-Henri Baron de Jomini, *Summary of the Art of War*, 1938

. . . As to physical courage, although sheer cowardice (i.e., a man thinking of his own miserable carcass when he ought to be thinking of his men) is fatal, yet, on the other hand, a reputation for not knowing fear does not help an officer in his war discipline: in getting his company to follow him as the Artillery of the Guard followed Drouot at Wagram. I noticed this first in Afghanistan in 1879 and have often since made the same observation. If a British officer wishes to make his men shy of taking a lead from him let him stand up under fire whilst they lie in their trenches as did the Russians on the 17th of July at the battle of Motienling. Our fellows are not in the least impressed by such bravado. All they say is, 'This fellow is a fool. If he cares so little for his own life, how much less will he care for ours.'

General Sir Ian Hamilton, *Soul and Body of an Army*, 1921

The well-known Spanish proverb, 'He was a brave man on such a day', may be applied to nations as to individuals. The French at Rossbach were not the same people as at Jena, nor the Prussian at Prenzlau as at Dennewitz.

Lieutenant General Antoine-Henri Baron de Jomini, *Summary of the Art of War*, 1838

One man with courage makes a majority.

American saying attributed to Andrew Jackson (1767–1845)

Courage is rightly esteemed the first of human qualities, because . . . it is the quality that guarantees all others.

Sir Winston S. Churchill, *Great Contemporaries*, 1937

No sane man is unafraid in battle, but discipline produces in him a form of vicarious courage.

General George S. Patton, Jr., *War As I Knew It*, 1947

On the field there is no substitute for courage, no other binding influence toward unity of action. Troops will excuse almost any stupidity; excessive timidity is simply unforgivable.

Brigadier General S. L. A. Marshall, *The Armed Forces Officer*, 1950

For without belittling the courage with which men have died, we should not forget those acts of courage with which men . . . have lived. The courage of life is often a less dramatic spectacle than the courage of a final moment; but it is no less a magnificent mixture of triumph and tragedy. A man does what he must – in spite of personal consequences, in spite of obstacles and dangers and pressures – and that is the basis of all human mortality.

John F. Kennedy, *Profiles in Courage*, 1956

I don't believe there is any man, who in his heart of hearts, wouldn't rather be called brave than have any other virtue attributed to him. And this elemental, if you like unreasoning, male attitude is a sound one, because courage is not merely a virtue; it is *the* virtue. Without it there are no other virtues. Faith, hope, charity, all the rest don't become virtues until it takes courage to exercise them. Courage is not only the basis of all virtue; it is its expression. True, you may be bad and brave, but you can't be good without being brave.

Field Marshal Viscount Slim, *Courage* and Other Broadcasts, 1957

. . . There is no better ramrod for the back of a senior who is beginning to buckle than the sight of a junior who has kept his nerve. Land battles, as to the fighting part, are won by the intrepidity of men in grade from private to captain mainly. Fear is contagious, but courage is not less so. The courage of any one man reflects in some degree the courage of all those who are within his vision. To the man who is in terror and bordering on panic, no influence can be more steadying than that of seeing some other man near him who is retaining self-control and doing his duty.

Brigadier General S. L. A. Marshall, *The Armed Forces Officer*, 1950

Generally speaking, there is no more sincere or inspiring form of courage than that which shuns display and seeks the modest unostentatious way. Under certain conditions, however, the most modest of commanders will be compelled to make a show of courage either by paying an unexpected visit under fire to the front line or by taking the lead when advancing upon the enemy's poisitions. Yet one point must constantly be borne in mind: courage comes of cool thought and knowledge, never of hot-headedness or lack of knowledge. The kind of courage needed by a commander is serious and purposeful, not rash or adventurous.

General Yigal Allon, *The Making of Israel's Army*, 1960

Courage is the greatest of all virtues, for without it there are no other virtues . . . As officers you will, of course, accept greater hazards than your men. You will be first in danger, but an officer must have more than the ordinary physical courage. Anyone can be brave for five minutes. You will not only be braver than the men you lead; you will be brave *for longer*. You will go on being brave when others falter; brave not only in danger, but brave in hardship, in loneliness and, perhaps most difficult of all, in those long periods of inactivity, of boredom that come at times to all soldiers. In failure, too, you will show your courage. We can all be brave when we are winning. I'm a hell of General when everybody is whooping along and the enemy's on the run. But you won't always be winning. If you had ever been a British General at the start of a war you would know what I mean. You'll find some day when things are bad, whether you're the Commanding General or the Platoon Commander, there will come a sudden pause when your men stop and look at you. No one will speak; they will just look at you – look at you and ask, dumbly, for leadership. Their courage is ebbing; you must make if flow back – and it is not easy. You will never have felt more alone in your life.

Field Marshal Viscount Slim, *Courage* and Other Broadcasts, 1957

There is nothing like seeing the other fellow run to bring back your courage.

Field Marshal Viscount Slim, *Unofficial History*, 1959

Courage in the commander is a prior condition for courage in his men and for their ability to carry out orders promptly and well. It takes courage and daring to overcome the fears of battle and the trials of armed combat. The commander must also have the courage of his convictions, defying when necessary entrenched, conventional ideas. He must dare to make his views heard in front of his superiors and among his colleagues when permitted to do so, notwithstanding the contrary opinions of those he is addressing.

General Yigal Allon, *The Making of Israel's Army*, 1960

There are two kinds of courage, physical and moral, and he who would be a true leader must have both. Both are the products of the character-forming process, of the development of self-control, self-discipline, physical endurance, of knowledge of one's job and, therefore, of confidence. These qualities minimize fear and maximize sound judgment under pressure – with some of that indispensable stuff called luck – often bring success from seemingly hopeless situations.

Putting aside impulsive acts of reckless bravery, both kinds of courage bespeak an untroubled conscience, a man at peace with God. An example is Colonel John H. Glenn who was asked after his first rocket flight if he had been worried, and who replied: 'I am trying to live the best I can. My peace had been made with my Maker for a number of years, so I had no particular worries.'

General Mathew B. Ridgway, 'Leadership', *Military Review*, 10/1966

COWARDICE/COWARDS

Why have you not launched the attack? Are you standing in chariots of water? Did you also turn to water? . . . If you were to fall down before him, you . . . would either kill him, or at least, frighten him! But instead you acted like a 'fairy'!

Hattusili I, 1650–1620 BC, letter to the commander of his army besieging the city of Ursa

---

The moment the Persian left went to pieces, under Alexander's attack and Darius, in his war-chariot, saw that it was cut off, he incontinently fled – indeed, he led the race for safety. Keeping to his chariot as long as there was smooth ground to travel on, he was forced to abandon it when ravines and other obstructions barred his way; then, dropping his shield and stripping off his mantle – even leaving his bow in the war-chariot – he leapt upon a horse and rode for his life. Darkness soon closed in; and that alone saved him from falling into the hands of Alexander, who, while daylight held, relentlessly pressed the pursuit; but when there was no longer light enough to see what he was coming to, he turned back – but not without taking possession of Darius' chariot together with his shield, mantle, and bow.

Flavius Arrianus Xenophon (Arrian), the flight of Darius at the Battle of Issus (333 BC), *The Campaigns of Alexander the Great*, c. AD 150, tr. de Selincourt

---

Be remiss in nothing whilst you are in the battle, for if any mark of fear or cowardice is revealed in you, even if you had a thousand lives you would be unable to save a single one, and the humblest person could overwhelm you. Cowardice results either in your being slain or in the besmirching of your name. Once you become notorious among men for poltroonery and for a display of sloth and feebleness in such circumstances as these, and for failing your comrades, you will be disgraced amongst your friends; amongst your contemporaries you will be stricken with shame. Death is preferable to such a life, and it is far better to die in good odour than to spend one's life in disrepute.

Kai Ka'us Ibn Iskander, *A Mirror For Princes*, AD 1082, tr. Levy

---

. . . A coward, when taught to believe, that if he breaks his Ranks, and abandons his Colours, will be punished with Death by his own party, will take his chance against the Enemy. But the man who thinks little of the one, and is fearful of the other, Acts from present feelings regardless of consequences.

General George Washington, 9 February 1776, to the President of Congress

---

Cowards do not count in battle;
They are there, but not in it.

Euripides (480–406 BC)

All men are timid on entering any fight. Whether it is the first fight or the last fight, all of us are timid. Cowards are those who let their timidity get the better of their manhood.

General George S. Patton, Jr., *War As I Knew It*, 1747

CRITICISM

. . . room for a military criticism as well as a place for a little ridicule upon some famous transactions of that memorable day . . . But why this censure when the affair is so happily decided? To exercise one's ill-nature? No, to exercise the faculty of judging . . . The more a soldier thinks of the false steps of those that are gone before, the more likely he is to avoid them.

Major General Sir James Wolfe (1727–1759), reflections on the Battle of Culloden, quoted in Liddell Hart, *Great Captains Unveiled*, 1927

There's a man for you! He is forced to flee from an army that he dares not fight, but he puts eighty leagues of devastation between himself and his pursuers. He slows down the march of the pursuing army, he weakens it by all kinds of privation – he knows how to ruin it without fighting it. In all of Europe, only Wellington and I are capable of carrying out such measures. But there is a difference between him and myself: In France . . . I would be criticized, whereas England will praise him.

Napoleon, 1810, conversation on Wellington's retreat to Portugal, Herold, ed., *The Mind of Napoleon*, 1955

We must remember that man remains man and that his heart does not change. Secondly, we must remember that the means of war do not change and that the intelligence of man must keep pace with these changes. We must keep minds subtle and active, and never let ourselves be hypnotized by traditions; we must criticize ourselves, and criticize our criticisms; we must experiment and explore.

Major General J. F. C. Fuller, *Sir John Moore's System of Training*, 1925

In an institution such as the Army the hierarchical system and the habit of subordination, most necessary in its right place, preclude criticism to a degree unknown in other professions, and that would be hardly comprehensible to a scientific mind. This reflection reminds me of the humorous, yet serious, warning of my beloved old chief: 'You'll learn that generals are as sensitive as prima donnas.' I have long learned to appreciate his wisdom. (February 1933)

Captain Sir Basil Liddell Hart, *Thoughts on War*, 1944

. . . I always remember the Japanese soldier who outraged the sense of patriotism and duty in his superior officer by saying, 'In Osaka I would get five yen for digging this gun pit; here I only get criticism.'

General Sir Ian Hamilton, *The Soul and Body of an Army*, 1921

Much military criticism is no better than 'wisdom after the event' – although not in the usual sense of this facile retort so popular with the friends of the commander whose conduct is called into question. Soldiers who ought to know better, are as addicted as laymen to the habit of judging by results, and of concentrating their attention on the way a plan works out. Even at the best, criticism of performance is difficult – the properties of the general's instrument and the conditions in which it operates are so variable. Moreover, accident plays such an immense part in any result. The truest basis of criticism lies in the general's appreciation and plan – in his conception rather than in its execution. For if we know the orbit of his thought we have a reasonable guide as to how far he was fitted, mentally, for his task. Here we have a real standard for military criticism, a standard created by many centuries of recorded experience. (August 1934)

Captain Sir Basil Liddell Hart, *Thoughts on War*, 1944

CUNNING

The term 'cunning' implies secret purpose. It contrasts with the straight-forward, simple, direct approach much as wit contrasts with direct proof. Consequently, it has nothing in common with methods of persuasion, of self-interest, or of force, but a great deal with deceit, which also conceals its purpose. It is itself a form of deceit, when it is completed; yet no deceit in the ordinary sense of the word, since no outright breach of faith is involved. The use of a trick or strategem permits the intended victim to make his own mistakes, which, combined in a single result, suddenly change the nature of the situation before his very eyes. It might be said, that, as wit juggles with ideas and beliefs, so cunning juggles with actions.

Major General Carl von Clausewitz, *On War*, 1832, tr. Howard and Paret

---

. . . the weaker the forces that are at the disposal of the supreme commander, the more appealing the use of cunning becomes. In a state of weakness and insignificance, when prudence, judgment, and ability no longer suffice, cunning may well appear the only hope. The bleaker the situation, with everything concentrating on a single desperate attempt, the more readily cunning is joined to daring. Released from all future considerations, and liberated from thoughts of later retribution, boldness and cunning will be free to augment each other to the point of concentrating a faint glimmer of hope into a single beam of light which may yet kindle a flame.

Major General Carl von Clausewitz, *On War*, tr. Howard and Paret

DANGER

The Romans also have an excellent method of encouraging young soldiers to face danger. Whenever any have especially distinguished themselves in a battle, the general assembles the troops and calls forward those he considers to have shown exceptional courage. He praises them first for their gallantry in action and for anything in their previous conduct which is particularly worthy of mention, and then he distributes gifts . . . These presentations are not made to men who have wounded or stripped an enemy in the course of a pitched battle, or at the storming of a city, but to those who during a skirmish or some similar situation in which there is no necessity to engage in single combat, have voluntarily and deliberately exposed themselves to danger.

Polybius, *The Rise of the Roman Empire*, c. 125 BC

Remember also that God does not afford the same protection in unprovoked as in necessary dangers.

Count Belisarius, April AD 531

I wish to have no Connection with any Ship that does not sail *fast*, for I intend *to go in harm's way*.

Admiral John Paul Jones, November 1778, letter

If I had been censured every time I have run my ship, or fleets under my command, into great danger, I should long ago have been *out* of the Service, and never *in* the House of Peers.

Admiral Lord Nelson, March 1805, letter to the Admiralty

To someone who has never experienced danger, the idea is attractive rather than alarming. You charge the enemy, ignoring bullets and casualties, in a surge of excitement. Blindly you hurl yourself toward icy death, not knowing whether you or anyone else will escape him. Before you lies that golden prize, victory, the fruit that quenches the thirst of ambition. Can that be so difficult? No, and it will seem even less difficult than it is. But such moments are rare; and even they are not, as is commonly thought, brief like a heartbeat, but come rather like a medicine, in recurring doses, the taste diluted by time.

Major General Carl von Clausewitz, *On War*, iv, 1832, tr. Howard and Paret

---

. . . headlong, dogged, or innate courage, overmastering ambition, or long familiarity with danger – all must be present to a considerable degree if action in this debilitating environment is not to fall short of achievements that in the study would appear as nothing out of the ordinary.

Major General Carl von Clausewitz, *On War*,, iv, 1832, tr. Howard and Paret

---

Danger is part of the friction of war. Without an accurate conception of danger we cannot understand war.

Major General Carl von Clausewitz, *On War*, iv, 1832, tr. Howard and Paret

---

Nothing in life is so exhilarating as to be shot at without result.

Sir Winston S. Churchill, *The Malakand Field Force*, 1898

Danger gleams like sunshine to a brave man's eyes.

Euripides, *Iphigenia in Tauris*, 412 BC

DEATH

As for you, who now survive them, it is your business to pray for a better fate, but to think it your duty to preserve the same spirit and warmth of courage against your enemies; not judging of the expediency of this from a mere harangue – where any man indulging in a flow of words may tell you what you yourselves know as well as he, how many advantages there are in fighting valiantly against your enemies – but, rather making the daily-increasing grandeur of this community the object of your thoughts and growing quite enamoured of it. And when it really appears great to your apprehensions, think again that this grandeur was acquired by brave and valiant men, by men who knew their duty, and in the moments of action were sensible of shame; who, whenever their attempts were unsuccessful, thought it not dishonour for their country to stand in need of anything their valour could do for it, and so made it the most glorious present. Bestowing thus their lives on the public, they have every one received a praise that will never decay, a sepulchre that will always be the most illustrious – not that in which their bones lie mouldering, but that in which their fame is preserved, to be on every occasion, when honour is the employ of either word or act, eternally remembered. For the whole earth is the sepulchre of illustrious men . . .

Pericles, 431 BC, funeral oration for the Athenian war dead at the beginning of the Peloponnesian War, quoted in Thucydides, *The History of the Peloponnesian War*, c. 404 BC

Both armies battled it out along the river banks –
they raked each other with hurtling bronze-tipped spears.
And Strife and Havoc plunged in the fight, and violent Death –
now seizing a man alive with fresh wounds, now one unhurt, now
hauling a dead man through the slaughter by the heels,
the cloak on her back stained red with human blood.
So they clashed and fought like living, breathing men
grappling each other's corpses, dragging off the dead.

Homer, *The Iliad*, xviii, c. 800 BC, tr. Fagles

The general should take thought for the burial of the dead, offering as a pretext for delay neither occasion nor time nor place nor fear, whether he happens to be victorious or defeated. Now this is both a holy act of reverence toward the dead and also a necessary example for the living. For if the dead are not buried, each soldier believes that no care will be taken of his own body, should he chance to fall, observing what happens before his own eyes, and thereby judging of the future, feeling that he, likewise, if he should die, would fail of burial, waxes indignant at the contemptuous neglect of burial.

Onasander, *The General*, AD 58

Remember, gentlemen, what a Roman emperor said: 'The corpse of an enemy always smells sweet.'

Napoleon, 1812, ed. Herold, *The Mind of Napoleon*, 1955

Death is nothing; but to live defeated and without glory is to die every day.

Napoleon, 1804, letter to General Lauriston, ed. Herold, *The Mind of Napoleon*, 1955

For hours the army gathered the dead; and to Blackford's amazement, Jackson stood over the working men as they laid out rows, with up to fifty bodies in each, and spread blankets or oilcloths over them. He looked about the field, sending men to pick up every scrap of cloth or grisly debris. Jackson was stranger even than his reputation, Blackford thought. But he saw at the end that the scene was much less depressing and that the numbers of casualties appeared much less.

'Why did you have the field cleaned like this?' the cavalryman asked Jackson.

'Because I am going to attack here presently, as soon as the fog rises, and it won't do to march troops over their own dead, you know. That's what I'm doing it for.'

Lieutenant General Thomas 'Stonewall' Jackson, description of Stonewall Jackson, 1 July 1862, during the Seven Days Battle, quoted in Davis, *They Called Him Stonewall*,

---

Death in combat is not the end of the fight but its peak, and since combat is a part, and at times the sum total of life, death which is the peak of combat, is not the destruction of life, but its fullest, most powerful expression.

General Moshe Dayan, address in honour of Natan Altermann, 1971

There should be no real neglect of the dead, because it has a bad effect on the living; for each soldier values himself as highly as though he were living in a good house at home.

General of the Army William T. Sherman, *Memoirs of W. T. Sherman*, 1875

One thing he [General Patton] said always stuck with me, for it was contrary to what I had believed up to that moment, but when I had been in combat only a short while, I knew he was right. Speaking to all of us late one afternoon as we assembled in the North African sunset, he said, 'Now I want you to remember that no sonuvabitch ever won a war by dying for his country. He won it making the other poor dumb bastard die for his country.'

General James M. Gavin, *On to Berlin*, 1978

To die with glory, if one has to die at all,
is still, I think, pain for the dier.

Euripides, *Rhesus*, c. 455–441 BC. tr. Lattimore

DECEPTION

When . . . his majesty had arrived at the locality south of the town of Shabtuna, there came two Shasu, to speak to his majesty as follows: 'Our brethren, who belong to the greatest of the families with the vanquished chief of Kheta, have made us come to his majesty, to say: "We will be subjects of Pharaoh . . . and . . . will flee from the vanquished chief of Kheta; for the vanquished chief of Kheta sits in the land of Aleppo . . . He fears because of Pharaoh . . . to come southward."' Now, these Shasu spake these words . . . falsely, for the vanquished chief of Kheta made them come to spy where his majesty was, in order to cause the army of his majesty not to draw up for fighting him, to battle with the vanquished chief of Kheta.

Lo, the vanquished chief of Kheta came with every chief of every country, their infantry and their chariotry, which he had brought with him by force, and stood, equipped, drawn up in line of battle behind Kadesh the Deceitful, while his majesty knew it not. Then his majesty proceeded northward and arrived on the north-west of Kadesh; and the army of his majesty made camp there.

Ramses II, the official record of the Battle of Kadesh, 1284 BC, quoted in Breasted, *Ancient Records of Egypt*, 1906

All warfare is based on deception.

Therefore, when capable, feign incapacity; when active inactivity.

When near, make it appear that you are far away; when far away, that you are near.

Offer the enemy a bait to lure him; feign disorder and strike him.

Now the crux of military operations lies in the pretence of accommodating one's self to the designs of the enemy.

Sun Tzu, *The Art of War*, c. 500 BC, tr. Griffith

I make the enemy see my strengths as weaknesses and my weaknesses as strengths while I cause his strengths to become weaknesses and discover where he is not strong . . . I conceal my tracks so that none can discern them; I keep silence so that none can hear me.

Ho Yen-hsi, Sung Dynasty commentator to Sun Tzu, *The Art of War*, c. 500 BC, tr. Griffith

---

It is very important to spread rumours among the enemy that you are planning one thing; then go and do something else. Your plans about major operations should not be made known to many, but to just a few and those very close to you.

The Emperor Maurice, *The Strategikon*, c. AD 600

When a delegation comes from the enemy, inquire about the leaders of the group, and on the arrival treat them very friendly, so their own people will come to suspect them.

The Emperor Maurice, *The Strategikon*, c. AD 600

In his movements the general should act like a good wrestler; he should feint in one direction to try to deceive his adversary and then make good use of the opportunities he finds, and in this way he will overpower the enemy.

The Emperor Maurice, *The Strategikon*, c. AD 600

Success in war is obtained by anticipating the plans of the enemy, and by diverting his attention from our own designs.

Francesco Guicciacardini (1483–1540)

Though fraud in other activities be detestable, in the management of war it is laudable and glorious, and he who overcomes an enemy by fraud is as much to be praised as he does so by force.

Niccolo Machiavelli, *Discourses*, 1517

Always mystify, mislead, and surprise the enemy, if possible; and when you strike and overcome him, never give up the pursuit as long as your men have strength to follow; for an army routed, if hotly pursued, becomes panic-stricken, and can then be destroyed by half their number.

Lieutenant General Thomas 'Stonewall' Jackson, quoted in Henderson, *Stonewall Jackson*, 1898

If I can deceive my own friends, I can make certain of deceiving the enemy.

Lieutenant General Thomas 'Stonewall' Jackson, 1862

To achieve victory we must as far as possible make the enemy blind and deaf by sealing his eyes and ears, and drive his commanders to distraction by creating confusion in their minds.

Mao Tse-tung, *On Protracted War*, 1938

'In wartime', I said, 'truth is so precious that she should always be attended by a bodyguard of lies.' Stalin and his comrades greatly appreciated this remark when it was translated, and upon this note our formal conference ended gaily.

Sir Winston S. Churchill, *The Second World War: Closing the Ring*, 1951

DECISION/DECISIVENESS

It is essential to be cautious and take your time in making plans, and once you come to a decision to carry it out right away without any hesitation or timidity. Timidity after all is not caution, but the invention of wickedness.

The Emperor Maurice, *The Strategikon*, c. AD 600

On the other hand, experience shows me that, in an affair depending on vigour and dispatch, the generals should settle their plan of operations so that no time may be lost in idle debate and consultations when the sword is drawn; that pushing on smartly is the road to success, and more particularly so in an affair of this sort; that nothing is to be reckoned an obstacle to your undertaking which is not found really so on trial; that in war something must be allowed to chance and fortune, seeing that it is in its nature hazardous, and an option of difficulties; that the greatness of an object should come under consideration, opposed to the impediments that lie in the way; that the honour of one's country is to have some weight; and that, in particular circumstances and times, the loss of a thousand men is rather an advantage to a nation than otherwise, seeing that gallant attempts raise its reputation . . . whereas the contrary appearances sink the credit of a country, ruin the troops, and create infinite uneasiness and discontent at home.

Major General Sir James Wolfe, quoted in Liddell Hart, *Great Captains Unveiled*, 1927

. . . The commander is compelled during the whole campaign to reach decisions on the basis of situations which cannot be predicted. All consecutive acts of war are, therefore, not executions of a premeditated plan, but spontaneous actions, directed by military tact. The problem is to grasp in innumerable special cases, the actual situation which is covered by the mist of uncertainty, to appraise the facts correctly and to guess the unknown elements, to reach a decision quickly and then to carry it out forcefully and relentlessly . . . It is obvious that theoretical knowledge will not suffice, but that here the qualities of mind and character come to a free, practical and artistic expression, although schooled by military training and led by experiences from military history or from life itself.

Field Marshal Helmuth Graf von Moltke, quoted in Earle, ed., *The Makers of Modern Strategy*, 1943

. . . Various incidents may happen which may necessitate a certain deviation from the original plan here and there. It will not be possible to ask the commander for orders in this case, since telegraph and other communications may not work. The corps commander will be faced with the necessity of arriving at a decision of his own. In order that this decision should meet the ideas of the commander-in-chief he must keep the corps commanders sufficiently informed, while, on the other hand, the latter must continuously strive to keep in mind the basic ideas of all the operations and to enter into the mind of the commander-in-chief.

Field Marshal Alfred Graf von Schlieffen, *Dienstschriften*, Vol. II

Once you have taken a decision, never look back on it.

Field Marshal Viscount Allenby of Meggido, quoted in Wavell, *Allenby, Soldier and Statesman*, 1946

Sometimes a commander who cannot make decisions will by fussing and fuming, try to persuade himself and others that he is taking an active part in the operation. He will give his personal attention to details of secondary importance, and satisfy his desire to leave his impress on the action by interfering in an aimless sort of way.

General Charles de Gaulle, *The Edge of the Sword*, 1932

And now the last point, and the most difficult of all. Our map problems generally close with a statement that it is now such an hour and call for a decision. We know, therefore, that the situation is such and such, that we have all the information we are going to get and that we must make a decision. The forgoing action clearly indicates that one of the most difficult things we have to do in war *is to recognize the moment for making a decision*. The information comes in by degrees. We never know but that the next minute will bring us further information that is fresh and vital. Shall we make a decision now or shall we wait a little longer? It is usually more difficult to determine the moment for making a decision than it is to formulate the decision itself.

Captain Adolf von Schell, *Battle Leadership*, 1933

In making an attack, there can be no vacillation or indecision, even though a bayonet charge be involved. Any vacillation will result in greater casualties, loss of victory, and general discouragement of the whole force.

General Lin Piao, 1946, quoted in Ebon, *Lin Piao*, 1970

Once you have made up your mind, stick to it; there is no longer any *if* or *but* . . .

Napoleon, 18 February 1812, to Marshal Marmont, *Correspondance*, No. 18503, Vol. XXIII, 1858–1870

War is waged only with vigour, decision, and unshaken will; one must not grope or hesitate.

Napoleon, 6 June 1813, to General Bertrand, *Correspondance*, No. 20090, Vol. XXV, 1858–1870

. . . The power of decision develops only out of practice. There is nothing mystic about it. It comes of a clear-eyed willingness to accept life's risks, recognizing that only the enfeebled are comforted by thoughts of an existence devoid of struggle.

Brigadier General S. L. A. Marshall, *The Armed Forces Officer*, 1950

It is customary to treat 'estimate of the situation' as if it were pure mathematical process, pointing almost infallibly to a definite result. But this is contrary to nature. The mind of man does not work that way, nor is it consistent with operational realities. Senior commanders are as prone as even the newest junior lieutenant to labor in perplexity between two opposing courses of action during times of crisis, and then make their decisions almost with the abruptness of an explosion. *It is post-decision steadiness more than pre-decision certitude which carries the day.* A large part of decision is intuitive; it is the by-product of the subconscious. In war, much of what is most pertinent lies behind a drawn curtain. The officer is therefore badly advised who would believe that a hunch is without value, or that there is something unmilitary about the simple decision to take some positive action, even though he is working in the dark.

Brigadier General S. L. A. Marshall, *The Armed Forces Officer*, 1950

Any man facing a major decision acts, consciously or otherwise, upon the training and beliefs of a lifetime. This is no less true of a military commander than of a surgeon who, while operating, suddenly encounters an unsuspected complication. In both instances, the men must act immediately, with little time for reflection, and if they are successful in dealing with the unexpected it is upon the basis of past experience and training.

Admiral of the Fleet Ernest J. King, *Fleet Admiral King, Naval Record*, 1952

He [the leader] cannot afford to be ambiguous.

Ever since I was a boy and read about Gettysburg, I've thought that ambiguity was the reason for Lee's losing the battle. Lee said, 'General Ewell was instructed to carry the hill occupied by the enemy, if he found it practicable . . . ' I call that leaning on a subordinate, most definitely. Ewell didn't find the attack practicable; so he didn't attack that late afternoon or early evening. During the night the Federals heavily reinforced, and were never driven from their ridges, but instead repelled every Confederate attack poised against them.

Lee left it up to Ewell to make a decision which I feel that great general should have made himself. And Lee was a great general. Figure it out if you can.

General Curtis LeMay, *Mission With LeMay*, 1965

The flow of battle may sometimes change conditions or create new tasks no amount of advance planning could have allowed for. Then the success or failure of the entire mission may well depend on the resourcefulness, the training, the alertness, the decisiveness of the leaders of even the smallest units. The ability to make prompt decisions and to execute them vigorously is best bred in men who, through confidence in their troops and in their superiors, have persuaded themselves that they are unbeatable. Unrelenting attention to detail, concern for the well-being of every member of every unit, painstaking checking out of each assignment with the men charged with performing it – all these help instill that self-confidence, that sense of belonging to a tightly organized and well-fed organization, the feeling that can give momentum to a whole force and make it truly unconquerable.

General Mathew B. Ridgway, *The Korean War*, 1967

When in field operations a man jellies on the pivot, anything moving in the what seems like the right direction pulls like a magnet.

Brigadier General S. L. A. Marshall, *Bringing Up the Rear*, 1979

. . . the first demand in war is decisive action.

German Army Field Service Regulations

The matter of 'decision' is vital. The modern tendency is to avoid taking decisions, and to procrastinate in the hope that things will come out all right in the wash. The only policy for the military leader is decision in action and calmness in crisis: no bad doctrine for the political leader as well.

Field Marshal Viscount Montgomery of Alamein, *The Memoirs of Field Marshal Montgomery*, 1958

When all is said and done the greatest quality required in a commander is 'decision'; he must be able to issue clear orders and have the drive to get things done. Indecision and hesitation are fatal in any officer; in a C-in-C they are criminal.

Field Marshal Viscount Montgomery of Alamein, *The Memoirs of Field Marshal Montgomery*, 1958

When a decision is taken belatedly, its execution inevitably leads to haste.

Marshal of the Soviet Union Vasili I. Chuikov, 1962, quoted in Leites, *The Soviet Style of War*, 1982

It was an example of inflexibility in the pursuit of previously conceived ideas that is, unfortunately, too frequent in modern warfare. Final decisions are made not at the front by those who are there, but many miles away by those who can but guess at the possibilities and potentialities.

General of the Army Douglas MacArthur, *Reminiscences*, 1964

Command is a pretty lonesome job. There are multiple decisions which you have to make entirely by yourself. You can't lean on anybody else. And a good commander, once he issues an order, must receive complete compliance. An indecisive commander cannot achieve instant compliance. Or one who is unable to make up his own mind and tries to lean on his subordinates will never achieve instant compliance either.

General Curtis LeMay, *Mission With LeMay*, 1965

DECORATIONS/AWARDS

At the storming of a city the first man to scale the wall is awarded a crown of gold. In the same way those who have shielded and saved one of their fellow-citizens of their allies is honoured with gifts from the consul, and the men whose lives they have saved present them of their own free will with a crown; if not, they are compelled to do so by the tribunes who judge the case. Moreover, a man who has been saved in this way reveres his rescuer as a father for the rest of his life and must treat him as if he were a parent. And so by means of such incentives even those who stay at home feel the impulse to emulate such achievements in the field no less than those who are present and see and hear what takes place. For the men who receive these trophies not only enjoy great prestige in the army and soon afterwards in their homes, but they are also singled out for precedence in religious processions when they return. On these occasions nobody is allowed to wear decorations save those who have been honoured for their bravery by the consuls, and it is the custom to hang up the trophies they have won in the most conspicuous places in their houses, and to regard them as proofs and visible symbols of their valour. So when we consider this people's almost obsessive concern with military rewards and punishments, and the immense importance which they attach to both, it is not surprising that they emerge with brilliant success from every way in which they engage.

Polybius, *The Rise of the Roman Empire*, vi, c. 125 BC

It is not titles that honour men, but men that honour titles.

Niccolo Machiavelli, *Discourses*, 1517

Show me a republic, ancient or modern, in which there have been no decorations. Some people call them baubles. Well, it is by such baubles that one leads men.

Napoleon, 19 May 1802, on establishing the Legion of Honour

Honour them with titles, present them with goods, and soldiers willingly come join you. Treat them courteously, inspire them with speeches, and soldiers willingly die. Give them nourishment and rest so that they do not become weary, make the code of rules uniform, and soldiers willingly obey. Lead them into battle personally, and soldiers will be brave. Record even a little good, reward even a little merit, and the soldiers will be encouraged.

Zhuge Liang (AD 180–234) commentary to Sun Tzu, *The Art of War*

A soldier will fight long and hard for a bit of coloured ribbon.

Napoleon, 15 July 1815, to the Captain, H.M.S. *Bellerophon*, upon going into exile

---

. . . a coward dressed as a brave man will change from his cowardice and, in nine cases out of ten, will on the next occasion demonstrate the qualities fortuitously emblazoned on his chest . . .

We must have more decorations and we must give them with no niggard hand . . .

War may be hell; but for John Doughboy there is a heaven of suggestion in anticipating what Annie Rooney will say when she sees him in his pink feather and his new medal.

General George S. Patton, Jr., October 1927, *The Patton Papers*, Vol. I, 1972–74

The result of decorations works two ways. It makes the men who get them proud and determined to get more, and it makes the men who have not received them jealous and determined to get some in order to even up. It is the greatest thing we have for building a fighting heart.

General George S. Patton, quoted in Semmes, *Portrait of Patton*, 1955

They recommended me for a Silver Star for that action, and back in Corps Headquarters at Noumea some jerk reduced it to a Bronze Star. What right have those people got to put their cotton-picking hands into things like that? They didn't see the action, and have no way on earth to judge. Wouldn't you think they could see what it does to morale? I can stand it. I've got enough damned medals. But what it does to the young kids is inexcusable.

Lieutenant General Lewis 'Chesty' Puller, quoted in Davis, *Marine*, 1962

No order taken by me while I was chief of staff gave me greater satisfaction than re-establishing the Order of the Purple Heart. It had not been in use for a century and a quarter.

This decoration is unique in several ways; first, it is the oldest in American history, and antedates practically all of the famous military medals in the world; second, it comes from the greatest of all Americans, George Washington, and thereby carries with it something of the reverence which haloes his great name; and third, it is the only decoration which is completely intrinsic in that it does not depend upon approval or favor by anyone. It goes only to those who are wounded in battle, and enemy action alone determines the award. It is a true badge of courage and every breast that wears it can beat with pride.

General of the Army Douglas MacArthur, *Reminiscences*, 1964

During the war I found that men who won the Medal of Honor – and survived – were so conscious of the distinction that when they returned to combat they showed even greater courage, took greater risks and were far more likely to be killed than their comrades. This tendency was so pronounced that I instituted a policy of sending such men back to the states to train and inspire our untried troops. Usually they would protest bitterly, but I never let myself be swayed.

General Mark Clark, 'Recapturing What Honor Really Means', *Rocky Mountain News*, 5 May 1984

DEFEAT

Those who reached my boundary, their seed is not; their heart and their soul are finished forever and ever. As for those who had assembled before them on the sea, the full flame was in their front, before the harbour-mouths, and a wall of metal upon the shore surrounded them. They were dragged, overturned, and laid low upon the beach; slain and made heaps from stern to bow of their galleys. While all their things were cast upon the water. Thus I turned back the waters to remember Egypt; when they mention my name in their land, may it consume them, while I sit upon the throne of Harakhte, and the serpent-diadem is fixed upon my head, like Re.

Ramses III, *c.* 1190 BC, record of the Northern War, quoted in Breasted, *Ancient Records of Egypt*, 1906

We do have Prayers, you know, Prayers for forgiveness,
daughters of mighty Zeus . . . and they limp and halt,
they're all wrinkled, drawn, they squint to the side,
can't look you in the eyes, and always bent on duty,
trudging after Ruin, maddening, blinding Ruin.
But Ruin is strong and swift –
She outstrips them all by far, stealing a march,
leaping over the whole wide earth to bring mankind to grief.
And the Prayers trail after, trying to heal the wounds.

Phoenix, the Charioteer, tutor and comrade to Achilles in Homer, *The Iliad*, ix, *c.* 800 BC, tr. Fagles

---

Once more may I remind you that you have beaten most of the enemy's fleet already; and, once defeated, men do not meet the same dangers with their old spirit.

Phormio, address to the Athenian Navy, 429 BC, quoted in Thucydides, *The Peloponnesian War*, *c.* 404 BC, tr. Jowett

---

Troops defeated in open battle should not be pampered or, even if it seems like a good idea, take refuge in a fortified camp or some other strong place, but while their fear is still fresh, they should attack again. By not indulging them they may with greater assurance renew the fighting.

The Emperor Maurice, *The Strategikon*, c. AD 600

A general of great merit should be said to be a man who has met with at least one great defeat. A man like myself who has gone his whole life with victories alone and suffered no defeats cannot be called a man of merit, even though he gains in years.

Asakura Soteki (1474–1555) *Soteki Waki*, c. 1550, in Wilson, *Ideals of the Samurai*, 1982

. . . Trophies apart, there is no accurate measure of loss of morale; hence in many cases the abandonment of the fight remains the only authentic proof of victory. In lowering one's colours one acknowledges that one has been at fault and concedes in this instance that both might and right lie with the opponent. This is shame and humiliation, which must be distinguished from all other psychological consequences of the transformation of the balance, is an essential part of victory.

Major General Carl von Clausewitz, *On War*, iv, 1832, tr. Howard and Paret

---

Man in war is not beaten, and cannot be beaten, until he owns himself beaten. Experience of all war proves this truth. So long as war persists as an instrument of policy, the objects of that policy can never be attained until the opponent admits his defeat. Total annihilation, even if it were possible, would recoil on the victor in the close-knit organization of the world's society. Moreover, it is not necessary to the military result. In all wars, and battles, decision is obtained at the moment when the survivors – normally the vast majority – realize that unless they yield their extinction has become inevitable.

The history of war shows that man tends to give way directly he recognizes a stronger – when it dawns upon his mind that he has no further hope of victory by continuing the fight. That is why in battle, contrary to popular impression, actual shock is the rarest episode. When an assault takes place, the weaker side – weaker in either numbers, morale, or momentum – have almost invariably surrendered or fled before the clash actually comes.

Captain Sir Basil Liddell Hart, *Thoughts on War*, 1944

In war there is no prize for the runner-up.

General of the Army Omar N. Bradley, in *Military Review*, 1950

In the final choice a soldier's pack is not so heavy a burden as a prisoner's chains.

General of the Army Dwight D. Eisenhower, Inaugural Address, 20 January 1953

As to being prepared for defeat, I certainly am not. Any man who is prepared for defeat would be half defeated before he commenced. I hope for success; shall do all in my power to secure it and trust to God for the rest.

Admiral David G. Farragut, 1864, letter to his wife

The finest theories and most minute plans often crumble. Complex systems fall by the wayside. Parade ground formations disappear. Our splendidly trained leaders vanish. The good men which we had at the beginning are gone. Then raw truth is before us.

Charles O'Daniel

Everything seems quiet in Berlin for the time being. We are rather in a mess here unfortunately [German Army on the Eastern Front]. We shall not be able to hold Riga.

The troops will fight no more.

General Max Hoffmann, 31 December 1918, *War Diaries and Other Papers*, vol 2, 1927

I have seen much war in my life time and I hate it profoundly. But there are worse things than war; and all of them come with defeat.

Ernest Hemingway, *Men at War*, 1942

Errors and defeats are more obviously illustrative of principles than successes are . . . Defeat cries aloud for explanation; whereas success, like charity, covers a multitude of sins.

Rear Admiral Alfred Thayer Mahan, *Naval Strategy*, 1911

. . . Any engagement includes the abstract possibility of defeat and there is no other means for reducing the possibility than the organized preparation of the engagement.

V. I. Lenin, quoted in *Voyenno istorichiskiy zhurnal*, 3/1986

A beaten general is disgraced forever.

Marshal of France Ferdinand Foch, *Precepts and Judgments*, 1919

I claim we got a hell of a beating. We got run out of Burma and it is humiliating as hell. I think we ought to find out what caused it, and go back and retake it.

General Joseph 'Vinegar Joe' Stilwell, *New York Times*, 26 May 1942

DEFENCE

Little minds try to defend everything at once, but sensible people look at the main point only; they parry the worst blows and stand a little hurt if thereby they avoid a greater one. If you try to hold everything, you hold nothing.

Frederick the Great, quoted in Foertsch, *The Art of Modern War*, 1940

What is the concept of defence? The parrying of a blow. What is its characteristic feature? Awaiting the blow. It is this feature which turns any action into a defensive one; it is the only test by which defence can be distinguished from attack in war. Pure defence, however, would be completely contrary to the idea of war, since it would mean that only one side was waging it . . . But if we are really waging war, we must return the enemy's blows; and these offensive acts in a defensive war come under heading of 'defence' – in other words, our offensive takes place within our own positions or theatre of operations. Thus, a defensive campaign can be fought with offensive battles, and in a defensive battle, we can employ our divisions offensively. Even in a defensive position awaiting the enemy assault, our bullets take the offensive. So the defensive form of war is not a simple shield, but a shield made up of well-directed blows.

Major General Carl von Clausewitz, *On War*, vi, 1832, tr. Howard and Paret

---

What is the object of defence? Preservation. It is easier to hold ground than take it. It follows that defence is easier than attack, assuming both sides have equal means. Just what is it that makes preservation and protection so much easier? It is the fact that time which is allowed to pass unused accumulates to the credit of the defender. He reaps where he did not sow. Any omission of attack – whether from bad judgment, fear, or indolence – accures to the defenders' benefit.

Major General Carl von Clausewitz, *On War*, vi, 1832, tr. Howard and Paret

---

He who stays on the defensive does not make war, he endures it.

Field Marshal Colmar Baron von der Goltz, *The Nation in Arms*, 1883

The defence of a river crossing is the worst of all assignments especially if the front that you are to defend is long; in this case defence is impracticable.

Frederick the Great, *Instructions to His Generals*, 1747

To me death is better than the defensive.

Field Marshal Prince Aleksandr V. Suvorov (1729–1800)

DEFENSIVE POSITIONS/ FORTIFICATIONS/FORTRESSES

By God's throat, even if that castle were all built of butter and not of iron and stone, I have no doubt it would defend me against him [King of France] and all his forces.

Richard I, 'The Lion Heart', King of England, 1198–9, of his new castle, Gaillard

---

When I hear talk of lines, I always think I am hearing talk of the walls of China. The good ones are those that nature has made and the good entrenchments are good dispositions and brave soldiers.

Field Marshal Maurice Comte de Saxe, *My Reveries*, 1732

The art of defending fortified places consists in putting off the moment of their reduction.

Frederick the Great, *Instructions to His Generals*, 1747

The only advantage a fortified line may afford is to render the situation of the enemy so difficult as to induce him to operate incorrectly, which in turn may cause him to be defeated with numerically inferior forces. Or if one is confronted by a capable general, the fortifications may have the effect of compelling him methodically to negotiate the obstacles that one has created with due deliberation. In this manner time is gained.

Napoleon, 1807, 'Notes sur la défense de l'Italie', *Correspondance*, No. 14704, Vol. XVIII, 1858–1870

MAXIM 17. In a war of march and manoeuvre, if you would avoid a battle with a superior army, it is necessary to entrench every night, and occupy a good defensive position. Those natural positions which are ordinarily met with, are not sufficient to protect an army against superior numbers without recourse to art.

MAXIM 40. Fortresses are equally useful in offensive and defensive warfare. It is true they will not in themselves arest an army, but they are an excellent means of retarding, embarrassing, weakening and annoying a victorious enemy.

Napoleon, *The Military Maxims of Napoleon*, 1831, tr. D'Aguilar

---

MAXIM 103. When they are thoroughly understood, field fortifications are always useful and never injurious.

Napoleon, *The Military Maxims of Napoleon*, 1827, ed., Burnod

---

Any position in which one accepts battle and makes use of terrain to protect oneself is a defensive position, and it makes no difference whether one's general attitude is mainly passive or mainly active.

Major General Carl von Clausewitz, *On War*, 1832

Intrenchments, or continuous lines, of which great use was made in the old wars, are adapted to absolutely defensive operations. They are open to the great inconvenience of spreading the means of defence along a considerable extent, and consequently, of being weak upon all points which the enemy may attack.

They offer moreover the great disadvantage of forcing an army to abandon them the instant any part of them is carried.

Marshal of France Michel Ney, Duc d'Elchingen, Prince de la Moskova, *The Memoirs of Marshal Ney*, 1834

It was one of Professor Mahan's maxims that the spade was as useful in war as the musket, and to this I will add the ax.

General of the Army William T. Sherman, *The Memoirs of General W. T. Sherman*, 1875

The trick expression, 'Dig or die', is much over-used and much misunderstood. Wars are not won by defensive tactics. Digging is primarily defensive. The only time it is proper for a soldier to dig is when he has reached his final objective . . . Personally, I am opposed to digging under such circumstances, as the chance of getting killed while sleeping normally on the ground is quite remote, and the fatigue from digging innumerable slit trenches is avoided. Also, the psychological effect on the soldier is bad, because if he thinks he has to dig he must think the enemy is dangerous, which he usually is not.

General George S. Patton, Jr., *War As I Knew It*, 1947

. . . Leave soldiers in defensive positions too long and they begin to scare themselves to death. Every rustle of the wind sends shivers up their spine, and a single enemy sniper can paralyze an entire battalion. Instead of taking action to eliminate the threat, they hold back, convinced that any such attempt would only make matters worse and trigger a terrible retribution.

Colonel Harry Summers, 13 January 1989, *Washington Times*

The person who is able fully to utilize all the equipment of military engineering and show a flexibility of mind and military inventiveness is always able to confront the enemy with unexpected events and unexpected dimensions of the defeat dealt it.

Marshal of the Soviet Union Mikhail N. Tukhachevskiy, April 1929, quoted in *Voyenno istorichiskiy zhurnal*, 2/1983

Anyone can realize that positions, however strong, cannot repel an attack, when once they are deprived of a sufficient living garrison.

General Aleksei A. Brusilov, A *Soldier's Notebook*, 1931

The Italians and Germans spent tremendous effort in time, labor, and money, building defensive positions. I am sure that just as in the case of the walls of Troy and the Roman walls across Europe the fact that they trusted to defensive positions reduced their power to fight. Had they spent one-third as much effort fighting as they did in building, we never could have taken the positions.

General George S. Patton, Jr., July 1943, *The Patton Papers*, Vol. II, 1974

DELEGATION OF AUTHORITY

According to my custom . . . I was present at the attack near the monastery of Svyanty Kryzh, but held my tongue, not wishing in the least to detract from the praiseworthy, skilfull and brave commands of my subordinates.

Field Marshal Prince Aleksandr V. Suvorov (1729–1800)

Major Briscoe, I am but the nominal commander. The President and Secretary of State have interfered with my intended operations, and I fear for the success of the day.

General William H. Winder, Commander of the American forces at the Battle of Blandensburg (1814), quoted in *Military Review*, 4/1981

Be content to do what you can for the well-being of what properly belongs to you; commit the rest to those who are responsible.

General Robert E. Lee, quoted in Fuller, *Grant and Lee: A Study in Personality and Generalship*, 1933

My interference in battle would do more harm than good . . . I have, then, to rely on my brigade and division commanders. I think and work with all my power to bring the troops to the right place at the right time; then I have done my duty. As soon as I order them forward into battle, I leave my army in the hands of God.

General Robert E. Lee, quoted in Fuller, *Grant and Lee: A Study in Personality and Generalship*, 1933

No army can be efficient unless it be a unit for action; and the power must come from above, not from below: The President usually delegates his power to the commander-in-chief, and he to the next, and so on down to the lowest actual commander of troops, however small the detachment. No matter how troops come together, when once united, the highest officer in rank is held responsible, and should be consequently armed with the fullest power of the Executive, subject only to law and existing orders. The more simple the principle, the greater the likelihood of determined action; and the less a commanding officer is circumscribed by bounds or precedent, the greater is the probability that he will make the best use of his command and achieve the best results.

General of the Army William T. Sherman, *The Memoirs of General W. T. Sherman*, 1875

The battalion officer cannot be four company commanders.

General Mikhial I. Dragomirov (1830–1905), quoted in *Voyenno istoricheskiy zhurnal*, 9/1985

I also said that he, as Supreme Commander, could not descend into the land battle and become a ground C-in-C; the Supreme Command has to sit on a very lofty perch and be able to take a detached view of the whole intricate problem – which involves land, sea, air, civil control, political problems, etc., etc. Someone must run the land battle for him; we had won a great victory in Normandy *because* of unified land control and *NOT in spite of it.*

Field Marshal Viscount Montgomery of Alamein, 1944, quoted in Hamilton, *Master of the Battlefield*, 1983

If the state of a command is such that the commander's hands are not free for the administration of his exterior tasks, it is an indication that the morale and leadership of the unit are based on incorrect principles and that insufficient authority has been delegated to subordinates. There can be no more miserable cause of failure though it is a weakness that is found in many conscientious officers. The man who cannot bring himself to trust the judgment and good faith of other men cannot command very long. He will soon break under the unnecessary strain he puts on himself. Sleeplessness, nervous irritation, and loss of self-control will be his lot until he is at last found totally unfit.

Brigadier General S. L. A. Marshall, *The Armed Forces Officer*, 1950

The (our company commanders) are now bound hand and foot by a whole new list of mandatory subjects . . . You've delegated to the commander only the choice of what he is going to catch hell for.

General Creighton Abrams (1914–1974), quoted in *Military Review*, 4/1985

Each commander's plan had to be clear so that every other commander would understand what his colleagues were doing – what their objectives were, and how they intended to accomplish them. By the time the order presentations and reviews were completed, they each knew the overall plan in detail, they knew how their own roles fit into the overall concept, they knew what their neighbors were trying to do and where they would be. They also knew that I trusted them to carry out their missions with an absolute minimum of interference. The field assignments were theirs, and my philosophy (as each commander understood) was that they would handle them best.

General Ariel Sharon, referring to the Battle of Agheila, 1967, *Warrior*, 1989

. . . A company commander can more hope to supervise directly the acts of several hundred men in battle without reposing large faith in his lieutenants than the general can expect good results to come of by-passing his staff and his corps commanders and dealing directly with his divisions. But there are generals who have failed because they did not learn this lesson as captains.

Brigadier General S. L. A. Marshall, *Men Against Fire*, 1947

DEMORALIZATION

. . . The men of both Lee's and Johnston's armies were like their brethren of the North, as brave as men can be; but no man is so brave that he may not meet such defeats and disasters as to discourage and dampen his ardor for any cause, no matter how just he deems it.

General of the Army Ulysses S. Grant, *Personal Memoirs of U.S. Grant,* 1885

The Principle of Demoralization. As the principle of endurance has, as its primary objejct, the security of the minds of men by shielding their morale against the shock of battle, inversely the principle of demoralization has as its object the destruction of morale: first, in the moral attack against the spirit and nerves of the enemy's nation and government; secondly against his nation's policy; thirdly against the plan of its commander-in-chief, and fourthly against the morale of the soldiers commanded by him. Hitherto the fourth, the least important of these objectives, has been considered by the traditionally minded soldier as the sole psychological objective of this great principle.

Major General J. F. C. Fuller, *The Reformation of War,* 1923

DEPENDABILITY

Even when in a battle that is beyond one's means to win, one should lift up his heart and be resolved that no one will surpass him in firmness. He should think to be another's strength and a man to be relied upon.

Shiba Yoshimasa (1350–1410), *The Chikubasho,* 1838, in Wilson, *Ideals of the Samurai,* 1982

During a critical phase of the Battle of the Bulge, when I commanded the 18th Airborne Corps, another corps commander just entering the fight next to me remarked: 'I'm glad to have you on my flank. It's character that counts.' I had long known him, and I knew what he meant. I replied: 'That goes for me, too.' There was no need for amplification. None was necessary. Each knew the other would stick however great the pressure; would extend help before it was asked, if he could; and would tell the truth, seek no self-glory, and everlastingly keep his word. Such feelings breed confidence and success.

General Mathew B. Ridgway, 'Leadership', *Military Review,* 10/1966

DESERTION/DESERTERS

To seduce the enemy's soldiers from their alliance and encourage them to surrender is of especial service, for an adversary is more hurt by desertion than by slaughter.

Flavius Vegetius Renatus, *Military Institutions of the Romans,* c. AD 378

Letters ought to be sent to deserters from our side who have joined the enemy in such a way that the letters will fall into the enemy's hands. These letters should remind the deserters of the prearranged time for their treachery, so that the enemy will become suspicious of them, and they will have to flee.

The Emperor Maurice, *The Strategikon,* c. AD 600

If any of the enemy's troops desert him and come over to you, it is a great acquisition – provided they prove faithful; for their loss will be more than that of those killed in battle, although deserters will always be suspected by their new friends and odious to their old ones.

Niccolo Machiavelli, *The Art of War,* 1521

Come, come, let us fight another battle today: if I am beaten, we will desert together tomorrow.

Frederick the Great (1712–1786), to a captured deserter, quoted in Campbell, *Frederick the Great,* 1843

A soldier who offers to quit his rank, or offers to flag, is to be instantly put to death by the officer who commands that platoon, or by the officer or sergeant in the rear of that platoon;

a soldier does not deserve to live who won't fight for his king and country.

Major General Sir James Wolfe, 1755, Order to the 20th Foot at Canterbury

---

Not upon a man from the colonel to the private in a regiment – both inclusive. We may pick up a marshal or two; but not worth a damn.

The Duke of Wellington, June 1815, before the Battle of Waterloo when asked if he anticipated any French desertions, quoted in Creevy, *The Creevy Papers*, 1934

No dangers can for me and mine be compared with the base desertion of such a noble cause – the honour of God, the peace of the provinces, the freedom of the Fatherland – and abandonment of the sacred and honourable side which I have up till now followed.

Prince William I of Orange, 1584

DESERT WARFARE

Desert operations should be like . . . wars at sea, 'in their mobility, their ubiquity, their independence of bases and communications, their lack of ground features, of strategic areas, of fixed directions, of fixed points'.

Colonel T. E. Lawrence, quoted in Liddell Hart, *Colonel Lawrence: The Man Behind the Legend*, 1934

Immobility and a rigid adherence to pattern are bad enough in European warfare; in the desert they are disastrous. Here everything is in flux; there are no obstructions, no lines, water or woods for cover; everything is open and incalulable; the commander must adapt and reorientate himself daily, even hourly, and retain his freedom of action. Everything is in motion; he must be constantly on the alert, all the time on the edge of capture or destruction by a more cunning, wide-awake or versatile enemy. There can be no conservatism in thought or action, no relying on tradition or resting on the laurels of previous victory. Speed of judgment, and action to create changing situations and surprises for the enemy faster than he can react, never making dispositions in advance, these are the fundamentals of desert tactics.

General Fritz Bayerlein, in *Rommel Papers*, 1953

Being barren, the desert cannot be conquered by armies. Of itself, as Winston Churchill once wrote, the desert yields nothing to them but hardship and suffocation. To occupy it is therefore purposeless and wasteful, unless occupation is an essential step toward ultimate security.

It is this very singleness of purpose about desert warfare that makes it unique. The arena is suitable for nothing except primordial combat, with death or survival as the issue. All movement is limited or regulated by the availability of water, and all maneuver seeks the destruction of the enemy force.

Brigadier General S. L. A. Marshall, 'The Desert: It's Different', *Infantry*, 12/1970

In desert warfare you do not necessarily go after a terrain objective. What you do is seek to destroy the enemy's center of gravity. And there are a lot of ways you can do that without ever having to worry about house-to-house, dig-'em-out fighting.

General H. Norman Schwarzkopf, 13 September 1990, interview, quoted in Pyle, *Schwarzkopf, In His Own Words*, 1991

Of all theatres of operations, it was probably in North Africa that the war took on its most advanced form. The protagonists on both sides were fully motorized formations, for whose employment the flat and obstruction-free desert offered hitherto undreamed-of possibilities. It was the only theatre where the principles of motorized and tank warfare, as they had been taught theoretically before the war, could be applied in full – and further developed. It was the only theatre where pure tank battles between major formations were fought.

Field Marshal Erwin Rommel, *The Rommel Papers*, 1953

DESPERATION

I have often myself witnessed occasions, and I have heard of many more from others where men who had been conquered by an enemy, having been driven quite to desperation, have renewed the fight, and retrieved their former disasters.

Themistocles, quoted in Herodotus (*c.* 484–*c.* 430 BC), *The Struggle for Greece*

It behooves you to reflect that the Lacedaemonians, if they be forced to relinquish the hope of life, will fight with desperation; and the divine powers, as it seems, often take delight in making the little great, and the great little.

Jason of Pherae, advice to the Thebans after the Battle of Leuctra, quoted by Xenophon (*c.* 438–*c.* 354 BC), *Hellenica*

. . . and there is nobody more terrible than the desperate.

Field Marshal Prince Aleksandr V. Suvorov, 1799, account of his icy Swiss campaign

---

I was cut off and surrounded, night and day we attacked the enemy, in front and in the rear, captured his guns, which we were obliged to throw off the precipices owing to the shortage of pack animals, and inflicted on him losses four times heavier than ours. Everywhere we forced our way through as victors . . .

Field Marshal Prince Aleksandr V. Suvorov, 1799, account of his icy Swiss campaign

DESTRUCTION OF THE ENEMY

The objective is not the occupation of a geographical position but the destruction of the enemy force.

General Piotr A. Rumyantsev (1725–1796), quoted in Parkinson, *Fox of the North*, 1976

. . . did you ever hear the story of the *Bismarck*? I forget the British Admiral's name. He told the story himself. He wirelessed Churchill. 'The *Bismarck* is a wreck. The crew has left her but she won't sink. And I've got just enough oil in my ship to get home.' The Prime Minister [Churchill] sent him a cable, 'You stay there until the *Bismarck* is on the bottom; and we'll send out and tow you in.'

General of the Army Dwight D. Eisenhower, *The Military Churchill*, 1970

In case opportunity for destruction of major portion of the enemy fleet is offered or can be created, *such destruction becomes the primary task.*

Admiral of the Fleet Chester W. Nimitz, quoted in Morison, *Leyte, June 1944–January 1945*, 1958

When the enemy is driven back, we have failed, and when he is cut off, encircled and dispersed, we have succeeded.

Field Marshal Prince Aleksandr V. Suvorov (1729–1800) quoted in Reznichenko, *Tactica*, 1987

Had we taken ten sail and allowed the eleventh to escape, being able to get at her, I could never have called it well done.

Admiral Lord Nelson (1758–1805)

The most important goal of our actions is destruction of the enemy to the last limit of possibility.

Field Marshal Prince Mikhail I. Kutuzov (1745–1813), quoted in Savkin, *Basic Principles of Operational Art and Tactics*, 1972

Only destruction of the enemy can be called victory.

Admiral Stepan O. Makarov, 1904, Port Arthur

Jomini's dictum that the organized forces of the enemy are the chief objective, pierces like a two-edged sword to the joints and marrow of many specious propositions . . .

Rear Admiral Alfred Thayer Mahan, *From Sail to Steam*, 1907

DETAILS

Appreciate all those details; they are not without glory. It is the first step that leads to glory.

Frederick the Great, *Art de la Guerre*, 1751

It is absolutely vital that a senior commander should keep himself from becoming immersed in details, and I always did so. I would spend many hours in quiet thought and reflection in thinking out the major problems. In battle a commander has got to think how he will defeat the enemy. If he gets involved in details he cannot do this since he will lose sight of the essentials which really matter; he will then be led off on side issues which will have little influence on the battle, and he will fail to be that solid rock on which he staff can lean. Details are their province. No commander whose daily life is spent in the consideration of details, and who has not time for quiet thought and reflection, can make a sound plan of battle on a high level or conduct large-scale operations efficiently.

Field Marshal Viscount Montgomery of Alamein, *The Memoirs of Field Marshal Montgomery*, 1958

DETERMINATION

In my fifth campaign the inhabitants of the cities . . . of Mount Nipu, a steep mountain, like the nest of the eagle, king of birds, – were not submissive to my yoke. I had my camp pitched at the foot of Mount Nipur and with my picked bodyguard and my relentless warriors, I, like a strong wild-ox, went before them. Gullies, mountain torrents and waterfalls, dangerous cliffs, I surmounted in my sedan chair. Where it was too steep for my chair, I advanced on foot. Like a young gazelle I mounted the highest peaks in pursuit of them. Whenever my knees gave out, I sat down on some mountain boulder and drank the cold water from the water skin to quench my thirst. To the summits of the mountains I pursued them and brought about their overthrow. Their cities I captured and I carried off their spoil; I destroyed, I devastated, I burned them with fire.

Sennacherib, King of Assyria (705–681 BC), quoted in Luckenbill, *Ancient Records of Assyria and Babylonia*, 1926

We will either find a way or make one.

Hannibal (247–183 BC)

I have not the particular shining bauble or feather in my cap for crowds to gaze or kneel to, but I have power and resolution for forces to tremble at.

Oliver Cromwell (1559–1658)

He who understands the nature of a decision in practical affairs, especially in war where it must be made under the pressure of great responsibility and in the midst of a thousand uncertainties and contradictions, will appreciate that decisions cannot be made without many doubts; and what may seem a very simple problem often cannot be decided without the exercise of great determination. Conceiving plans for military operations is therefore the easiest part . . .

Major General Carl von Clausewitz, *Campaign of 1799 in Italy and Switzerland*, 1906

Determination in a single instance is an expression of courage; if it becomes characteristic, a mental habit. But here we are referring not to physical courage but to the courage to accept responsibility, a courage in the face of moral danger. This has often been called *courage d'esprit*, because it is created by the intellect. That, however, does not make it an act of the intellect: it is an act of temperament. Intelligence alone is not courage; we often see that the most intelligent people are irresolute. Since in the rush of events a man is governed by feelings rather than by thought, the intellect needs to arouse the quality of courage, which then supports and sustains it in action.

Looked at in this way, the role of determination is to limit the agonies of doubt and the perils of hesitation when the motives for action are inadequate . . .

Major General Carl von Clausewitz, *On War*, i, 1832, tr. Howard and Paret

---

If the entire Union Army comes across here, I will kill them all.

Lieutenant General James Longstreet, 13 December 1862, at Marye's Heights during the Battle of Fredericksburg

---

My troops may fail to take a position, but are never driven from one.

Lieutenant General Thomas 'Stonewall' Jackson (1824–1863)

True wisdom for a general is in vigorous determination.

Napoleon, 'Précis des guerres de Frédéric II', *Correspondance,* No. 209, Vol. XXXII, 1858–1870

Tell the men to fire faster and not to give up the ship. Fight her till she sinks.

Captain James Lawrence, 1813, order, as he lay dying, during the battle between his USS *Chesapeake* and HMS *Shannon*

My centre is giving way, my right is pushed back, situation excellent, I am attacking.

Marshal of France Ferdinand Foch, August 1914, at the First Battle of the Marne, quoted in Liddell Hart, *Reputations Ten Years After*, 1928

. . . Simply to exhalt the will as a force without limit tends only to create delusions about what may be accomplished by the exercise of it. These delusions in turn destroy the confidence of troops, then as confidence flags, they become unresponsive to the bidding of the commander. The quality of mind most worth seeking is not power of will but, as said by Marcus Aurelius, 'Freedom of will and undeviating steadiness of purpose'.

Brigadier General S. L. A. Marshall, *Men Against Fire*, 1947

It is fatal to enter any war without the will to win it.

General of the Army Douglas MacArthur, July 1952, speech to the Republican National Convention

---

. . . It is true that in war determination by itself may achieve results, while flexibility without determination in reserve, cannot, but it is only the blending of the two that brings final success. The hardest test of generalship is to hold this balance between determination and flexibility.

Field Marshal Viscount Slim, *Defeat Into Victory*, 1963

I think it would be better to order up some artillery and defend the present location.

General of the Army Ulysses S. Grant, 1864, when members of his staff advised retreat at the Battle of the Wilderness

---

I propose to fight it out on this line, if it takes all summer.

General of the Army Ulysses S. Grant, 11 May 1864, dispatch to Washington during the Battle of Spotsylvania Courthouse

THE DIRECT APPROACH

. . . to move directly on an opponent consolidates his balance, physical and psychological, and by consolidating it increases his resisting power. For in the case of an army it rolls the enemy back towards his reserves, supplies, and reinforcements, so that as the original front is driven back and worn thin, new layers are added to the back. At most, it imposes a strain rather than producing a shock.

Captain Sir Basil H. Liddell Hart, *Strategy*, 1954

DISCIPLINE

When the troops continually gather in small groups and whisper together the general has lost the confidence of the army.

Too frequent rewards indicate that the general is at the end of his resources; too frequent punishments that he is in acute distress.

If the officers treat the men violently and later are fearful of them, the limit of indiscipline has been reached.

Sun Tzu, *The Art of War*, *c.* 500 BC, tr. Griffith

If troops are punished before their loyalty is secured they will be disobedient. If not obedient, it is difficult to employ them. If troops are loyal, but punishments are not enforced, you cannot employ them.

Thus, command them with civility and imbue them uniformly with martial ardour and it may be said that victory is certain.

If orders which are consistently effective are used in instructing the troops, they will be obedient. If orders which are not consistently effective are used in instructing them, they will be disobedient.

When orders are consistently trustworthy and observed, the relationship of a commander with his troops is satisfactory.

Sun Tzu, *The Art of War*, *c.* 500 BC, tr. Griffith

Maintain discipline and caution above all things, and be on the alert to obey the word of command. It is both the noblest and the safest thing for a great army to be visibly animated by one spirit.

Archidamus II, King of Sparta, to the Spartan expeditionary force departing for Athens, 431 BC, quoted in Thucydides, *The History of the Peloponnesian War*, *c.* 404 BC, tr. Jowett

---

The Illyrians, to those who have no experience of them, do indeed at first sight present a threatening aspect. The spectacle of their numbers is terrible, their cries are intolerable, and the brandishing of their spears in the air has menacing effect. But in action they are not the men they look, if their opponents will only stand their ground; for they have no regular order, and therefore are not ashamed of leaving any post in which they are hard pressed; to fly and to advance being alike honourable, no imputation can be thrown on their courage. When every man is his own master in battle, he will readily find a decent excuse for saving himself.

Brasidas of Sparta, to the Spartan Army, 423 BC, quoted in Thucydides, *History of the Peloponnesian War*, *c.* 404 BC

The strength of an army lies in strict discipline and undeviating obedience to its officers.

Thucydides, *History of the Peloponnesian War*, ii, *c.* 404 BC

. . . where there is no one in control nothing useful or distinguished can be done. This is roughly true of all departments of life, and entirely true where soldiering is concerned. Here it is discipline that makes one feel safe, while lack of discipline has destroyed many people before now.

Xenophon, speech to the Greek officers after the defeat of Cyrus at Cunaxa, 401 BC, *The Persian Expedition* (*Anabasis*)

In military operations, order leads to victory. If rewards and penalties are unclear, if rules and regulations are unreliable, and if signals are not followed, even if you have an army of a million strong it is of no practical benefit.

An orderly army is one that is mannerly and dignified, one that cannot be withstood when it advances and cannot be pursued when it withdraws. Its movements are regulated and directed; this gives it security and presents no danger. The troops can be massed but not scattered, can be deployed but not worn out.

Zhuge Liang (AD 180–234), *The Way of the General*

The expense of keeping up good or bad troops is the same; but it depends wholly on you, most August Emperor, to restore the excellent discipline of the ancients and to correct the abuses of later times. This is a reformation the advantages of which will be equally felt by ourselves and our posterity.

Flavius Vegetius Renatus, *Military Institutions of the Romans*, c. AD 378

Punishment, and fear thereof, are necessary to keep soldiers in order in quarters; but in the field they are more influenced by hope and rewards.

Flavius Vegetius Renatus, *Military Institutions of the Romans*, c. AD 378

He ought not to be easy with those who have committed offences out of cowardice or carelessness in the hope of being regarded as a good leader, for a good leader does not encourage cowardice or laziness. On the other hand, he ought not to punish hastily and without a full investigation just to show he can act firmly. The first leads to contempt and disobedience; the other to well-deserved hatred with all its consequences. Both of these are extremes. The better course is to join fear with justice, that is, impose a fitting punishment upon offenders after proof of guilt. This, for reasonable men, is not punishment, but correction, and aids in maintaining order and discipline.

The Emperor Maurice, *The Strategikon*, c. AD 600

Few men are brave by nature, but good order and experience make many so. Good order and discipline in any army are to be more depended upon than courage alone.

Niccolo Machiavelli, *The Art of War*, 1521

After the organization of troops, military discipline is the first matter that presents itself. It is the soul of armies. If it is not established with wisdom and maintained with unshakeable resolution you will have no soldiers. Regiments and armies will be only contemptible, armed mobs, more dangerous to their own country than to the enemy.

It is a false idea that discipline, subordination and slavish obedience debase courage. It has always been noted that is with those armies in which the severest discipline is enforced that the greatest deeds are performed.

Many generals believe that they have done everything as soon as they have issued orders, and they order a great deal because they find many abuses. This is a false principle; proceeding in this fashion, they will never reestablish discipline in an army in which it has been lost or weakened. Few orders are best, but they should be followed up with care; negligence should be punished without partiality and without distinction of rank or birth; otherwise, you will make yourself hated. One can be exact and just, and be loved at the same time as feared. Severity must be accompanied with kindness, but this should not have the appearance of pretence, but of goodness.

Field Marshal Maurice Comte de Saxe, *My Reveries*, 1732

The commander should practise kindness and severity, should appear friendly to the soldiers, speak to them on the march, visit them while they are cooking, ask them if they are well cared for, and alleviate their needs if they have any. Officers without experience in war should be treated kindly. Their good actions should be praised. Small requests should be granted and they should not be treated in an overbearing manner, but severity is maintained about everything regarding the service. The negligent officer is punished; the man who answers back is made to feel your severity by being reprimanded with the authoritative air that superiority gives; pillaging or argumentative soldiers, or those whose obedience is not immediate should be punished.

Frederick the Great, *Instructions to His Generals*, 1747

. . . do we not see that every Nation under the Sun find their acct. therein; and without it no Order no regularity can be observed?

Why then shou'd it be expected from us, (who are all young and inexperienced,) to govern, and keep up a proper spirit of Discipline with't Laws; when the best, and most Experienced, can scarcely do it with. Then if we consult our Interest, I am sure it is loudly called for.

General George Washington, 11 October 1775, to Robert Dinwiddie

---

Discipline is the soul of an army. It makes small numbers formidable; procures success to the weak, and esteem to all.

General George Washington, 29 July 1759, Letter of Instructions to the Captains of the Virginia Regiments

---

The firmness requisite for the real business of fighting is only to be obtained by a constant course of discipline and service. I have never yet been witness to a single instance that would justify a different opinion . . .

General George Washington, *The Writings of George Washington*, Vol. 20, 1931–44

To bring men to a proper degree of subordination is not the work of a day, a month, or a year.

General George Washington, quoted in Marshall, *The Armed Forces Officer*, 1950

In camp, and especially in the presence of an active enemy, it is much easier to maintain discipline than in barracks in time of peace. Crime and breaches of discipline are much less frequent, and the necessity for courts-martial less. The captain can usually inflict all the punishment necessary, and the colonel *should* always.

General of the Army William T. Sherman, *The Memoirs of General W. T. Sherman*, 1875

Discipline is not made to order, cannot be created off-hand; it is a matter of the institution of tradition. The Commander must have absolute confidence in his right to command, must have the habit of command, pride in commanding. It is this which gives a strong discipline to Armies commanded by an aristocracy, whenever such a thing exists.

Colonel Charles Ardnant du Picq, quoted in Hamilton, *The Soul and Body of an Army*, 1921

. . . be strict in your discipline; that is, to require nothing unreasonable of your officers and men, but see that whatever is required be punctually complied with. Reward and punish every man according to his merit, without partiality or prejudice; hear his complaints; if well founded, redress them; if otherwise, discourage them, in order to prevent frivolous ones. Discourage vice in every shape, and impress upon the mind of every man, from the lowest to the highest, the importance of the cause, and what it is they are contending for.

General George Washington, letter of 10 November 1775, *The Writings of George Washington*, Vol. 4, 1931–1944

. . . with severity, kindness is needed, or else severity is tryanny. I am strict in maintaining the health of the soldiers and a true sense of good conduct; kind soldierly strictness, and then general brotherhood. To me strictness by whim would be tryanny.

Field Marshal Prince Aleksandr V. Suvorov (1729–1800), *Dokumenty*, 1952

The officers of companies must attend to the men in their quarters as well as on the march, or the army will soon be no better than a banditti.

The Duke of Wellington, General Order of 19 May 1800

As the severity of military operations increases, so also must the sternness of the discipline. The zeal of the soldiers, their warlike instincts, and the interests and excitements of war may ensure obedience of orders and the cheerful endurance of perils and hardships during a short and prosperous campaign. But when fortune is dubious or adverse; when retreats as well as advances are necessary; when supplies fail, arrangements miscarry and disasters impend, and when the struggle is protracted, men can only be persuaded to accept evil things by the lively realization of the fact that greater terrors await their refusal.

Sir Winston S. Churchill, *The River War*, 1899

Discipline is a reciprocal question, that is, it is strong only where it exists not only from the bottom up but also from the top down, for the very law which imposes certain duties on the soldier also protects him from unjust violations and superiors who allow themselves such violations are the violators of both the law and discipline.

General Mikhail I. Dragomirov (1830–1905), quoted in *Voyenno istorichiskiy zhurnal*, 9/1985

The discipline that makes the soldiers of a free country reliable in battle is not to be gained by harsh or tyrannical treatment. On the contrary, such treatment is far more likely to destroy than to make an army. It is possible to impart instruction and give commands in such a manner and in such a tone of voice as to inspire in the soldier no feeling but an intense desire to obey, while the opposite tone of voice cannot fail to excite strong resentment and a desire to disobey.

Lieutenant General John M. Schofield (1831–1906), Address to the Corps of Cadets at West Point, and now inscribed in the Sally Port

---

Popularity, however desirable it may be to individuals, will not form, or feed, or pay an army; will not enable it to march, and fight; will not keep it in a state of efficiency for long and arduous service.

The Duke of Wellington, 8 April 1811, letter from Portugal

They should be made to understand that discipline contributes no less to their safety than to their efficiency . . . Let officers and men be made to feel that they will most effectively secure their safety by remaining steadily at their posts, preserving order, and fighting with coolness and vigor.

General Robert E. Lee, 1865, Circular to Troops, Army of Northern Virginia

---

To be disciplined does not mean that one commits no offence against discipline; such a definition might suffice for the rank and file, but it is quite insufficient for a leader of whatever rank.

To be disciplined does not mean either that one executes orders received only in such measure as seems proper or possible, but it means that one enters freely into the thought and aims of the chief who has ordered, and that one takes every possible means to satisfy him.

To be disciplined does not mean to keep silent, to do only what one thinks can be done without risk of being compromised, the art of avoiding responsibilities, but it means *acting* in the spirit of the orders received, and to that end assuring by thought and planning the possibility of carrying out such orders, assuring by strength of character the energy to assume the risks necessary in their execution. The laziness of the mind results in lack of discipline as much as does insubordination. Lack of ability and ignorance are not either excuses, for knowledge is within reach of all who seek it.

Marshal of France Ferdinand Foch, *Principle of War*, 1913

Discipline, to which officer and private alike are subjected, was, in my opinion, the only basis on which an army could be effectively trained for war. Such a training could be acquired through long service. It is only what discipline makes second nature in a man that is lasting, and survives even the demoralizing impressions of the battle-field and the psychological changes wrought by a long campaign.

General Erich Ludendorff, *My War Memories, 1914–1918*, 1919

It is a mistaken idea that precision of movement and smartness of appearance, which for the popular mind are often the whole meaning of the word 'discipline', can only be obtained by a Prussian discipline, where the individuality of man is ground out until only a robot-like body is left. A century ago, Sir John Moore . . . introduced a new discipline based on intelligence and comradeship instead of on sullenness and fear. His work was justified by its fruits, and not only did the results show that this discipline of the thinking man instead of automata produces more efficient soldiers for war, but also that the former can rival the latter in order and smartness. Intelligent men whose minds are disciplined in the best sense can acquire these qualities as well as, and more quickly than, the barrack-square product. But the reverse order of progress from the Frederican discipline to initiative is not possible. (September 1925.)

Captain Sir Basil Liddell Hart, *Thoughts on War*, 1944

. . . We had no discipline in the sense in which it was restrictive, submergent of individuality, the Lowest Common Denominator of men. In peace-armies discipline meant the hunt, not of an average but of an absolute; the hundred per cent standard in which the ninety-nine were played down to the level of the weakest man on parade. The aim was to render the unit a unit, the man a type; in order that their effort might be calculable, and the collective output even in grain and bulk. The deeper the discipline, the lower was the individual excellence; also the more sure the performance.

By this substitution of a sure job for a possible masterpiece, military science made a deliberate sacrifice of capacity in order to reduce the uncertain element, the bionomic factor, in enlisted humanity.

Colonel T. E. Lawrence, *The Seven Pillars of Wisdom*, 1926

. . . for it had seemed to me that discipline, or at least formal discipline, was a virtue of peace: a character or stamp by which to mark off soldiers from complete men, and obliterate the humanity of the individual. It resolved itself easiest into the restrictive, the making men not do this or that: and so could be fostered by a rule severe enough to make them despair of disobedience. It was a process of the mass, an element of the impersonal crowd, inapplicable to one man, since it involved obedience, a duality of will. It was not to impress upon men that their will must actively second the officer's, for then the momentary pause for thought transmission, or digestion; for the nerves to resolve the relaying private will into active consequence. On the contrary, each regular Army sedulously rooted out this significant pause from its companies on parade. The drill-instructors tried to make obedience an instinct, a mental reflex, following as instantly on the command as though the motor power of the individual wills had been invested in the system.

This was well, so far as it increased quickness: but it made no provision for casualties, beyond the weak assumption that each subordinate had his will-motor not atrophied, but reserved in perfect order, ready at the instant to take over his late superior's office; the efficiency of direction passing smoothly down the great hierarchy till vested in the senior of the two surviving privates.

Colonel T. E. Lawrence, *The Seven Pillars of Wisdom*, 1926

Commanding and obeying are the characteristics of the army, and the one is as hard as the other. Both are simplified in proportion as orders are given with prudence and intelligence and obedience is rendered with perception and confidence. Human nature demands compulsion when many are to be united for one purpose. Thus discipline becomes an inseparable feature of the army, and its nature and degree are the true measure of the army's efficiency. The more voluntary the nature of the discipline the better, but only a discipline that has become habit and matter of course can survive the test in the hour of danger.

Colonel General Hans von Seekt, *Thoughts of a Soldier*, 1930

[Although soldiers] carry within themselves a thousand and one seeds of diversity . . . men, in their hearts, can no more do without being controlled than they can live without food, drink or sleep. [Discipline is thus] the basic constituent of all armies [but must be] shaped by the conditions and moral climate of the times.

Charles de Gaulle, *The Edge of the Sword,* 1932

It is the basic constituent of all armies. By virtue of discipline something resembling a contract comes into being between the leader and the subordinates. It is an understood thing that the latter owe obedience to the former, and that each single component of an organized body must, to the best of its ability, carry out the orders transmitted to him by a higher authority. In this way a fundamental attitude of good will is created which guarantees a minimum degree of cohesion. But it is not enough for a chief to bind his men into a whole through the the medium of impersonal obedience. It is on the inner selves that he must leave the imprint of his personality. If he is to have a genuine and effective hold on his men, he must know how to make their wills part and parcel of his own, and so to inspire them as something of their own choosing. He must increase and multiply the effects of mere discipline and implant in those under him a sort of moral suggestion which goes far beyond reasoning, and crystallize round his own person all their potentialities of faith, hope, and devotion.

Charles de Gaulle, *The Edge of the Sword,* 1932

The idea that real discipline is best instilled by fear of punishment is a delusion – especially in these days of open-order fighting. For such discipline kills the highest military qualities – initiative, and intelligent fulfilment of the superior's intentions. Moreover, the common notion that discipline is produced by drill is a case of putting the cart before the horse. A well-drilled battalion has often proved a bad one in the field. By contrast a *good* battalion is often good at drill – because if the spirit is right, it likes to do all things well. Here is the real sequence of causation. (June 1935)

Captain Sir Basil Liddell Hart *Thoughts on War,* 1944

Military discipline, in the narrowest sense of the term, is the enforcement of instant obedience to orders through threat of punishment, supplemented by the control reflexes established through drill. In the wider sense it is bound up with the soldierly spirit. So far as this is susceptible to analysis, its components appear to be pride of manhood, the pride of arms (of being an initiate in the martial cult), the confidence that comes from skill at arms, the sense of comradeship, the sense of duty, and the sense of loyalty – to comrades, commanders, corps, and country. (January 1936)

Captain Sir Basil Liddell Hart, *Thoughts on War,* 1944

In all armies, obedience of the subordinates to their superiors must be exacted . . . but the basis for soldier discipline must be the individual conscience. With soldiers, a discipline of coercion is ineffective, discipline must be self-imposed, because only when it is, is the soldier able to understand completely why he fights and how he must obey. This type of discipline becomes a tower of strength within the army, and it is the only type that can truly harmonize the relationship that exists between officers and soldiers.

Mao Tse-tung, *On Protracted War,* 1938

. . . You cannot be disciplined in great things and undisciplined in small things . . . Brave, undisciplined men have no chance against the discipline and valor of other men. Have you ever seen a few policemen handle a crowd?

General George S. Patton, Jr., May 1941, *The Patton Papers,* Vol. II, 1974

All human beings have an innate resistance to obedience. Discipline removes this resistance and, by constant repetition, makes obedience habitual and subconscious. Where would an undisciplined football team get? The players react subconsciously to the signals. They must, because the split second required for thought would give the enemy the jump.

Battle is much more exigent than football. No sane man is unafraid in battle, but discipline produces in him a form of vicarious courage which, with his manhood, makes for victory. Self-respect grows directly from discipline. The Army saying, 'Who ever saw a dirty soldier with a medal?' is largely true. Pride, in turn, stems from self-respect and from the knowledge that the soldier is an American. The sense of duty and obligation to his comrades and superiors comes from a knowledge of reciprocal obligation, and from the sharing of the same way of life. Self-confidence, the greatest military virtue, results from the demonstrated ability derived from the acquisition of all preceding qualities and from exercise in the use of weapons.

General George S. Patton, Jr., *War As I Knew It*, 1947

It is absurd to believe that soldiers who cannot be made to wear the proper uniform can be induced to move forward in battle. Officers who fail to perform their duty by correcting small violations and in enforcing proper conduct are incapable of leading.

General George S. Patton, Jr., April 1943, *The Patton Papers*, Vol. II, 1974

Discipline can only be obtained when all officers are so imbued with the sense of their awful obligation to their men and to their country that they cannot tolerate negligence. Officers who fail to correct errors or to praise excellence are valueless in peace and dangerous misfits in war.

General George S. Patton, Jr., *War As I Knew It*, 1947

One of the primary purposes of discipline is to produce alertness. A man who is so lethargic that he fails to salute will fall an easy victim to the enemy.

General George S. Patton, Jr., *War As I Knew It*, 1947

Soldiers can endure hardship. Most of their training is directed toward conditioning them for unusual privation and exertion. But no power on earth can reconcile them to what common sense says is unnecessary hardship which might have been avoided by greater intelligence in their superiors. The more intelligent the soldier, the more likely it is that he will see that as a sign of indiscipline up above and will answer it in the same way.

Brigadier General S. L. A. Marshall, *Men Against Fire*, 1947

Between these two things – discipline in itself and a personal faith in the military value of discipline – lies all the difference between military maturity and mediocrity. A salute from an unwillingly soldier is as meaningless as the moving of a leaf on a tree; it is a sign only that the subject has been caught by a gust of wind. But a salute from the man who takes pride in the gesture because he feels privileged to wear the uniform, having found the service good, is an act of the highest military virtue.

Brigadier General S. L. A. Marshall, *Men Against Fire*, 1947

It is said that our men are somewhat deficient in the usual forms of military discipline. But against that they have the virtues of responsibility and courage. Any loss due to lack of military discipline is more than made up for by the self-reliance, the initiative and the spirit of our men.

General Yisrael Galili, 1947, quoted in Allon, *The Making of Israel's Army*, 1961

Discipline destroys the spirit and working loyalty of the general force when it is pitched to the minority of malcontented, undutiful men within the organization, whether to punish or appease them.

Brigadier General S. L. A. Marshall, *The Armed Forces Officer*, 1950

He [the soldier] recognizes a right and reasonable discipline as such, even though it causes him personal inconvenience, because he has acquired a sense of military values. But if it is either unduly harsh or unnecessarily lax, he likewise knows it and wears it as a hairshirt, to the undoing of his morale. Though the man, like the group, can be hurt by being pushed beyond sensible limits, his spirit will suffer even more sorely if no real test is put upon his abilities and moral powers. The greater his intelligence, the stronger will be his resentment. That is a law of nature. The enlightened mind has always the greatest measure of self-discipline but it also has a higher sense of what constitutes justice, fair play and a reasonable requirement in the performance of duty. If denied these things, he will come to hold his chief, his job, and himself in contempt. The greater part of man's satisfactions comes of activity and only a very small remnant comes of passive enjoyment. Forgetting this rather obvious fact in human nature, social reformers aim at securing more leisure, rather than at making work itself more satisfactory. But it need not be forgotten in the military service.

Brigadier General S. L. A. Marshall, *The Armed Forces Officer*, 1950

That's the crux of the matter. Discipline is something that is enforced, either by fear or by understanding. Even in an Army, it is not merely a question of giving orders; there is more to a soldier's discipline than blind obedience. To take men into your confidence is not a new technique invented in the last war. Good generals were doing that long before you and I got into khaki to save the world.

Field Marshal Viscount Slim, *Courage* and Other Broadcasts, 1957

It is only discipline that enables men to live in a community and yet retain individual liberty . . . You can have discipline without liberty but cannot have liberty without discipline . . . The self-discipline of the strong is the safeguard of the weak.

Field Marshal Viscount Slim, *Courage* and Other Broadcasts, 1957

The more modern war becomes, the more essential appear the basic qualities that from the beginning of history have distinguished armies from mobs. The first of these is discipline. We very soon learned in Burma that strict discipline in battle and in bivouac was vital, not only for success, but for survival. Nothing is easier in jungle or dispersed fighting than for a man to shirk. If he has no stomach for advancing, all he has to do is flop into the undergrowth; in retreat, he can slink out of the rear guard, join up later, and swear he was the last to leave. A patrol leader can take his men a mile into the jungle, hide there, and return with any report he fancies. Only discipline – not punishment – can stop that sort of thing; the real discipline that a man holds to because it is a refusal to betray his comrades. The discipline that makes a sentry, whose whole body is tortured for sleep, rest his chin on the point of his bayonet because he knows, if he nods, he risks the lives of the men sleeping behind him. It is only discipline, too, that can enforce the precautions against disease, irksome as they are, without which an army would shrivel away. At some stage in all wars armies have let their discipline sag, but they have never won victory until they made it taut again; nor will they. We found it a great mistake to belittle the importance of smartness in turnout, alertness of carriage, cleanliness of person, saluting, or precision of movement, and to dismiss them as naive, unintelligent parade-ground stuff. I do not believe that troops can have unshakeable battle discipline without showing those outward and formal signs, which mark the pride men take in themselves and their units and the mutual confidence and respect that exist between them and their officers. It was our experience in a tough school that the best fighting units, in the long run, were not necessarily those with the most advertised reputations, but those who when they came out of battle at once resumed a more formal discipline and appearance.

Field Marshal Viscount Slim, *Courage* and Other Broadcasts, 1957

When we speak of conscious discipline, it means that it is built up on the basis of political consciousness of the officers and men, and the most important method of maintaining discipline is education and persuasion, thus making the army men of their own accord, respect and remind each other to observe discipline. When we speak of strict discipline, it means that everyone in the army, regardless of rank or office must observe discipline and no infringements are allowed.

General Vo Nguyen Giap, *People's War, People's Army*, 1961

There was no tactical organization other than the tribal group, and few preparations were made for a campaign. There was a total lack of planning; battles were headlong assaults in rough phalangial order in which the warriors rapidly exhausted themselves and became disorganized; courage shattered itself upon the rocks of discipline.

Major General J. F. C. Fuller, on the military organization of the Gauls compared to the Romans, *Julius Caesar: Man, Soldier, Tyrant*, 1965

Within our system, that discipline is nearest perfect which assures to the individual the greatest freedom of thought and action while at all times promoting his feeling of responsibility toward the group. *These twin ends are convergent and interdependent for the exact converse of the reason that it is impossible for any man to feel happy and successful if he is in the middle of a failing institution.*

Brigadier General S. L. A. Marshall, *The Armed Forces Officer*, 1950

DISLOCATION

Remember also that in battle, and battles are tests of military structure, the object of each side is not to kill for the sake of killing, but for the sake of disorganizing, for military strength does not reside in individuals, but in the co-operation of individuals and masses. Co-operation depends on control; and the endurance of force depends on maintenance. Every plan must have a threefold base; it must permit of the existing structure of an army remaining unaltered, or as unaltered as possible; it must permit of the existing system of control working without friction; and it must permit of the administrative units carrying out their duties without let or hinderance.

Major General J. F. C. Fuller, *The Foundations of the Science of War*, 1926

How is the strategic dislocation produced? In the physical, or 'logistical', sphere it is the result of a move which (a) upsets the enemy's dispositions and, by compelling a sudden 'change of front', dislocates the distribution and organization of his forces; (b) separates his forces; (c) endangers his supplies; (d) menances the route or routes by which he could retreat in case of need and reestablish himself in his base or homeland.

Captain Sir Basil Liddell Hart, *Strategy*, 1954

In the psychological sphere, dislocation is the result of the impression on the commander's mind of the physical effects which we have listed. The impression is strongly accentuated if his realization of his being at a disadvantage is *sudden*, and if he feels that he is unable to counter the enemy's move. *Psychological dislocation fundamentally springs from this sense of being trapped.*

This is the reason why it has most frequently followed a physical move on to the enemy's rear. An army, like a man, cannot properly defend its back from a blow without turning round to use its arms in a new direction. 'Turning' temporarily unbalances an army as it does a man, and with the former the period of instability is inevitably much longer. In consequence, the brain is much more sensitive to any menace to its back.

Captain Sir Basil Liddell Hart, *Strategy*, 1954

DISTRACTION

. . . The purpose of . . . 'distraction' is to *deprive the enemy of his freedom of action*, and it should operate in both the physical and psychological spheres. In the physical, it should cause a distention of his forces or their diversion to unprofitable ends, so that they are too widely distributed, and too committed elsewhere, to have the power of interfering with one's own decisively intended move. In the psychological sphere, the same effect is sought by playing upon the fears of, and by deceiving, the opposing command. 'Stonewall' Jackson aptly expressed this in his strategical motto – 'Mystify, mislead, and surprise'. For to mystify and to mislead constitutes 'distraction', while surprise is the essential cause of 'dislocation'. It is through the 'distraction' of the commander's mind that the distraction of his forces follows. The loss of freedom of action is the sequel to the loss of his freedom of conception.

Captain Sir Basil Liddell Hart, *Strategy*, 1954

DOCTRINE

A *doctrine* of war consists first in a common way of objectively approaching the subject; second, in a common way of handling it, by adapting without reserve the means to the goal aimed at, to the object.

Marshal of France Ferdinand Foch, *Precepts and Judgments*, 1919

A unified military doctrine is that concept accepted in the army which establishes the character of the construction of the armed forces of the country, the methods of combat training of the troops and their guidance on the basis of views held by those ruling in the state concerning the character of the problems lying before them and the means of their solution, with such means springing from the class character of the state and determined by the level of development of the productive forces of the state.

Mikhail V. Frunze (1885–1925)

The central idea of an army is known as its doctrine, which to be sound must be based on the principles of war, and which to be effective must be elastic enough to admit of mutation in accordance with change in circumstances. In its ultimate relationship to the human understanding this central idea or doctrine is nothing else than common sense – that is, action adapted to circumstances.

Major General J. F. C. Fuller, *The Foundations of the Science of War*, 1926

Military doctrine refers to the point of view from which military history is understood and its experience and lessons understood. Doctrine is the daughter of history . . . Doctrine is needed so that in the realm of military thought an army does not represent human dust, but a cohesive whole . . . should be predatory and stern, ruthless toward defeat and the defeated.

General A. A. Svechin (1878–1938)

The crumbling of the whole system of doctrines and organization, to which our leaders had attached themselves, deprived them of their motive force. A sort of moral inhibition made them suddenly doubtful of everything, and especially of themselves. From then on the centrifugal forces were to show themselves rapidly.

General Charles de Gaulle, *The Call to Honour*, 1955

An inner battle took place in me . . . At the beginning of the operation when we . . . had not yet felt to the end what destruction lengthy combat in this area might entail, I gave the order for encirclement.

[Then] the thought matured in me that we were obliged to seize the Silesian industrial area . . . whole, that is, that we had to leave the Hitlerites out of this trap . . . On the other hand, it is precisely encirclement which is the highest form of operational art . . . Then how could I . . . renounce it? It was not easy for me, a professional military person . . . to go against established doctrine . . . This was a difficult pyschological problem.

Marshal of the Soviet Union Ivan S. Konev, 1970, quoted in Leites, *The Soviet Style of War*, 1982

DO OR DIE

Let no one prefer hunger, that unmanageable and distressing evil, to the walls and bodies of the enemy, which they will yield to bravery, to the sword, to despair. Our situation at the moment is so pressing that nothing can be postponed till to-morrow, but this very day must decide for us either a complete victory or an honourable death.

Mark Anthony, address to his troops at the First Battle of Phillipi in 42 BC, quoted in Appian

When men find they must inevitably perish, they willingly resolve to die with their comrades and with their arms in their hands.

Flavius Vegetius Renatus, *Military Institutions of the Romans*, c. AD 378

We must neither suffer ourselves to be confounded nor lose courage since it is in the midst of afflictions that valiant men become more resolute and strengthen themselves with new hopes . . . Let us apply ourselves to seek out the means of taking, with usury, our revenge upon the enemy, so that, with the favour of Heaven, this place may remain in the hands of the Spaniards only as many days as our ancestors left it years in possession of the English.

Henry IV (Henry of Navarre), King of France, (1553–1610), address to the French Army at the siege of Calais

Some commanders have forced their men to fight by depriving them of all means of saving themselves except victory; this is certainly the best method of making them fight desperately. This resolution is commonly heightened either by the confidence they put in themselves, their arms, armour, discipline, good order, and lately-won victories, or by the esteem they have for their general . . . or it is a result of the love their country, which is natural to all men. There are various ways of forcing men to fight, but that is the strongest and most operative; it leaves men no other alternative but to conquer or die.

Niccolo Machiavelli, *The Art of War*, 1521

There is one certain means by which I can be sure never to see my country's ruin: I will die in the last ditch.

William III (William of Orange), King of England (1650–1702)

Saburov returned from the meeting at Protsenko's to his battalion with a feeling that it was urgent to do something big in the next few days, something he would remember all his life. What they had already done and what they had still to do did not seem heroic. The men defending Stalingrad had developed an unfailing and stubborn resistant strength which grew out of many different reasons: because the longer they fought the more impossible it became to retreat; because to retreat now meant to die uselessly while you were retreating; because the nearness of the enemy and the unceasing danger gave some of them an actual taste for danger and others a feeling that it inescapable; and finally because all of them, crowded on this little piece of ground, had grown to know each other; with all their vices and their virtues, better than anywhere else on the front. All these circumstances together gradually created that stubborn force which came to be called 'the men of Stalingrad.' The heroic strength of this phrase was understood throughout the country, all around them, much earlier than it was in Stalingrad itself.

Konstantin Simonov, *Days and Nights*, 1945

If we had not driven them into hell . . . hell would have swallowed us.

Field Marshal Prince Aleksandr V. Suvorov, 1787, of the Battle of Kinburn

It so often happens that, when men are convinced that they have to die, a desire to bear themselves well and to leave life's stage with dignity conquers all other sensations.

Sir Winston S. Churchill, *Savarola*, 1900

Throw the troops into a position from which there is no escape and even when faced with death they will not flee. For if prepared to die, what can they not achieve? Then officers and men together put forth their utmost efforts. In a desperate situation they fear nothing; when there is no way out they stand firm.

Sun Tzu, *The Art of War*, c. 500 BC, tr. Griffith

. . . there is a necessity for you to be brave, and, since all between victory and death is broken off from you by inevitable despair, either to conquer, or if fortune should waver, to meet death rather in battle than in flight. If this be well fixed and determined in the minds of you all, I will repeat, you have already conquered; no stronger incentive to victory has been given to man by the immortal gods.

Hannibal, address to his troops upon crossing the Alps into Italy, 218 BC

Go, therefore, to meet the foe with two objects before you, either victory or death. For men animated by such a spirit must always overcome their adversaries, since they go into battle ready to throw away their lives.

Scipio Africanus, address to the Roman Army before the Battle of Zama in 202 BC, quoted in Polybius, *The Rise of the Roman Empire*, c. 125 BC

Now the field of battle is a land of standing corpses; those determined to die will live; those who hope to escape with their lives will die.

Wu Ch'i (430–381 BC), commentary in Sun Tzu, *The Art of War*, c. 500 BC

Alexander and Victory!

The battle cry of Alexander the Great as quoted in Plutarch (*c.* AD 46 – after AD 119)

I want to make it clear to all of you that I shall never give the order 'abandon ship', the only way you can leave the ship is if she sinks beneath your feet.

Admiral Earl Mountbatten, 1949, order to his crew while commander of HMS *Kelly*.

It is essential to impress on all officers that determined leadership will be very vital in this battle, as in any battle. There have been far too many unwounded prisoners taken in this war. We must impress on our officers, N.C.O.s and men that when they are cut off or surrounded, and there appears to be no hope of survival, they must organize themselves into a defensive locality and hold out where they are. By so doing they will add enormously to the enemy's difficulties; they will greatly assist the development of our own operations; and they will save themselves from spending the rest of the war in a prison camp.

Nothing is ever hopeless so long as troops have stout hearts, and have weapons and ammunition.

Field Marshal Viscount Montgomery of Alamein, 6 October 1942, instructions for Operation 'Lightfoot', the Battle of El Alamein, quoted in de Guingand, *Operation Victory*, 1947

With our backs to the wall, and believing in the justice of our cause, each one of us must fight on to the end. The safety of our Homes and the Freedom of mankind alike depend on the conduct of each one of us at this critical moment.

Field Marshal Earl Haig, Order of the Day, 12 April 1918, at the height of the German *Friedenstürm* offensive that nearly broke the British Army

DOGMA

To dogmatize upon that which you have not practised is the prerogative of ignorance; it is like thinking that you can solve, by an equation of the second degree, a problem of transcendental geometry which would have daunted Legrange or Laplace.

Napoleon (1769–1821)

It will be better to offer certain considerations for reflection than to make sweeping dogmatic assertions.

Rear Admiral Alfred Thayer Mahan (1840–1914)

In itself, the danger of a doctrine is that it is apt to ossify into a dogma, and to be seized upon by mental emasculates who lack virility of judgment, and who are only too grateful to rest assured that their actions, however inept, find justification in a book, which, if they think at all, is, in their opinion, written in order to exonerate them from doing so. In the past many armies have been destroyed by internal discord, and some have been destroyed by the weapons of their antagonists, but the majority have perished through adhering to dogmas springing from their past successes – that is, self-destruction or suicide through inertia of mind.

Major General J. F. C. Fuller, *The Reformation of War*, 1923

Adherence to dogmas has destroyed more armies and lost more battles and lives than any other cause in war. No man of fixed opinions can make a good general; consequently, if this series of lectures, however hypothetical many of its contentions may seem, succeeds in unfixing dogmas, then certainly it will not have been written in vain.

Major General J. F. C. Fuller, *Lectures on F. S. R. III*, 1932

DRILL AND CEREMONIES

Drill is necessary to make a soldier steady and skillful, but it does not warrant exclusive attention. Among all the elements of war it even is the one that deserves the least.

Field Marshal Maurice Comte de Saxe, My *Reveries*, 1732

I had never looked at a copy of tactics from the time of my graduation. My standing in that branch of the studies had been near the foot of the class. In the Mexican war in the summer of 1846, I had been appointed regimental quartermaster and commissary and had not been at a battalion drill since.

General of the Army Ulysses S. Grant, *The Personal Memoirs of U.S. Grant*, 1885

Drill may be beautiful: but beauty is not perceptible when you are expecting a punishment every moment for not doing it well-enough. Dancing is beautiful: – because it's the same sort of thing, without the sergeant-major and the 'office'. Drill in the R.A.F. is always punitive: – it is always practice-drill, never exercise-drill or performance-drill. Airmen haven't the time to learn combined rhythm. If they did learn it, their (necessarily) individual work with screwdrivers and spanners would suffer. Rhythm takes months to acquire and years to lose.

Colonel T. E. Lawrence, 6 November 1928, letter to Robert Graves; Lawrence's perspective at this time was that of an enlisted man, having resigned his wartime commission and enlisted in the RAF; ed., Brown, *The Letters of T. E. Lawrence*, 1988

Combat experience has proven that ceremonies, such as formal guard mounts, formal retreat formations, and regular supervised reveille formations, are a great help, and, in some cases, essential, to prepare men and officers for battle, to give them that perfect discipline, that smartness of appearance, that alertness without which battles cannot be won.

General George S. Patton, Jr., *War As I Knew It*, 1947

The military manifestations of discipline are many and various. At one end of the scale may be placed the outward display, such as saluting and smartness of drill, the meaning and value of which are often misunderstood and misused both inside the Army and outside. Saluting should be in spirit the recognition of a comrade in arms, the respect of a junior for a senior – a gesture of brotherhood on both sides. Good drill should either be a ceremony for the uplifting of the spirit or a time-saver for some necessary purpose – never mere formalism or pedantry. No one who has participated in it or seen it well done should doubt the inspiration of ceremonial drill.

Field Marshal Earl Wavell, *Soldiers and Soldiering*, 1953

DUTY

You tell me to put my trust in birds,
flying off on their long wild wings? Never.
I would never give them a glance, a second thought,
whether they fly on the right towards the dawn and sunrise
or fly on the left toward the haze and coming dark!
No, no, put our trust in the will of mighty Zeus,
King of the deathless gods and men who die.
Bird signs!
Fight for your country – that is the best, the only omen!

Hector, rebuking Polydamus for reliance on seers in Homer, *The Iliad*, xii, c. 800 BC, tr. Fagles

---

So fight by the ships, all together. And that comrade
who meets his death and destiny, speared or stabbed,
Let him die! He dies fighting for fatherland –
no dishonour there!
He'll leave behind wife and sons unscathed,
His house and estates unharmed – once these Argives sail for home,
the fatherland they love.

Hector, rallying the Trojans to storm the Greek camp in Homer, *The Iliad*, xv, c. 800 BC, tr. Fagles

---

Our business in the field of fight
Is not to question, but to fight.

Homer, *The Iliad*, xx, c. 800 BC, tr. Pope

And therefore the general who in advancing does not seek personal fame, and in withdrawing is not concerned with avoiding punishment, but whose only purpose is to protect the people and promote the best interests of his sovereign, is the precious jewel of the state.

Sun Tzu, *The Art of War*, c. 500 BC, tr. Griffith

Go tell the Spartans, thou who passest by,
That here, obedient to their laws, we lie.

Simonides of Keos (c. 556 BC–c. 468 BC), of the Spartans who fell at Thermopylae

---

The sense of duty makes the victor.

Flavius Vegetius Renatus, *Military Institutions of the Romans*, c. AD 378

I thank all for the advice which you have given me. I know that my going out of the city might be of some benefit to me, inasmuch as all that you foresee might really happen. But it is impossible for me to go away! How could I leave the churches of our Lord, and His servants the clergy, and the throne, and my people in such a plight? What would the world say about me? I pray you, my friends, in future do not say to me anything else but, 'Nay, sire, do not leave us!' Never, never will I leave you! I am resolved to die here with you!

The Emperor Constantine XI, upon the advice to flee Constantinople before the Turkish siege of 1453, quoted in Majatovich, *Constantine, The Last Emperor of the Greeks*, 1892

Three things prompt men to a regular discharge of their Duty in time of Action: natural bravery, hope of reward, and fear of punishment. The first two are common to the untutor'd, and the Disciplin'd Soldiers; but the latter, most obviously distinguishes the one from the other. A Coward, when taught to believe, that if he breaks his Ranks, and abandons his Colours, he will be punished by Death by his own party, will take his chance against the Enemy.

General George Washington, letter to the President of Congress, 9 February 1776, *The Writings of George Washington*, Vol. 4, 1931–1944

Firstly you must always implicitly obey orders, without attempting to form any opinion of your own respecting their propriety. Secondly, you must consider every man your enemy who speaks ill of your king; and thirdly you must hate a Frenchman as you hate the devil.

Admiral Viscount Nelson, 1793, to a midshipman under his command in HMS *Agamemnon*

England expects that every man will do his duty.

Admiral Viscount Nelson, 21 October 1805, order to the fleet before the Battle of Trafalgar

---

Thank God, I have done my duty.

Admiral Viscount Nelson, 21 October 1805, dying of his wounds at Trafalgar

---

I am nimmukwallah, as we say in the east, that is, I have eaten of the King's salt and, therefore, conceive it to be my duty to service with unhesitating zeal and cheerfulness where and whenever the King and his Government may think proper to employ me.

The Duke of Wellington, 1805, upon accepting an obscure posting of brigade command in a military backwater after his victories in India

---

The brave man inattentive to his duty, is worth little more to his country, than the coward who deserts her in the hour of danger.

Major General Andrew Jackson, 8 January 1815, to troops who had abandoned their lines during the Battle of New Orleans

---

Duty is the sublimest word in our language. Do your duty in all things. You cannot do more. You should never do less.

General Robert E. Lee (1807–1870)

Duty is ours; consequences are God's.

Lieutenant General Thomas 'Stonewall' Jackson, 1862

Defeat is a common fate of a soldier and there is nothing to be ashamed of in it. The great point is whether we have performed our duty. I cannot but express admiration for the brave manner in which the officers and men of your vessels fought in the late battle for two days continuously. For you, especially, who fearlessly performed your great task until you were seriously wounded, I beg to express my sincerest respect and also my deepest regrets. I hope you will take great care of yourself and recover as soon as possible.

Admiral Marquis Togo Heihachiro, 3 June 1905, words to the wounded Admiral Rodzhesvensky at his bedside after the Battle of Tshushima

---

You and I were long side by side, and, like you, I was only unpopular with those soldiers who did not do their duty . . .

Lieutenant General D. H. Hill, *c.* 1870, letter to Jubal A. Early

---

The higher the soldier rises on the military ladder, the graver becomes his duty; not in itself, for it only changes form – and no man can do more than his duty – but because to his own duty and his own honour is added the responsibility for the duty and the honour of his subordinates. Responsibility grows to immensity; at one time the lives and the honour of hundreds, of thousands, are at stake; at another, the security of the state itself.

Colonel General Hans von Seekt, *Thoughts of a Soldier*, 1930

If I do my full duty, the rest will take care of itself.

General George S. Patton, Jr., 8 November 1942, diary entry before the North African landings, *The Patton Papers*, Vol. II, 1974

'Not in vain' may be the pride of those who survived and the epitaph of those who fell.

Sir Winston S. Churchill, 28 September 1944, speech in the House of Commons in tribute to the British 1st Airborne Division's sacrifice at the Battle of Arnhem

What a strange position I find myself in now having to make a decision diametrically opposed to my private opinion, with no choice but to push full speed in pursuance of that decision. Is that, too, fate? And what a bad start we've made.

Admiral Yamamoto Isoroku, November 1941, letter to a friend as the Imperial Japanese Navy strike force departed for its attack on Pearl Harbor, quoted in Toland, *The Rising Sun*, 1970

Any commander who fails to attain his objective, and who is not dead or wounded, has not done his full duty.

General George S. Patton, Jr., *War As I Knew It*, 1947

Our honour lies in doing our duty toward our people and our fatherland as well as in the consciousness of our mutual obligation to keep faith with one another, so we can depend on each other. We must remember that, even in our technological age, it is a man's fighting spirit that ultimately decides between victory and defeat.

Major General Hasso von Manteuffel, quoted in von Mellenthin, *German Generals of World War II*, 1977

. . . if it were accepted in principle that when a commander realized that the military situation was hopeless, it was his duty to advocate the conclusion of peace, even if that involved unconditional surrender, there would always be a danger that the struggle might be given up prematurely. As history shows, in war even a seemingly all but hopeless situation can sometimes be radically altered by unexpected political developments and similar occurrences.

Admiral Karl Dönitz, *Memoirs, Ten Years and Twenty Days*, 1959

I had missed the boat in the war we had been told would end all wars. A soldier's place was where the fighting went on. I hadn't fully learned the basic lesson of the military – that the proper place for a soldier is where he is ordered by his superiors.

General of the Army Dwight D. Eisenhower, *At Ease: Stories I Tell My Friends*, 1967

ECONOMY OF FORCE

Another such victory over the Romans, and we are undone.

Pyrrhus (319–272 BC), King of Epirus, quoted in Plutarch (c. AD 46 – after c. AD 119), *Lives of the Noble Grecians and Romans*

One must know how to accept a loss when advisable, how to sacrifice a province (he who tries to defend everything saves nothing) and meanwhile march with *all* one's forces against the other forces of the enemy, compel them to battle, spare no effort for their destruction, and turn them against the others.

Frederick the Great (1712–1786), quoted in Foch, *The Principles of War*, 1913

The art of war consists in having always more forces than the opponent, with an army weaker than his, at the point where one attacks, or where one is attacked.

Napoleon, quoted in Foch, *The Principles of War*, 1913

One of these simplified features, or aids to analysis, is always to make sure that all forces are involved – always to ensure that no part of the whole force is idle. If a segment of one's force is located where it is not sufficiently busy with the enemy, or if troops are on the march – that is, idle – while the enemy is fighting, then these forces are being managed uneconomically. In this sense they are being wasted, which is even worse than using them inappropriately. When the time for action comes, the first requirement should be that all parts must act: even the least appropriate task will occupy some of the enemy's forces and reduce his overall strength, while completely inactive troops are neutralized for the time being.

Major General Carl von Clausewitz, *On War*, iii, 1832, tr. Howard and Paret

The principle of economy of forces consist in throwing *all* one's forces at a given time on one point, in using there all one's troops, and, in order to render such a thing possible, having them always in communication among themselves instead of splitting them and of giving to each a fixed and unchangeable purpose.

Marshal of France Ferdinand Foch, *The Principles of War*, 1913

To me, an unnecessary action, or shot, or casualty, was not only waste but sin.

Colonel T. E. Lawrence (1888–1935)

Economy of force rightly means, not a mere husbanding of one's resources in man-power, but the employment of one's force, both weapons and men, in accordance with economic laws, so as to yield the highest possible dividend of success in proportion to the expenditure of strength. While husbanding the lives of the troops is a matter of common sense, full economy of force demands organized common sense – the habit of thinking scientifically, and weighing up tactical values and conditions. Economy of force is the supreme law of successful war.

Surprise is the psychological and Concentration the physical means to thoroughly *economic* 'offensive action' – blows that will attain the maximum result at the minimum cost. Similarly, *information* and a correct *distribution* of our forces are the economic means to Security. *Flexibility* and *co-operation* increase Mobility – just as by choosing and steering the best course and by oiling and tuning-up the mechanism we get the best speed out of a motor-car. (October 1924)

Captain Sir Basil Liddell Hart, *Thoughts on War*, 1944

In cold truth our small band, which at the most comprised some 300 Europeans and about 11,000 Askari, had occupied a very superior enemy force for the whole war. According to what English officers told me, 137 Generals had been in the field, and in all about 300,000 men had been employed against us. The enemy's losses in dead would not be put too high at 60,000 for an English Press notice stated that about 20,000 Europeans and Indians alone had died or been killed, and to that must be added the large number of black soldiers who fell. The enemy had left 140,000 horses and mules behind in the battle area. Yet in spite of the enormously superior numbers at the disposal of the enemy, our small force, the rifle strength of which was only about 1,400 at the time of the armistice, had remained in the field always ready for action and possessed of the highest determination.

Major General Paul von Lettow-Vorbeck, *East African Campaigns*, 1957 Note: The war (1914–1918) in East Africa is undoubtedly the greatest economy-of-force operation in modern military history – PGT.

The Principle of Economy of Force. Economy of force may be defined as the efficient use of all means: physical, moral and material, towards winning a war. Of all the principles of war it is the most difficult to apply, because of its close interdependence on the ever changing conditions of war. In order to economize the moral energy of his men, a commander must not only be in spirit one of them, but must ever have fingers on the pulse of the fighters. What they feel he must feel, and what they think he must think; but while they feel fear, exerience discomfort and think in terms of easy victory or disaster, though he must understand what all these mean to the men themselves, he must in no way be obsessed by them. To him economy of force first means planning a battle which his men *can* fight, and secondly, adjusting this plan according to the psychological changes without forgoing his objective. This does not only entail his possessing judgment, but also foresight and imagination. His plan must never crystallize, for the energy of the battle front is always fluid. He must realize that a fog, or shower of rain, a cold night or unexpected resistance may force him to adjust his plan, and in order to enable him to do so, the grand tactical economy of force rests with his reserves, which form the staying power of the battle and the fuel of all tactical movement.

Major General J. F. C. Fuller, *The Reformation of War*, 1923

ENDURANCE/PHYSICAL FITNESS/HEALTH

Now lead the way, wherever your fighting spirit bids you.
All of us right behind you, hearts intent on battle.
Nor do I think you'll find us short on courage,
long as our strength will last. Past his strength
no man can go, though he's set on mortal combat.

Paris to his brother Hector, in Homer, *The Iliad*, xiii, c. 800 BC, tr. Fagles

---

What can a soldier do who charges when out of breath?

Flavius Vegetius Renatus, *Military Institutions of the Romans*, c. AD 378

When you are at war, if your army has suffered a setback, it is imperative to examine the physical and mental health of the soldiers. If they are healthy, then inspire them to fight; if they are run down and low in spirits, then nurture their health for the time being, until they are again fit for service.

Liu Ji (1310–1375), *Lessons of War*

. . . for bad indeed is the condition of a general when he has a sickness among his men and an enemy to contend with at the same time. But nothing is more conduicive to keeping an army in good health and spirits than exercise; the ancients used to exercise their troops every day. Proper exercise, then, is surely of great importance for it preserves your health in camp and secures you victory in the field.

Niccolo Machiavelli, *The Art of War*, 1521

. . . without arrogance or the smallest deviation from the truth, it may be said that no history now extant, can furnish an instance of an Army's suffering such uncommon hardships as ours have done, and bearing them with the same patience and fortitude. To see men without clothes to clothe their nakedness, without blankets to lie on, without shoes, by which their marches might be traced by the blood from their feet, and almost as often without provisions as with; marching through frost and snow, and at Christmas taking up their winter quarters within a day's march of the enemy, without a house or hut to cover them till they could be built, and submitting to it without a murmur, is a mark of patience and obedience which in my opinion can scarce be paralleled.

General George Washington, 1777, at Valley Forge

Remember that victory depends on the legs; the hands are only the instruments of victory.

Field Marshal Prince Aleksandr V. Suvorov (1729–1800), quoted in Blease, *Suvorof*, 1920

The soldier's health must come before economy or any other consideration.

Napoleon, 1813, letter, ed. Herold, *The Mind of Napoleon*, 1955

War is the realm of physical exertion and suffering. These will destroy us unless we can make ourselves indifferent to them, and for this birth or training must provide us with a certain strength of body and soul. If we do possess those qualities, then even if we have nothing but common sense to guide them we shall be well equipped for war: it is exactly these qualities that primitive and semi-civilized peoples usually possess.

Major General Carl von Clausewitz, *On War*, i, 1832, tr., Howard and Paret

---

The Principle of Endurance. Springing directly from the principle of determination is the principle of endurance. The will of the commander-in-chief and the will of his men must endure, that is they must continue in the same state. It is the local conditions, mental and material, which continually weaken this state and in war often threaten to submerge it. To the commander endurance consists, therefore, in power of overcoming conditions – by foresight, judgment and skill. These qualities cannot be cultivated at a moment's notice, and the worst place to seek their cultivation is on the battlefield itself. The commander-in-chief must be, therefore, a mental athlete, his dumb-bells, clubs, and bars being the elements of war and his exercises the application of the principles of war to the conditions of innumerable problems.

Major General J. F. C. Fuller, *The Reformation of War*, 1923

Collectively, in an army, endurance is intimately connected with numbers, and, paradoxial as it may seem, the greater the size of an army the less is its psychological endurance. The reason for this is a simple one: one man has one mind; two men have three minds – each his own and a crowd mind shared between them; a million men have millions and millions and millions of minds. If a task which normally requires a million men be carried out by one man, this one man possesses psychologically an all but infinitely higher endurance than any single man out of the million. Man, I will again repeat, is an encumbrance on the battlefield, psychologically as well as physically; consequently, endurance should not be sought in numbers, for one Achilles is worth a hundred hoplites.

Major General J. F. C. Fuller, *The Reformation of War*, 1923

Good health and a robust consitution are invaluable in a general . . . In a sick body, the mind cannot possibly remain permanently fresh and clear. It is stunted by the selfish body from the great things to which it should be entirely devoted.

Field Marshal Colmar Baron von der Goltz, *The Nation in Arms* 1887

In this situation the real leaders of the Army stood forth in bold contrast to the ordinary clay. Men who had sustained a reputation for soldierly qualities, under less trying conditions proved too weak for the ordeal and became pessimistic calamity howlers. Their organizations were quickly infected with the same spirit and grew ineffective unless a more suitable commander was given charge. It was apparent that the combination of tired muscles, physical discomforts, and heavy casualties weakened the backbone of many. Officers of high rank who were not in perfect physical condition usually lost the will to conquer and took an exceedingly gloomy view of the situation.

General of the Army George C. Marshall, *Memoirs of My Service in the World War*, written 1923–1926, first published 1976

The great advantage of the gain in moral force through all forms of physical training is that it is an unconscious gain. Will power, determination, mental poise, and muscle control all march hand-in-hand with the general health and well-being of the man. Fatigue will beat men down as quickly as any other condition, for fatigue brings fear with it. There is no quicker way to lose a battle than to lose it on the road for lack of adequate hardening of the troops. Such a condition cannot be redeemed by the resolve of a commander who insists on driving troops an extra mile beyond their general level of endurance. Extremes of this sort make men rebellious and hateful of command, and thus strike at tactical efficiency from two directions at once. For when men resent a commander, they will not fight as willingly for him, and when their bodies are spent, their nerves are gone. In this state the soldier's every act is mechanical. He is reduced to that automatism of mind which destroys his physical response. His courage is killed. His intellect falls asleep.

Brigadier General S. L. A. Marshall, *Men Against Fire*, 1947

The civil comparison to war must be that of a game, a very rough and dirty game, for which a robust mind and body are essential.

Field Marshal Earl Wavell, quoted in Marshall, *The Armed Forces Officer*, 1950

Health in a general is, of course, most important, but it is a relative quality only. We would all of us, I imagine, sooner have Napoleon sick on our side than many of his opponents whole. A great spirit can rule in a frail body, as Wolfe and others have showed us. Marlborough during his great campaigns would have been ploughed by most modern medical boards.

Field Marshal Earl Wavell, *Soldiers and Soldiering*, 1953

In war, it sometimes happens that the strain to which the commanders are subjected in critical situations becomes too severe and necessitates a change. This is not only the case with commanders at the front but also, and perhaps to an even greater degree, the higher command which has to make grave decisions alone and far from the theatre of war, from which conflicting and incomplete reports are received. This requires sure intuition, realistic imagination, and great strength of will, and entails a strain on the physical and mental forces which is sometimes overwhelming. If the troops at this time were exhausted, especially in the most exposed sectors, it is obvious that both the higher and lower commanders' endurance had also been severely tested.

Field Marshal Carl Baron von Mannerheim, *The Memoirs of Marshal Mannerheim*, 1953

See also: TRAINING (Physical Training)

THE ENEMY

Therefore I say: 'Know the enemy and know yourself; in a hundred battles you will never be in peril.

When you are ignorant of the enemy but know yourself, your chances of winning or losing are equal.

If ignorant both of your enemy and of yourself, you are certain in every battle to be in peril.'

Sun Tzu, *The Art of War*, c. 500 BC, tr. Griffith

That general is wise who before entering into war carefully studies the enemy, and can guard against his strong points and take advantage of his weaknesses. For example, the enemy is superior in cavalry; he should destroy his forage. He is superior in number of troops; cut off their supplies. His army is composed of diverse peoples; corrupt them with gifts, favours, and promises. There is dissention among them; deal with their leaders. This people relies on the spear; lead them into difficult terrain. This people relies on the bow; line up in the open and force them into close, hand-to-hand fighting. Against Scythians or Huns launch your assault in February or March when their horses are in wretched condition after suffering through the winter . . . If they march or make camp without proper precautions, make unexpected raids on them by night and by day. If they are reckless and undisciplined in combat and not inured to hardship, make believe you are going to attack, but delay and drag things out until their ardour cools, and when they begin to hesitate, then make your attack on them. The foe is superior in infantry; entice him into the open, not too close, but from a safe distance hit him with javelins.

The Emperor Maurice, *The Strategikon*, c. AD 600

He who does not carefully compare his own forces with those of the enemy will come to a disastrous end.

Things which are unexpected or sudden frighten the enemy, but they pay little attention to things to which they are accustomed.

It is better to avoid a tricky opponent than one who never lets up. The latter makes no secret of what he is doing, whereas it is difficult to find out what the other is up to.

The Emperor Maurice, *The Strategikon*, c. AD 600

A general in all of his projects should not think so much about what he wishes to do as what his enemy will do; that he should never underestimate this enemy, but he should put himself in his place to appreciate difficulties and hindrances the enemy could not interpose; that his plans will be deranged at the slightest event if he has not foreseen everything and if he has not devised means with which to surmount the obstacles.

Frederick the Great, *Instructions to His Generals*, 1747

Let no one imagine that it is sufficient merely to move an army about, to make the enemy regulate himself according to your movements. A general who has too presumptious confidence in his skill runs the risk of being grossly duped. War is not an affair of chance.

Frederick the Great, *Instructions to His Generals*, 1747

What design would I be forming if I were the enemy?

Frederick the Great, *General Principles of War*, 1748

The one who is to draw up a plan of operations must possess a minute knowledge of the power of his adversary and of the help the latter may expect from his allies. He must compare the forces of the enemy with his own numbers and those of his allies so that he can judge which kind of war to lead or undertake.

Frederick the Great, 1748, letter

Never despise your enemy, whoever he is. Try to find out about his weapons and means, how he uses them and fights. Research into his strengths and weaknesses.

Field Marshal Prince Aleksandr V. Suvorov, 1789 or 1790, letter, quoted in Longworth, *The Art of Victory*, 1966

Fight the enemy with the weapons he lacks.

Field Marshal Prince Aleksandr V. Suvorov, quoted in Ossipov, *Suvorov*, 1945

In war one sees his own troubles and not those of the enemy.

Napoleon, 30 April 1809, to Eugene, *Correspondance*, No. 15144, Vol. XVIII, 1858–1970

The practical measures that we take are always based on the assumption that our enemies are not unintelligent.

Archidamus II, King of Sparta, quoted in *The History of the Peloponnesian War*, c. 404 BC

For every one must confess that there is no great proof of the abilities of a general than to investigate, with the utmost care, into the character and natural abilities of his opponent.

Polybius, *The Histories*, c. 125 BC

That march would have been a rash thing in front on Conde's headquarters, but . . . I was familiar enough with the Spanish service to know that before the Archduke would hear of it and notify the Prince de Conde and hold his council of war, I should have returned to my camp.

General Marshal Vicomte de Turenne (1611–1675), quoted in Weygand, *Turenne, Marshal of France*, 1930

There is still another factor that can bring military action to a standstill: imperfect knowledge of the situation. The only situation a commander can know fully is his own; his opponent's he can know only from unreliable intelligence. His evaluation, therefore, may be mistaken and can lead him to suppose that the initiative lies with the enemy when in fact it remains with him. Of course such faulty appreciation is as likely to lead to ill-timed action as to ill-timed inaction, and is no more conducive to slowing down operations than it is to speeding them up. Nevertheless, it must rank among the natural causes which, *without entailing inconsistency, can bring military activity to a halt.* Men are always more inclined to pitch their estimate of the enemy's strength too high than too low, such is human nature. Bearing this in mind, one must admit that partial ignorance of the situation is, generally speaking, a major factor in delaying the progress of military action . . .

Major General Carl von Clausewitz, *On War*, i, 1832, tr. Howard and Paret

MAXIM 8. A general-in-chief should ask himself frequently in the day, What should I do if the enemy's army appeared now in my front, or on my right, or on my left? If he have any difficulty in answering these questions he is ill posted, and should seek to remedy it.

Napoleon, *The Military Maxims of Napoleon*, 1831, tr. D'Aguilar

---

MAXIM 16. It is an approved maxim of war, never to do what the enemy wishes you to do, for this reason alone, that he desires it. A field of battle, therefore, which he has previously studied and reconnoitred, should be avoided, and double care should be taken where he has had time to fortify or entrench. One consequence deducible from this principle is, never to attack a position in front which you can gain by turning.

Napoleon, *The Military Maxims of Napoleon*, 1831, tr. D'Aguilar

---

The natural disposition of most people is to clothe a commander of a large army whom they do not know with almost superhuman abilities. A large part of the National army, for instance, and most of the press of the country, clothed General Lee with just such qualities, but I had known him personally, and knew that he was mortal; and it was just as well that I felt this.

General of the Army Ulysses S. Grant, *The Memoirs of U.S. Grant*, 1885

While dispositions taken in peace times can be weighted at length, and infallibly lead to the result desired, such is not the case with the use of forces in war, with operations. In war, hostilities are begun, our will soon encounters the independent will of the enemy. Our dispositions strike against the freely-made dispositions of the enemy.

Field Marshal Helmuth Graf von Moltke, quoted in Foch, *Principles of War*, 1913

If we try to give an exact answer to the question who deserves the most credit for the victory at Tannenberg, we must also briefly consider the conduct of the enemy, without whose blunders the success would not have been possible.

Major General Max Hoffman, *War Diaries and Other Papers*, Vol. II, 1929

. . . I have always said what I here expressly repeat in print, that the German people exhibited such astounding energy, doggedness, sturdy patriotism, courage, endurance, discipline, and readiness to die for their country, that, as a soldier, I cannot but salute them. They fought with lion-like bravery and an amazing tenacity against the whole world. The German soldier – and this means the German people – deserves unstinted respect.

General Aleksei A. Brusilov, *A Soldier's Notebook*, 1931

A dead enemy always smells sweet.

General Kemal Mustafa Ataturk, 1922, at the conquest of Smyrna (variously attributed to Roman emperors)

The more energetic and resourceful the enemy, the more difficult it will be to predict the further development of events. Only by ever bolder and faster action can one be confident of putting the enemy to rout on the main thrust line and hence of going on to further offensive action at high speed. Only in this way can one trump the enemy's cards and remain absolute master of the situation.

Marshal of the Soviet Union Mikail N. Tukhachevskiy, 1924, quoted in Simpkin, *Deep Battle*, 1987

I hope you have kept the enemy always in the picture. War-books so often leave them out.

Colonel T. E. Lawrence, 9 February 1928, letter to Colonel A. P. Wavell

---

I was in a furious rage. The Turks should never, by the rules of sane generalship, have ventured back to Tafileh at all. It was simple greed, a dog-in-the-manger attitude unworthy of a serious enemy, just the sort of hopeless thing a Turk would do. How could they expect a proper war when they gave us no chance to honour them? Our morale was continually being ruined by their follies, for neither could our men respect their courage, nor our officers respect their brains. Also, it was an icy morning, and I had been up all night and was Teutonic enough to decide that they should pay for my changed mind and plan.

Colonel T. E. Lawrence, *The Seven Pillars of Wisdom*, 1926

An important difference between a military operation and a surgical operation is that the patient is not tied down. But it is a common fault of generalship to assume that he is. (May 1934)

Captain Sir Basil Liddell Hart, *Thoughts on War*, 1944

I have today written General Bradley that if he can give me 2,000 tons of ammunition . . . and then guarantee a like amount daily, I can initiate an attack any time. I have learned from you to fight on a shoestring, and I realize further that the enemy is fighting on a shoestring which is even more frayed than ours.

General George S. Patton, Jr., 19 October 1944, letter to Eisenhower, *The Patton Papers*, Vol. II, 1974

However absorbed a commander may be in the elaboration of his own thoughts, it is sometimes necessary to take the enemy into account.

Sir Winston S. Churchill (1874–1965)

Remember that these enemies, who we shall have the honor to destroy, are good soldiers and stark fighters. To beat such men, you must not despise their ability, but you must be confident in your own superiority.

General George S. Patton Jr., December 1941, *Patton Papers*, II, 1974

When speaking to a junior about the enemy confronting him, always understate their strength. You do this because the person in contact with the enemy invariably overestimates their strength to himself, so, if you understate it, you probably hit the approximate fact, and also enhance your junior's self-confidence.

General George S. Patton, Jr., *War As I Knew It*, 1947

I want to emphasize the importance of focusing on the enemy when planning and conducting combat operations. First, you must know your enemy. Second, you must develop your plan keeping the enemy foremost in mind. Third, you must wargame your plan to enhance your ability to develop or adjust the plan once enemy contact is made.

General H. Norman Schwarzkopf, 'Food for Thought', *How They Fight*, 1988

A battle plan is good only until enemy contact is made. From then on, your ability to execute the plan depends on what the enemy does. If your assessment of the enemy is good, then you probably have anticipated his actions or reactions and will be able to quickly develop or adjust your plan accordingly. If your assessment is bad, due to poor understanding of how the enemy fights or poor battlefield intelligence, then you will probably get surprised, react slowly, lose the initiative, and cause your unit to be defeated.

General H. Norman Schwarzkopf, 'Food for Thought', *How They Fight*, 1988

ENERGY

Generals wage war through the armed forces, the armed forces fight by energy. Energy prevails when it is drummed up. If you can energize your troops, don't do it too frequently, otherwise their energy will easily wane. Don't do it at too great a distance either, otherwise their energy will be easily exhausted. You should drum up the energy of your soldiers when enemies are within a calculated critical distance, having your troops fight at close range. When enemies wane and you prevail, victory over them is assured.

The rule is 'Fight when full of energy, flee when drained of energy.'

Liu Ji (AD 1310–1375), *Lessons of War*

If you wage war, do it energetically and with severity. This is the only way to make it shorter, and consequently less inhuman.

Napoleon, 1799, letter to General Hedouville, ed. Herold, *The Mind of Napoleon*, 1955

If we read history with an open mind, we cannot fail to conclude that, among all the military virtues, *the energetic conduct of war* has always contributed most to glory and success.

Major General Carl von Clausewitz, *On War*, iii, 1832, tr. Howard and Paret

---

The true speed of war is not headlong precipitance, but the unremitting energy which wastes no time.

Rear Admiral Alfred Thayer Mahan (1840–1914)

The organization of supplies, the command of men, anything in any way constructive requires more than intellect; it requires energy and drive and an unrelenting will to serve the cause, regardless of one's personal interests. Academic soldiers mostly look on war as a pure intellectual problem and demand energy and drive only from those whom they somewhat contemptuously refer to as 'Troupier', but from themselves, never. They rest content with their professional qualifications, which were attested for them by others of the same ilk, and regard themselves as the source of all good and the 'Troupier' as the source of all evil.

Field Marshal Erwin Rommel, *The Rommel Papers*, 1953

Mental conception must be followed by immediate execution. This is a matter of energy and initiative. What the soldier needs is a combination of realist intellect and energy. Whatever is attempted must be carried through. The young officer must understand from the outset of his training that just as much energy is required of him as mental ability. The sensational victory is, more often than not, largely a victory for the energy of the officers.

Field Marshal Erwin Rommel, *The Rommel Papers*, 1953

The conservation of men's powers, not the spending thereof, is the object of main concern to the truly qualified commander.

Brigadier General S. L. A. Marshall, *The Armed Forces Officer*, 1950

ENGINEERS

Good engineers are so scarce, that one must bear with their humours and forgive them because we cannot be without them.

Lord Galway, 1704, report from Spain to the Board of Ordnance, quoted in Chandler, *The Art of War in the Age of Marlborough*, 1976

Their science demands a great deal of courage and spirit, a solid genius, perpetual study and consumate experience in all the arts of war.

Marshal of France Sébastien de Vauban, *A Manual of Siegecraft*, 1740

MAXIM 85. A general of engineers who must conceive, propose and direct all the fortifications of an army, needs good judgment and a practical mind above all.

Napoleon, *The Military Maxims of Napoleon*, 1827, ed., Burnod

---

Clearly, we would need great skill and even greater courage. But our first need was for men and equipment. It will be seen why the creation of engineer units was a top priority. In just over two years we succeeded in creating and training almost 40 engineer battalions, some of them highly specialized. It was our biggest coup and the foundation of our success. In his memoirs, the Israeli Chief of Staff, General David Elazaar, records that during a discussion in the Israeli High Command of the possibility of Egyptian forces attempting to cross the Canal, General Dayan, the Israeli Defence Minister, said: 'To cross the Canal the Egyptians would need the support of both the American and the Soviet engineer corps.' I do not hold Dayan's dismissive remark against him. I accept the tribute. I know how hard we worked.

General Saad El Shazly, *The Crossing of the Suez*, 1980

No Corps was more constantly in demand, so much master of so many tasks.

General Sir David Fraser, Royal Engineers, *And We Shall Shock Them*, 1983

'Essayons' (Let's try!)

The motto of the US Army Corps of Engineers

ENTHUSIASM/EXCITEMENT

I do not like the enthusiastic man.

Cornelius Scipio 'The Younger' Africanus Aemilianus (185–129 BC), quoted in Cicero, *De Oratore*

. . . in all men there is an innate excitability and drive which is kindled by the heat of the fight, and it is the function of the general not to quench but to heighten the excitement. There is sound sense in the ancient practice of sounding trumpet blasts on all sides and raising a battle cry from all throats; these things, they thought, serve both to terrify the enemy and to heighten the ardour of one's own men.

Julius Caesar, *The Gallic War*, c. 51 BC, tr. Hadas

Military virtues should not be confused with simple bravery, and still less with enthusiasm for a cause. Bravery is obviously a necessary component. But just as bravery, which is part of the natural make-up of a man's character, can be developed in a soldier – a member of an organization – it must develop differently in him than in other men. In the soldier the natural tendency for unbridled action and outbursts of violence must be surbordinated to demands of a higher kind: obedience, order, rule, and method. An army's efficiency gains life and spirit from enthusiasm for the cause for which it fights, but such enthusiasm is not indispensible.

Major General Carl von Clausewitz, *On War*, ii, 1832, tr. Howard and Paret

---

I already knew that even in the ordinary condition of mind enthusiasm is a potent element with soldiers, but what I saw that day convinced me that if it can be excited from a state of despondency its power is almost irresistible. I said nothing except to remark, as I rode among those on the road: 'If I had been with you this morning this disaster would not have happened. We must face the other way; we will go back and recover our camp.'

General of the Army Philip H. Sheridan, of how his army turned itself around in the midst of the rout of the Battle of Winchester (19 September 1864) at his appearance on the field, *Personal Memoirs of P. H. Sheridan*, 1888

ENVELOPMENT

It is an invariable axiom of war to secure your own flanks and rear and endeavour to turn those of your enemy.

Frederick the Great, *Instructions to His Generals*, 1747

A general should show boldness, strike a decided blow, and manoeuvre upon the flank of his enemy. The victory is in his hands.

Napoleon, *Maxims of War*, 1831

. . . the greater conditions of warfare have remained unchanged. The battle of extermination may be fought today according to the same plan as elaborated by Hannibal in long forgotten times. The hostile front is not the aim of the principle attack. It is not against that point that the troops should be massed and the reserves disposed; the essential thing is to crush the flanks.

Field Marshal Alfred Graf von Schlieffen, *Cannae*, 1913

It may be seen from all the battles, won or lost by Frederick the Great, that his aim was attack from the very beginning a flank or even the rear of the enemy, to push him, if possible against an insurmountable obstacle and then to annihilate him by enveloping one or both of his flanks.

Field Marshal Alfred Graf von Schlieffen, *Cannae*, 1913

Envelopment. An enemy may be enveloped and so placed at a severe disadvantage. Envelopment, whether accomplished by converging or overlapping, presupposes a flank, a flank which may be tactically rolled up, or, if turned, will expose the command and lines of communication behind it. The attack by envelopment is a very common action in war, which more often than not has led to victory.

Major General J. F. C. Fuller, *The Reformation of War*, 1923

If I had worried about flanks, I could never have fought the war.

General George S. Patton, Jr., *War As I Knew It*, 1947

The deep envelopment based on surprise, which severs the enemy's supply lines, is and always has been the most decisive maneuver of war. A short envelopment which fails to envelop and leaves the enemy's supply system intact, merely divides your own forces and can lead to heavy loss and even jeopardy.

General of the Army Douglas MacArthur, 23 August 1950, conference before Inchon

EQUILIBRIUM

The fate of a battle is a question of a single moment, a single thought . . . the decisive moment arrives, the moral spark is kindled, and the smallest reserve force settles the argument.

Napoleon, quoted in Las Cases, *Mémorial de Sainte Hélène*, 1923

There is a moment in engagements when the least manoeuvre is decisive and gives the victory; it is the one drop of water which makes the vessel run over.

Napoleon, 'Précis des guerres de J. César', *Correspondance*, Vol. XXXII, 1858–1870

All in all, loss of moral equilibrium must not be underestimated merely because it has no absolute value and does not always show up in the final balance. It can attain such massive proportions that it overpowers everything by its irresistible force. For this reason it may in itself become a main objective of the action . . .

Major General Carl von Clausewitz, *On War*, iv, 1832, tr. Howard and Paret

---

War is in itself only a matter of maintaining harmonious proportion between the spiritual and material elements. Fundamentally that is so. If no such proportion is attained, however excellent an army may be, it can do nothing against its adversaries.

Marshal of France Ferdinand Foch, quoted in Liddell Hart, *Foch, the Man of Orleans*, 1931

ESPRIT DE CORPS

All that can be done with the soldier is to give him *esprit de corps* – i.e., a higher opinion of his own regiment than all the other troops in the country.

Frederick the Great, *Military Testament*, 1768

Personal bravery of a single individual does not decide on the day of battle, but the bravery of the unit, and the latter rests on the good opinion and the confidence that each individual places in the unit to which he belongs. The exterior splendour, the regularity of movements, the adroitness, and at the same time the firmness of the mass – all this gives the individual soldier the safe and calming conviction that nothing can withstand his particular regiment or battalion.

Field Marshal François Comte de Guibert (1744–1790), quoted in Keegan, *Soldiers*, 1986

An army that maintains its cohesion under the most murderous fire; that cannot be shaken by imaginary fears and resists well-founded ones with all its might; that, proud of its victories, will not lose the strength to obey orders and its respect and trust for its officers even in defeat; whose physical power, like the muscles of an athlete, has been steeled by training, privation and effort; a force that regards such efforts as a means to victory rather than a curse on its cause; that is mindful of all these duties and qualities by virtue of the single powerful idea of the honour of its arms – such an army is imbued with the true military spirit.

Major General Carl von Clausewitz, *On War*, iii, 1832, ed. Howard and Paret

---

By inuring armies to labour and fatigue, by keeping them from stagnation in garrison in times of peace, by inculcating their superiority over their enemies (without despising the latter too much), by inspiring a love for great exploits – in a word, by exciting their enthusiasm by every means in harmony with their tone of mind, by honouring courage, punishing weakness, and disgracing cowardice – we may except to maintain a high military spirit.

Lieutenant General Antoine-Henri Baron de Jomini, *Summary of the Art of War*, 1838

. . . (*esprit de Corps*) which does not forget itself in the heat of action and which alone makes true combatants. Then we have an army; and it is no longer difficult to explain how men carried away by passion, even men who know how to die without flinching, without turning pale, really strong in the presence of death, but without discipline, without solid organization, are vanquished by others individually less valiant, but firmly, jointly, and severally knit into a fighting unit.

Colonel Ardnant du Picq, *Battle Studies*, 1880

My success had been so uninterrupted that the men thought that victory was chained to my standard. Men who go into a fight under the influence of such feelings are next to invincible, and generally are the victors before it begins.

Colonel John S. Mosby, *Mosby's War Reminiscences*, 1887

Consider your corps as your family; your commander as your father; your comrade as your brother; your inferior as a young relative. Then all will be happy and friendly and easy. Don't think of yourself, think of your comrades; they will think of you. Perish yourself, but save your comrades.

General Mikhail I. Dragomirov, *Notes for Soldiers*, 1890 (Note: The last sentence is a quote from Suvorov – PGT)

I have alluded before to the *esprit de corps*, founded as it was upon the sentiment of saving life – sentiment to which appeal has never failed. Other factors went to strengthen it. It was braced by a high standard of results demanded, by determination to make good in spite of partial first success. But the strongest element in it was the faith in our weapon – the machine necessary to supplement the other machines of war, in order to break the stalemate produced by the great German weapon, the machine-gun – our mobile offensive answer to the immobile defensive man-killer.

Major General Hugh Elles, Commander of the British Tank Corps in World War I, in preface to Ellis, *The Tank Corps*, 1919

They tell the tale of an American lady of notable good works, much esteemed by the French, who, at the end of June 1918, visited one of the field-hospitals behind Degoutte's French Sixth Army. Degoutte was fighting on the face of the Marne salient, and the 2nd American Division, then in action around the Bois de Belleau, northwest of Chateau Thierry, was under his orders. It happened that occasional casualties of the Marine Brigade of the 2nd American Division, wounded toward the flank where Degoutte's own horizon-blue infantry joined on, were picked up by French stretcher-bearers and evacuated to French hospitals. And this lady, looking down a long, crowded ward, saw on a pillow a face unlike the fiercely whiskered Gallic heads there displayed in rows. She went to it.

'Oh,' she said. 'Surely you are an American!'

'No, ma'am', the casualty answered, 'I'm a Marine.'

Colonel John W. Thompson, *Fix Bayonets!* 1926

Esprit . . . is the product of a thriving mutual confidence between the leader and the led, founded on the faith that together they possess a superior quality and capability. The failure of the spirit of any military organization is less frequently due to what men have forgotten than to what they can't forget. No 'imperishable record' of past greatness can make men serve with any greater vigor if they are being served badly. Nor can it sustain the fighting will of the organization so much as one mile beyond the radius within which living associations enable men to think great thoughts and act with nobility toward their fellows. Unless the organization's past conveys to its officers a sense of having been especially chosen, and unless they respond to this trust by developing a complete sense of duty toward their men, the old battle records might as well be poured down the drain, since they will not rally a single man in the hour of danger.

Brigadier General S. L. A. Marshall, *The Armed Forces Officer*, 1950

Given such officers, the organization comes to possess a sense of unity and of fraternity in its routine existence which express itself as the force of cohesion in the hour when all ranks are confronted by a common danger. It is not because of mutual enthusiasm for an honored name but because of mutual confidence in one another that the ranks of old regiments or the blue jackets serving a ship with a great tradition are able to convert their esprit into battle discipline. Under stress they move and act together because they have imbibed the great lesson, and experience has made its application almost instinctive, that only in unity is their safety. They believe that they can trust their comrades and commanders as they would trust their next of kin. They have learned the necessity of mutual support and a common danger serves but to bind the ranks closer.

Brigadier General S. L. A. Marshall, *The Armed Forces Officer*, 1950

Well, the main thing – that I have remembered all my life – is the definition of *esprit de corps*. Now my definition – the definition I was taught, that I've always believed in – is that *esprit de corps* means love for one's military legion, in my case the United States Marine Corps. I also learned that this loyalty to one's Corps travels both ways, up and down.

Lieutenant General Louis 'Chesty' Puller, 13 August 1956, quoted in Davis, *Marine*, 1962

I wanted to imbue my crews with enthusiasm and a complete faith in their arm and to instil in them a spirit of selfless readiness to serve in it. Only those possessed of such a spirit could hope to succeed in the grim realities of submarine warfare. Professional skill alone would not survive.

Admiral Karl Dönitz, *Memoirs, Ten Years and Twenty Days*, 1959

With the 82nd and 101st living side by side, a new factor was added to the prejudice of good order and discipline: the growing rivalry between the members of the divisions which, as the war went on, became famous throughout the Army. Sometimes it was wholesome and constructive, sometimes it was partisan and acrimonious, but in whatever form it created many problems in unit and individual relationships. Actually it was the rivalry of twin brothers who were so much alike in temperament as to be indistinguishable to innocent bystanders who soon discovered to their sorrow how quickly they would put aside family quarrels to turn on outsiders. Having served in both divisions in the course of the war, I could never have chosen between them except for the inevitable bias which came from having commanded the 101st. One's own division is necessarily number one.

General Maxwell Taylor, *Swords and Ploughshares*, 1972

EXAMPLE

I charged all countries, while I was alone, my infantry and my chariotry having forsaken me. Not one among them stood to turn about. I swear, as Re Loves me, as my father, Atum, favors me, that as for every matter which his majesty has stated, I did it in truth, in the presence of my infantry and my chariotry.

Ramses II, 1284 BC, in his own words of the Battle of Kadesh when the Hittites surprised his camp and cut him off from even his escort; he rallied his men by charging single-handed into the mass of Hittite chariotry to cut a way through, quoted in Breasted, *Ancient Records of Egypt*, 1906.

If the general is in haste to finish some enterprise that he has on hand, he should not hesitate to be prominent in the work, for soldiers are not forced to activity so much by the threats of their immediate superiors as by the influence of men of higher rank. For a soldier seeing his officer the first to put his hand to the task not only realizes the need of haste too but also is ashamed not to work, and afraid to disobey orders; and the rank and file no longer feel that they are being treated as slaves under orders but are moved as though urged by one on the same footing as themselves.

Onasander, *The General*, AD 58

According to the code of generalship, generals do not say they are thirsty before the soldiers have drawn from the well; generals do not say they are hungry before the soldiers' food is cooked; generals do not say they are cold before the soldiers' fires are kindled; generals do not say they are hot before the soldiers' canopies are drawn. Generals do not use fans in summer, do not wear leather in winter, do not use umbrellas in the rain. They do as everyone does.

Zhuge Liang (AD 180–234), *The Way of the General*

. . . Clearchos . . . marched with the army in battle order, himself commanding the rearguard. They came on ditches and canals full of water, which they could not cross without bridges; but they managed by making bridges of fallen date-palms or others which they cut down. Then one could learn what a commander Clearchos was. In his left hand he held the spear, a stick in his right; and if he thought there was any shirking, he picked out the right man and gave him one, and also he lent a hand himself, jumping into the mud, so that all were ashamed not to work as hard as he did. Those of thirty years and under were told off for the job; but when they saw how keen Clearchos was, the older men helped.

Xenophon (*c.* 431 BC–*c.* 352 BC), describing Clearchos, the original commander of the Ten Thousand in *The March Up Country* (*Anabasis*), *c.* 379 BC–*c.* 365 BC, tr. Rouse

---

Quiet!
What if one of the men gets wind of your brave plan?
No one should ever let such nonsense pass his lips,
no one with any skill in fit and proper speech –
and least of all yourself, a sceptred king.
Full battalions hang on your words, Agamemnon –
look at the countless loyal fighters you command!
Now where's your sense? You fill me with contempt –
what are you saying? With the forces poised to clash
you tell us to haul our oar-swept vessels out to sea?

Just so one more glory can crown these Trojans –
god help us, they have beaten us already –
and the scales of headlong death can drag us down.
Achaean troops will never hold the line, I tell you,
not while the long ships are being hauled to sea.
They'll look left and right – where can they run? –
and fling their lust for battle to the winds. Then,
commander of armies, our plan will kill us all!

Odysseus, berrating Agamemnon for wanting to put to sea as the Trojans stormed the Greek camp, in Homer, *The Iliad*, xiv, *c.* 800 BC, tr. Fagles

---

. . . All these soldiers have their eyes on you. If they see you are discouraged they will all be cowards; but if you show that you are making preparations against the enemy, and if you call on them, you may be sure they will follow you and try to imitate you. Perhaps it is fair to expect you to be a bit better than they are. You are captain, you see, and you are in command of troops and companies, and while there was peace, you had more wealth and honour; then now when war has come, we must ask you to be better than the mob, and to plan and labour for their behalf, if necessary.

Xenophon (*c.* 431 BC–*c.* 352 BC), rallying the few surviving officers of the Ten Thousand to action, after most of the officers were murdered under truce by the Persians in 401 BC, in *The March Up Country* (*Anabasis*), *c.* 370 BC–*c.* 365 BC, tr. Rouse

---

In carrying out very critical operations the general ought not set himself apart as though such labour was beneath him, but he should begin the work and toil along with his troops as much as possible. Such behaviour will lead the soldier to be more submissive to his officers, even if only out of shame, and he will accomplish more.

The Emperor Maurice, *The Strategikon*, *c.* AD 600

The general should set an example of how things ought to be done for his subordinates, training himself in the highest ideals, doing what is right and refraining from those things his soldiers should refrain from.

The Emperor Maurice, *The Strategikon*, c. AD 600

In the course of battle, as long as you are able to advance on foot, never take a step backward. Even when you are hemmed in amongst the enemy, never cease the struggle; you may with your bare fist knock the enemy out of the fight. And as long as they see activity, proving you to be in good fettle, they will stand in awe of you. At a time like this reconcile your heart with death. Under no conditions be afraid, but be bold; for a short blade grows longer in the hands of the brave.

Kai Ka'us Ibn Iskander, *A Mirror for Princes*, 1082, tr. Levy

---

Essential to generalship is to share the pleasures and pains of the troops. If you encounter danger, do not abandon the troops to save yourself, do not seek personal escape from difficulties confronting you. Rather, make every effort to protect the troops, sharing their fate. If you do this, the soldiers will not forget you.

The rule is 'When you see danger and difficulty, do not forget the troops' ('Sima's Art of War').

Liu Ji (1310–1375), *Lessons of War*

I used to say to them, 'go boldly in among the English', and then I used to go boldly in myself.

Joan of Arc (*c.* 1412–1431) quoted in Creasy, *Decisive Battles of the World*, 1899

To form a perfect general, the courage, the fortitude and activity of Charles XII, the penetrating glance and policy of Marlborough, the vast plains and art of Eugene, the strategems of Luxembourg, the wisdom, order, and foresight of Montecuccoli, and the grand art, which Turenne possessed, of seizing the critical moment should be united. Such a

Phoenix will with difficulty be engendered. Some pretend that Alexander was the model on which Charles XII patterned himself. If that be true, it is no less true that the successor of Charles is Prince Edward, and if unfortunately the latter should serve as an example to anyone, the copy at best can only be a Don Quixote.

But what right have I to judge the most celebrated and the greatest generals? Have I myself observed the precepts I have just described? I can only reply that the faults of others come into view with the slightest effort of memory, and that we glide lightly over our own.

Frederick the Great, 'Reflections on Charles XII', *Posthumus Works of Frederick II, King of Prussia,* 1789; quoted in Luvaas, *Frederick the Great and the Art of War,* 1966

. . . drill your soldiers well, and give them a pattern yourself.

Alexander V. Suvorov (1729–1800), quoted in Blease, *Suvorof*, 1920

Is it not the manner in which the leaders carry out the tasks of command, of impressing their resolution in the hearts of others, that makes them warriors, far more than all other aptitudes or facilities which theory may expect of them?

General Gerhardt von Scharnhorst, quoted in Marshall, *Infantry in Battle*, 1939

As if the forces in one individual after another become prostrated, and can no longer be excited and supported by an effort of his own will, the whole inertia of the mass gradually rests its weight on the will of the commander: by the spark in his breast, by the light of his spirit, the spark of purpose, the light of hope, must be kindled afresh in others.

Major General Carl von Clausewitz, *On War,* 1832

The officers should feel the conviction that resignation, bravery, and faithful attention to duty are virtues without which no glory is possible and no army is respectable, and that firmness amid reverses is more honourable than enthusiasm in success – since courage alone is necessary to storm a position whereas it requires heroism to make a difficult retreat before a victorious and enterprising enemy, always opposing to him a firm and unbroken front.

Lieutenant General Antoine-Henri Baron de Jomini, *Summary of the Art of War,* 1838

Officers should never seek for houses, but share the conditions of the men.

General of the Army William T. Sherman, *Memoirs of General W. T. Sherman*, 1875

I will not send troops to danger which I will not myself encounter.

The Duke of Marlborough, quoted in Churchill, *Marlborough*, 1933

I am now at an age when I find no heat in my blood that gives me temptation to expose myself out of vanity; but as I would deserve and keep the kindness of this army, I must let them see that when I expose them, I would not exempt myself.

The Duke of Marlborough, quoted in Churchill, *Marlborough*, 1933

One should take care in his activities so that he will be without negligence. In the *Shih Chi* is says, 'If the master acts correctly, his retainers will perform well, even if given no commands. But if the master acts incorrectly, even though he gives commands they will not be followed.

Takeda Nobushige (1525–1561), *Opinions in 99 Articles,* 1558, in Wilson, *Ideals of the Samurai*, 1982

Thus we see how surely the physical is the foundation of the moral, and how these physical defects, for defects they are in war, react upon a general's moral sense by subordinating it to intellectual achievements. More and more do strategical, administrative and tactical details occupy his mind and pinch out the moral side of his nature. Should he be a man of ability, he becomes a thinker rather than a doer, a planner rather than a leader, until morally he is as far removed from his men as a chess player is from the chessmen on his board. The more he is thrown back up on the intellectual side of war, the more sedentary he becomes, until a kind of military scholasticism enwarps his whole life.

The repercussion of such generalship on subordinate command has always been lamentable, because whatever a general may be, he is always the example which the bulk of his subordinate commanders will follow. If he becomes an officer soldier, they become officer soldiers; not only because his work makes their work, but because his morale makes their morale: how can he order them into danger if he remains in safety? If the general-in-chief does not face discomfort and danger neither will they; if they do not, neither will their subordinates, until the repercussion exhausts itself in a devitalized firing line.

Major General J. F. C. Fuller, *Generalship: Its Diseases and Their Cure*, 1933

Be an example to your men, both in your duty and in private life. Never spare yourself, and let the troops see that you don't, in your endurance of fatigue and privation. Always be tactful and well mannered and teach your subordinates to be the same. Avoid excessive sharpness or harshness of voice, which usually indicates the man who has shortcomings of his own to hide.

Field Marshal Erwin Rommel, 1938, remarks to graduating cadets of the Wiener Neustadt Military School

---

I am getting on for sixty, and this old body of mine is no longer worth caring for. But you are all young men with futures before you, so take care of yourselves and continue living in order to serve your country.

Admiral Marquis Togo Heihachiro, 1905, to his staff at the Battle of Tsushima, quoted in Warner, *The Tide at Sunrise*, 1974

Gentlemen, at least we've all been blooded now. I don't want you to be mooning over the losses and feeling sorry for yourselves or taking all the blame on your shoulders. We've all got to leave this world some day; we're all in the same pickle. And there are worse things than dying for your country. Some things about our action in the last four days I want you to remember forever. There are some we'd all like to forget – but they'll be in your mind's eye as long as you live. I hope we've all learned something. Now take care of your men, and make yourselves ready. We haven't seen anything yet.

One other thing. Back there on the hillside at Mount Austen, I had trouble getting company officers up. I hope you saw what that cost us in casualties. Never do I want to see that again in my command. I want to see my officers leading. I want you to know that you're leaders, and not simply commanders. You cannot operate a military force in the field under these conditions with commanders alone. Civilians wouldn't know what I was talking about, but you've found out now that it's true. There are many qualities in a man, but one that is absolutely necessary in an infantry leader is stark courage. Give that idea to your men in your own way.

Lieutenant General Louis 'Chesty' Puller, to his officers at Guadalcanal, 1943, quoted in Davis, *Marine*, 1962

Fear is always worse when man is isolated; it is least prominent when following an example. This consideration points to the true cellule of combat on which tactics should be built up. The little compact group formed of men who have long shared their training and recreation, following the leader they know and who has trained them, is the ideal formation for battle. Example is better, as an antidote to fear, than the companionship of equals.

Captain Sir Basil Liddell Hart, *Thoughts on War*, 1944

To the young officer who is conscious of his own reserve and is anxious to do something about it, I can suggest nothing better than to make a habit of full physical participation.

That is, instead of watching the squad or platoon work out a problem and either directing or criticizing its action, let him pick up a weapon, relieve one man in the group, then let himself be one of the commanded until the conclusion of the operation. Is this course of conduct beneath the dignity of an American officer? Certainly not!

Brigadier General S. L. A. Marshall, *Men Against Fire*, 1947

In cold weather, General Officers must be careful not to appear to dress more warmly than the men.

General George S. Patton, Jr., *War As I Knew It*, 1947

The small unit commander who practises self-exposure to danger in the hope of having a good moral effect on men, instead, frays the nerves of troops and most frequently succeeds in getting himself killed under conditions which do no earthly good to the army. Troops expect to see their officers working and moving *with* them; morale is impaired when they see that their leaders are shirking danger. But they do not care to see them play the part of a mechanical rabbit, darting to the front so as to tease on the hounds. In extreme emergencies, when the stakes are high and the failure of others to act has made the need imperative, such acts are warranted. But their value lies largely in their novelty. A commander cannot rally his men by a spectacular intervention in the hour when they have lost their grip if they have grown accustomed to seeing him run unnecessary risks in the average circumstances of battle.

Brigadier General S. L. A. Marshall, *Men Against Fire*, 1947

Since it was through the young officer that we could shape the kind of army we wanted, I would use the occasion of a graduation parade at an Officers' course whenever I had something special to say. I remember one such occasion at the end of May 1955, and also what I said when I addressed the cadets on whom I had just pinned officers' insignia. A few days earlier, I had had the unpleasant duty of terminating the service of a young career officer who had ordered a soldier to proceed on a dangerous action while he himself sat in safety. A vehicle of ours was stuck close to the border of the Gaza Strip and was under heavy fire from the Egyptians. The officer in charge sent a driver to retrieve it, while he himself lay behind cover and issues directions from there. I told the cadets: 'I would not have dismissed this officer if he had decided that the danger was too great and it was better to abandon the vehicle rather than endanger lives. But if he decided to take daring action and save the vehicle, he should have advanced with his troops and laid his own life on the line together with theirs. Officers of the Israeli army do not send their men into battle. They lead them into battle.'

General Moshe Dayan, *The Story of My Life*, 1976

The power of example is very important to people under stress. For one thing it affords an outlet for hero-worship, to which there seems to be an important and deep-rooted inclination in men. The person under stress is aware of inadequacies. He sees someone else apparently less burdened in this way. To some extent he identifies with that other person. This gives him some release. He is then likely to be grateful and become even more biddable. He will be even more open to the influence of suggestion and example than he was before.

General Sir John Hackett, *The Profession of Arms*, 1983

Our leaders will discover also that part of the their duty involves teaching. Every great leader I have known has been a great teacher, able to give those around him a sense of perspective and to set the moral, social, and motivational climate among his followers. This takes wisdom and discipline and requires both the sensitivity to perceive philosophical disarray in one's charges and the knowledge of how to put things in order. A leader must aspire to a strength, a compassion, and a conviction several octaves above that held sufficient by the workaday world. He must be at home under pressure and can never settle for the lifestyle or the outlook of that sheltered man on the street whom Joseph Conrad characterized as '. . . skimming over the years of existence to sink gently into a placid grave, ignorant of life to the last, without ever being made to see all it may contain of perfidy, or violence, or terror'.

Admiral James B. Stockdale, 'Educating Leaders', *Washington Quarterly*, Winter 1983

How the hell are you going to lead troops if they're freezing their butts off beneath shelter halves and you're dozing in a warm tent? A warm man can't understand a cold man's problems.

Lieutenant Colonel Jack Nix, US Army, quoted in *Newsweek*, 18 April 1988

. . . optimism and pessimism are infectious and they spread more rapidly from the head downward than in any other direction . . . I firmly determined that my mannerisms and speech in public would always reflect the cheerful certainty of victory – that any pessimism and discouragement I might ever feel would be reserved for my pillow.

General of the Army Dwight D. Eisenhower, quoted in *Military Review*, 10/1990

FAILURE

He who in war fails to do what he undertakes, may always plead the accidents which invariably attend military affairs: but he who declares a thing to be impossible, which is subsequently accomplished, registers his own incapacity.

The Duke of Wellington (1769–1852), quoted in Fraser, *Words of Wellington*, n.d.

The service cannot afford to keep a man who does not succeed.

Lieutenant General Thomas 'Stonewall' Jackson (1824–1863), quoted in Henderson, *Stonewall Jackson*, 1898

On the heels of the South African War came the sleuth-hounds pursuing the criminals, I mean the customary Royal Commissions. Ten thousand words of mine stand embedded in their Blue Books, cold and dead as so many mammoths in glaciers. But my long spun-out intercourse with the Royal Commissioners did have living issue – my Manchurian and Gallipoli notes. Only constant witnesses could have shown me how fallible is the unaided military memory or have led me by three steps to a War Diary:

(1) There is nothing certain about war except that one side won't win.

(2) The winner is asked no questions – the loser has to answer for everything.

(3) Soldiers think of nothing so little as failure and yet, to the extent of fixing intentions, orders, facts, dates firmly in their own minds, they ought to be prepared.

Conclusion: – in war, keep your own counsel, preferably in a note-book.

General Sir Ian Hamilton, *Gallipoli Diary*, 1920

FATIGUE AND REST

I have decided to engage successively and without halt one body of troops after the other, until harassed and worn out the enemy will be unable to further resist.

Mohammed II, 1453, Ottoman Sultan, at the Siege of Constantinople, quoted in Fuller, *A Military History of the Western World*, 1954

Gentlemen, the dispositions for tomorrow, or rather for today since midnight has gone, cannot be altered. You have heard them read out. We shall do our duty. But before a battle, there is nothing more important than to sleep well. Gentlemen, let us take some rest.

Field Marshal Prince Mikhail I. Kutuzov, 1 December 1805, the evening before the Battle of Austerlitz, quoted in Parkinson, *Fox of the North*, 1976

In military, public or administrative affairs there is a need for deep thought as well as deep analysis, and also for an ability to concentrate on subjects for a long time without sleep.

Napoleon, *Correspondance*, No. 337, Vol. I, 1858–1870

You must not needlessly fatigue the troops.

Napoleon, 29 July 1806, *Correspondance*, No. 10563, Vol. XII, 1858–1870

It is at night when a commander must work: if he tires himself to no purpose during the day, fatigue overcomes him at night . . . A commander is not expected to sleep.

Napoleon, conversation with Gourgaud, *Journal inédit de 1815 à 1818*, Vol. II, n.d.

Oh God! Perhaps the rain of the 17th of June had more to do with the loss of Waterloo than we think! If I had not been so exhausted I would have spent the whole night on horseback. The apparently most trivial circumstances have often the greatest effect.

Napoleon, ed. Kircheisen, *Memoirs of Napoleon I*, n.d.

If no one had the right to give his views on military operations except when he is frozen, or faint from heat and thirst, or depressed from privation and fatigue, objective and accurate views would be even rarer than they are.

Major General Carl von Clausewitz, *On War*, i, 1832, tr. Howard and Paret

---

I have often thought that their fierce hostility to me was more on account of the sleep I made them lose than the number we killed or captured.

Colonel John S. Mosby, *Mosby's War Reminiscences*, 1887

That men should be prepared to fight to the last is neither surprising nor unusual; it is counted upon, in fact, in establishing any plan of operations; but that men who have been retreating continuously for ten days and are lying about half-dead with exhaustion can pick up their arms and go into the attack with bugles blowing is something that had never entered into our calculations, never been taught as even a remote possibility in our military schools.

General Alexander Von Kluck, 1920, quoted in Blond, *The Marne*, 1967

Fatigue, exertion, and privation constitute a separate destructive factor in war – a factor not essentially belonging to combat, but more or less intricately involved in it, and pertaining especially to the realm of strategy. This factor is also present in tactical situations, and possibly in its most intense form; but since tactical actions are of shorter duration, the effects . . . will be limited. On the strategic plane, however, where the dimensions of time and space are enlarged, the effects are always perceptible, and often decisive . . .

Major General Carl von Clausewitz, *On War*, iii, 1832, tr. Howard and Paret

---

After a certain period or event the victor becomes tired of war; and the more civilized a people is, the more quickly will this weakness become more apparent.

Field Marshal Colmar Baron von der Goltz, *The Conduct of War*, 1883

No human being knows how sweet sleep is but a soldier.

Colonel John S. Mosby, *Mosby's War Reminiscences*, 1887

. . . There are two things I will not put up with in war; one of them is 'nerves', and the other a consequence of the first, the maintenance of units in a constant state of tension and readiness and alarm, since it is insupportable for men and horses. I have so often seen it during the war. I have always come to the conclusion that in cases of uncertainty the commanders had simply quite uselessly shifted the burden of preparedness on to the shoulders of their subordinates, instead of carrying it themselves with the help of their staff. Even in moments of gravity, I struggled with myself for a long time before dragging out an order for increased wariness and a state of alarm.

Marshal Joseph Pilsudski, *The Memoirs of a Polish Revolutionary and Soldier*, 1931

The tanks travelled, in a straight line, perhaps 570 kilometers . . . But, Andreya Lavrentevich, their speedometers show more than 2,000. A man has no speedometer and nobody knows what wear and tear has taken place.

General Mikhail Ye. Katukov, 1945, quoted in Ryan, *The Last Battle*, 1966

There are more tired corps and division commanders than there are tired corps and divisions.

General George S. Patton, Jr., *War As I Knew It*, 1947

Fatigue makes cowards of us all.

General George S. Patton, Jr., *War As I Knew It*, 1947

. . . In battle, whatever wears out the muscles reacts on the mind and whatever impairs the mind drains physical strength.

Tired men take flight more easily.

Frightened men swiftly tire.

The arrest of fear is as essential to the recovery of physical vigor by men as is rest to the body which has been spent by hard marching or hard work.

We are therefore dealing with a chain reaction. Half of control during battle comes of the commander's avoiding useless expenditure of the physical resources of his men while taking action to break the hold of fear. The other half comes from sensible preparation beforehand.

Brigadier General S. L. A. Marshall, *The Soldier's Load*, 1950

Fear and fatigue impacted on the body in the same way, draining it of energy. That being true, the overloaded soldier became more susceptible to fear; and the more heavily fear began to oppress him, the less strength he had to sustain his burdens. Overloading plus fear – result, mass panic under fire.

Brigadier General S. L. A. Marshall, *Bringing Up the Rear*, 1979

Yet historians and tacticians do not get it. If they did, there would be less nonsense written on the subject of vigorous pursuit, such as the current criticism of General Meade for allowing the beaten Confederate Army to slip away after Gettysburg. The Union Army was spiritually and physically spent after three days of battle ending in victory. Getting away, the southerners were homeward bound with no need to be urged.

Brigadier General S. L. A. Marshall, *Bringing Up the Rear*, 1979

If there's no escape, you don't experience combat fatigue.

You may get the soul scared out of you – or pretty well shot out – but you won't be suffering from combat fatigue, so-called.

General Curtis LeMay, *Mission With LeMay*, 1965

FEAR

Fear makes men forget, and skill which cannot fight is useless.

Brasidas of Sparta, 429 BC, to the Peloponnesian Fleet before action in the Criseaen Gulf, quoted in Thucydides, *The Peloponnesian War*, c. 404 BC, tr. Jowett

---

On the battlefield the real enemy is fear and not the bayonet or bullet.

Robert Jackson, *A Systematic View of the Formation, Discipline, and Economy of Armies*, 1804

Fear is the beginning of wisdom.

General of the Army William T. Sherman (1820–1891)

War is not at all such a difficult art as people think . . . In reality it would seem that he is vanquished who is afraid of his adversary and that the whole secret of war is this.

Napoleon, 1807, comment to Alexander I at Tilsit

All soldiers run away, madam.

The Duke of Wellington, when asked whether British soldiers ever ran away

---

You tell me I can't stand up to monstrous Ajax?
I tell you I never cringe at war and thundering horses!
But the will of Zeus will always overpower the will of men,
and tears away his triumph, all in a lightning flash,
and at other times he will spur a man to battle.
Come on, my friend, stand by me, watch me work!
See if I prove a coward dawn to dusk – your claim –
or I stop some Argive, blazing in all his power,
from fighting on to shield Patroculus' corpse!

Hector's reply to Glaucus for chiding him for not seeking out Ajax

Homer, *The Iliad*, xvii, *c.* 800 BC, tr. Fagles

One of the greatest qualities which we have is the ability to produce in our enemy the fear of the unknown. Therefore, we must always keep moving, do not sit down, do not say 'I have done enough', keep on, see what else you can do to raise the devil with the enemy . . .

General George S. Patton, Jr., May 1941, *The Patton Papers*, Vol. II, 1974

You are not all going to die. Only two per cent of you here, in a major battle would die. Death must not be feared. Every man is frightened at first in battle. If he says he isn't he is a God damn liar. Some men are cowards, yes, but they fight just the same or get the hell scared out of them watching men who do fight and who are just as scared as they. The real hero is the man who fights even though he is scared. Some get over their fright in a few minutes under fire, some take hours, some take days. The real man never lets the fear of death over-power his honor, his duty to his country and his innate manhood.

General George S. Patton, Jr., June 1944, speech to the Third Army

---

The exceptional man may not feel fear, but the great mass of men do – their nervous control alone stands between them and a complete yielding to fear.

This nervous control may be upset in two principal ways. It may be worn thin by a long-continued strain – it may be shattered in a single instant by sudden shock. Usually it gives under a combination of these influences. The control if worn away imperceptibly by the anxiety and suspense of waiting for the enemy's blow, by the noise and concussive effect of shellfire, and by loss of the sleep that renovates the tired will. Then without warning the shock of a sudden surprise danger snaps the fine drawn thread of the will to resist. Stubborn resistance changes in a moment to panic-stricken flight. Fear becomes uncontrollable terror. (July 1921)

Captain Sir Basil Liddell Hart, *Thoughts on War*, 1944

When the infantryman's mind is gripped by fear, his body is captured by inertia, which is fear's Siamese twin. 'In an attack half of the men on a firing line are in terror and the other half are unnerved.' So wrote Major General J. F. C. Fuller when a young captain. The failure of the average soldier to fire is not in the main due to conscious recognition of the fact that the act of firing may entail increased exposure. It is a result of a paralysis which comes of varying fears. The man afraid wants to do nothing; indeed, he does not care even to think of taking action.

Getting him on his way to the doing of one positive act – the digging of a foxhole or the administering of first-aid to a comrade – persuading him to make any constructive use of his muscle power, and especially putting him at a job which he can share with other men, may become the first step toward getting him to make appropriate use of his weapons under combat conditions. The man who finds that he can still control his muscles will shortly begin to use them. But if he is to make a rapid and complete recovery, he requires help from others.

Brigadier S. L. A. Marshall, *Men Against Fire*, 1947

Fear unhinges the will, and by unhinging the will it paralyses the reason; thoughts are dispersed in all directions in place of being concentrated on one definite aim. Fear, again, protects the body; it is the barometer of danger; is danger falling or rising, is it potent or weak? Fear should answer these questions, especially physical fear, and, thus knowing that danger confronts us, we can secure ourselves against it. Whilst moral fear is largely overcome by courage based on reason, physical fear is overcome by courage based on physical means.

Major General J. F. C. Fuller, *The Foundations of the Art of War*, 1926

Everybody gets frightened. This is basic. I do not believe that many soldiers are frightened of death. Most people are frightened of dying and everybody is frightened of being hurt. The pressures of noise, of weariness, of insecurity lower the threshold of a man's resistance to fear. All these sources of stress can be found in battle, and others too – hunger, thirst, pain, excess of heat or cold and so on. Fear in war finds victims fattened for the sacrifice.

General Sir John Hackett, *The Profession of Arms*, 1983

Never look a frightened man in the eye if you want to keep him going. Of course, it occurs to me now that this can operate the other way round too. One can also be found out oneself in this way.

General Sir John Hackett, *The Profession of Arms*, 1983

FIGHTING/KILLING

Italian commander: 'If you are a great general, come down and fight me.'
Marius: 'If you are a great general, come and make me fight you.'

Gaius Marius (*c.* 157 BC–86 BC), Adcock, *The Roman Art of War Under the Republic*, 1940

You see my sword dripping with blood and my horse with sweat. It is thus that the Moors are beaten in the field of battle.

El Cid (Rodrigo Diaz de Vivar), 1094, after routing the Almoravides before Valencia, quoted in *Cronica del Cid Ruy Diaz*, 1498

Essentially war is fighting, for fighting is the only effective principle in the manifold activities generally designated as war. Fighting, in turn, is a trial of moral and physical forces through the medium of the latter. Naturally moral strength must not be excluded, for phsychological forces exert a decisive influence on the elements involved in war.

Major General Carl von Clausewitz, *On War*, ii, 1832, tr. Howard and Paret

---

War means fighting . . . The business of the soldier is to fight . . . to find the enemy and strike him; to invade his country, and do him all possible damage in the shortest time.

Lieutenant General Thomas 'Stonewall' Jackson (1824–1863)

War means fighting, and fighting means killing.

Lieutenant General Nathan Bedford Forrest (1821–1877)

We must never forget our mission, as soldiers, is to kill, even though being killed ourselves. We must not shut our eyes to this fact. To make war – to kill without being killed is an illusion. To fight – to be killed without killing

– that is clumsy, inept. One must know how to kill while being ready to be killed oneself. The man who is dedicated to death is terrible. Nothing can stop him if he is not shot down on the way.

General Makhail I. Dragomirov (1830–1905)

The object of a good general is not to fight, but to win. He has fought enough if he gains a victory.

The Duke of Alva, *c.* 1560

To imagine that it is possible to perform great military deeds without fighting is just empty dreams.

Napoleon (1769–1821)

Hard pounding, this, gentlemen; try who can pound the longest.

The Duke of Wellington, 18 June 1815, to his staff at the Battle of Waterloo, quoted in Longworth, *Years of the Sword*, 1969

War is a savage business, a form of legalized murder, a business where killing one's fellow men without mercy is a duty and sometimes a form of sport. Nobody enjoys killing their fellow creatures, but in war one's likes and dislikes must take second place to defeat and survival. In this book there is much killing and I make no excuse for recording cases with satisfaction and often with relish, whilst on reflection one is shocked at depriving others, just as good as oneself, of their lives and for no better reason than that both of us are obeying orders and performing an unpleasant duty.

Colonel Richard Meinertzhagen, *Army Diary 1899–1925*, 1960

The object of fighting is to kill without getting killed. Don't disperse your force; you can't punch with an open hand; clench your fist; keep your command together.

Fight when holding, advancing or retiring: always fight or be ready to fight.

Aim at surprise; see without being seen. If you meet a man in a dark room, you jump; you should always try to make your enemy jump, either by day or night. A jumping man can't hit.

Never remain halted without a look-out. Sentries must be posted, no matter what troops are supposed to be in front of you.

Guard your flanks and keep touch with neighbouring units. Try to get at the enemy's flanks.

Send information back to your immediate Commander. Negative information is as important as positive. State time and place of your message. You cannot expect assistance from your superiors unless you tell them where you are and how you are situated.

Hold what you have got and what you gain. Never withdraw from a position until ordered to do so.

WHEN IN DOUBT, FIGHT IT OUT. (1918)

Major General J. F. C. Fuller, *Memoirs of an Unconventional Soldier*, 1936

FIREPOWER

Battles are won by superiority of fire.

Frederick the Great, *Military Testament*, 1768

. . . But by concentrating our howitzers and cannon at a single point we will gain the local superiority, and perhaps be able to beat them. The real difficulty is to make the hole in the enemy line, but once we have done that we will overcome the remaining obstacles soon enough.

Frederick the Great, 11 June 1778, letter to Prince Henry, *Politische Correspondenz Friedrichs des Grossen*, 1879–1939

There is but one means to extenuate the effects of enemy fire: it is to develop a more violent fire oneself.

Marshal of France Ferdinand Foch, *Precepts and Judgments*, 1919

No endurance, no physical force, no gregarious instinct and solidarity of the mass struggle could give preponderance in the era of high-speed small-calibre rifles, machine guns, intricate technical devices on vessels, and the loose formation in ground battles.

V. I. Lenin, quoted in Savkin, *Basic Principles of Operational Art and Tactics*, 1972

It is fire-power, and fire-power that arrives at the right time and place, that counts in modern war – not man-power. (March 1924)

Captain Sir Basil Liddell Hart, *Thoughts on War*, 1944

Errors of conception were to cost more than any errors of execution. Lessons of the South Africa War that went wider than the selection of leaders have been overlooked. Read in the light of 1914–1918, the 'Evidence taken before the Royal Commission on the War in South Africa' offers astonishing proof of how professional vision may miss the wood for the trees. There is little hint, among those who were to be the leaders in the next war, that they had recognized the root problem of the future – the dominating power of fire defence and the supreme difficulty of crossing the bullet-swept zone.

Captain Sir Basil Liddell Hart, *A History of the World War 1914–1918*, 1935

The tendency towards under-rating firepower . . . has marked every peace interval in modern military history.

Captain Sir Basil Liddell Hart, *Thoughts on War*, 1944

MAXIM 92. In battle like in siege, skill consists in converging a mass of fire on a single point: once the combat is opened, the commander who is adroit will suddenly and unexpectedly open fire with a surprising mass of artillery on one of these points, and is sure to seize it.

Napoleon, *Military Maxims of Napoleon*, 1827, ed. Burnod

I think if we should say that 'Fire is the Queen of Battles', we should avoid arm arguments and come nearer telling the truth. Battles are won by fire and movement. The purpose of the movement is to get the fire in a more advantageous place to play on the enemy. This is from the flank or rear.

General George S. Patton, Jr., *War As I Knew It*, 1947

The greater the visible effect of fire on the attacking infantry, the firmer grows the defenders' morale, whilst a conviction of the impregnability of the defence rapidly intensifies in the mind of the attacking infantryman.

Captain Sir Basil Liddell Hart, *Thoughts on War*, 1944

The best protection against the enemy's fire is a well directed fire from our own guns.

Admiral David G. Farragut, 14 March 1863, General Order for the attack on Port Hudson

---

Fire kills.

Marshal of France Henri-Phillipe Pétain (1856–1951)

A superiority of fire, and therefore a superiority in directing and delivering fire and in making use of fire, will become the main factors upon which the efficiency of a force will depend.

Marshal of France Ferdinand Foch, *Precepts and Judgments*, 1919

. . . Battles are won by frightening the enemy. Fear is induced by inflicting death and wounds on him. Death and wounds are produced by fire. Fire from the rear is more deadly and three times more effective than fire from the front, but to get fire behind the enemy, you must hold him by frontal fire and move rapidly around his flank . . .

General George S. Patton, Jr., *War As I Knew It*, 1947

FIRST BATTLE

. . . Only a fool will claim that he is as calm in his first battle as in his tenth . . . I know that my heart was pounding when reveille sounded on the morning of that memorable day.

Frederick the Great, referring to the Battle of Mollwitz, 10 April 1741, quoted in Duffy, *The Military Life of Frederick the Great*, 1986

War loses a great deal of its romance after a soldier has seen his first battle. I have a more vivid recollection of the first than the last one I was in. It is a classical maxim that it is sweet and becoming to die for one's country; but whoever has seen the horrors of a battlefield feels that is far sweeter to live for it.

Colonel John S. Mosby, *Mosby's War Reminiscences*, 1887

I must have been unconscious for sometime. When I came to, Sergeant Bentele was working over me. French shell and shrapnel were striking intermittently in the vicinity . . . A quarter of an hour later, buglers sounded 'regimental call' and 'assembly'. From all sides parts of the regiment worked their way toward the area west of Bleid. One after the other the different companies came in. There were many gaps in their ranks. In its first fight the regiment had lost twenty-five per cent of its officers and fifteen per cent of its men in dead, wounded, and missing. I was deeply grieved to learn that two of my best friends had been killed . . .

Field Marshal Erwin Rommel, of the first battle of the 124th Infantry Regiment, August 1914, *Infantry Attacks*, 1937

Many of you have not been fortunate enough to have engaged in combat, and owing to the foolish writings of sob-sisters and tear-jerkers, you may have erroneous ideas of what battle is like. You will read of men – imaginary men – who on the eve of battle sit around the camp fire and discuss their mothers, and their sisters, and their sweet-hearts, and talk regretfully of their past life and fear foolishly for their future. No one has a higher or more respectful devotion to women than I have; but the night before battle you do not sit around a fire . . . You go to sleep and have to be kicked in the butt in the morning so as to start the war. You have not dreamed of dying or worried about your boyhood. You have slept the sleep of fighting males eager for the kill.

General George S. Patton, December 1941, *The Patton Papers*, Vol. II, 1974

The first fight, the first contact with war! I do not know what this held for others, but for me it had as much moving poetry as my first youthful love affair, my first kisses. (1918)

Marshal Joseph Pilsudski, *The Memoirs of a Polish Revolutionary and Soldier*, 1931

. . . Far as training went, our pilots were coming directly from basic trainers right into B-17s. They knew nothing about formation flying. We merely prayed to get 'em off the ground and get 'em down again.

You'd go to bed at night and think: *How could anybody ever have the gall to bring a rabble like this into battle?*

And then you'd say to yourself: *You, too. How will you stand up? You've never been shot at, you don't know how you'll feel. Maybe you know more about the business of being a pilot; maybe you know more about navigation and bombardment, and even gunnery, than your men. But what do you know about how it feels to be in combat? Will you stand up? Will you have the nerve to ask them to stand up to it?*

Things like that you'd keep thinking. Then you'd be too exhausted to go to sleep.

General Curtis LeMay, *Mission With LeMay*, 1968

FLEXIBILITY

Plans and counter-strategems for victory that are originated at the very moment of battle are sometimes preferable to those which are conceived and contrived by generals in anticipation and before the engagement, and they are sometimes more worthy of remark, in the case of those made by men who are skilled in military science, though they are things which cannot be reduced to rules or planned beforehand. For just as pilots for their voyages, before sailing from the harbour, fit their ship out with everything that a ship requires; yet when a storm blows up they do, not what they wish, but what they must, boldly staking their fortunes against the driving peril of chance and calling to their aid no memory of their past practice but assistance appropriate to the existing circumstances; just so generals will prepare their armies as they believe will be best, but when the storm of war is at hand repeatedly shattering, overthrowing, and bringing varied conditions, the sight of present circumstances demands expedients based on the exigencies of the moment, which the necessity of chance rather than the memory of experience suggests.

Onasander, *The General*, AD 58

One must not always use the same modes of operation against the enemy, even though they seem to be working out successfully. Often enough the enemy will become used to them, adapt to them, and inflict disaster on us.

The Emperor Maurice, *The Strategikon*, c. AD 600

The Duke being asked how it was that he had succeeded in beating Napoleon's Marshals, one after another, said 'I will tell you. They planned their campaigns just as you might make a splendid set of harness. It looks very well; and answers very well; until it gets broken; and then you are done for. Now I made my campaigns of ropes. If anything went wrong, I tied a knot; and went on.'

The Duke of Wellington, quoted in Fraser, *Words on Wellington*, n.d.

Strategic development can, and must be, planned far ahead. Battles in a war of positions demand similar treatment, but in the war of movement and the actions incidental to it the situations which the commander has to visualize follow one another in motley succession. He has to decide in accordance with his instinct. Thus soldiering becomes an art, and the soldier a strategist.

General Erich Ludendorff, *My War Memories, 1914–1918*. 1919

. . . every Red Army commander must firmly grasp the fact that slavery to routine and extreme enthusiasm for some specific plan or some specific method are the most dangerous thing of all for us . . . nothing can be absolute or solidly fixed; everything flows and changes, and any means, any methods might be used in a certain situation.

Mikhail V. Frunze (1885–1925), quoted in *Voyenno istoricheskiy zhurnal*, 11/1984

The success of a commander does not arise from following rules or models. It consists in absolutely new comprehension of the dominant facts of the situation at the time, and all the forces at work. Every great operation of war is unique. What is wanted is a profound appreciation of the actual event. There is no surer road to disaster than to imitate the plans of bygone heroes and fit them to novel situations.

Sir Winston S. Churchill, *Marlborough*, 1933–1938

. . . strategy is an art, and one who practises it must be supple and cunning and know how to make a timely alteration at every turn of it. For there is a time when it is not shameful to flee, if the occasion allows, and again to pursue relentlessly, each according to one's advantage; where success would seem more by cunning than by force, risking everything is to be deprecated. Since many and various matters lead toward one end, victory, it is a matter of indifference which one uses to reach it.

General John Kinnamos, *Deeds of John and Manuel Comnenus*, 1176

No jealousies, no counter-marches, no demonstrations which are only child's play . . . it is a question of a month. One need only be on one's guard against the bottomless pit of systematic rules.

Field Marshal Prince Aleksandr V. Suvorov (1729–1800), quoted in Blease, *Suvorof*, 1920

Unhappy the general who comes on the field of battle with a system.

Napoleon (1769–1821)

Let us now discuss flexiblity. What is flexibility? It is the concrete realization of the initiative in military operations; it is the flexible employment of armed force . . . the central task in directing a war, a task most difficult to perform well. In addition to organizing and educating the army and the people, the business of war consists in the employment of troops in combat, and all these things are done to win the fight. Of course it is difficult to organize an army, etc., but it is even more difficult to employ it, particularly when the weak are fighting the strong. To do so requires subjective ability of a very high order and requires the overcoming of the confusion, obscurity and uncertainty peculiar to war and the discovery of order, clarity and certainty in it; only thus can flexibility in command be realized.

Mao Tse-tung, *On Protracted War*, May 1938

Renouncing this counter-strike . . . was insistently required by a change of the situation. And that decision, in my view, expresses one of the characteristic traits of the gift for military leadership of Nikolai Fedorovich Vatutin: the ability to precisely capture the smallest changes in the situation, to infer the further development of events from them . . . not even shrinking from basic changes in plans made earlier.

Marshal of the Soviet Union Kirill S. Moskalenko (1902–1985), speaking of the counter-strike of tank units on the Voronezh Front foreseen for 4 July 1943, quoted in Leites, *The Soviet Style in War*, 1982

Ensure that both plan and dispositions are flexible – adaptable to circumstances. Your plan should foresee and provide for a next step in case of success or failure, or partial success – which is the most common case in war. Your dispositions (or formation) should be such as to allow this exploitation or adaptation in the shortest possible time.

Captain Sir Basil Liddell Hart, *Strategy*, 1954

In battle, the art of command lies in understanding that no two situations are ever the same; each must be tackled as a wholly new problem to which there will be a wholly new answer.

Field Marshal Viscount Montgomery of Alamein, *The Memoirs of Field Marshal Montgomery*, 1958

I have used the word 'flexibility' for the last time. It seems that it is a large general purpose tent under which chaos, confusion, and incompetency are kept well hidden from the public.

General Creighton Abrams (1914–1974), quoted in *Military Review*, 4/1985

The mark of a good military unit is versatility in combat situation, not simply the ability to memorize and execute a set piece.

Lieutenant General Arthur S. Collins, Jr., *Common Sense Training*, 1978

FOG OF WAR

In war obscurity and confusion are normal. Late, exaggerated or misleading information, surprise situations and counterorders are to be expected.

General of the Armies George C. Marshall, *Infantry in Battle*, 1939

. . . the general unreliability of all information presents a special problem in war: all action takes place, so to speak, in a kind of twighlight, which like fog or moonlight, often tends to make things seem grotesque and larger than they really are.

Whatever is hidden from full view in this feeble light has to be guessed at by talent, or simply left to chance. So once again for lack of objective knowledge one has to trust to luck.

Major General Carl von Clausewitz, *On War*, ii, 1832, tr. Howard and Paret

---

Frederick: That was a diabolical day. Did you understand what was going on?
Catt: Your Majesty, I had a good grasp of the preliminary march, and the first arrangements for the battle. But all the rest escaped me. I could make no sense of the various movements.
Frederick: You were not the only one, my dear friend. Console yourself you weren't the only one!

Frederick the Great, speaking of the Battle of Zorndorff, 1758, quoted in Catt, *Unterhaltungen mit Friedrich dem Grossen*, 1884

Errors of judgment there must be in war, and few would cavil at them, especially those due to the fog of war. But it is different when the fog is self-created by confused thought and limited study.

Captain Sir Basil Liddell Hart, *Thoughts on War*, 1944

FOLLOW-UP

The principal task of the general is mental, large projects and major arrangements. But since the best dispositions become useless if they are not executed, it is essential that the general should be industrious to see whether his orders are executed or not.

Frederick the Great, *Instructions to His Generals*, 1747

If you want anything done well, do it yourself.

The Duke of Wellington, quoted in Fraser, *Words of Wellington*, n.d.

No man of action, no commander, has finished when he has taken his decision and embodied it in an order. He remains to the last moment responsible for its execution in the way he intended and for the manifestation of his will in every stage of its accomplishment . . . One evening before a battle I was taking steps to discover whether our order had reached all the quarters concerned, and I received the brief answer in an honest Berlin accent, '*Ick greife an*',* He had understood, and that was the essential thing.

* *Ich greife an*, I attack

Colonel General Hans von Seekt, *Thoughts of a Soldier*, 1930

Many generals believe that they have done everything as soon as they have issued orders, and they order a great deal because they find many abuses. This is a false principle; proceeding in this fashion, they will never reestablish discipline in an army in which it has been lost or weakened. Few orders are best, but they should be followed up with care; negligence should be punished without partiality and without distinction of rank or birth; otherwise you will make yourself hated. One can be exact and just, and be loved at the same time as feared. Severity must be accompanied with kindness, but this should not have the appearance of pretence, but of goodness.

Field Marshal Maurice Comte de Saxe, *My Reveries*, 1732

Commanders must remember that . . . the issuance of an order, or the devising of a plan, is only abut 5 per cent of responsibility of command. The other 95 per cent is to ensure, by personal observation, or through interposing of staff officers that the order is carried out.

General George S. Patton, Jr., quoted in *Military Review*, 9/1980

It is a mistake to think that once an order is given there is nothing more to be done; you have got to see that it is carried out in the spirit which you intended. Once he has decided on his outline plan and how he will carry it out, the commander should himself draft the initial operational order or directive, and not allow his staff to do so. His staff and subordinates then begin their more detailed work, and this is based on the written word of the commander himself. Mistakes are thus reduced to a minimum. This was my method, beginning from the days when I commanded a battalion.

Field Marshal Viscount Montgomery of Alamein, *The Memoirs of Field Marshal Montgomery*, 1958

I paid a lot of surprise visits at night, mostly driving alone. I wanted to check whether units were in a constant state of readiness, ensure that there was always a responsible senior officer in every command headquarters, and talk to the soldiers returning from a night exercise or from guard duty at an outpost. Whenever there was an operational problem, I would see the head of the Operations Branch, the unit commander, and his junior platoon commanders. I wanted to learn what had happened, if it was after an action, or what special problems were envisaged, if it was before an operation. I wanted to hear things from them at first hand, without intermediaries, and I believed that the young officer should hear what I had to say directly from me, in my own words and in my own style.

General Moshe Dayan, *The Story of My Life*, 1976

A unit does well only what the commander checks.

Old US Army saying

FORCE

Force is only justifiable in extremes; when we have the upper hand, justice is preferable.

Napoleon, *Political Aphorisms*, 1848

Force, to counter opposing force, equips itself with the inventions of art and science. Attached to force are certain self-imposed, imperceptible limitations hardly worth mentioning, known as international law and custom, but they scarcely weaken it. Force – that is, physical force, for moral force has no existence save as expressed in the state and the law – is thus the *means* of war; to impose our will on the enemy is its object. To secure that object we must render the enemy powerless; and that, in theory, is the true aim of warfare . . .

Major General Carl von Clausewitz, *On War*, i, 1832, tr. Howard and Paret

---

The right use of the sword is that it should subdue the barbarians while lying gleaming in its scabbard. If it leaves its sheath it cannot be said to be used rightly. Similarly the right use of military power it that it should conquer the enemy while concealed in the breast. To take the field with an army is to be found wanting in the real knowledge of it. Those who hold the office of Shogun are to be particularly clear on this point.

Tokugawa Ieyasu (1543–1616), quoted in Sadler, *The Maker of Modern Japan: The Life of Tokugawa Ieyasu*

An efficient military body depends for its effect in war – and in peace – less upon its position than upon its concentrated force.

Rear Admiral Alfred Thayer Mahan, *Naval Administration and Warfare*, 1906

Not believing in force is the same as not believing in gravitation.

Leon Trotsky (1879–1940)

The more I reflect on the experience of history the more I come to see the instability of solutions achieved by force, and to suspect even those instances where force has had the appearance of resolving difficulties.

Captain Sir Basil Liddell Hart, *Thoughts on War*, 1944

Once you decide to use force, you had better make sure you have plenty of it. If you need a battalion to do a job, it's much better to have the strength of a division. You probably won't suffer any casualties at all in that way.

General of the Army Dwight D. Eisenhower, *The New York Times*, 10 May 1965

Between two groups that want to make inconsistent kinds of worlds, I see no remedy but force.

Justice Oliver Wendall Holmes, Jr. (1841–1935)

FORESIGHT/ANTICIPATION

After agreeing upon a treaty or a truce with the enemy, the commander should make sure that his camp is guarded more strongly and more closely. If the enemy chooses to break the agreement, they will only gain a reputation for faithlessness and the disfavour of God, while we shall remain in safety and be true to our word. A general should not have to say: 'I did not expect it.'

The Emperor Maurice, *The Strategikon*, c. AD 600

. . . as it turned out, they wished to do it all in safety and to fall upon us by night, when we were not watching. And as I was so much on my guard, they found me standing in front of their thoughts.

Hernan Cortes, 1519, on his entry into the Valley of Mexico, in Madariaga, *Hernan Cortes*, 1942

While Rommel was leading his troops in person against our strongly-held defensive positions on the Alam Halfa ridge, Montgomery was planning the battle of Alamein. That was the difference between the two.

Lieutenant General Sir Brian Horrocks, *Escape to Action*, 1961

. . . It is a cardinal responsibility of a commander to foresee insofar as possible where and when crises affecting his command are likely to occur. It starts with his initial estimate of the situation – a continuing mental process from the moment of entering the combat zone until his unit is pulled out of the line. Ask yourself these questions. What are the enemy's capabilities? What shall I do, or what could I do, if he should exercise that one of his capabilities which would be most dangerous to me, or most likely to interfere with the accomplishment of my missions?

General Mathew B. Ridgway, 'Leadership', *Military Review*, 10/1966

. . . the job of a commander is to think ahead. A GHQ must be dealing with events foreseen two and preferably three or more days away. The imagination and foresight with which he manages that is one of the inescapable tests of a commander: because if there is one rule about a battle it is, 'What is possible today may not be possible tomorrow.'

General Saad El Shazly, *The Crossing of the Suez*, 1980

MAXIM 8. A general-in-chief should ask himself frequently in the day, What should I do if the enemy's army appeared now in my front, or on my right, or on my left? If he has any difficulty in answering these questions he is ill posted, and should seek to remedy it.

Napoleon, *The Military Maxims of Napoleon*, 1831, tr. D'Aguilar

---

Think to a finish!

Field Marshal Viscount Allenby of Meggido, 1902, upon taking command of the 5th Lancers

---

The vital point in actual warfare is to apply to the enemy what we do not wish to be applied to ourselves and at the same time not to let the enemy apply it to us. Therefore, it is most important that what we consider would embarrass the enemy we should apply to them before they can do the same to us; we must always forestall them.

Admiral Marquis Togo Heihachiro, February 1905, to the officers of the Japanese Fleet, quoted in Ogasawara, *Life of Admiral Togo*, 1934

Again for the small unit commander, there is need to seek the true meaning of the counsel insistently given by Major General Stuart Heintzelman, Major General Frank A. Ross, and others: 'Anticipation is 60 per cent of the art of command.'

But like ten years in the penitentiary, it is very easy to say 'Anticipate!' and very hard to do it.

Brigadier General S. L. A. Marshall, *Men Against Fire,* 1947

. . . a commander must think two stages ahead.

Field Marshal Viscount Montgomery of Alamein, *The Memoirs of Field Marshal Montgomery*, 1958

FORTUNE

In war we must always leave room for strokes of fortune and accidents that cannot be foreseen.

Polybius, *Histories,* c. 125 BC

For it is necessary to pray to Fortune to do its share, but not to consider that Fortune has entire control. Stupid are those who make disasters chargeable to Fortune alone, rather than to the negligence of commanders, as well as those who attribute successes to her, and to the skill of the generals.

Onasander, *The General,* AD 58

It is better to subdue an enemy by famine than by sword, for in battle, *fortuna*, has often a much greater share than *virtu*.

Niccolo Machiavelli, *The Art of War,* 1521

There is nothing so subject to inconstancy of fortune as war.

Miguel de Cervantes, *Don Quixote,* 1615

When a general conducts himself with all prudence, he still can suffer ill fortune; for how many things do not cooperate at all with his labours! Weather, harvest, the officers, the health or sickness of his troops, blunders, the death of an officer on whom he counts, discouragement of the troops, exposure of your spies, negligence of the officers who should reconoitre the enemy, and finally betrayal. These are the things that should be kept continually before your eyes so as to be prepared for them and so that good fortune will not blind us.

Frederick the Great, *Instructions to His Generals*, 1747

. . . rule fortune . . .

Field Marshal Prince Aleksandr V. Suvorov (1729–1800), quoted in Blease, *Suvorof*

. . . the contingent element inseparable from the waging of war [which] gives to that activity both its difficulty and its grandeur.

Charles de Gaulle, *The Edge of the Sword,* 1932

FRICTION

Friendly relations contribute a lot to diminish the friction of war. Comradeship is the oil in the machine. If 'Bill' appeals to his friend 'Dick' he is more sure of good and prompt support than if an overtired O.C. 9th Battalion sends a formal message to an overburdened O.C. 999th Battery. It is thus important to lose no opportunity of creating a co-operative atmosphere by making touch. The military saying that 'time spent in reconnaissance is rarely wasted' applies just as clearly to the more personal kind of reconnaissance within your own front. (July 1933)

Captain Sir Basil Liddell Hart, *Thoughts on War,* 1944

Everything in war is very simple, but the simplest thing is difficult. The difficulties accumulate and end by producing a kind of friction that is inconceivable unless one has experienced war . . . Countless minor incidents – the kind you can never really foresee – combine to lower the general level of performance, so that one always falls far short of the intended goal. Iron will-power can overcome friction; it pulverizes every obstacle, but of course it wears down the machine as well.

Friction is the only concept that more or less corresponds to the factors that distinguish real war from war on paper. The military machine . . . is composed of individuals, every one of whom retains his potential for friction . . . the least important of whom may chance to delay things or somehow make them go wrong.

Major General Carl von Clausewitz, *On War*, i, 1832, tr. Howard and Paret

GENERALS/GENERALSHIP

There are five qualities which are dangerous in the character of a general.

If reckless, he can be killed.
If cowardly, captured.
If quick-tempered you can make a fool of him.
If he has too delicate a sense of honour you can calumniate him.
If he is of a compassionate nature you can harass him.

Now these traits of character are serious faults in a general and in military operations are calamitous.

The ruin of the army and death of the general are inevitable results of these shortcomings. They must be deeply pondered.

Sun Tzu, *The Art of War*, c. 500 BC, tr. Griffith

A good general not only sees the way to victory; he also knows when victory is not possible.

Polybius, *The Rise of the Roman Empire*, c. 125 BC

The general must know how to get his men their rations and every other kind of stores needed for war. He must have imagination to originate plans, practical sense, and energy to carry them through. He must be observant, untiring, shrewd, kindly and cruel, simple and crafty: a watchman and a soldier; lavish and miserly; generous and yet tight-fisted; both rash and conservative. All these and other qualities, natural and acquired, he must have. He should also, as a matter of course, know his tactics; for a disorderly mob is no more an army than a heap of building material is a house.

Socrates (470 BC–399 BC), quoted by Xenophon, (c. 431 BC–c. 352 BC), *Memorabilia*

It was Caesar's hope that he could finish the business without fighting or casualties because he had cut his adversaries off from food supply. Why lose men, even for victory? Why expose soldiers who deserved so well of him to wounds? Why even tempt fortune. Victory through policy is as much a mark of the good general as victory by the sword.

Julius Caesar, *The Civil War*, c. 45 BC, tr. Hadas

There are five skills and four desires involved in generalship.

The five skills are: skill in knowing the disposition and power of enemies, skill in knowing the ways to advance and withdraw, skill in knowing how empty or how full countries are, skill in knowing nature's timing and human affairs, and skill in knowing the features of terrain.

The four desires are: desire for the extraordinary and unexpected in strategy, desire for thoroughness in security, and desire for calm among the masses, and desire for unity of hearts and minds.

Zhuge Liang (AD 180–234), *The Way of the General*

A ship cannot cross the sea without a helmsman, nor can one defeat an enemy without tactics and strategy. With these and the aid of God it is possible to overcome not only an enemy force of equal strength but even one greatly superior in numbers. For it is not true, as some inexperienced people believe, that wars are decided by courage and numbers of troops, but, along with God's favour, by tactics and generalship, and our concern should be with these rather than wasting our time in mobilizing large numbers of men. The former provide security and advantage to men who know how to use them well, whereas the other brings trouble and financial ruin.

The Emperor Maurice, *The Strategikon*, c. AD 600

It is safer and more advantageous to overcome the enemy by planning and generalship than by sheer force; in the one case the results are achieved without loss to oneself, while in the other some price has to be paid.

The Emperor Maurice, *The Strategikon*, c. AD 600

Pay well, command well, hang well.

General Sir Ralph Hopton (1596–1652), *Maxims for the Management of an Army*, 1643

. . . the knowledge of a real general must be varied. He must have an accurate idea of politics in order to be informed of the intention of princes and the forces of states and of their communications; to know the number of troops that the princes and their allies can put in the field; and to judge the condition of their finances. Knowledge of the country where he must wage war serves as the base for all strategy. He must be able to imagine himself in the enemy's shoes in order to anticipate all the obstacles that are likely to be placed in his way. Above all, he must train his mind to furnish him with a multitude of expedients, ways and means in case of need. All this requires study and exercise. For those who are destined for the military profession, peace must be a time of meditation, and war the period where one puts his ideas into practice.

Frederick the Great, 'Réflexions sur les projets de campagne', *Oeuvres de Frédéric le Grand*, 1846–56, in Luvaas, ed., *Frederick the Great on the Art of War,* 1966

A perfect general, like Plato's republic, is a figment. Either would be admirable, but it is not characteristic of human nature to produce beings exempt from human weaknesses and defects. The finest medallions have reverse sides. But in spite of this awareness of our imperfections it is not less necessary to consider all the different talents that are needed by an accomplished general . . .

It is essential that a general should dissemble while appearing to be occupied, working with the mind and working with the body, ceaselessly suspicious while affecting tranquility, saving of his soldiers and not squandering them except for the most important interests, informed of everything, always on the lookout to deceive the enemy and careful not to be deceived himself. In a word he should be more than an industrious, active, and indefatigable man, but one who does not forget one thing to execute another, and above all who does not despise those little details which pertain to great projects.

Frederick the Great, *Instructions to His Generals,* 1747

MAXIM 81. It is exceptional and difficult to find all the qualities of a great general combined in one man. What is most desirable and distinguishes the exceptional man, is the balances of intelligence and ability with character or courage. If courage is predominant, the general will hazard far beyond his conceptions; and on the contrary, he will not dare to accomplish his conceptions if his character or his courage are below his intelligence.

Napoleon, *The Military Maxims of Napoleon,* 1827, ed. Burnod

---

MAXIM 82. With a great general there is never a continuity of great actions which can be attributed to chance and good luck; they always are the result of calculation and genius.

Napoleon, *The Military Maxims of Napoleon,* 1827, ed. Burnod

---

The most essential quality of a general is firmness of character and the resolution to conquer at any price.

Napoleon, quoted in Gourgaud, *Journal inédit de 1815 à 1818*, Vol. II, n.d.

Intelligent and fearless generals assure the success of affairs.

Napoleon, to Marshal Mortier, 29 November 1806, *Correspondance*, No. 113255, XIII, 1858–1870

The greatest general is he who makes the fewest mistakes.

Napoleon (1769–1821)

A general's principal talent consists in knowing the mentality of the soldier and in gaining his confidence.

Napoleon (1769–1821)

The most essential qualities for a general will always be: *first*, a high moral courage, capable of great resolution; *second*, a physical courage which takes no account of danger. His scientific or military acquirements are secondary to these. It is not necessary that he should be a man of vast erudition; his knowledge may be limited but it should be thorough, and he should be perfectly grounded in the principles at the base of the art of war.

Next in importance come the qualities of his personal character. A man who is gallant, just, firm, upright, capable of esteeming merit in others instead of being jealous of it, and skillful in making his merit add to his own glory, will always be a good general and may even pass for a great man . . . *finally*, the union of wise theory with great character will constitute the great general.

Lieutenant General Antoine-Henri Baron de Jomini, *Summary of the Art of War*, 1838

A complete battle of Cannae is rarely met in history. For its achievement, a Hannibal is needed on one side, and a Terrentius Varro, on the other, both cooperating for the attainment of the great goal.

A Hannibal must possess, if not a superiority in numbers, the knowledge how to create one. It is desirable for this purpose that the general combine in himself something of a Scharnhorst, a Frederick William, or William I, to weld together a strong army, of a Moltke, to assemble it solely against the principal enemy, of a Frederick the Great, to bring all his guns and rifles into action, of a Frederick the Great or a Napoleon, to direct the principal attack against the flank or rear, of a Frederick the Great or a Moltke to replace the absent Hasdrubal by a natural obstacle or the frontier of a neutral state. Lastly, there are needed subordinate commanders, well trained in their profession, and able to comprehend the intentions of their chiefs.

A Terrentius Varro has a great army, but does not do his best to increase and train it. He does not assemble his forces against the principal enemy. He does not wish to vanquish by fire superiority from several sides, but by the weight of masses in narrow and deep formations, selecting for attack the hostile front as being the side most capable of resistance.

All these desirable conditions will not be found combined on either side. A few of Hannibal's qualities and some of the means at his disposal were possessed by other generals. Terentius Varro, on the other side, was always the product of the school . . .

Field Marshal Alfred Graf von Schlieffen, *Cannae*, 1913

A general has much to bear and needs strong nerves. The civilian is too inclined to think that war is only like the working out of an arithmetical problem with given numbers. It is anything but that. On both sides it is a case of wrestling with powerful, unknown physical and psychological forces, a struggle which inferiority in numbers makes all the more difficult. It means working with men of varying force of character and with their own views. The only quality that is known and constant is the will of the leader.

General Erich Ludendorff, *My War Memories, 1914–1918*, 1919

If our western adversaries failed to obtain any decisive results in the battle from 1915 to 1917 it must mainly be ascribed to a certain unimaginativeness in their generalship. The necessary superiority in men, war material and ammunition was certainly not lacking, nor can it be suggested that the quality of the enemy troops would not have been high enough to satisfy the demands of a more vigorous and ingenious leadership. Moreover, in view of the highly-developed railway and road system, and the enormous amount of transport at their disposal, our enemies in the West had free scope for far greater strategic subtlety. However, the enemy commander did not make full use of these possibilities, and our long resistance was to be attributed, apart from other things, to a certain barrenness of the soil in which the enemy's plans took root . . .

Field Marshal Paul von Hindenburg, *Out of My Life*, 1920

The ever-victorious general is rare and there have been very few of these in history, but what is necessary is that our generals should have studied the art of war and paid attention to its rules; it is then that, with this wisdom tempered by courage, our military leaders will have better chances of success.

Mao Tse-tung, *On the Study of War*, 1936

Here are six points which are important for successful generalship. A general must:

(a) Know his 'stuff' thoroughly.
(b) Be known and recognized by the troops.
(c) Ensure the troops are given tasks that are within their powers. Success means mutual confidence – failure the reverse.
(d) See that subordinate commanders are disturbed as little as possible in carrying out their tasks.
(e) Command by personal contact.
(f) Be human and study the human factor.

Major General Sir Francis de Guingand, *Operation Victory*, 1947

To exercise high command successfully one has to have an infinite capacity for taking pains and for careful preparation; and one has also to have an inner conviction which at times will transcend reason. Having fought, possibly over a prolonged period, for the advantage and gained it, there then comes the moment for boldness. When that moment comes, will you throw your bonnet over the mill and soar from the known to seize the unknown? In the answer to this question lies the supreme test of generalship in high command.

Field Marshal Viscount Montgomery of Alamein, *The Memoirs of Field Marshal Montgomery*, 1958

Let me give you some advice, Lieutenant. Don't become a general. Don't ever become a general. If you become a general you just plain have too much to worry about.

General of the Army Dwight D. Eisenhower, 9 May 1965

GENERAL STAFF

The most highly trained general staff, with the best of ideas and principles cannot guarantee good leadership in an army by itself. It must be backed by a great general who serves as its leader and counterbalance, who from time to time prevents the staff from entangling itself in its own red tape. A good staff on the other hand is an indispensable aid for a general.

Major General Carl von Clausewitz, *The Campaign of 1799 in Italy and Switzerland*, 1906

Great achievements, small display: more reality than appearance.

Field Marshal Helmuth Graf von Moltke (1800–1891), motto of the German General Staff, quoted in Guderian, *Panzer Leader*, 1953

A trained general staff is even more necessary to a commander today than when Clausewitz wrote, because of the greater masses of men involved in operations today. There is little danger that such a staff will, as with the opponents of Napoleon, 'entangle itself in its own red tape'; but there is another danger which has resulted from the highly technical character of general staff training. This is the tendency to overemphasize detail. We must take care that the mass of minutiae required in modern-day planning does not overshadow the spirit of the operation itself, and cause us to judge the operation by the excellence or insufficiency of its staff work, rather than by its scope and daring, as we should. To use Boyen's words, we must see that 'industry without talent is not unduly rewarded'. There is no better way to prevent this than to study the leadership of great generals of the past and endeavour to follow the workings of their minds. Only in this way can we gain for ourselves some part of the insight and determination they possessed.

General der Infantrie Hugo Baron von Freytag-Loringhoven, *The Power of Personality in War*, 1911

The difficulty of always selecting a good general has led to the formation of a good general staff, which being near the general may advise him and thus exercise a beneficial influence over the operations. A well-instructed general staff is one of the most useful organizations, but care must be observed to prevent the introduction into it of false principles.

Lieutenant General Antoine-Henri Baron de Jomini, *Summary of the Art of War*, 1838

The General Staff officer was, so to speak, a man apart. As the war become more technical, his duties became more arduous. It was no longer sufficient for him to have a general knowledge of all arms and their employment. He had to be a good artilleryman and, in addition, to possess a sound knowledge of the use of aircraft, signalling, supply questions and a thousand other things, while he had to master many details which the divisional commander had no time to settle. In spite of every effort to keep them brief, the orders which he had to draft grew ever longer and more complicated. The more technical the war became, the more did these orders grow into veritable works of art, involving infinite skill and knowledge. There was no other way, if things were to go smoothly. The variety of his functions often compelled the General Staff officer to keep many things in his own hands. Care had to be taken that the independence of other services did not suffer on this account, and that the commander, too, was not 'shelved'. I could never have allowed either of these developments.

General Erich Ludendorff, *My War Memories, 1914–1918*, 1919

The General Staff was certainly one of the most remarkable structures within the framework of our German Army. Side by side with the distinctly heirarchical form of the commands it constituted a special element which had its foundation in the great intellectual prestige of the Chief of Staff of the Army, Field Marshal Count von Moltke. The peace training of the General Staff officer offered a guarantee that in case of war all the commanders in the field should be controlled from a single source, and all their plans governed by a common aim. The influence of the General Staff on those commanders was not regulated by any binding order. It depended far more on the military and personal qualities of the individual officer. The first requirement of the General Staff officer was that he should keep his own personality and actions entirely in the background. He had to work out of sight, and therefore be more than he seemed.

Field Marshal Paul von Hindenburg, *Out of My Life*, 1920

History has often shown that, where a particular aim has an instinctive appeal, a General Staff appreciation may have more resemblance to counsel's address to the jury on behalf of the client than to a judicial summing-up. (May 1933.)

Captain Sir Basil Liddell Hart, *Thoughts on War*, 1944

No soldier can doubt the immense value of a general staff if it is the general's servant, and not the general's gaoler. I have said that the staff has no responsibilites; it has none, though it has duties; because it has no powers of decision or command. It can suggest, but it has no responsibility for actions resulting; the general alone is responsible, therefore the general alone should and must decide, and, more than this, he must elaborate his own decisions and not merely have them thrust upon him by his staff like a disc upon a gramophone.

Major General J. F. C. Fuller, *Generalship: Its Diseases and Their Cure,* 1936

My general staff training revolted against an operation of this sort. If old von Moltke thought I had planned this offensive he would turn in his grave.

Field Marshal Gerd von Rundstedt (1875–1953), referring to the plan for the Ardennes Offensive of December 1944

Prerequisites for appointment to the General Staff were integrity of character and unimpeachable behaviour and way of life both on and off duty. Next came military competence; a man had to have proved himself at the front, had to have an understanding for tactical and technical matters, a talent for organization and powers of endurance both physical and mental; he had also to be industrious, of sober temperament, and determined.

In selecting officers from this point of view it is possible that intellectual ability was sometimes overvalued in comparison to strength of character and particularly to warmth of personality; but these last two qualities are much less easily estimated, particularly since they do not by their very nature tend to be spectacular.

Colonel General Heinz Guderian, *Panzer Leader*, 1953

The German General Staff was a human institution, and it, too, had human frailties and weaknesses, which contributed, at least in part, to its two major defeats in this century. Yet the performance of that General Staff, and of the Army it designed and built in both of those disastrous wars, was comparable in terms of military excellence to Napoleon and Hannibal at their best. Perhaps, in this sense, it is not too much to say that in striving to institutionalize excellence in military affairs, the German General Staff can be said to have institutionized military genius itself.

Colonel Trevor N. Dupuy, *A Genius for War: The German General Staff and the German Army*, 1977

General Staff officers have no names.

Colonel General Hans von Seekt, quoted in Foertsch, *The Art of Modern War,* 1940

If any one aspect of military performance was emphasized more than any other in the General Staff and in all German military training, it was encouragement of initiative. American General Albert C. Wedemeyer, the last American officer who attended the war Academy before World War II, remembers this emphasis. It was summarized, he recalls, in the concept: 'When in doubt, attack!'

Colonel Trevor N. Dupuy, *A Genius For War: The German General Staff and the German Army*, 1977

GENIUS

. . . they [the Romans] failed to reckon with the talents of Archimedes or to foresee that in some cases the genius of one man is far more effective than superiority in numbers. This lesson they now learned by experience.

Polybius, of the Roman attack on Syracuse in 214 BC, *The Rise of the Roman Empire, c.* 125 BC

A genius paramount in force and originality manifested in the character of a general, animates the operations of the army with one impulse . . . It impresses an opinion of superiority on the mass: everyone views himself in the splendour of his commander, assimilates, in idea, with his excellence, and, being superior in opinion, soon becomes so in reality. Hence, it is not the dry mechanical wisdom of the plan of battle, so much as the animating spirit of the leader, which may be considered as the pledge of success in war . . .

It may be observed, in this place that good officers abound among all nations who cultivate the art of war; the genuine military genius is of rare occurrence: no power of industry can produce it; and no one can define the path in which it moves. The genius which leads to conquest, like the genius of a poet, is original: it is a first impression, improved by study in the book of nature only . . . the spirit which knows not to submit, which retires from no danger because it is formidable, is the soul of a soldier. The soldier fixes victory in his eye, as a passion rather than a reason. This forms what may be considered as genius – a paramount genius, which domineers and conquers and enslaves.

Robert Jackson, *A Systematic View of the Formation, Discipline, and Economy of Armies*, 1804

Of the conquerors and eminent military characters who have at different times astonished the world, Alexander the Great and Charles the Twelfth of Sweden are two of the most singular; the latter of whom was the most heroic and the most extraordinary man of whom history has left any record. An army which had Alexander or Charles in its eye was different from itself in its simple nature, it imbibed a share of their spirit, became insensible of danger, heroic in the extreme.

Robert Jackson, *A Systematic View of the Formation, Discipline, and Economy of Armies*, 1804

Great geniuses have a sort of intuitive knowledge; they see at once the cause, and its effect, with the different combinations, which unite them: they do not proceed by common rules, successively from one idea to another, by slow languid steps, no: the *Whole*, with all its circumstances and various combinations, is like a picture, all together present to their mind; they want no geometry, but an age produces few of this kind of men: and in the common run of generals, geometry, and experience, will help them to avoid gross errors.

Major General Henry Lloyd, *History of the Late War in Germany*, 1766

Achilles was the son of a goddess and a mortal: he symbolizes the genius of war. Its divine half is composed of all those elements which depend on moral factors – on character, on talent, on the interests of your adversary, on public opinion, and on the morale of your troops, who are either strong and victorious or weak and beaten, depending on which they think they will be. Armaments, entrenchments, positions, order of battle, everything pertaining to the combinations of material factors – these make up the earthly half.

Napoleon, Dictation at Saint Helena, ed., Herold, *The Mind of Napoleon*, 1955

Genius is sometimes only an instinct which is incapable of being perfected. In most cases the art of judging correctly is perfected only through observation and experience. A good thought is not always associated with good judgment, but good judgment always presupposes a good thought.

Napoleon, ed., Kircheisen, *The Memoirs of Napoleon*, n.d.

There is no genius who tells me suddenly and in secret what I must say or do in any circumstance unexpected by others, it is reflection, meditation.

Napoleon, quoted in Foch, *Principles of War*, 1913

Genius is not extravagant; it is ardent, and it conceives great projects; but it knows beforehand how to attain the result, and it uses the simplest means, because its faculties are essentially, calculating, industrious, and patient. It is creative, because its knowledge is vast; it is quick and preemptory, not because it is presumptious, but because it is well-prepared.

General Sir William Napier (1782–1853), quoted in Henderson, *Stonewall Jackson*, 1898

Every special calling in life . . . requires special qualities of intellect and temperament. When these are of a high order, and manifest themselves by extraordinary achievements, the mind to which they belong is termed genius . . . a superior mental capacity for certain activities . . . The essence of military genius is to bring under consideration all of the tendencies of the mind and soul in combination towards the business of war . . . We say 'in combination', for military genius is not just one single quality bearing upon war – as, for instance courage . . . – but it is a harmonious combination of powers, in which one or the other may predominate, but one must be in opposition.

Major General Carl von Clausewitz, *On War*, i, 1832

Genius is, of course, an exception to all rules and is in a category by itself, as we see from the personality of Frederick, whose versatility is unequalled among military men. The Comte de Segur, as quoted by Pierron, says very pertinently: 'Tremendous intellect, plus a vast amount of good sense, coupled with great strength of character add up to genius. The first two of these characteristics make it possible to see the critical point in any affair; the third enables one to employ all his force to reach it.'

General Lieutenant Hugo Baron von Freytag-Loringhoven, *The Power of Personality in War*, 1911

Despite every wish to give full weight to the genius of the Commander, we must fain recognize that there are immutable bounds set to what any individual can achieve. We know what a Lee can do against the good, ordinary General; yet – there are the appointed limits – as Gettysburg showed. A thrust is well-timed and deadly; the over-tasked blade breaks in the hand of the master. The stroke remains a master stroke; *it did not penetrate,* that was all!

General Sir Ian Hamilton, *The Soul and Body of an Army*, 1921

The military genius is he who can produce original combinations out of the forces of war; he is the man who can take all these forces and so attune them to the conditions which confront him that he can produce startling and, frequently, incomprehensible results. As an animal cannot explain the instincts which control it, neither can a man of genius explain the powers which control him. He acts on the spur of the moment, and he acts rightly, because this power is in his control. That some explanation exists cannot be doubted, but so far science has not revealed it, though the psychologist is working towards its fringe.

Major General J. F. C. Fuller, *The Foundations of the Science of War*, 1926

As genius is a personal gift, so is imitation a collective instinct. One man possessed of genius may alter the course of history, in fact, such a man has always altered the course of history, when alteration has been rapid. Three men of genius, working as a committee, could not do this, and still less so a crowd of normal men.

Major General J. F. C. Fuller, *The Foundations of the Science of War*, 1926

Great war leaders have always been aware of the importance of instinct. Was not what Alexander called his 'hope', Caesar his 'luck', and Napoleon his 'star' simply the fact that they knew they had a particular gift of making contact with realities sufficiently closely to dominate them? For those who are greatly gifted, this facility often shines through their personalities. There may be nothing in itself exceptional about what they say or their way of saying it, but other men in their presence have the impression of a natural force destined to master events.

Charles de Gaulle, *The Edge of the Sword*, 1932

Generals have been legion; artists of war few. Many more evidences of military genius can be traced in the scantier records of irregular and guerrilla forces. The natural explanation is that the natural gifts of the leaders who have emerged straight from the womb of conflict, instead of a professional incubation chamber, have not been cramped or warped by convention . . . it does cast a reflection upon the customary method of training leaders in organized armies. It is significant that in the American Civil War, which was the most fruitful field of generalship in modern times, most of the outstanding leaders were men who after an apprenticeship in the Regular Army had gone into some field of civil activity, which in varying degree freed their minds from the fetters of military convention and routine, while they retained a useful experience of the functioning of the military machine. (March 1937)

Captain Sir Basil Liddell Hart, *Thoughts on War*, 1944

'Genius' is a tiresome and misleading word to apply to the military art, if it suggests, as it does to many, one so gifted by nature as to obtain his successes by inspiration rather than through study. Nor does the definition of genius attributed to Carlyle as 'an infinite capacity for taking pains' suit the great commander, as it suggests the pedant or martinet. Good generals, unlike poets, are made rather than born, and will never reach the first rank without much study of their profession; but they must have certain natural gifts – the power of quick decision, judgment, boldness and, I am afraid, a considerable degree of toughness, almost callousness, which is harder to find as civilization progresses.

Field Marshal Earl Wavell, *Soldiers and Soldiering*, 1953

The stroke of genius that turns the fate of a battle? I don't believe it. A battle is a complicated operation, that you prepare laboriously. If the enemy does this, you say to yourself I shall do that. If such and such happens, these are the steps I shall take to meet it. You think out every possible development and decide on the way to deal with the situation created. One of these developments occurs; you put in operation your pre-arranged plan, and everyone says, 'What genius to have thought of that at the critical moment!' Whereas the credit is really due to the labour of preparation done beforehand.

Marshal of France Ferdinand Foch, *Daily Mail*, 19 April 1919

GLORY

Mother tells me,
the immortal goddess Thetis with her glistening feet,
that two fates bear me on to the day of death.
If I hold out here and I lay siege to Troy,
my journey home is gone, but my glory never dies.
If I voyage back to the fatherland I love,
my pride, my glory dies . . .
true, but the life that's left me will be long . . .

Achilles, in Homer, *The Iliad*, ix, c. 800 BC, tr. Fagles

The nearest way to glory – a short cut as it were – is to strive to be what you wish to be thought to be.

Socrates, quoted in Cicero, *De Officius*

. . . my own assessment of myself is based on the extent not of my life but of my glory. I could have been content with my father's inheritance, and within Macedonia's bounds have enjoyed a life of ease as I awaited an old age without renown or distinction (though even inactive men cannot control their destiny, and those who believe a long life is the only good are often overtaken by a premature death). But no – I count my victories, not my years and, if I accurately compute fortune's favours to me, my life has been long.

Alexander the Great, quoted in Curtius, *The History of Alexander*, c. AD 41, tr. Yardley

---

. . . my greatest happiness is to serve my gracious King and Country, and I am envious only of glory; for if it be a sin to covet glory, I am the most offending soul alive.

Admiral Lord Nelson, 18 February 1800, letter to Lady Hamilton

---

The art of war is the most difficult of all arts; therefore military glory is universally considered the highest, and the services of warriors are rewarded by a sensible government in a splendid manner and above all other services.

Napoleon (1769–1821)

Glory can only be won where there is danger.

Napoleon, 14 November 1806

A sailor does not have a difficult or an easy path. There is just one, the glorious path.

Admiral P. S. Nakhimov, quoted in Ablamonov, *Admiral*, 1986

Stand firm; for well you know that hardship and danger are the price of glory, and that sweet is the savour of a life of courage and of deathless renown beyond the grave.

Alexander the Great, 324 BC, in India when the Macedonians refused to go on, quoted in Arrian, *The Campaigns of Alexander*, v, c. AD 150, tr. de Selincourt

---

It is men who endure toil and dare dangers that achieve glorious deeds, and it is a lovely thing to live with courage and to die leaving behind an everlasting renown.

Alexander the Great (356–323 BC), quoted in Plutarch (*c.* AD 46–after AD 119), *The Lives of the Noble Grecians and Romans*

The desire for glory clings to men even longer than any other passion.

Cornelius Tacitus (*c.* AD 56–*c.* AD 120), *Annals*

There must be a beginning of any great matter, but the continuing unto the end until it be thoroughly finished yields the true glory.

Sir Francis Drake, 17 May 1587, to Lord Walsingham

In the hearts of most men there lurks a sentiment which they carefully try to hide from their fellow men. This foolish sentiment is that which causes man to aspire constantly to immortality. Not all men, however, succeed in inscribing their names on the walls of the temple of glory.

General Antonio Lopez de Santa Anna, *The Eagle: The Autobiography of Santa Anna*, c. 1870

The love of glory, the ardent desire for honorable distinction by honorable deeds, is among the most potent and elevating of military motives.

Rear Admiral Alfred Thayer Mahan, *Life of Nelson*, 1897

A nation without glory is like a man without courage, a woman without virtue. It takes the first place in our human estimate of national fame. All States long for it, and certainly it is a big factor in the consciousness of national strength which commands the respect of both friends and enemies. It is a national heirloom of priceless value to the people to whom the world accord it and who are ready to fight rather than risk its loss. When the nation to whom it was once universally conceded begins to sneer at it as unimportant, and to ridicule its worth, the tide of the nation's greatness has surely turned . . .

Glory to the nation is what sunlight is to all human beings. Without it the State dwindles in size and grows weak in strength, as the man in a dark dungeon becomes daily whiter, until at last his whiteness passes into the colourlessness of death.

Field Marshal Viscount Wolseley, *The Story of a Soldier's Life*, Vol. II, 1903

Common men endure these horrors and overcome them, along with the insistent yearnings of the belly and reasonable promptings of fear; and in this, I think, is glory.

Colonel John W. Thompson, Jr. *Fix Bayonets*, 1926

There is a true glory and a true honor, the glory of duty done, the honor of the integrity of principle.

General Robert E. Lee, *Southern Historical Society Papers*, XI

THE GOLDEN BRIDGE

Do not thwart an enemy returning home.
To surround an enemy you must leave a way of escape.
Do not press an enemy at bay.

Sun Tzu, *The Art of War,* c. 500 BC, tr. Griffith

For nothing makes men so brave as the fear of what ills they will suffer if they surrender; indeed the expectation of the evils which will ensue from their subjection produces a terrible pertinacity in danger. Moreover, fighting is dangerous against desperate men, who expect from surrender no amelioration of the fate which will be theirs if they continue to fight, and therefore prefer, if they can inflict much harm, also to suffer much.

Onasander, *The General,* AD 58

The maxim of Scipio, that a golden bridge should be made for a flying enemy, has much been commended. For when they have free room to escape they think of nothing but how to save themselves by flight, and the confusion becoming general, great numbers are cut to pieces. The pursuers can be in no danger when the vanquished have thrown away their arms for greater haste. In this case the greater the numbers of the flying army, the greater the slaughter. Numbers are of no significance where troops once thrown into consternation are equally terrified at the sight of the enemy as at their weapons. But on the contrary, men when penned in, although weak and few, become a match for the enemy from the very fact that they have no resource but in despair. The conquer'd's safety is to hope for none.

Flavius Vegetius Renatus, *Military Institutions of the Romans,* c. AD 378

When an enemy army is in flight, you must either build a golden bridge for it or stop it with a wall of steel.

Napoleon, ed. Herold, *The Mind of Napoleon,* 1955

When the enemy is surrounded, it is well to leave a gap in our lines to give them the opportunity to flee, in case they judge that flight is better than remaining and taking their chances.

The Emperor Maurice, *The Strategikon,* c. AD 600

When the two sides are too evenly matched to offer a reasonable chance of early success to either, the statesman is wise who can learn something from the psychology of strategy. It is an elementary principle of strategy that, if you find your opponent in a strong position costly to force, you should leave him a line of retreat – as the quickest way of loosening his resistence. It should, equally, be a principle of policy, especially in war, to provide your opponent with a ladder by which he can climb down.

Captain Sir Basil Liddell Hart, *Strategy*, 1954

We did not set ourselves the task of cutting the last path of retreat of the Hitlerites. If we had done this, it would then have been necessary to root them out there at length, and we would doubtlessly destroyed the city [Cracow]. Tempting as it was to create a ring of encirclement, we did not do that, though it was possible for us to do so. Confronting the enemy with a real threat of envelopment, our troops pushed him out of the city through the straight strike of infantry and tanks.

Marshal of the Soviet Union Ivan S. Konev, 1970, quoted in Leites, *The Soviet Style in War,* 1982

GREAT CAPTAINS/GREATNESS

MAXIM 18. A general of ordinary talent occupying a bad position, and surprised by a superior force, seeks his safety in retreat; but a great captain supplies all deficiencies by his courage, and marches boldly to meet the attack. By this means he disconcerts his adversary, and if this last shows any irresolution in his movements, a skilful leader profiting by his indecision may even hope for victory, or at least employ the day in manoeuvring – at night he entrenches himself, or falls back to a better position. By this determined conduct he maintains the honour of his arms, the first essential to all military superiority.

Napoleon, *The Military Maxims of Napoleon*, 1831, tr. D'Aguilar

---

The right appreciation of their opponents . . ., the audacity to leave for a short space of time, a small force only before them, energy in forced marches, boldness in sudden attacks, the intensified activity which great souls acquire in the moment of danger, these are the ground of such victories . . .

Major General Carl von Clausewitz, *On War*, 1832

Many a time has the soldier's calling exhausted strong characters, and that surprisingly quickly. The fine intellect and resolute will of one year give place to the sterile imagings and faint heart of the next. That is perhaps the tragedy of military greatness.

Field Marshal Paul von Hindenburg, *Out of My Life*, 1920

If I may hazard to set down the qualifications of the great captain, then I should say that they are:

(i) Imagination operating through reason.
(ii) Reason operating through audacity.
(iii) And audacity operating through rapidity of movement.

The first creates unsuspected forms of thought; the second establishes original forms of action; and the third impels the human means at the disposal of the command to accomplish his purpose with the force and rapidity of a thunderbolt. From the mind, through the soul, we thus gain our ends by means of the body.

Major General J. F. C. Fuller, *The Foundations of the Science of War*, 1926

The road to failure is the road to fame – such apparently must be the verdict on posterity's estimate of the world's greatest figures. The flash of the meteor impresses the human imagination more than the remoter splendour of the star, fixed immutably in the high heavens. Is it that final swoop earthwards, the unearthly radiance ending in the common dust, that, by its evidence of the tangible or the finite, gives to the meteor a more human appeal? So with lumanaries of the human system, provided that the ultimate fall has a dramatic note, the memory of spectacular failure eclipses that of enduring success. Again, it may be that the completeness of his course lends individual emphasis to the great failure, throwing his work into clearer relief, whereas the man whose efforts are crowned with permanent success builds a stepping-stone by which others may advance still farther, and so merges his own fame in that of his successors.

Captain Sir Basil Liddell Hart, *A Greater Than Napoleon*, 1926

It may well be that the greatest soldiers have possessed superior intellects, may have been thinkers; but this was not their dominant characteristic . . . [they] owed their successes to indomitable wills and tremendous energy in execution and achieved their initial hold upon the hearts of their troops by acts of demonstrated valor . . . the great leaders are not our responsibility, but God's.

General George S. Patton, Jr., October 1927, *The Patton Papers*, Vol. I, 1972

He must accept the loneliness which according to Faguet, is the 'wretchedness of superior beings'. Contentment and tranquillity and the simple joys which go by the name of happiness are denied to those who fill positions of great power. The choice must be made, and it is a hard one: whence that vague sense of melancholy which hangs about the skirts of majesty, in things no less than in people. One day somebody said to Napoleon, as they were looking at an old and noble monument: 'How sad it is!' 'Yes,' came the reply, 'as sad as greatness.'

Charles de Gaulle, *The Edge of the Sword*, 1932

History shows that rather than resign himself to a direct approach, a Great Captain will take even the most hazardous indirect approach – if necessary, over mountains, deserts, or swamps, with only a fraction of his force, even cutting himself from his communications. Facing, in fact, every unfavourable condition rather than the risk of stalemate. (October 1928)

Captain Sir Basil Liddell Hart, *Thoughts on War*, 1944

It goes without saying that the successes achieved by great men have always depended on their possessing many different faculties. Character alone, if unsupported by other qualities, results only in rashness and obstinacy. On the other hand, purely intellectual gifts, even of the highest order, are not sufficient. History is filled with examples of men who, though they were gifted beyond the ordinary, saw their labours brought to nothing because they were lacking in character. Whether serving, or betraying, their masters in the most expert fashion, they were entirely uncreative. Notable they may have been, but famous never.

Charles de Gaulle, *The Edge of the Sword*, 1932

'Enlightened Views and Supreme Wisdom' . . . are all a matter of intuition and character which no decree can compel, no instruction can impart. Only flair, intelligence and above all the latent eagerness to play a part which alone enables a man to develop ability and strength of character, can be of service. It all comes to this, that nothing great will ever be achieved without great men, and men are great only if they are determined to be so.

Charles de Gaulle, *The Edge of the Sword*, 1932

The commander with the imagination – the genius, in fact – to use the new forces may have his name written among the 'great captains'. But he will not win that title lightly or easily; consider for a moment the qualifications he will require. On the ground he will have to handle forces moving at a speed and ranging at a distance far exceeding that of the most mobile cavalry of the past; a study of naval strategy and tactics as well as those of cavalry (tanks) will be essential to him. Some ideas on his position in battle and the speed at which he must make his decisions may be derived from the battle of Jutland; not much from Salisbury Plain or the Long Valley. Needless to say, he must be able to handle Air Forces with the same knowledge as forces on land.

Field Marshal Earl Wavell, *Generals and Generalship*, 1941

Many times during the war I have tried to analyse the ingredients of the 'big man'. The following are the points that I consider important:

(a) He should be able to sit back and avoid getting immersed in the detail.
(b) He must be a good 'picker' of men.
(c) He should trust those under him, and let them get on with their job without interference.
(d) He must have the power of clear decision.
(e) He should inspire confidence.
(f) He must not be petty.
(g) He should not be pompous.

Major General Sir Francis de Guingand, *Operation Victory*, 1947

1805

At Viscount Nelson's lavish funeral,
  While the mob milled and yelled about
  St Paul's,
A General chatted with an Admiral:

'One of your Colleagues, Sir, remarked today
  That Nelson's *exit*, though to be lamented,
Falls not inopportunely, in its way.'

'He was a thorn in our flesh,' came the reply –
  The most bird-witted, unaccountable,
Odd little runt that ever I did spy.

'One arm, one peeper, vain as Pretty Poll,
  A meddler, too, in foreign politics
And gave his heart in pawn to a plain moll.

'He would dare lecture us Sea Lords, and then
Would treat his ratings as though men of honour
And play at leap-frog with his midshipmen!

'We tried to box him down, but up he popped,
And when he'd banged Napoleon at the Nile
Became too much the hero to be dropped.

'You've heard that Copenhagen "blind eye" story?
We'd tied him to Nurse Parker's apron-strings –
By G-d, he snipped them through and snatched the glory!'

'Yet,' cried the General, 'six-and-twenty sail
Captured or sunk by him off Trafalgar –
That writes a handsome *finis* to the tale.'

'Handsome enough. The seas are England's now.
The fellow's foibles need no longer plague us.
He died most creditably, I'll allow.'

'And, Sir, the secret of his victories?'
'By his unServicelike, familiar ways, Sir,'
he made the whole Fleet love him, damn his eyes!'

Robert Graves, *New Collected Poems*, 1977

HERO/HEROISM

Though the actions of our hero shine with great brilliancy, they must not be imitated, except with peculiar caution. The more resplendent they are, the more easily may they seduce the youthful, headlong, and angry warrior, to whom we cannot often enough repeat that valour without wisdom is insufficient, and that the adversary with a cool head, who can combine and calculate, will finally be victorious over the rash individual.

Frederick the Great, 'Reflections on Charles XII', *Posthumus Works of Frederick II, King of Prussia*, tr. Holcroft, 1789; in ed. Luvaas, *Frederick the Great on the Art of War*, 1966

They say: 'Exalted is the Sun of our land!' We were lost in the land daily in the darkness, which King Ramses III has expelled. The lands and countries are stripped, and brought to Egypt as slaves; gifts gathered together for her gods' satiety, provisions, supplies, are a flood in the Two Lands. The multitude rejoices in this land, none is sad, for Amon has established his son upon his throne, all the circuit of the sun is united in his grasp; the vanquished Asiatics and the Tehenu. Taken are those who were spoiling the condition of Egypt. The land had been exposed in continual extremity, since the former kings. They were desolated, the gods as well as people. There was no hero to seize them when they retreated. Lo, there was a youth like a gryphon, like a bull ready for battle upon the field. His horses were like hawks . . . roaring like a lion terrible in rage . . . His name is a flame, the terror of him is in the countries.

Ramses III, Egyptian Pharaoh, c. 1193 BC, description of the First Libyan War, quoted in Breasted, *Ancient Records of Egypt*, 1906

High and dangerous action teaches us to believe as right beyond dispute things for which our doubting minds are slow to find words of proof; out of heroism grows faith in the worth of heroism.

Justice Oliver Wendell Holmes, Jr. (1841–1935)

At the grave of the hero we end, not with sorrow at the inevitable loss, but with the contagion of his courage; and with a kind of desperate joy we back to the fight.

Justice Oliver Wendell Holmes, Jr. (1841–1935)

War is, or anyhow should be, an heroic undertaking; for without heroism it can be no more than an animal conflict, which in place of raising man through an ideal, debases him through brutality.

Major General J. F. C. Fuller, *Generalship: Its Diseases and Their Cure*, 1933

Soldiers, all men in fact, are natural hero-worshippers. Officers with a flame for command realize this and emphasize in their conduct, dress and deportment the qualities they seek to produce in their men . . . the influence one man can have on thousands is a never ending source of wonder to me . . .

General George S. Patton, Jr., 6 June 1944, *The Patton Papers*, Vol. II, 1974

It is an unfortunate and, to me, tragic fact that, in our attempts to prevent war, we have taught our people to belittle the heroic qualities of the soldier. They do not realize that, as Shakespeare put it, the pursuit of 'The bubble reputation even at the cannon's mouth' is not only a good military characteristic, but also very helpful to the young man when bullets and shells are whistling and cracking around him. Much more could be done if the women of America would praise their heroes . . .

General George S. Patton, Jr., *War As I Knew It*, 1947

On a trip to the West Coast, President Kennedy was asked by a little boy, 'Mr. President, how did you become a war hero?' 'It was involuntary. They sank my boat.'

President John F. Kennedy, quoted in Adler, *The Kennedy Wit*, 1964

It is in disaster, not success, that the heroes and the bums really get sorted out.

Admiral James B. Stockdale, 'In War, In Prison, In Antiquity', *Parameters*, 12/1987

HIERARCHY/SENIORITY

. . . the military hierarchical system must come to the aid of subordination, even of ideas.

Field Marshal Helmuth Graf von Moltke, 'The Italian Campaign of 1859', quoted in von Freytag-Loringhoven, *Generalship in the Great War*, 1920

It follows that the stuffed dummies of the hierarchy can never enjoy prestige, for they are parasites who take everything and give nothing in return, weak-kneed creatures forever trembling in their shoes, jumping jacks who will turn their coats without scruple at the first opportunity. They can often safeguard their official careers, their rank if they are soldiers, their portfolios if they are ministers. They even, on occasion, receive the deference which custom and convention accords to their office . . . But such cold and shrewd intelligences can never command the confidence and enthusiasm of others.

Charles de Gaulle, *The Edge of the Sword*, 1932

It is a military convention that infallibility is the privilege of seniority. (May 1933)

Captain Sir Basil Liddell Hart, *Thoughts on War*, 1944

The pretence to infallibility is instinctive in a hierarchy. If military commanders are prone to it that is only because of the prominent part that authority plays in the military profession. But to understand the cause is not to underrate the harm that this pretence has produced – in every sphere. We learn from history that the critics of authority have always been rebuked in self-righteous tones – if no worse fate has befallen them – yet have often been justified by history. (January 1934)

Captain Sir Basil Liddell Hart, *Thoughts on War*, 1944

HISTORY

The more an army lacks war experience the more it needs to make use of the history of war for its instruction. Although the history of war is no substitute for actual experience it can be a foundation for such experience. In peace times it becomes the true method of learning war and of determining the invariable principles of the art of war.

General de Peuker, quoted in Foch, *The Principles of War*, 1913

It is not only Experience and Practice which maketh a Soldier worthy of his Name: but the knowledge of the manifold Accidents which arise from the variety of humane actions, which is best, and most speedily learned by reading History: for upon the variety of Chances that are set forth, he may meditate on the Effects of other men's adventures, that their harms may be his warnings, and their happy proceedings his fortunate Directions in the Art Military.

Anonomous Captain J. S., *Military Discipline – Or the Art of War,* 1689

I had it from Xenophon, but our friends here are astonished at what I have done because they have read nothing.

Major General James Wolfe, of his use of light infantry at the siege of Louisburg, 1758, quoted in Liddell Hart, *Great Captains Unveiled,* 1927

MAXIM 78. Peruse again and again the campaigns of Alexander, Hannibal, Caesar, Gustavus Adolphus, Turenne, Eugene and Frederick. Model yourself upon them. This is the only means of becoming a great captain, and of acquiring the secret of the art of war. Your own genius will be enlightened and improved by this study, and you will learn to reject all maxims foreign to the principles of the great commanders.

Napoleon, *The Military Maxims of Napoleon,* 1831, tr. D'Aguilar

---

The knowledge of higher leadership can only be acquired by the study of military history and actual experience. There are no hard and fast rules; everything depends on the plans of the general, the condition of his troops, the season of the year, and a thousand other circumstances, which have the effect that no one case will ever resemble another.

Napoleon, quoted in von Freytag-Loringhoven, *Generalship in the World War,* 1920

. . . The study of military history thus becomes very instructive in a strategical point of view; whilst, on the other hand, in endeavoring to apply . . . the tactics of the ancients to our modern armies, errors of the gravest character might be committed. Every servile imitation in this latter case . . . from not having weighed the enormous difference . . . between . . . present day . . . firearms, and the weapons for like purpose used by the ancients . . . is greatly to be deprecated . . . Misapprehension of the value of a new agent renders all celerity, that secret of success, impossible.

Dennis Hart Mahan, *Advance Guard, Outpost and Detachment Service of Troops, with the Essential Principles of Strategy and Tactics,* 1864

Only the study of military history is capable of giving those who have no experience of their own a clear picture of what I have just called the friction of the whole machine.

Major General Carl von Clausewitz, *Principles of War,* 1812

Military history, accompanied by sound criticism, is indeed the true school of war.

Lieutenant General Antoine-Henri Baron de Jomini, *Summary of the Art of War,* 1838

It is in military history that we are to look for the source of all military science. In it we shall find those exemplifications of failure and success by which alone the truth and value of the rules of strategy can be tested.

Dennis Hart Mahan, *Advance Guard, Outpost and Detachment Service of Troops, with the Essential Principles of Strategy and Tactics*, 1864

The smallest detail, taken from the actual incident in war, is more instructive to me, a soldier, than all the Thiers and Jominis in the world. They speak, no doubt, for the heads of states and armies but they never show me what I wish to know – a battalion, a company, a squad in action.

Colonel Charles Ardnant du Picq, *Battle Studies*, 1880

Although the story of a campaign is made up of many details which cannot be omitted, since they are essential to the truth as well as the interest of the account, it is of paramount importance that the reader should preserve a general idea . . . To appreciate the tale it is less necessary to contemplate the wild scenes and stiring incidents, than thoroughly to understand the logical sequence of incidents which all tend ultimately to culminate in a decisive trial of strength.

Sir Winston S. Churchill, *The River War*, 1899

It is now accepted with naval and military men who study their profession, that history supplies the raw material from which they are to draw their lessons, and reach their working conclusions. Its teachings are not, indeed, pedantic precedents; but they are the illustrations of living principles.

Rear Admiral Alfred Thayer Mahan, *From Sail to Steam*, 1907

To every one who desires to become a commander, there is available a book entitled 'military history' . . . This reading matter, I must admit, is not always highly favoured. One will have to work through a mass of ingredients anything but palatable. But behind it all one arrives just the same at the facts, often most gratifying facts; at the bottom lies the knowledge as to how everything happened, how it had to happen, and how it will happen again . . .

Field Marshal Alfred Graf von Schlieffen, 15 October 1910, speech at the centenary celebration of the Kriegsacademie, quoted in von Freytag-Loringhoven, *Generalship in the Great War*, 1920

. . . the most effective means of teaching war during peace.

Field Marshal Helmuth Graf von Moltke (1800–1891)

We cannot draw our inspiration indifferently from Turenne, Condé, Prince Eugene, Villars, or Frederick the Great, even less from the tottering theories and degenerate forms of the last century. The best of these doctrines answered a situation and needs which are no longer ours.

Marshal of France Ferdinand Foch, *The Principles of War*, 1913

History. To keep the brain of an army going in time of peace, to direct it continually towards its task of war, there is no book more fruitful to the student than that of history. If war, in its just aspect, is but a struggle between two wills more or less powerful and more or less informed, then the accuracy of decisions arrived at in war will always depend upon the same considerations as those of the past. The same errors reappear, leading to the same checks. The art of war is always to be drawn from the same sources.

Marshal of France Ferdinand Foch, *Precepts and Judgments*, 1919

Unless history can teach us how to look at the future, the history of war is but a bloody romance.

Major General J. F. C. Fuller, *British Light Infantry in the Eighteenth Century*, 1925

What a fine mirror history is . . . In great events which pass to the bosom of history how clearly do the conduct and acts of those who take an active part in these events show their moral character.

General Kemal Mustapha Ataturk, quoted in Moorehead, *Gallipoli*, 1956

For the first time in my life I have seen 'History' at close quarters, and I know that its actual process is very different from what is presented to posterity.

General Max Hoffmann, *War Diaries and Other Papers*, 1929

. . . The value of the knowledge acquired by study must not be overestimated. The soldier faced with the necessity for independent decision must not mentally search the pages of his professional encyclopaedia nor seek to remember how the great generals of history, from Alexander to Zeiten, would have acted in a similar case. Such knowledge as that derived from the study of the history of war is only of living practical value when it has been digested, when the permanent and the important has been extracted from the wealth of detail and has been incorporated with a man's own mental resources – and it is not every man who has the gift for this.

Colonel General Hans von Seekt, *Thoughts of a Soldier*, 1930

If I am doubted, then again must I ask the student to turn to military history, and not merely examine one or two incidents as I have done, but read and re-read the campaigns of the great captains and study operations of the great fools, for not only are these latter folk in the majority, but their art is immensely instructive. What will the student's verdict be? I imagine that it will agree with mine: namely, that we soldiers are mostly alchemists, and many of us little more than military sorcerers.

Major General J. F. C. Fuller, *The Foundations of the Science of War*, 1926

The discovery of uncomfortable facts had never been encouraged in armies, who treated their history as a sentimental treasure rather than a field of scientific research. (June 1936)

Captain Sir Basil Liddell Hart, *Thoughts on War*, 1944

To make war without a thorough knowledge of the history of war is on a par with the casualness of a doctor who prescribes medicine without taking the trouble to study the history of the case he is treating. (December 1933)

Captain Sir Basil Liddell Hart, *Thoughts on War*, 1944

A military philosopher may recognize the inevitability of gradual evolution, but a military historian must point out how often it has proved fatal to nations. The money spent on armies that failed to adapt themselves to changing conditions has proved too literally a sinking fund – acting as a millstone around the investor's neck when he was plunged into the deep waters of war. (March 1934)

Captain Sir Basil Liddell Hart, *Thoughts on War*, 1944

There is a modern, and too common, tendency to regard history as a specialist subject. On the contrary, it is the corrective to specialization. Viewed right, it is the broadest of studies, embracing every aspect of life. It lays the foundation of education by showing how mankind repeats its follies, and what those follies are . . . It is universal experience – infinitely longer, wider, and more varied than any individual's experience. How often do we hear people claim a knowledge of the world and of life because they are sixty or seventy years old. Most of them might be described as a 'young sixty, or seventy'. There is no excuse for any literate person if he is less than three thousand years old in mind.

Captain Sir Basil Liddell Hart, *Through the Fog of War*, 1938

More than most professions the military is forced to depend upon intelligent interpretation of the past for signposts charting the future. Devoid of opportunity, in peace, for self-instruction through actual practice of his profession, the soldier makes maximum use of historical record in assuring the readiness of himself and his command to function efficiently in emergency. The facts derived from historical analysis he applies to conditions of the present and the proximate future, thus developing a synthesis of appropriate method, organization, and doctrine.

General of the Army Douglas MacArthur, *Annual Report of the Chief of Staff for FY June 30, 1935*

It is true that full study of war will not seriously assist a subaltern on picket duty; but when it comes to understanding present war conditions and the probable origins of the next war, a deep and impartial knowledge of history is essential. Further still, as it is not subalterns or generals who make wars, but governments and nations, unless the people as a whole have some understanding of what war meant in past ages, their opinions on war . . . today will be purely alchemical.

Major General J. F. C. Fuller, *Decisive Battles: Their Influence Upon History and Civilization,* 1939

The real way to get value out of the study of military history is to take particular situations, and as far as possible get inside the skin of the man who made a decision, realize the conditions in which the decision was made, and then see in what way you could have improved upon it.

Field Marshal Earl Wavell, 1930, lecture to officers at Aldershot

---

Study the human side of military history, which is not a matter of cold-blooded formulas of diagrams, or nursery-book principles, such as: Be good and you will be happy. Be mobile and you will be victorious. Interior lines at night are the general's delight. Exterior lines at morning are the general's warning. And so on.

Field Marshal Earl Wavell, 1930, lecture to officers at Aldershot

---

To be a successful soldier, you must know history. Read it objectively . . . You must read biography and especially autobiography . . .

General George S. Patton, Jr., 6 June 1944, letter to his son, *The Patton Papers*, Vol. II, 1974

The value of military history is the creative perception of the experience and lessons of the past in the capability to disclose the regular laws of the development of methods for the conduct of war, in its boundless capabilities for the expansion of the military world outlook and military thinking of officers and generals.

Marshal of the Soviet Union Andrei Grechko, *Voyenno-istoricheskiy zhurnal*, 2/1961

In my view, the single most important foundation for any leader is a solid academic background in history. That discipline gives perspective to the problems of the present and drives home the point that there is really very little new under the sun. Whenever a policy maker starts his explication of how he intends to handle a problem with such phrases as 'We are at the take-off point of a new era . . . ' You know you are heading for trouble. Starting by ignoring the natural yardstick of 4,000 years of recorded history, busy people, particularly busy opportunists, have a tendency to see their dilemmas as so unique and unprecedented that they deserve to make exceptions to law, custom, or morality in their own favor to get around them . . .

Admiral James B. Stockdale, 'Educating Leaders', *The Washington Quarterly*, Winter 1983

HONOUR

The honourable thing, that which makes the real general is to have clean hands.

Aristides (*c.* 530 BC – *c.* 468 BC), to Themistocles, quoted in Plutarch, (*c.* AD 46–after AD 119) *Lives of the Noble Grecians and Romans*

In no way should a sworn agreement made with the enemy be broken.

The Emperor Maurice, *The Strategikon*, *c.* AD 600

Your Majesty may send me to attack the enemy batteries, and I will readily obey. But to act contrary to my honour, oath and duty is something which my will and conscience do not permit me to do.

General F. C. von Saldern, 1761, response to Frederick the Great's order to commit the atrocity of destroying the royal palaces of Saxony, quoted in Duffy, *The Military Life of Frederick the Great,* 1986

The soldier trade, if it is to mean anything at all, has to be anchored to an unshakeable code of honour. Otherwise, those of us who follow the drums become nothing more than a bunch of hired assassins walking around in gaudy clothes . . . a disgrace to God and mankind.

Major General Carl von Clausewitz, *On War*, 1832

I would lay down my life for America, but I cannot trifle with my honour.

Admiral John Paul Jones, 4 September 1777, letter to A. Livingston

---

The bravest man feels an anxiety *circa praecordia* as he enters the battle; but he dreads disgrace more.

Admiral Lord Nelson, quoted in Mahan, *The Life of Nelson*, 1897

What is life without honour? Degradation is worse than death. We must think of the living and of those who are to come after us, and see that by God's blessing we transmit to them the freedom we have ourselves inherited.

Lieutenant General Thomas 'Stonewall' Jackson, 1862

The muster rolls on which the name and oath were written were pledges of honour redeemable at the gates of death. And they who went up to them, knowing this, are on the lists of heroes.

Major General Joshua Lawrence Chamberlain, *The Passing of the Armies*, 1915

Never give in, never give in, never, never, never – in nothing great or small, large or petty – never give in except in convictions of honour and good sense.

Sir Winston S. Churchill, 29 October 1941, address at Harrow School

---

An officer must teach his men to honour the country and the uniform and to respect the symbols of these things. But one way in which to do them honour is to begin to understand the limits in which they operate as forces influencing the conduct and shaping the fortunes of combat troops. The rule for the soldier should be that given the Australian mounted infantryman when he asked the Sphinx for the wisdom of the ages: 'Don't expect too much!'

Brigadier General S. L. A. Marshall, *Men Against Fire*, 1947

A man has honour if he holds himself to a course of conduct because of a conviction that it is in the general interest, even though he is well aware that it may lead to inconvenience, personal loss, humilitation, or grave personal risk.

Brigadier General S. L. A. Marshall, *The Armed Forces Officer*, 1950

What I want you to preserve is honour, and not a few planks of wood.

Napoleon, 1804, letter to the Minister of Marine, in Herold, ed., *The Mind of Napoleon*, 1955

Whoever prefers death to ignominity will save his life and live in honour, but he who prefers life will die and cover himself with disgrace.

Napoleon, 1809, letter to General Clarke, in Herold, ed., *The Mind of Napoleon*, 1955

The honour of a general consists in obeying, in keeping subalterns under his orders on the honest path, in maintaining good discipline, devoting oneself solely to the interests of the State and the sovereign, and in scorning completely private interests.

Napoleon, 8 June 1811, letter to Marshal Bertier, *Correspondance*, No. 17782, Vol. XXII, 1858–1870

HUMAN NATURE/ THE HUMAN HEART/MAN

The same troops, who if attacking would have been victorious, may be invariably defeated in entrenchments. Few men have accounted for it in a reasonable manner, for it lies in human hearts and one should search for it there. No one has written of this matter which is the most important, the most learned, and the most profound of the profession of war. And without a knowledge of the human heart, one is dependent upon the favour of fortune, which sometimes is very inconstant.

Field Marshal Maurice Comte de Saxe, *My Reveries*, 1732

The human character is the subject of the military officer's study; for it is upon man that his trials are made. He must, therefore, know, in the most precise manner, what man can do, and what he cannot do; he must also know the means by which his exertions are to be animated to the utmost extent of exertion. The general's duty is consequently an arduous duty; the capacity of learning it is the gift of Nature; the school is in the camp and the cottage rather than the city and the palace; for a man cannot know things in their foundations till he sees them without disguise; as he cannot judge of the hardships of service till he has felt them in experience . . .

Robert Jackson, *A Systematic View of the Formation, Discipline, and Economy of Armies*, 1804

Man, not men, is the most important consideration.

Napoleon, *Maxims of War*, 1831

Sentiment rules the world, and he who fails to take that into account can never hope to lead.

Napoleon (1769–1821)

Remember also that one of the requisite studies for an officer is *man*. Where your analytical geometry will serve you once, a knowledge of men will serve you daily. As a commander, to get the right man in the right place is one of the questions of success or defeat.

Admiral David G. Farragut, 13 October 1864, letter to his son

The man whose profession is arms should calm his mind and look into the depths of others. Doing so is likely the best of the martial arts.

Shiba Yoshimasa (1350–1410), *Chikubasho*, 1380, in Wilson, *Ideals of the Samurai*, 1982

Men are your castles
Men are your walls
Sympathy is your ally
Enmity your foe.

Takeda Shingenm (1521–1573), in Wilson, *Ideals of the Samurai*, 1982

In all things, think with one's starting-point in man.

Nabeshima Naoshige (1538–1618), in Wilson, *Ideals of the Samurai*, 1982

There is a soul to an army as well to the individual man, and no general can accomplish the full work of his army unless he commands the souls of his men, as well as their bodies and legs.

General of the Army William T. Sherman, *Memoirs of General W. T. Sherman*, 1875

Man is the first weapon of battle: let us then study the soldier in battle, for it is he who brings reality to it. Only the study of the past can give us a sense of reality, and show us how the soldier will fight in the future.

Colonel Charles Ardnant du Picq, *Battle Studies*, 1880

On foot, on horseback, on the bridge of ä vessel, at the moment of danger, the same man is found. Anyone who knows him well, deduces from his action in the past what his future action will be.

Colonel Charles Ardnant du Picq, *Battle Studies*, 1880

. . . I will again sound the note of warning against that plausible cry of the day which finds *all* progress in material advance, disregarding that noblest sphere in which the mind and heart of man, in which all that is god-like in man, reign supreme; and against the temper which looks not to man, but to his armor.

Rear Admiral Alfred Thayer Mahan, *Naval Administration and Warfare*, 1911

Give me soldiers prepared to fight to the death, and I will look after the tactics. Men, men, and again men. Always men – they are the first and best instrument of battle.

General Mikhail I. Dragomirov (1830–1895)

It is the leader who reckons with the human nature of his troops, and of the enemy, rather than with their mere physical attributes, numbers, armament and the like, who can hope to follow in Napoleon's footsteps.

Colonel G. F. R. Henderson, quoted in Marshall, *The Armed Forces Officer*, 1950

We will have to say that in any cause the decisive role does not belong to technology – behind technology there is always a living person without whom the technology is dead.

Mikhail Frunze (1885–1925), quoted in Gareyev, *Frunze, Military Theorist*, 1985

Wars may be fought with weapons, but they are won by men. It is the spirit of the men who follow and the man who leads that gains the victory.

General George S. Patton, Jr., in *The Cavalry Journal*, 9/1933

This is the so-called theory that 'weapons decide everything', which constitutes a mechanical approach to the question of war and a subjective and one-sided view. Our view is opposed to this; we see not only weapons but also people. Weapons are an important factor in war, but not the decisive factor; it is people, not things, that are decisive. The contest of strength is not only a contest of military and economic power, but also a contest of human power and morale.

Mao Tse-tung, *On Protracted War*, May 1938

The art of leading, in operations large or small, is the art of dealing with humanity, of working diligently on behalf of men, of being sympathetic with them, but equally, of insisting they make a square facing toward their own problems.

Brigadier General S. L. A. Marshall, *Men Against Fire*, 1947

A commander should have a profound understanding of human nature, the knack of smoothing out troubles, the power of winning affection while communicating energy, and the capacity for ruthless determination where required by circumstances. He needs to generate an electrifying current and to keep a cool head in applying it. (October 1933)

Captain Sir Basil Liddell Hart, *Thoughts on War*, 1954

An army must be as hard as steel in battle and can be made so; but, like steel, it reaches its finest quality only after much preparation and only provided the ingredients are properly constituted and handled. Unlike steel, an army is a most sensitive instrument and can easily become damaged; its basic ingredient is men and, to handle an army well, it is essential to understand human nature. Bottled up in men are great emotional forces which have got to be given an outlet in a way which is positive and constructive, and which warms the heart and excites the imagination. If the approach to the human factor is cold and impersonal, then you achieve nothing. But if you gain the confidence and trust of your men, and they feel their best interests are safe in your hands, then you have in your possession a priceless asset and the greatest achievements become possible.

Field Marshal Viscount Montgomery of Alamein, *The Memoirs of Field Marshal Montgomery*, 1958

It is essential to understand that battles are won primarily in the hearts of men.

Field Marshal Viscount Montgomery of Alamein, *The Memoirs of Field Marshal Montgomery*, 1958

I was blamed for excessive demands and exactness which I consider to be a necessary quality of any commander – bolshevik. Looking at this part, I realize that I was sometimes excessively demanding and I did not always control myself when offences of my subordinates were involved. I used to lose control over myself because of carelessness in any work and in the behaviour of the military men. Some of them did not understand that and I probably wasn't soft enough toward human weaknesses.

Marshal of the Soviet Union Georgi K. Zhukov, quoted in Gareyev, *Frunze, Military Theorist*, 1985

. . . the essential nature of war has not changed. Wars are fought by men, and there has been no discernible difference in the fundamental nature of man over the past five thousand years of recorded history. Because the nature of man has not changed, neither has his basic objective when he turns to war: the employment of lethal instruments to force his will upon other men with opposing points of view.

Colonel N. T. Dupuy, *Understanding War*, 1987

HUMANITY

Do no put a premium on killing.

Li Ch'uan, Tang Dynasty commentator on Sun Tzu, *The Art of War*, c. 500 BC, tr. Griffith

---

No man is justified in doing evil on the ground of expediency.

President Theodore Roosevelt, *The Strenuous Life*, 1900

As we turned back it began to snow; and only very late, and by a last effort did we get our hurt men in. The Turkish wounded lay out, and were dead next day. It was indefensible, as was the whole theory of war: but no special reproach lay on us for it. We risked our lives in the blizzard (the chill of victory bowing us down) to save our own fellows; and if our rule was not to lose Arabs to kill even many Turks, still less might we lose them to save Turks.

Colonel T. E. Lawrence, *The Seven Pillars of Wisdom*, 1926

The art of leading, in operations large or small, is the art of dealing with humanity.

Brigadier General S. L. A. Marshall, *Men Against Fire*, 1947

. . . quite contrary to trivial opinion, all professional military men do not walk blind and brutal. I have known some who demonstrated as much pity as they did courage, and they showed a lot of that.

When you are dealing firsthand with the quivering element of life and death – When you are trying to figure out the best manner in which to save certain lives as well as to take others, and in the same operation – You do not necessarily become calloused. Neither does a surgeon.

General Curtis LeMay, *Mission With LeMay*, 1965

Kind-hearted people might of course think there was some ingenious way to disarm or defeat an enemy without too much bloodshed, and might imagine this is the true goal of the art of war. Pleasant as it sounds, it is a fallacy that must be exposed: war is such a dangerous business that the mistakes which come from kindness are the very worst.

Major General Carl von Clausewitz, *On War*, i, 1832, tr. Howard and Paret

---

Men who take up arms against one another in public do not cease on this account to be moral beings, responsible to one another and to God.

U.S. Army General Order No. 100, 1863

The greatest kindness in war is to bring it to a speedy conclusion.

Field Marshal Helmuth Graf von Moltke, 11 December 1800

HUMOUR

To speak of the importance of a sense of humor would be futile, if it were not that what cramps so many men isn't that they are by nature humorless as that they are hesitant to exercise what humor they possess. Within the military profession, this is as unwise as to let the muscles go soft or to spare the mind the strain of original thinking. Great humor has always been in the military tradition.

Brigadier General S. L. A. Marshall, *The Armed Forces Officer*, 1950

Humor is an effective but tricky technique in command and leadership, beneficial when used wisely and with skill, but it can backfire into a dangerous booby-trap if overworked or crudely employed.

Major General Aubrey 'Red' Newman, *Follow Me*, 1981

Mix empathy with your humor; consider how the jest you have in mind will affect others before you utter it.

Major General Aubrey 'Red' Newman, *Follow Me*, 1981

Like profanity, sarcasm and irony often leave men in doubt as to exactly what the leader means. Even a bantering tone should not be used often . . . At the same time, any wise leader will know that in some circumstances a certain amount of joking is helpful. During periods of exhaustion and discouragement, humor may impart confidence or relieve tension. Often humor is well received as a means implying sympathy and understanding or cooperation in the midst of difficulty. This method is very effective when employed by those leaders who display great dignity.

US Army Field Manual (FM) 22–100 Leadership, 1983

People are afraid of a leader who has no sense of humor. They think that he is not capable of relaxing, and as a result of this there is a tendency for that leader to have a reputation for pomposity, which may not be the case at all. Humor has a tendency to relax people in times of stress.

Louis H. Wilson, quoted in Montor, *Naval Leadership: Voices of Experience*, 1987

HUNGER

. . . he [the general] should not hesitate to order the first meal at sunrise, lest the enemy, by a prior attack force his men to fight while still hungry. On the whole, this matter must not be considered of slight importance nor should a general neglect to pay attention to it; for soldiers who have eaten moderately, so as not to put too great a load into their stomachs, are more vigorous in battle; armies have often been overpowered for just this reason, their strength failing for lack of food – that is, whenever the decision rests, not on a moment's fighting, but when the battle lasts throughout the entire day.

Onasander, *The General*, AD 58

Famine makes greater havoc in an army than the enemy, and is more terrible than the sword.

Flavius Vegetius Renatus, *Military Institutions of the Romans*, c. AD 378

The general achieves the most who tries to destroy the enemy's army more by hunger than by force of arms.

The Emperor Maurice, *The Strategikon*, c. AD 600

Few victories are won on an empty belly.

Sir John Hawkwood (1320–1394)

The greatest secret of war and the masterpiece of a skillful general is to starve his enemy. Hunger exhausts men more surely than courage, and you will succeed with less risk than by fighting.

Frederick the Great, *Instructions to His Generals*, 1747

The first art of a military leader is to deprive the enemy of subsistence.

Field Marshal Prince Aleksandr V. Suvorov (1729–1800)

A starving army is actually worse than none. The soldiers lose their discipline and their spirit. They plunder even in the presence of their officers.

The Duke of Wellington, August 1809, letter from Spain

The person selected to feed the army was a metaphysical dyspeptic, who it was said, lived on rice-water, and had a theory that soldiers could do the same. A man, to fill such a position well, should be in sympathy with hungry men, on the principle that he who drives fat oxen must himself be fat.

Colonel John S. Mosby, *Mosby's War Reminiscences*, 1887

The waning *morale* at home was intimately connected with the food situation. In the daily food the human body did not receive the necessary nourishment, especially albumen and fats, for the maintenance of physcial and mental vigour. In wide quarters a certain decay of bodily and mental powers of resistance was noticeable, resulting in an unmanly and hysterical state of mind which under the spell of enemy propaganda encourage the pacifist leanings of many Germans. In the summer of 1917 my first glimpse of this situation gave me a great shock. This state of mind was a tremendous element of weakness. It was all a question of human nature. It could be eliminated to some extent by strong patriotic feeling, but in the long run could only be finally overcome by better nourishment. More food was needed.

General Erich Ludendorff, *My War Memories, 1914–1918*, 1919

Nothing undermines morale more decisively than hunger; quickest of all is the effect of any digestive upset. That was strikingly demonstrated in the late summer of 1918, when the morale slump of the German troops became most marked at a moment when stomach disorders, due to bad food, were rife among them. The old saying that 'an army marches on its stomach' has a wider and deeper application than has yet been given to it. An army fights on its stomach, and falls if its stomach is upset. (May 1939)

Captain Sir Basil Liddell Hart, *Thoughts on War*, 1944

See also: LOGISTICS

IDEALISM/IDEALS

It is, indeed, an observable fact that all leaders of men, whether as political figures, prophets, or soldiers, all those who can get the best out of others, have always identified themselves with high ideals, and this has given added scope and strength to their influence. Followed in their lifetime because they stand for greatness of mind rather than self-interest, they are later remembered less for the usefulness of what they have achieved than for the sweep of their endeavours. Though sometimes reason may condemn them, feeling clothes them in an aura of glory.

Charles de Gaulle, *The Edge of the Sword*, 1932

Men who are infused with a faith, even a false one, will beat men who have no faith; only a good one can withstand the impact. Those who complain of the younger generation's lack of patriotism should, rather, reproach themselves for their failure to define and teach patriotism in higher terms than the mere preservation of a geographical area, its inhabitants, and their material interests. Such material appeal offers no adequate inspiration, nor cause for sacrifice to the young. Those who are concerned with practical questions of defence ought to realize the practical importance of ideals. (August 1936)

Captain Sir Basil Liddell Hart, *Thoughts on War*, 1944

Military ideals are not different from the ideals that make any man sound in himself and in his relation to others. They are called military ideals only because the proving ground is a little more rugged in the Service than elsewhere. But they are all founded in hard military experience; they did not find expression because some admiral got it in his head one day to set an unattainable goal for his men, or because some general wished to turn a pious face to the public, professing that his men were aspiring to greater virtue than anything the public knew.

Brigadier General S. L. A. Marshall, *The Officer as a Leader,* 1966

IMAGINATION

. . . without that great gift [imagination] only a very inferior order of ambition in any walk of life can be satisfied, and certainly without it no one can ever become a renowned leader of armies. How largely it was possessed by Moses, Xenophon, Hannibal, Caesar, Turenne, Marlborough, Napoleon, and Wellington! It is said to rule the world . . . And yet, whilst imagination may convert into a poet the man of poor physique, it would not of itself make an able general of him. He who aspires to lead soldiers in war should be not only a thorough master of the soldier's science, but he must possess a healthy strength of body, an iron nerve, calm determination, and be instinct with that electric power which causes men to follow the leader who possesses it, as readily, as surely, as iron filings do the magnet . . . It is the necessity for this rare, this exceptional combination of mental gifts with untiring physical power and stern resolution that accounts for the fact that the truly great commander is rare indeed among God's creations.

Field Marshal Viscount Wolseley, *The Story of a Soldier's Life,* 1869

The development of the power of imagination and its various ramifications is an essential part of general staff training, and an indispensable requisite for leaders of large forces distributed over a considerable area. The ability to form accurate mental pictures of a situation quickly is especially important today when the higher commander cannot hope to see his troops with his physical eyes . . .

General der Infantrie Hugo Baron von Freytag-Loringhoven, *The Power of Personality in War,* 1911

Imagination forms one of the four parts of genius and has itself two clearly marked attributes: the one, fancy or the power of ornamenting facts as, for instance, making the wolf speak Little Red Riding Hood or describing the Battle of Le Cateau; the other, inventions; not fairy tales but machines. Inventions do not often make their first bow to armies on the battle-field. They have been in the air for some time; hawked about the ante-chambers of the men of the hour; spat upon by common sense; cold-shouldered by interests vested in what exists; held up by stale functionaries to whom the sin against the Holy Ghost is to 'make a precedent' – until, one day, arrives a genius who by his imagination sees; by his enthusiasm moves; by his energy keeps moving; by his courage cuts the painter of tradition.

General Sir Ian Hamilton, *The Body and Soul of an Army,* 1921

The most indispensable attribute of the great captain is imagination.

General of the Army Douglas MacArthur, 1959, letter to Liddell Hart

IMPEDIMENTA

When an army brings along a useless crowd, unnecessary baggage, luxury items, expensive equipment which serves no purpose, then a journey of one day will not be completed in four. In crossing deep or swampy rivers, moreover, or passing over bridges the army will suffer no little delay because of the useless crowd. Then, too, there is the food supply which ought to be for the use of the fighting men along with a moderate number of necessary servants. But a useless crowd will devour that and swiftly reduce the army to want and force it to return home without having achieved anything. And so, often what even a strong force of the enemy has been unable to accomplish is brought about by want alone which stems from thoughtlessness. They are the very people who, growing weary of the work and fearful of the enemy, hasten to spread rumours and harmful stories and they devise whatever tricks they can to overturn the plans which are in the interest of the emperor and to bring about a swift return.

General Nikephorus Ouranos, *Campaign Organization and Tactics,* c. AD 994, in Dennis, *Three Byzantine Military Treatises,* 1985

It is necessary from time to time to inspect the baggage and force the men to throw away useless gear. I have frequently done this. One can hardly imagine all the trash they carry with them year after year . . . It is no exaggeration to say that I have filled twenty wagons with rubbish which I have found in the review of a single regiment.

Field Marshal Maurice Comte de Saxe, *My Reveries*, 1732

We can get along without anything but food and ammunition. The road to glory cannot be followed with much baggage.

Lieutenant General Richard S. Ewell, 1862, orders during the Valley Campaign

---

. . . an army is efficient for action and motion exactly in the inverse ratio of its *impedimenta.*

General of the Army William T. Sherman, *Memoirs of General W. T. Sherman,* 1875

A serious handicap on mobility lies in the tendency of armies to increase their requirement. The desire for stronger armament causes a demand for more transport; the desire for better communications leads to more complex communications; the desire for superior organization creates an additional organization to deal with the development. In building up its powers to overcome the enemy's resistance an army is apt to set up an internal resistance to its advance. Thus what has been gained by the advent of mechanized means of movement is largely offset by the growth of impedimenta. It is not a necessary growth, but military evolution has always tended to be a process of accretion, tacking new means on to the old body. The idea of remodelling design is foreign to tradition. And the result today is the paradoxial one that armies have become less mobile as their limbs have become more mobile. The hare dons the shell of the tortoise.

Captain Sir Basil Liddell Hart, *Thoughts on War,* 1944

INCOMPETENCE

Many badly conceived enterprises have the luck to be successful because the enemy has shown an even smaller degree of intelligence.

Thycidides, *The History of the Peloponnesian War,* c. 404 BC

Those whose performance is consistently poor should not be entrusted even with just ordinary responsibilities.

The Emperor Maurice, *The Strategikon,* c. AD 600

One must obey these old fellows who, never having studied their profession, obsessed by an antiquated routine which they call experience, and taking advantage of a long existence which they consider a long life, set out to traduce, pull to pieces, and ridicule budding genius which they detest, because they are compelled to value it more than themselves.

Major General Henry Lloyd, *Introduction of the History of the War in Germany*, 1766

Normally it is not possible for an army simply to dismiss incompetent generals. The very authority which their office bestows upon generals is the first reason for this. Moreover, the generals form a clique, tenaciously supporting each other, all convinced that they are the best possible representatives of the army. But we can at least give them capable assistants. Thus the General Staff officers are those who support incompetent generals, providing the talents that might otherwise be wanting among leaders and commanders.

General Gerhard von Scharnhorst (1755–1813), quoted in Dupuy, *A Genius for War*, 1977

There are field marshals who would not have shone at the head of cavalry regiments and vice versa.

Major General Carl von Clausewitz, *On War*, 1832

There are plenty of small-minded men who, in time of peace, excel in detail, are inexorable in matter of equipment and drill, and perpetually interfere with the work of their subordinates.

They thus acquire an unmerited reputation, and render the service a burden, but they above all do mischief in preventing development of individuality, and in retarding the advancement of independent and capable spirits.

When war arises the small minds, worn out by attention to trifles, are incapable of effort, and fail miserably. So goes the world.

The Archduke Albert (1817–1895), quoted in Fuller, *Generalship: Its Diseases and Their Cure*, 1933

There is an enemy greater than the hospital: the damned fellow who 'doesn't know'. The hint-dropper, the riddle-poser, the deceiver, the word-spinner, the prayer-skimper, the two-faced, the mannered, the incoherent. The fellow who 'doesn't know' has caused a great deal of harm . . . One is ashamed to talk about him. Arrest for the officer who 'doesn't know' and house arrest for the field or general officer.

Field Marshal Prince Aleksandr V. Suvorov, *The Science of Victory*, 1796

Presumably victory is our object. This war is a business proposition; it therefore requires ability. Nevertheless, our Army is crawling with 'duds'; though habitual offenders, they are tolerated because of the camaraderie of the old Regular Army: an Army so small as to permit of all its higher members being personal friends. Good-fellowship ranks with us above efficiency; the result is a military trade union which does not declare a dividend.

Major General J. F. C. Fuller, 1918, *Memoirs of an Unconventional Soldier*, 1936

An ignorant man cannot be a good soldier. He may be brave and audacious, and, in the hand-to-hand struggles of the past, his ignorance may have appeared but a small defect, since he could rapidly clinch with danger. But to-day this defect has grown big; the stout arm of Cannae, of Crécy, or even of Inkerman, demands at least a cunning brain. Fighting intervals and distances have increased, and there is more room for ignorance to display its feathers, and the corridors of fear are long and broad.

Major General J. F. C. Fuller, *The Foundations of the Science of War*, 1926

The incompetence of the authorities and of G.H.Q. is greater on the other side than with us; and that is saying a good deal.

General Max Hoffmann, November 1915, *War Diaries and Other Papers*, Vol. I, 1929

I have known him for a long time. I am very fond of him, but he has not yet gone as far as *Caesar's Commentaries* in studying the history of war since he forgot the history he learned at West Point.

John J. Pershing, 1918, comments on a deficient officer, quoted in Vandiver, *Black Jack*, 1977

More and more does the 'system' tend to promote to *control*, men who have shown themselves efficient *cogs* in the machine. It is especially so in the Army to-day. There are few commanders in our higher commands. And even these, since their chins usually outweigh their foreheads, are themselves outweighted by the majority – of commanders who are essentially staff officers. These tend to be *desperately* conventional in organizing the Army and moulding its doctrines for war. And in war they would almost certainly prove *recklessly* cautious. (May 1936)

Captain Sir Basil Liddell Hart, *Thoughts on War*, 1944

It is always a bad sign in an army when scapegoats are habitually sought out and brought to sacrifice for every conceivable mistake. It usually shows something very wrong in the highest command. It completely inhibits the willingness of junior commanders to take decisions, for they will always try to get chapter and verse for everything they do, finishing up more often than not with a miserable piece of casuistry instead of the decision which would spell release. The usual result is that the man who never does more than supinely pass on the opinion of his seniors is brought to the top, while the really valuable man, the man who accepts nothing ready-made but has an opinion of his own, gets put on the shelf.

Field Marshal Erwin Rommel, *The Rommel Papers*, 1953

It takes a war to separate the men from the boys. To many serving officers, the army is simply a better 'ole. They do duty routinely in peace time and develop a keen eye for a secure spot where they will have minimum problems and maximum comfort.

Though some so-called line officers are in this category, their wearing of the crossed rifles or cannon is a masquerade; they are not interested in combat, and if they think of it at all, they are caught between chill and sweat. They have no intention of ever going into battle. Should the accident of assignment force them to do so, they invariably fall apart and are relieved for cause, provided they live long enough. This type of officer usually knows the regulations to the letter. To his mind, they are as inviolable as the tablets from Mount Sinai were intended to be. Being what he is, he cannot understand that regulations are nothing more than a general guidance to conduct, written to protect and advise the soldier, but never intended to fetter him and circumscribe his action when great daring and initiative are required in coping with some unprecedented problem.

Brigadier General S. L. A. Marshall, *Bringing Up the Rear*, 1979

INDEPENDENCE OF COMMAND

MAXIM 72. A general-in-chief has no right to shelter his mistakes in war under cover of his sovereign, or of a minister, when they are both distant from the scene of operation, and must consequently be either ill informed or wholly ignorant of the actual state of things.

Hence it follows that every general is culpable who undertakes the execution of a plan which he considers faulty. It is his duty to represent his reasons, to insist upon a change of plan; in short, to give in his resignation rather than allow himself to become the instrument of his army's ruin. Every general-in-chief who fights a battle in consequence of superior orders, with the uncertaintly of losing, is equally blameable.

Napoleon, *Military Maxims of Napoleon*, 1831, tr. D'Aguilar

TO FIELD MARSHAL ROMMEL

In the situation in which you find yourself there can be no other thought but to stand fast and throw every gun and every man into the battle. The utmost efforts are being made to help you. Your enemy, despite his superiority, must also be at the end of his strength. It would not be the first time in history that a strong will has triumphed over the bigger battalions. As to your troops, you can show them no other road than that to victory or death.

ADOLF HITLER

This order demanded the impossible. Even the most devoted soldier can be killed by a bomb.

Erwin Rommel, quoting Hitler's order from East Prussia to stand fast at El Alamein in North Africa and Rommel's own comment, *The Rommel Papers*, 1953

There are commands of the sovereign which must not be obeyed.

Sun Tzu, *The Art of War*, c. 500 BC

THE INDIRECT APPROACH

Force has no place where there is need of skill

Herodotus, (c. 484 BC–c. 420 BC), *The Histories of Herodotus*

Generally, in battle, use the normal force [direct approach] to engage; use the extraordinary [indirect approach] to win.

Now the resources of those skilled in the use of extraordinary forces are as infinite as the heavens and earth; as inexhaustible as the flow of great rivers.

For they end and recommence; cyclical, as are the movements of the sun and moon. They die away and are reborn; recurrent, as are the passing seasons.

The musical notes are only five in number but their melodies are so numerous that one cannot hear them all.

The primary colours are only five in number but their combinations are so infinite that one cannot visualize them all.

The flavours are only five in number but their blends are so various that one cannot taste them all.

In battle there are only the normal and extraordinary forces, but their combinations are limitless; none can comprehend them all.

For these two forces are mutually reproductive; their interaction as endless as that of interlocked rings. Who can determine where one ends and the other begins.

Sun Tzu, *The Art of War*, c. 500 BC, tr. Griffith

. . . three men behind the enemy are worth more than fifty in front of him.

Frederick the Great (1712–1786)

MAXIM 16. It is an approved maxim of war, never do what the enemy wishes you to do, for this reason alone, that he desires it. A field of battle, therefore, which he has previously studied and reconnoitred, should be avoided, and double care should be taken where he has had time to fortify or entrench. One consequence deducible from this principle is, never to attack a position in front which you can gain by turning.

Napoleon, *The Military Maxims of Napoleon*, 1831, tr. D'Aguilar

---

Should one ask: 'How do I cope with a well-ordered enemy host about to attack me?' I reply: 'Seize something he cherishes and he will conform to your desires.

Sun Tzu, *The Art of War*, c. 500 BC, tr. Griffith

It is well to hurt the enemy by deceit, by raids, or by hunger and never be enticed into a pitched battle, which is a demonstration more of luck than of bravery.

The Emperor Maurice, *The Strategikon*, c. AD 600

To see correctly a general must understand the nature of the changes which take place in war. The enemy does not attack him physically, but mentally; for the enemy attacks his ideas, his reason, his plan. The physical pressure directed against his men react on him through compelling him to change his plan, and changes in his plan react on his men by creating a mental confusion which weakens their *morale*. Psychologically, the battle is opened by a physical blow which unbalances the commander's mind, which in its turn throws out of adjustment the *morale* of his men, and leads to their fears impeding the flow of *his* will. If the blow is a totally unexpected one, the will of the commander may cease altogether to flow, and, the balance in the moral sphere being utterly upset, self-preservation fusing with self-assertion results in panic.

Major General J. F. C. Fuller, *The Foundations of the Science of War*, 1926

Throughout history the direct approach has been the normal form of strategy, and a purposeful indirect approach the rare exception. It is curious how often generals have adopted the latter, not as their initial strategy, but as a last resource. Yet it has brought them a decision where the direct approach had brought them failure – and thereby left them in a weakened condition to attempt the indirect. A decisive success obtained in such deteriorated conditions acquires all the greater significance. (October 1928)

Captain Sir Basil Liddell Hart, *Thoughts on War*, 1944

The art of the indirect approach can only be mastered, and its full scope appreciated, by study of and reflection upon the whole history of war. But we can at least crystallize the lessons into two simple maxims, one negative, and the other positive. The first is that in face of the overwhelming evidence of history no general is justified in launching his troops to a *direct* attack upon an enemy *firmly* in position. The second, that instead of seeking to upset the enemy's equilibrium by one's attack, it must be upset *before* a real attack is, or can be, successfully launched . . . (November 1928)

Captain Sir Basil Liddell Hart, *Thoughts on War*, 1944

Battle should no longer resemble a bludgeon fight, but should be a test of skill, a manoeuvre combat, in which is fulfilled the great principle of surprise by striking 'from an unexpected direction against an unguarded spot'.

Captain Sir Basil Liddell Hart, *Thoughts on War*, 1944

You can tell a man's character by the way he makes advances to a woman. Men like you, for example – when the fleet's in port and you go off to have a good time, you seem to have only two ways of going about things. First, you put it straight to the woman: 'Hey, how about a lay?' Now, any woman, even the lowest whore, is going to put up at least a show of refusing if she's asked like that. So what do you do next? You either act insulted and get rough, or you give up immediately and go off to try the same thing on the next woman. That's all you're capable of. But take a look at Western men – they're quite different. Once they've set their sights on a woman, they invite her out for a drink, or to dinner, or to go dancing. In that way they gradually break down her defences until, in the end, they get what they want, and in style at that. Where achieving a particular aim is concerned, that's surely a far wiser way of going about things. At any rate, they're the kind of men you'd be dealing with if there were a war, so you'd better give it some thought.

Admiral Yamamoto Isoroku, 1939, to his junior officers, quoted in Agawa, *The Reluctant Admiral*, 1979

Use streamroller strategy; that is, make up your mind on course and direction of action, and stick to it. But in tactics, do not streamroller. Attack weakness. Hold them by the nose and kick them in the pants.

General George S. Patton, Jr., *War As I Knew It*, 1947

Never attack where the enemy expects you to come. It is much better to go over difficult ground where you are not expected than it is over good ground where you are expected.

General George S. Patton, Jr., *War As I Knew It*, 1947

The enemy wanted to concentrate their forces. We compelled them to disperse. By successfully launching strong offensives on the points they had left relatively unprotected, we obliged them to scatter their troops all over the place in order to ward off our blows, and thus created favourable conditions for the attack at Dien Bien Phu, the most powerful entrenched camp in Indo–China, considered invulnerable by the Franco–American general staff. We decided to take the enemy by the throat at Dien Bien Phu.

General Vo Nguyen Giap, *People's War, People's Army*, 1961

INFANTRY

Those who go on foot or on a war-horse, with the mettle to take on a hundred men, who are skilled in the use of close-range weapons, swords, and spears, are called infantry generals.

Zhuge Liang (AD 180–234), *The Way of the General*

The infantry must ever be regarded valued as the very foundation and nerve of an army.

Niccolo Machiavelli, *Discourses*, 1517

Infantry is the Queen of Battles.

General Sir William Napier (1785–1860)

MAXIM 93. The better the infantry is, the more it should be used carefully and supported with good batteries. Good infantry is, without doubt, the sinew of an army; but if it is forced to fight for a long time against a very superior artillery, it will become demoralized and will be destroyed.

Napoleon, *The Military Maxims of Napoleon*, 1827, ed., Burnod

---

In the end of ends, infantry is the deciding factor in every battle. I was in the infantry myself and was body and soul an infantry man. I told my sons to join the infantry. They did so, but, as happened to so many of our young men, the freedom of the air drew them from the trenches. But the fine saying of the old 'Directions for Infantry Exercise', remain true in war: 'The infantry bears the heaviest burden of a battle and requires the greatest sacrifice; so also it promises the greatest renown.'

General Erich Ludendorff, My *War Memories, 1914–1918*, 1919

Let us remember the great part that is played by the infantry soldier in war. The artillery help us, the cavalry help us, and the engineers are there to confirm our success and overcome obstacles, but it is the infantry soldier, officer and man, who must bear the great stress of battle. He has no immovable gun to serve, and no horse to carry him forward; of his own initiative he advances against the enemy position with a fixed determination to drive him back and defeat him. It is not generally considered that he requires any very high training, and yet his training in the correct use of ground and of his rifle, in the dire stress of battle, is more complicated and more difficult than that of any other arm of the service. He is more influenced than any other soldier by those characteristics of the human mind which are adverse to success; ground is never the same and can never be treated in the same way, and therefore we can give no fixed rules to work by.

Captain R. C. B. Haking, *Company Training*, 1917

It is admitted by all military men that infantry is the great lever of war, and that the artillery and cavalry are only indispensable accessories . . .

Two essential conditions constitute the strength of infantry:

That the men be good walkers and inured to fatigue.

That the firing be well-executed.

The physical constitution, and the national composition of the French armies, fulfill the former most advantageously; the vivacity and intelligence of the soldiers ensure the success of the latter.

Marshal of France Michel Ney, Duc d'Elchingen, Prince de la Moskova, *Memoirs of Marshal Ney*, 1834

Infantrymen are they who in the frosts and storms of night watch over the sleep of camps, climb under fire the highest crests, fight and die, without their voluntary sacrifice receiving the reward of heroism.

General Francesco Franco, *Diario de Una Bandera*, 1922

I love the infantrymen because they are the underdogs. They are the mud-rain-and-wind boys. They have no comforts, and they learn to live without the necessities. And in the end they are the guys that wars can't be won without.

Ernie Pyle, *New York World Telegram,* 5 May 1943

Look at an infantryman's eyes and you can tell how much war he has seen.

William 'Bill' Mauldin, *Up Front,* 1944

. . . the least spectacular arm of the Army, yet without them you cannot win a battle. Indeed, without them, you can do nothing. Nothing at all, nothing.

Field Marshal Viscount Montgomery of Alamein, quoted in *Military Review,* 5/1981

When the smoke cleared away, it was the man with the sword, or the crossbow, or the rifle, who settled the final issue on the field.

General of the Army George C. Marshall, 3 February 1939, speech before the National Rifle Association

INFORM THE TROOPS

The great impediment to action is, in our opinion, not discussion, but the want of that knowledge which is gained by discussion prepatory to action. For we have a peculiar power of thinking before we act and of acting too, whereas other men are courageous from ignorance but hesitate upon reflection. And they are surely to be esteemed the bravest spirits who, having the clearest sense of both the pains and pleasures of life, do not on that account shrink from danger.

Pericles, 431 BC, funeral oration for the first Athenians to fall in the Peloponnesian War, quoted in Thucydides, *The Peloponnesian War,* c. 404 BC, tr. Jowett

---

I have many a time crept forward to the skirmish-line to avail myself of the cover of the pickets' 'little fort', to observe more closely some expected result; and always talked familiarly with the men, and was astonished to see how well they comprehended the general object, and how accurately they were informed of the state of facts existing miles away from their particular corps. Soldiers are very quick to catch the general drift and purpose of a campaign, and are always sensible when they are well commanded or well cared for. Once impressed with this fact, and that they are making progress, they bear cheerfully any amount of labor and privation.

General of the Army William T. Sherman, *Memoirs of General W. T. Sherman,* 1875

The power to command has never meant the power to remain mysterious, but rather to communicate, at least to those who immediately execute our orders, the idea which animates our plans. If any one ever had the chance of playing the mysterious role in war it was Napoleon. For this authority was beyond question, and he had taken upon himself to think out everything and decide everything for his army. Yet in his correspondence he always put his views and his programme for several days to come before the Commanders of his army corps. And if we call to mind a number of his proclamations we shall see that his very troops were made aware of the manoeuvre he intended. Souvarov said exactly the same thing. Every soldier should understand the manoeuvre in which he is engaged. He was convinced that one can get anything out of a force to which one speaks frankly, because such a force will understand what is asked of it and will then itself ask no better than to do what is required of it.

Marshal of France Ferdinand Foch, *Precepts and Judgments,* 1919

. . . the led put the worst construction on the silence of the leaders, that they assume no news to be bad news, despite all the proverbs.

Captain Sir Basil Liddell Hart, *A Greater Than Napoleon: Scipio Africanus,* 1926

To lie to American troops to cover up a blunder in combat rarely serves any valid purpose. They have a good sense of combat and an uncanny instinct for ferreting out the truth when anything goes wrong tactically. They will excuse mistakes but they will not forgive being treated like children.

Brigadier General S. L. A. Marshall, *The Armed Forces Officer*, 1950

But it is not enough to have a worthy object; you have got to convince everyone in the party that *is* a worthy object. What might have seemed obvious to me, sitting in Army HQs surrounded by maps, reports and returns, might not be so self-evident to the orderly at my door who hadn't seen his wife for four years, or to the wet, hungry soldier up there in the jungle who was being shot at. It may not be so plain to a lot of people now.

Field Marshal Viscount Slim, *Courage and Other Broadcasts*, 1957

We found that the best way to convince men that what they were doing was worthwhile was to tell them yourself. The spoken word, delivered in person – not over the wireless – is the greatest instrument. An occasional talk by the man who holds the responsiblity for the show counts a lot. It doesn't need an orator. Any man who holds control over others should be able to talk to them, provided he has two qualifications. First, that he is clear in his own mind about what he wants to put over, and secondly that he believes it himself. That last is important.

Field Marshal Viscount Slim, *Courage and Other Broadcasts*, 1957

I have always remembered Cromwell's saying that a soldier fights better when he knows what he is fighting for. Ever since I commanded twenty men in the first Desert Patrol in 1931, I have constantly collected all those under my command and explained to them every detail of the situation in which we have found ourselves. I have then invited questions and discussion.

Lieutenant General Sir John Glubb, *A Soldier With the Arabs*, 1957

The need for truth is not always realized. A leader must speak the truth to those under him; if he does *not* they will find it out and then their confidence in him will decline. I did not always tell *all* the truth to the soldiers in the war; it would have compromised secrecy and it was not necessary.

I told them all they must know for the efficient carrying out of their tasks. But what I did tell them was always true and they knew it; that produced a mutual confidence between us.

Field Marshal Viscount Montgomery of Alamein, *The Memoirs of Field Marshal Montgomery*, 1958

All commanders use directives to a degree, but I tried a novel approach. Mine were, for a start, irregular, each one inspired by a specific incident, a particular mistake. But I tried to go beyond mere correction of error, important as that was. I tried to instill new ideas and new concepts. Every word was chosen with care. I put everything I had into their composition. The papers had differing grades of secrecy. Some were distributed only at brigade level, others to battalion; but most were directed to company commanders with instructions that they should pass on their messages to their men. Whenever I visited the forces, I made a point of questioning all levels to see if they knew the applicable directives. I was astonished by the results. One of the happiest moments of my career came at the Suez front on October 8 [1973], two days after our crossing. Wherever I went, the troops cheered and shouted: 'Directive 41, we did it!' In Directive 41, I had laid down how our infantry divisions would cross the Canal.

General Saad el Shazly, *The Crossing of the Suez*, 1980

INITIATIVE: The Operational Dynamic

Leaving aside the failure of our allies, the Russians can credit their great victories to the fact that they have applied standard German command principles: Zhukov as military commander enjoys complete freedom within the framework of the task assigned to him; the Russians have adopted German tactics and German strategic doctrines. In the meantime, we have borrowed from the Russians their earlier system of rigidly laying down the law on virtually everything and going into the tiniest details, and therein lies the blame for our defeats. German military leaders who can think and act independently are discouraged – indeed both such qualities can lead to court-martial. Thus we have forfeited one of the fundamental requirements of a successful and versatile military command. We have become benumbed, and are incapable of strategic action . . . (1942)

General Reinhard Gehlen, *The Service*, 1972

I have never given a damn what the enemy was going to do or where he was. What I have known is what I have intended to do and then have done it. By acting in this manner I have always gotten to the place he expected me to come about three days before he got there.

General George S. Patton, Jr., August 1944, *The Patton Papers*, Vol. II, 1974

During war the ball is always kicking around loose in the middle of the field and any man who has the will may pick it up and run with it.

Brigadier General S. L. A. Marshall, *The Armed Forces Officer*, 1950

When the enemy is at ease, be able to weary him; when well fed, to starve him; when at rest, to make him move.

Appear at places to which he must hasten; move swiftly where he does not expect you.

Sun Tzu, *The Art of War*, c. 500 BC, tr. Griffith

The art of war is, in the last result, the art of keeping one's freedom of action.

Xenophon (431 BC–c. 352 BC)

A rule that I practise myself and which I have always found good is that in order to have rest oneself it is necessary to keep the enemy occupied. This throws them back on the defensive, and once they are placed that way they cannot rise up again during the entire campaign.

Frederick the Great, *Instructions to His Generals*, 1747

It is an incontestable truth that it is better to forestall the enemy, than to find yourself anticipated by him.

Frederick the Great, 'Anti-Machiavelli', *Posthumous works of Frederick II, King of Prussia*, 1789, tr. Holcroft, quoted in Luvaas, *Frederick the Great on the Art of War*, 1966

. . . every one of us knows from his personal experience how an opponent taking the initiative, though he be much weaker, confounds all calculations of his enemy, ruins his plans . . .

Mikhail Frunze, 1922

. . . The initiative . . . means an army's freedom of action as distinguished from an enforced loss of freedom. Freedom of action is the very life of an army and, once it is lost, the army is close to defeat or destruction . . .

Mao Tse-tung, *On Protracted War*, May 1938

I determined to make no-man's land *our* land. We set out to besiege the besiegers.

Lieutenant General Leslie Morshead, his principle of defence as Commander of the Tobruk garrison and the 9th Australian Division, 1941, quoted in Horner, *The Commanders*, 1984

It was to be considered that hostile troops would allow themselves to be held if we attacked, or at least threatened, the enemy at some really sensitive point. It was further to be remembered that, with the means available, protection of the Colony could not be ensured even by purely defensive tactics, since the total length of land frontier and coast-line was about equal to that of Germany. From these considerations it followed that it was necessary, not to split up our small available forces in local defence, but, on the contrary, to keep them together, to grip the enemy by the throat and force him to employ his forces for self-defence . . .

Major General Paul von Lettow-Vorbeck, describing the defence of German East Africa in World War I, in which German forces of fewer than 15,000 drew into the theatre a third of a million Allied troops, *East African Campaigns*, 1957

INITIATIVE: The Individual

Now in war there may be one hundred changes in each step. When one sees he can, he advances; when he sees that things are difficult, he retires. To say that a general must await commands of the sovereign in such circumstances is like informing a superior that you wish to put out a fire. Before the order to do so arrives the ashes are cold.

To put a rein on an able general while at the same time asking him to suppress a cunning enemy is like tying up the Black Hound of Han and then ordering him to catch elusive hares. What is the difference?

Ho Yen-hsi, commentator to Sun Tzu, *The Art of War*, c. 500 BC

When the signal, No. 39, was made, the Signal Lieutenant reported to him. He continued his walk, and did not appear to take notice of it. The Lieutenant meeting his Lordship at the next turn asked, whether he should repeat it. Lord Nelson answered, 'No, acknowledge it.' On the officer returning to the door, his Lordship called after him, 'Is No. 16 [for close action] still hoisted?' The lieutenant answering in the affirmative, Lord Nelson said, 'Mind you keep it so.' He now walked the deck considerably agitated, which was always known by his moving the stump of his right arm. After a turn or two, he said to me, in a quick manner. 'Do you know what's shown on board the Commander-in-Chief, No. 39?' On asking him what that meant, he answered, 'Why, to leave off action.' 'Leave off action!' he repeated, and then added, with a shrug, 'Now damn me if I do.' He also observed, I believe, to Captain Foley, 'You know, Foley, I have only one eye – I have a right to be blind sometimes:' and with an archness peculiar to his character, putting the glass to his blind eye, he exclaimed, 'I really do not see the signal.'

Colonel William Parker, describing Admiral Nelson's refusal to break off a successful action despite orders to do so during the Battle of Copenhagen on 2 April 1801, quoted in Mahan, *The Life of Nelson*, 1987

But, in case signals can neither be seen or perfectly understood, no captain can do very wrong if he places his ship alongside that of the enemy.

Admiral Lord Nelson, 9 October 1805, memorandum to the fleet before Trafalgar

---

All in all it seems . . . that an unusual spirit, of independence of those above and acceptance of responsibility, has grown up throughout the Prussian officer corps as it has in no other army . . . Prussian officers will not stand for being hemmed in by rules and stereotypes as happens in Russia, Austria or Britain . . . We follow the more natural course of giving scope to every individual's talent, of using a looser rein. We back up every success as a matter of course – even when it runs counter to the intentions of the commander-in-chief . . . the subordinate commander exploits every advantage by taking initiatives off his own bat, without his superior's knowledge or approval.

Prince Frederick of Prussia, 1860 essay, quoted in Simpkin, *Race to the Swift*, 1985

The [German Army] Field Service Regulations [of 1911] . . . require 'every officer, under all conditions, to exercise initiative to the maximum extent, without fear of the consequences. Commanding officers must encourage and require this.' This requirement is especially important during long periods of peace, when the necessity for maintaining discipline has a tendency to discourage independent action. All commanding officers, especially the higher ones, must, as their duty, make every effort to develop initiative in their subordinates, and oppose indifference and routine service. They must also restrain their supervisory activities and avoid too frequent visits to troop instruction, which sometimes causes embarrassment.

General der Infantrie Hugo Baron von Freytag-Loringhoven, *The Power of Personality in War*, 1911

You must frequently act without orders from higher authority. Time will not permit you to wait for them. Here again enters the importance of studying the work of officers above you. If you have a comprehensive grasp of the entire situation and can form an idea of the general plan of your superiors, that and your previous emergency training will enable you to determine that the responsibility is yours and to issue the necessary orders without delay.

Major C. A. Bach, 1917, farewell instructions to graduating student officers at Fort Sheridan, Wyoming

---

My aim was to turn highly-disciplined troops into responsible men possessed of initiative. Discipline is not intended to kill character, but to develop it. The purpose of discipline is to bring about the uniformity in co-operating for the attainment of a common goal, and this uniformity can only be obtained when each one sets aside the thought of his own personal interests. This common goal is – Victory.

General Erich Ludendorff, *My War Memories, 1914–1918*, 1919

I do not propose to lay down for you a plan of campaign: but simply to lay down the work it is desirable to have done and leave you free to execute it in your own way.

General of the Army General Ulysses S. Grant, 1864, mission orders to General Sherman for the destruction of Johnston's Army

---

A favourable situation will never be exploited if commanders wait for orders. The highest commander and the youngest soldier must always be conscious of the fact that omission and inactivity are worse than resorting to the wrong expedient.

Field Marshal Helmuth Graf von Moltke (1800–1891)

The magazine rifle with its low trajectory and smokeless powder, spoke volumes to the captain of 1899–1902. It told him he could still conduct his company into the zone of aimed fire, but that, having got them there, he must either:

(1) Keep his direct command at the cost of double losses.
(2) Let each little group understand the common objective. Then leave them to the promptings of their own consciences of what was right rather than to the dread of doing wrong.

General Sir Ian Hamilton, *The Soul and Body of an Army*, 1921

If, in the opinion of the leader, the plan has, through change in conditions, become inoperative, then he ceases to be a leader and becomes, for the time being, an independent commander and must act as if he were a general-in-chief. That is to say, he must replace the inoperative plan by an operative one – that is, one which will permit of the economical expenditure of force. To carry on a plan which manifestly has failed is the act of a fool, whether he be the general-in-chief or a private soldier. Once again we come back to our starting-point, namely, intelligence.

Major General J. F. C. Fuller, *The Foundations of the Science of War*, 1926

There are cases . . . of initiative exaggerated to the point of violating discipline and interfering with the concentration of the combined effort . . . In the last analysis, an exaggerated display of initiative is due in the first place to an absence or weakness of decision higher up. A spirit of enterprise in a commander is never a danger in itself.

. . . so it comes about that the authorities dread any officer who has the gift of making decisions and cares nothing for routine and soothing words. 'Arrogant and undisciplined' is what the mediocrities say of him, treating the thoroughbred with a tender mouth as they would a donkey which refuses to move, not realizing that asperity is, more often than not, the reverse side of a strong character, that you can only lean on something that offers resistance, and that resolute and inconvenient men are to be preferred to easy-going natures without initiative.

Charles de Gaulle, *The Edge of the Sword*, 1932

I wish to assure all officers and men that I shall never criticize them or go back on them for having done too much but that I shall certainly relieve them if they do nothing. You must keep moving . . .

General George S. Patton, Jr., May 1941, *The Patton Papers*, Vol. II, 1974

People must try to use their imagination and when orders fail to come, must act on their own best judgment. A very safe rule to follow is that in case of doubt, push on just a little further and then keep on pushing . . .

General George S. Patton, Jr., July 1941, *The Patton Papers*, Vol. II, 1974

Never tell people *how* to do things. Tell them *what* to do and they will surprise you with their ingenuity.

General George S. Patton, Jr., *War As I Knew It*, 1947

. . . improvisation is the essence of initiative in all combat just as initiative is the outward showing of the power of decision.

Brigadier General S. L. A. Marshall, *Men Against Fire*, 1947

As the situation demanded, Captain Mueller was given the greatest freedom of action. If any promising objective presented itself during the march, he was to decide without hesitation what his best course was. I would bring up our main body and intervene unconditionally in his support, and, in any case, I would accept the situation he had created. The main thing was that he should not wait for special orders and instructions. I realized that in acting thus I was in a large measure placing conduct of our operations in the hands of a subordinate officer. It was only possible because that subordinate officer possessed a very sound tactical judgment and great initiative.

Major General Paul von Lettow-Vorbeck, describing the campaign in German East Africa, 1914–1918, *East African Campaigns*, 1957

Initiative I gave as the third of the qualities of an officer. Initiative simply means that you do not sit down and wait for something to happen. In war, if you do, it will happen all right, but it will be almighty unpleasant. Initiative for you means that you keep a couple of jumps ahead, not only of the enemy, but of your own men.

Field Marshal Sir William Slim, *Courage and Other Broadcasts*, 1957

It is better to struggle with a stallion when the problem is how to hold it back, than to urge on a bull which refuses to budge.

General Moshe Dayan (1915–1918)

You must be able to underwrite the honest mistakes of your subordinates if you wish to develop their initiative and experience.

General Bruce C. Clarke

To make perfectly clear that action contrary to orders was not considered as disobedience or lack of discipline, German commanders began to repeat one of Moltke's favorite stories, of an incident observed while visiting the headquarters of Prince Frederick Charles. A major, receiving a tongue-lash from the Prince for a tactical blunder, offered the excuse that he had been obeying orders, and reminded the Prince that a Prussian officer was taught that an order from a superior was tantamount to an order from the King.

Frederick Charles promptly responded: 'His Majesty made you a major because he believed you would know when *not* to obey his orders.' This simple story became guidance for all following generations of German officers.

Colonel Trevor N. Dupuy, *A Genius For War: The Army and the German General Staff, 1807–1945*, 1977

The capacity for independent action of low-level infantry sub-units has long been our weakest point. The cry of 'I'm waiting for orders', which really meant 'I'm doing nothing', was the real scourge of our activities in the field.

Marshal of the Soviet Union Mikhail N. Tukhachevskiy, 1931–2, quoted in Simpkin, *Deep Battle*, 1987

INNOVATION/INVENTION/ ORIGINALITY

. . . qualifications a general ought to possess . . . abilities to strike out something new of his own occasionally. For no man excelled in his profession who could not do that, and if a ready and quick invention is necessary and honourable in any profession, it must certainly be so in the art of war above all others.

Niccolo Machiavelli, *The Art of War*, 1521

The mind is not interested by routine forms of duty; and, as it is important to success that the mind should be interested, it is useful . . . to endeavour to give a new cast, consequently a new force of impression, to military exercises and military forms of evolution, without changing the principles of such practices as are laid on a basis of truth. New modes of military exercise interest the individual by their novelty; they even not infrequently communicate an animating energy to the arm of the actor, which goes beyond the limits of ordinary calculation: they seldom fail to intimidate the enemy as striking him by surprise. If this be so, it belongs to military genius to change the appearances of things, with a view to animate one part and to intimidate another. But, while this be done, especial care is to be taken that the fundamental principles of military tactic be not rashly violated.

Robert Jackson, *A Systematic View of the Formation, Discipline, and Economy of Armies*, 1804

No soldier who has made himself conversant with the resources of his art, will allow himself to be trammelled by an exclusive system.

Dennis Hart Mahan (1802–1871)

Changes in tactics have not only taken place after changes in weapons which necessarily is the case, but the interval between such changes has been unduly long. An improvement of weapons is due to the energy of one or two men, while changes in tactics have to overcome the inertia of a conservative class.

Rear Admiral Alfred Thayer Mahan (1840–1914)

I was told by Mr. Charles King, when President of Columbia College, that he had been present in company with [Admiral Stephen] Decatur at one of the early experiments in steam navigation. Crude as the appliances still were, demonstration was conclusive; and Decatur, whatever his prejudices, was open to conviction. 'Yes,' he said gloomily, to King, 'it is the end of our business; hereafter any man who can boil a tea-kettle will be as good as the best of us.'

Rear Admiral Alfred Thayer Mahan, *From Sail to Steam*, 1907

In war novelties of an avatistic nature are generally horrible; nevertheless, in the public mind, their novelty is their crime; consequently, when novelties of a progressive character are introduced on the battlefield, the public mind immediately anathematizes them, not necessarily because they are horrible but because they are new. Nothing insults a human being more than an idea his brains are incapable of creating. Such ideas detract from his dignity for they belittle his understanding. In April 1915, a few hundred British and French soldiers were gassed to death; gas being a novelty, Europe was transfixed with horror. In the winter of 1918–1919, the influenza scourge accounted for over 10,000,000 deaths, more than the total casualties in killed throughout the Great War; yet the world scarcely twitched an eyelid, though a few people went so far as to sniff eucalyptus.

Major General J. F. C. Fuller, *The Reformation of War*, 1923

Originality is the most vital of all military virtues, as two thousand years of war attest. In peace it is at a discount, for it causes the disturbances of comfortable ways without producing dividends, as in civil life. But in war originality bears a higher premium than it can ever do in a civil profession. For its application can overthrow a nation and change the course of history in the proverbial twinkling of an eye. (August 1930)

Captain Sir Basil Liddell Hart, *Thoughts on War*, 1944

One of the most important talents of a general we would call that of a 'creative mind' . . . Originality, not conventionality, is one of the main pillars of generalship. To do something that the enemy does not expect, is not prepared for, something which will surprise him and disarm him morally. To be always thinking ahead and to be always peeping around corners. To spy out the soul of one's adversary, and to act in a manner which will astonish and bewilder him, this is generalship. To render the enemy's general ridiculous in the eyes of his men, this is the foundation of success . . .

Field Marshal Colmar Baron von der Goltz, quoted in Fuller, *Generalship: Its Diseases and Their Cure*, 1933

The past had its inventions and when they coincided with a man who staked his shirt on them the face of the world changed. Scythes fixed to the axles of the war chariots; the moving towers which overthrew Babylon; Greek fire; the short bow, the cross-bow, the Welsh long bow and the ballista; plate armour; the Prussian needle gun; the Merrimac and Ericsson's marvellous coincidental reply. The future is pregnant with inventions . . .

General Sir Ian Hamilton, *Soul and Body of an Army*, 1921

Accusing as I do without exception all the great Allied offensives of 1914, 1916 and 1917, as needless and wrongly conceived operations of infinite cost, I am bound to reply to the question – what else could have been done? And I answer it, pointing to the Battle of Cambrai, 'This could have been done'. This in many variants, this in larger and better forms ought to have been done, and would have been done if only the generals had not been content to fight machine-gun bullets with the breasts of gallant men, and think that was waging war.

Sir Winston Churchill, *The World Crisis*, 1923

We must hold our minds alert and receptive to the application of unglimpsed methods and weapons. The next war will be won in the future, not in the past. We must go on, or we will go under.

General of the Army Douglas MacArthur, 1931

The best hope of tilting the scales and of overcoming the resistance inherent in conflict lies in originality – to produce something unexpected that will paralyze the opponent's freedom of action . . . But in historical fact, such creative intelligence has been rare in the operations of war. And rarer still in preparation for war. Military history is filled with the record of military improvements that have been resisted by those who have profited richly from them. Between the development of new weapons or new tactics and their adoption there has always been a time-lag, often of generations. And that time-lag has often decided the fate of nations. (June 1936)

Captain Sir Basil Liddell Hart, *Thoughts on War*, 1944

When we study the lives of the great captains, and not merely their victories and defeats, what do we discover? That the mainspring within them was *originality*, outwardly expressing itself in unexpected actions. It is in the mental past in which most battles are lost, and lost conventionally, and our system teaches us how to lose them, because in the schoolroom it will not transcend the conventional. The soldier who thinks ahead is considered, to put it bluntly, a damned nuisance.

Major General J. F. C. Fuller, *Generalship: Its Diseases and Their Cure*, 1933

Prejudice against innovation is a typical characteristic of an Officer Corps which has grown up in a well-tried and proven system.

Field Marshal Erwin Rommel, *The Rommel Papers*, 1953

INSPECTIONS

The great mistake in inspections is that you officers amuse yourselves with God knows what buffooneries and never dream in the least of serious service. This a source of stupidity which would become most dangerous in case of a serious conflict. Take shoemakers and tailors and make generals of them and they will not commit worse follies.

Colonel Charles Ardnant du Picq, *Battle Studies*, 1880

When a unit has been alerted for inspection do not fail to inspect it and inspect it thoroughly. Further do not keep it waiting. When soldiers have gone to the trouble of getting ready to be inspected, they deserve the compliment of a visit. Be sure to tell the unit commander publically that his unit was good, if such is the case. If it is bad, tell him privately and in no uncertain terms. Be sure to speak to all enlisted men who have decorations, or who have been wounded, and ask how they got the decoration or how they were wounded.

General George S. Patton, Jr., *War As I Knew It*, 1947

Inspection is more important in the face of the enemy than during training because a fouled piece may mean a lost battle, an overlooked sick man may infect a fortress and a mislaid message can cost a war. In virtue of his position, every junior leader is an inspector, and the obligation to make certain that his force at all times is inspection proof is unremitting.

Brigadier General S. L. A. Marshall, *The Armed Forces Officer*, 1950

A unit that does well only those things the boss checks will have great difficulty.

Major General E. S. Leland, Commander of the National Training Center, Fort Erwin, California, quoted in *Parameters*, 9/1987

INSPIRATION

Hence the difference between a mechanic and a man of genius entrusted with the command of an army. The one operates mechanically by the impulse of fear on the slavish passions of man; the other insensibly insinuates and incorporates himself with his soldiers, forming them into heroes . . . hence the same instruments, independent of the mechanical mode of application, move forward to victory or recoil to defeat, according to the mode in which they are animated.

Robert Jackson, *A Systematic View of the Formation, Discipline, and Economy of Armies*, 1804

He who seeks, finds, if he does not lose heart; and to me, continuously seeking, came from within the suggestion that control of the sea was an historic factor which had never been systematically appreciated and expounded. Once formulated consciously, this thought became the nucleus of all my writing for twenty years then to come . . .

Rear Admiral Alfred Thayer Mahan, *From Sail to Steam*, 1907

It is this animation which so largely constitutes the art of war, and of which it is so difficult to write. It is not one soul lighting another – this is mere fanaticism – but rather one mind illuminating many minds, by one heart causing thousands to beat in rhythm which, like a musical instrument, accompanies the mind in control. It is a union between intelligence and heart; between the will of the general and the willingness of his men; that fusion of the mental and the moral spheres.

Major General J. F. C. Fuller, *The Foundations of the Science of War*, 1926

It is . . . an essential element in leadership. A commander has to work, not with a piece of clockwork which 'goes' when it is wound up, but with 'tool-using men'. These men are afraid of death, and suffer from hunger, thirst, lack of sleep, and bad weather. Some are brave, some are not. Some are slow in the uptake, others mentally alert. Some are trusting and loyal, others jealous and insubordinate. In short, they carry within themselves a thousand and one seeds of diversity. To make them act as one it is not enough for a commander to decide in his own mind what should be done, nor even to limit himself to issuing orders for the carrying out of his intentions. He must be able to create a spirit of confidence in those under him. He must be able to exert his authority.

Charles de Gaulle, *The Edge of the Sword*, 1932

It is said of Caesar that he never lacked a pleasant word for his soldiers. He remembered the face of anyone who had done a gallant deed and, when not in the presence of the enemy, joined his men in soldier games. Such little human acts as these inspired his legionnaires with the devotion which went so far to account for his success as a great captain.

General Maxwell D. Taylor, 21 January 1965, speech to the cadets of the Citadel

---

My soldiers were always like that throughout the war – an unfailing tonic for senior officers condemned to spend most of their time in planning and making the decisions upon which the lives of these men depended. In my experience the history books that depicted the role of the general as being that of galvanizing his men into action were all wrong. I found more often than not that I went up to the front lines not to urge on the troops but to escape the worries of the command post where all battle noises sounded like the doings of the enemy and where it was easy for the commander to give way to dire imaginings. A visit to the men of the 101st under fire never failed to send me back to the command post, assured that the situation was well in hand, and that there was no cause for worry.

General Maxwell Taylor, *Swords and Ploughshares*, 1972

Study the human side of history . . . To learn that Napoleon in 1796 with 20,000 beat combined forces of 30,000 by something called economy of force or operating on interior lines is a mere waste of time. If you can understand how a young unknown man inspired a half-starved, ragged, rather Bolshie, crowd; how he filled their bellies; how he outmarched, outwitted, outbluffed and defeated men who had studied war all their lives and waged it according to the text-books of their time, you will have learnt something worth knowing.

Field Marshal Earl Wavell (1883–1950)

. . . when all the units had been briefed by their officers, I would stand out on the porch of the headquarters and watch the preparations. Soldiers would be going and coming, checking their weapons and equipment, loading trucks, talking to each other and their officers. The camp would be a beehive of activity, alive with purpose. Each one knew precisely what his job was, and how he was going to do it. Each had been readied by months of the hardest training. I could see the determination in their eyes, and invariably I would feel a surge of assurance. It was a reciprocal process, a flow of confidence from them to me and from me to them. A commander has to inspire his men, but it was always clear to me that they inspire him as well.

General Ariel Sharon, *Warrior*, 1989

INSTINCT/INTUITION

Circumstances vary so enormously in war, and are so indefinable, that a vast array of factors has to be appreciated – mostly in the light of probabilities alone. The man responsible for evaluating the whole must bring to his task the quality of intuition that perceives the truth at every point. Otherwise a chaos of opinions and considerations would arise, and fatally entangle judgment . . .

Major General Carl von Clausewitz, *On War*, i, 1832, tr. Howard and Paret

---

The consecutive achievements of a war are not premeditated but spontaneous and guided by military instinct.

Field Marshal Helmuth Graf von Moltke, *Instructions for the Commanders of Large Formations*, 1869

Great war leaders have always been aware of the importance of instinct. Was not what Alexander called his 'hope', Caesar his 'luck' and Napoleon his 'star' simply the fact that they knew they had a particular gift of making contact with realities sufficiently closely to dominate them? For those who are greatly gifted, this faculty often shines through their personalities. There may be nothing in itself exceptional about what they say or their way of saying it, but other men in their presence have the impression of a natural force destined to master events.

Charles de Gaulle, *The Edge of the Sword*, 1932

The greatest commander is he whose intuitions most nearly happen.

Colonel T. E. Lawrence, 'The Science of Guerrilla Warfare', *Encyclopaedia Britannica*, 1929

The fighting instinct is necessary to success on the battlefield – although even here the combatant who can keep a cool head has an advantage over the man who 'sees red' – but should always be ridden on a tight rein. The statesman who gives that instinct its head loses his own; he is not fit to take charge of the fate of a nation.

Captain Sir Basil Liddell Hart, *Strategy*, 1954

INTEGRITY

False words are not only evil in themselves, but they infect the soul with evil.

Socrates (*c.* 470 BC–399 BC)

No matter how lacking a man may be in humanity, if he would be a warrior, he should first of all tell no lies. It is also basic that he be not the least bit suspicious, that he habitually stand on integrity, and that he know a sense of shame. The reason being that when a man who has formerly told lies and acted suspiciously participates in some great event, he will be pointed at behind his back and neither his allies nor his enemies will believe in him, regardless of how reasonable his words may be. One should be very prudent about this.

Asakura Soteki (1474–1555), *Soteki Waki*, *c.* 1550, in Wilson, *Ideals of the Samurai*, 1982

If you choose godly, honest men to be captains of horse, honest men will follow them.

Oliver Cromwell, September 1643, letter to Sir William Springe

---

A man may cease to be lucky, for that is beyond his control; but he should not cease to be honest.

Charles XII, King of Sweden, 1714, to Axel von Loewen

I know how to hold my tongue when it is necessary; but I do not know how to lie.

Field Marshal Prince Eugene of Savoy, 1728, to the English diplomat, Lord Waldegrave

---

I have just been offered two hundred fifty thousand dollars and the most beautiful woman I have ever seen to betray my trust. I am depositing the money with the Treasury of the United States and request immediate relief from this command. They are getting close to my price.

Lieutenant General Arthur MacArthur, (1845–1912), quoted by his son Douglas MacArthur, *Reminiscences*, 1964

Labor to keep alive in your breast that little spark of celestial fire called conscience.

General George Washington (1732–1799)

I hold the maxim no less applicable to public than to private affairs, that honesty is the best policy.

General George Washington, 1796, Farewell Address

Fidelity . . . because it comes of personal decision, is the jewel within reach of every officer who has the will to possess it. It is the epitome of character, and fortunately no other quality in the individual is more readily recognized and honored, by one's military associates.

Brigadier General S. L. A. Marshall, *The Armed Forces Officer*, 1950

He [a man] has integrity if his interest in the good of the Service is at all times greater than his personal pride, and when he holds himself to the same line of duty when unobserved as he would follow if all of his superiors were present.

Brigadier General S. L. A. Marshall, *The Armed Forces Officer*, 1950

Integrity is one of those words that many people keep in that desk drawer labeled 'too hard'. It is not a topic for the dinner table or the cocktail party. When supported with education, one's integrity can give a person something to rely on when perspective seems to blur, when rules and principles seem to waver, and when faced with a hard choice of right and wrong. To urge people to develop it is not a statement of piety but of practical advice. Anyone who has lived in an intense extortion environment [POW realizes that the most potent weapon an adversary can bring to bear is manipulation, the manipulation of a prey's shame. A clear conscience is one's only protection.

Admiral James B. Stockdale, 'Educating Leaders', *Washington Quarterly*, Winter 1983

INTELLECT

We pay a high price for being intelligent. Wisdom hurts.

Euripides, *Electra*, 413 BC

Men with sharpness of mind are to be found only among those with a penchant for thought.

Shiba Yoshimasa (1350–1410), *The Chiku-basho*, 1383, in Wilson, *Ideals of the Samurai*, 1982

What distinguishes a man from a beast of burden is thought, and the faculty of bringing ideas together . . . a pack mule can go on the campaigns with Prince Eugene of Savoy, and still learn nothing of tactics.

Frederick the Great (1712–1786)

We only wish to represent things as they are, and to expose the error of believing that a mere bravo without intellect can make himself distinguished in war.

Major General Carl von Clausewitz, *On War*, 1832

The instruction given by leaders to their troops, by professors of military schools, by historical and tactical volumes, no matter how varied it may be, will never furnish a model that need only be reproduced in order to beat the enemy . . . It is with muscles of the intellect, with something like cerebral reflexes that the man of war decides, and it is with his qualities of character that he maintains the decision taken. He who remains in abstractions falls into formula; he concretes his brain; he is beaten in advance.

General Cordonnier, 1917, French Corps Commander, quoted in Marshall, ed., *Infantry in Battle*, 1939

I feel a crippling incuriousness about our officers. Too much body and too little head.

Colonel T. E. Lawrence (1888–1935)

It is true that military men, exaggerating the relative powerlessness of the intelligence, will sometimes neglect to make use of it at all. Here the line of least resistance comes into its own. There have been examples of commanders avoiding all intellectual effort and even despising it on principle. Every great victory is usually followed by this kind of mental decline. The Prussian Army after the death of Frederick the Great is an instance of this. In other cases, the military men note the inadequacy of knowledge and therefore trust to inspiration alone or to the dictates of fate. That was the prevalent state of the French Army at the time of the Second Empire: 'We shall muddle through, somehow.'

Charles de Gaulle, *The Edge of the Sword*, 1932

Marked intellectual capacity is the chief characteristic of the most famous soldiers. Alexander, Hannibal, Caesar, Marlborough, Washington, Frederick, Napoleon, Wellington, and Nelson were each and all of them something more than fighting men. Few of their age rivalled them in strength of intellect. It was this, combined with the best qualities of Ney and Blücher, that made them masters of strategy, and lifted them high above those who were tacticians and nothing more . . .

Colonel George F. R. Henderson, *Stonewall Jackson and the American Civil War*, 1898

With soldiers, instinctive qualities play the chief part in action. The battlefield is not an intellectual sphere. But the organization and training of armies demand, above all, intellect based upon knowledge. Yet in the military system, these essentially different activities are treated as interchangeable. Here is the reason for so many failures. An army will produce a number of good fighting commanders, especially if their tactical sense is not atrophied by routine and discipline. But no army produces equally good strategists, because such can only be developed through a scientific approach to problems, and this is incompatible with the loyalty to other interests which the service demands. (November 1934)

Captain Sir Basil Liddell Hart, *Thoughts on War*, 1944

Physical exuberance does not go with mental fitness. A man who takes a lot of exercise rarely exercises his mind adequately – in developing his body he leaves his mind undeveloped. Pain may intensify the feelings but can hardly clarify the thought.

The mind works best when the body is quiescent – when one is totally unaware of the body: as if one were detached from it altogether for the time. If the body is either unwell or too active it impinges on the mind.

We ought to take account of this factor when weighing the mental form of commanders. Allowance must be made for the physical effect of active service conditions, and also for the intense cultivation of physical fitness by excessive exercise. (May 1935)

Captain Sir Basil Liddell Hart, *Thoughts on War*, 1944

Creative intelligence is always has been the supreme requirement in the commander – coupled with moral courage. War is a two-party affair, and thus conventionality of thought inevitably tends to confirm this equality of kind, conducing to stalemate. (June 1936)

Captain Sir Basil Liddell Hart, *Thoughts on War*, 1944

The Army, for all its good points, is a cramping place for a *thinking* man. As I have seen too often, such a man chafes and goes – or else decays. And the root of the trouble is the Army's root fear of the truth. Romantic fiction is the soldierly taste. The heads of the Army talk much of developing character yet deaden it – frowning on younger men who show promise of the personality necessary for command, and the originality necessary for surprise. (October 1935.)

Captain Sir Basil Liddell Hart, *Thoughts on War*, 1944

There has been no illustrious captain who did not possess taste and feeling for the heritage of the human mind. At the root of Alexander's victories one will always find Aristotle.

Charles de Gaulle, *The Army of the Future*, 1941

JUDGEMENT

Thus those unable to understand the dangers inherent in employing troops are equally unable to understand the advantageous ways of doing so.

Sun Tzu, *The Art of War*, c. 500 BC, tr. Griffith

There are some roads not to follow; some troops not to strike; some cities not to assault; and some ground which should not be contested.

Sun Tzu, *The Art of War*, c. 500 BC, tr. Griffith

Reason and judgment are the qualities of a leader.

Cornelius Tacitus, (c. AD 56–c. AD 120), *Annals*

Look at the essence of a thing, whether it be a point of doctrine, of practice, or interpretation.

The Emperor Marcus Aurelius (AD 121–AD 180), *Meditations*

That quality which I wish to see the officers possess, who are at the head of the troops, is a cool, discriminating judgement when in action, which will enable them to decide with promptitude how far they can go and ought to go, with propriety; and to convey their orders, and act with such vigour and decision, that the soldier will look to them with confidence in the moment of action, and obey them with alacrity.

The Duke of Wellington, General Order, 15 May 1811

Judgement, and not headlong courage, is the true arbiter of war.

Count Belisarius (c. AD 505–c. AD 565)

The general should not be hasty in placing confidence in people who promise to do something; if he does, almost everyone will think he is light headed.

The Emperor Maurice, *The Strategikon*, c. AD 600

Everyone complains of his memory, but no one complains of his judgement.

François, Duc de La Rochefoucauld, *Réflexions ou sentences et maximes morales*, 1665

When I am without orders, and unexpected occurrences arise, I shall always act as I think the honour and glory of my King and Country demand.

Admiral Lord Nelson, November 1804, letter to Hugh Elliot

If I am to be hanged for it, I cannot accuse a man who I believe has meant well, and whose error was one of judgement and not of intention.

The Duke of Wellington, 31 July 1809. On the subject of General Crauford's handling of the Light Division, 24 July 1809

In reviewing the whole array of factors a general must weigh before making his decision, we must remember that he can gauge the direction and value of the most important ones only by considering numerous other possibilities – some immediate, some remote. He must *guess*, so to speak: guess whether the first shock of battle will steel the enemy's resolve and stiffen his resistance, or whether, like a Bologna flask, it will shatter as soon as its surface is scratched; guess the extent of debilitation and paralysis that the drying up of particular sources of supply and the severing of certain lines of communication will cause in the enemy; guess whether the burning pain of an injury he has dealt will make the enemy collapse with exhaustion or, like a wounded bull, arouse his rage; guess whether and which political alliances will be dissolved or formed. When we realize that he must hit upon all this and much more by means of his discreet judgement, as a marksman hits a target, we must admit that such an accomplishment of the human mind is no small achievement. Thousands of wrong turns running in all directions tempt his perception; and if the range, confusion and complexity of the issues are not enough to overwhelm him, the dangers and responsibilities may.

Major General Carl von Clausewitz, *On War*, viii, 1832, tr. Howard and Paret

---

If you take a chance, it usually succeeds, presupposing good judgement.

Lieutenant General Sir Giffard Martel

The core of leadership is the ability and willingness to exercise judgement. It is the judgement which sits at the centre of the German army's *Innere Fuehrung*, at the centre of the Mongol Yassak, and the Roman Army's rules of the soldier, and it is judgement which sits at the centre of leadership today,. If judgement is vital to leadership, then it is clear that the code of values of the officer can never be the equivalent of bureaucratic rules.

Lieutenant Colonel Richard Gabriel, 'The Environment of Military Leadership', *Military Review*, 7/1980

JUNIOR OFFICERS

A young officer should never think he does too much; they are to attend to the looks of the men, and if any are thinner or paler than usual, the reason of their falling off may be enquired into.

Major General Sir James Wolfe (1727–1759), *Instructions for Young Officers*

The lieutenant, in the absence of the captain, commands the company, and should therefore make himself acquainted with the duties of the station; he must also be perfectly acquainted with the duties of the non-commissioned officer and soldier, and see them performed with the greatest exactness . . . he should endeavour to gain the love of his men, by his attention to every thing which may contribute to their health and convenience. He should often visit them at different hours; inspect into their manner of living; see that their provisions are good and well cooked, and as far as possible oblige them to take their meals at regulated hours. He should pay attention to their complaints, and when well founded, endeavour to get them redressed; but discourage them from complaining on every frviolous occasion . . . He must not suffer the soldiers to be ill treated by the non-commissioned officers through malevolence, or from any pique or resentment; but must at the same time be careful that proper degree of subordination is kept up between them.

Major General Friedrich Baron von Steuben, *Revolutionary War Drill Manual*, 1794

But the *platoon commander* – that is a special item. This by no means a trifle; this the commander, the leader, the head of the basic military group – the platoon. It is impossible to build an edifice with loose sand. One must have good building material, one must have a good platoon, and this means – a good, reliable, class-conscious, confident platoon commander.

Leon Trotsky, 1 April 1922, 'Our Current Basic Military Tasks', *Military Writings*, 1969

Modern battles are fought by platoon leaders. The carefully prepared plans of higher commanders can do no more than project you to the line of departure at the proper time and place, in proper formation, and start you off in the right direction.

General of the Army George C. Marshall, *Selected Speeches and Statements*, 1945

As to the way in which some of our ensigns and lieutenants braved danger – the boys just come out from school – it exceeds all belief. They ran as at cricket.

The Duke of Wellington, quoted in Rodgers, *Recollections*, 1859

The characteristics which are required in the minor commander if he is to prove capable of preparing men for and leading them through the shock of combat with high credit may therefore be briefly described:

(1) Diligence in the care of men.
(2) Administration of all organizational affairs such as punishments and promotions according to a standard of resolute justice.
(3) Military bearing.
(4) A basic understanding of the simple fact that soldiers wish to think of themselves as soldiers and that all military information is nourishing to their spirits and their lives.
(5) Courage, creative intelligence, and physical fitness.
(6) Innate respect for the dignity of the position and the work of other men.

Brigadier General S. L. A. Marshall, *The Armed Forces Officer*, 1950

Said a newly arrived lieutenant to an old sergeant of the 12th Cavalry: 'You've been here a long time, haven't you?' 'Yes sir,' replied the sergeant. 'The troop commanders, they come and go, but it don't hurt the troop.'

Brigadier General S. L. A. Marshall, *The Armed Forces Officer*, 1950

The pious Greek, when he had set up altars to all the great gods by name, added one more altar, 'To the Unknown God'. So whenever we speak and think of the great captains and set up our military altars to Hannibal and Napoleon and Marlborough and such-like, let us add one more altar, 'To the Uknown Leader', that is, to the good company, platoon, or section leader who carries forward his men or holds his post, and often falls unknown. It is these who in the end do most to win wars.

Field Marshal Earl Wavell, *Soldiers and Soldiering*, 1953

JUSTICE

The commander should be severe and thorough in investigating charges against his men, but merciful in punishing them. This will gain him their good will.

The Emperor Maurice, *The Strategikon*, c. AD 600

When certain offences are common among soldiers, moderation is called for. Do not judge and punish all indiscriminately. Widespread resentment might draw them all together, and discipline would become even worse. It is wiser to punish just a few of the ring leaders.

The Emperor Maurice, *The Strategikon*, c. AD 600

He [the general] should have a good disposition free from caprice and be a stranger to hatred. He should punish without mercy, especially those who are dearest to him, but never from anger. He should always be grieved when he is forced to execute the military rules and should have the example of Manlius constantly before his eyes. He should discard the ideas that it is he who punishes and should persuade himself and others that he only administers the military laws. With these qualities, he will be loved, he will be feared, and, without doubt, obeyed.

Field Marshal Maurice Comte de Saxe, *My Reveries*, 1732

First, it is very well known to all men of experience and judgment in matters of arms that all such great captains as have been lieutenants general to emperors, kings, or formed commonwealths, or that with regiments of their own nation have served foreign princes as mercenaries, knowing that justice is the prince of all order and government both in war and peace, by the which God is honoured and served and magistrates and officers obeyed, have at the first forming of their armies or such regiments, by great advice of counsel, established sundry laws both politic and martial, with officers for the superintending and due execution of the same. These laws have been notified to all their men of war, as also, at every encamping or lodging, they have been set, written, or printed in certain tables in convenient places for all soldiers and men of war to behold, to the intent that none might transgress the same through ignorance.

John Smythe (*c.* 1580–1631), *Certain Discourses Military*

A policy of rewards and penalties means rewarding the good and penalizing wrongdoers. Rewarding the good is to promote achievement; penalizing wrongdoers is to prevent treachery.

It is imperative that rewards and punishments be fair and impartial. When they know rewards are to be given, courageous warriors know what they are dying for; when they know penalties are to be applied, villains know what to fear.

Therefore, reward should not be given without reason, and penalties should not be applied arbitrarily. If rewards are given for no reason, those who have worked hard in public service will be resentful; if penalties are applied arbitrarily, upright people will be bitter.

Zhuge Liang (1310–1375), *The Way of the General*

No offence must go unpunished, for nothing can cause the men so much harm as lax discipline.

Field Marshal Prince Aleksandr V. Suvorov (1729–1800), quoted in Ossipov, *Suvorov*, 1945

Measured military punishment, together with a short and clear explanation of the offence, touches the ambitious soldier more than brutality which drives him to despair.

Field Marshal Prince Aleksandr V. Suvorov (1729–1800), quoted in Longworth, *The Science of Victory*, 1966

The presence of one of our regular civilian judge-advocates in an army in the field would be a first-class nuisance.

General of the Army William T. Sherman, *Memoirs of General W. T. Sherman*, 1875

Too many courts-martial in any command are evidence of poor discipline and inefficient officers.

General of the Army William T. Sherman, *Memoirs of General W. T. Sherman*, 1875

One of the great defects in our military establishment is the giving of weak sentences for military offenses. The purpose of military law is administrative rather than legal. As the French say, sentences are for the purpose of encouraging the others. I am convinced that, in justice to other men, soldiers who go to sleep on post, who go absent for an unreasonable time during combat, who shirk in battle, should be executed; and that Army Commands or Corps Commanders should have the authority to approve the death sentence. It is utterly stupid to say that General officers, as a result of whose orders thousands of gallant and brave men have been killed, are not capable of knowing how to remove the life of one miserable poltroon.

General George S. Patton, Jr., *War As I Knew It*, 1947

He [a man] has justice if he acknowledges the interests of all concerned in any particular transaction rather than serving his own particular interests.

Brigadier General S. L. A. Marshall, *The Armed Forces Officer*, 1950

KNOW YOUR MEN

Nothing is harder to see than into people's natures. Though good and bad are different, their conditions and appearances are not always uniform. There are some people who are nice enough but steal. Some people are outwardly respectful while inwardly making fools of everyone. Some people are brave on the outside yet cowardly on the inside. Some people do their best but are not loyal.

Hard though it be to know people, there are ways.

First is to question them concerning right and wrong, to observe their ideas.

Second is to exhaust all their arguments, to see how they change.

Third is to consult with them about strategy, to see how perceptive they are.

Fourth is to announce that there is trouble, to see how brave they are.

Fifth is to get them drunk, to observe their nature.

Sixth is to present them with the prospect of gain, to see how modest they are.

Seventh is to give them a task to do within a specific time, to see how trustworthy they are.

Zhuge Liang (AD 180–AD 234) *The Way of the General*

If the general knows the inclinations and tendencies of each officer and soldier, he will know better what duties should be assigned to each one.

The Emperor Maurice, *The Strategikon*, c. AD 600

A man with intelligence and a firm heart will be able to put others to use. People's ways are variant, and to use a man to whom one has taken a liking for all things – for example, to use a military man for literary matters, or a man untalented in speech as a messenger, or a slow-thinking man in a place where a quick wit is necessary – may bring about failure and even cause a man's life to be ruined. A man should be put to use in the line with which he seems most familiar.

When men are put to use in the same way that curved wood is used for the wheel and straight wood for the shaft, there will be no one without value.

Shiba Yoshimasa, *Chikubasho*, 1380, in Wilson, *Ideals of the Samurai*, 1982

He has great skill in fencing, no doubt, but he can't discriminate between those who need it and those who don't. A lord of the Empire, for instance, or a *daimyo* need not cut people down with his own hand, for if he is in a tight place in a fight he can call on his own men to do it for him. What he needs first of all is the capacity to judge the capacity of the people he uses.

Tokugawa Ieyasu (1543–1616), commenting on the qualifications of a fencing master, quoted in Sadler, *The Maker of Modern Japan: The Life of Tokugawa Ieyasu*

A general's principal talent consists in knowing the mentality of the soldier and in winning his confidence. And, in these two respects, the French soldier is more difficult to lead than any other. He is not a machine to be put in motion but a reasonable being that must be directed.

Napoleon, ed. Herold, *The Mind of Napoleon*, 1955

Know your men, and be constantly on the alert for potential leaders – you never know how soon you may need them. During my two years in command of the 82nd Airborne Division in Wold War II, I was in close and daily touch with every regimental and most battalion commanders. Before acceding to command of the division, and while I was General Omar Bradley's assistant division commander, I had learned to call by name every infantry officer in the division.

Later, by frequent exchange of views with the infantry regimental commanders and the divisional artillery commander, I knew in advance whom they had earmarked for battalion command. I do not recall any instance where I thought the regimental commander had not picked the right man. The payoff came in Normandy. I went in with 12 battalion commanders – four regiments – and I had 14 new ones when we came out, for some battalions lost as many as three commanders during the 33 days we were in that fight.

General Mathew B. Ridgway, 'Leadership', *Military Review*, 10/1966

I had always been one of those field commanders passionately concerned with minor tactics: the battlefield skills of the smallest section, platoon or company. The best plan in the world is useless if the young officer or his men do not have the training or will to carry it out. Every commander, however exalted, whatever armies he disposes, must keep in touch wih his 'poor bloody infantry' – know what they can do, sense what they can feel, imbue them in turn with what he feels and what he demands. Without that rapport he is no more than a 'classroom commander', a master of maps maybe, but lost on the field.

General Saad el Shazly, *The Crossing of the Suez*, 1980

It is not enough that we see the service in a different light, for it will acquire altogether different significance in service when we know how to act toward different people. We must not take uniformly identical steps against all and, under the guise of encouragement, beat everyone without distinction, with words and canes. Such uniformity in the actions of a commander shows that he has nothing in common with his subordinates and that he has no understanding at all of his countrymen. And this is very important.

Admiral P. S. Nakhimov (1802–1855), quoted in Danchenko and Vydrin, *Military Pedagogy*, 1973

THE LEADER/LEADERSHIP

An army of deer led by a lion is more to be feared than an army of lions led by a deer.

Attributed to Chabrias (410–375 BC) and to Philip II of Macedon (382–336 BC)

The leader must himself believe that willing obedience always beats forced obedience, and that he can get this only by really knowing what should be done. Thus he can secure obedience from his men because he can convince them that he knows best, precisely as a good doctor makes his patients obey him. Also he must be ready to suffer more hardships than he asks of his soldiers, more fatigue, greater extremes of heat and cold.

Xenophon (431 BC–c. 352 BC), *Cyropaedia*

A leader is a dealer in hope.

Napoleon (1769–1821)

The more the leader is in the habit of demanding from his men, the surer he will be that his demands will be answered.

Major General Carl von Clausewitz, *On War*, 1832

When you join your organization you will find there a willing body of men who ask from you nothing more than the qualities that will command their respect, their loyalty, and their obedience.

They are perfectly ready and eager to follow you so long as you can convince them that you have those qualities. When the time comes that they are satisfied you do not possess them you might as well kiss yourself goodbye. Your usefulness to that organization is at an end.

Major C. A. Bach, 1917, farewell instructions to the graduating student officers, Fort Sheridan, Wyoming

---

To be a highly successful leader in war four things are essential, assuming that you possess good common sense, have studied your profession and are physically strong.

When conditions are difficult, the command depressed and everyone seems critical and pessimistic, you must be especially cheerful and optimistic.

When evening comes and all are exhausted, hungry and possibly dispirited, particularly in unfavourable weather at the end of a march or in battle, you must put aside any thought of personal fatigue and display marked energy in looking after the comfort of your organization, inspecting your lines for tomorrow.

Make a point of extreme loyalty, in thought and deed, to your chiefs personally; and in your efforts to carry out their plans or policies, the less you approve the more energy you must direct to their accomplishment.

The more alarming and disquieting the reports received or the conditions viewed in battle, the more determined must be your attitude. Never ask for the relief of your unit and never hesitate to attack.

I am certain in the belief that the average man who scrupulously follows this course of action is bound to win great success. Few seemed equal to it in this war, but I believe this was due to their failure to realize the importance of so governing their course.

General of the Army George C. Marshall, 5 November 1920, letter to General John Mallory

---

The real secret of leadership in battle is the domination of the mass by a single personality. Influence over subordinates is a matter of suggestion. Discipline acquired during peace and the power of personal example are both used to exact great sacrifices.

General der Infantrie Hugo Baron von Freytag-Loringhoven, *The Power of Personality in War*, 1911

The moral equilibrium of the man is tremendously affected by an outward calmness on the part of the leader. The soldier's nerves, taut from anxiety of what lies ahead, will be soothed and healed if the leader sets an example of coolness. Bewildered by the noise and confusion of battle, the man feels instinctively that the situation cannot be so dangerous as it appears if he sees that his leader remains unaffected, that his orders are given clearly and deliberately, and that his tactics show decision and judgment. However 'jumpy' the man feels, the inspiration of his leader's example shames him into swallowing his own fear. But if the leader reveals himself irresolute and confused, the more even than if he shows personal fear, the infection spreads instantly to his men. (July 1921)

Captain Sir Basil Liddell Hart, *Thoughts on War*, 1944

The quality of leadership needs, above all, spirit, intelligence, and sympathy. Spirit is needed to fire men to self-sacrificing achievements; intelligence, because men will only respect and follow a leader whom they feel knows his profession thoroughly; sympathy, to understand the mentality of each individual in order to draw out the best that is in him. Given these qualities, men will conquer fear to follow a leader. (July 1921)

Captain Sir Basil Liddell Hart, *Thoughts on War*, 1944

At first blush one would scarcely expect to find the behaviour of a piece of cooked spaghetti an illustration of successful leadership in combat . . . it scarcely takes demonstration to prove how vastly more easy it is to PULL a piece of cooked spaghetti in a given direction along a major axis than it is to PUSH it in the same direction. Further, the difficulty increases with the size of either the spaghetti or the command.

General George S. Patton, Jr., January 1928, *The Patton Papers*, Vol. I, 1972

What, above all else, we look for in a leader is the power to dominate events, to leave his mark on them, and to assume responsibility for the consequences of his actions. The setting up of one man over his fellows can be justified only if he can bring to the common task the drive and certainty which comes of character. But why, for that matter, should a man be granted, free gratis and for nothing, the privilege of domination, the right to issue orders, the pride of seeing them obeyed, the thousand and one tokens of respect, unquestioning obedience, and loyalty which surround the seat of power? To him goes the greater part of the honour and glory. But this is fair enough, for he makes the best repayment that he can by shouldering risks. Obedience would by intolerable if he who demands it did not use it to produce effective results, and how can he do so if he does not possess the qualities of daring, decision, and initiative?

Charles de Gaulle, *The Edge of the Sword*, 1932

Leadership in war is an art, a free creative activity based on a foundation of knowledge. The greatest demands are made on the personality.

German Army Field Service Regulations, 1933

The real leader displays his quality in his triumphs over adversity, however great it may be.

General of the Army George C. Marshall, 18 September 1941, address to 1st Officer Candidate Class, Fort Benning, Georgia

---

War is much too brutal a business to have room for brutal leading; in the end, its only effect can be to corrode the character of men, and when character is lost, all is lost. The bully and the sadist serve only to further encumber an army; their subordinates must waste precious time clearing away the wreckage that they make. The good company has no place for an officer who would rather be right than loved, for the time will quickly come when he walks alone, and in battle no man may succeed in solitude.

Brigadier General S. L. A. Marshall, *Men Against Fire*, 1947

No man is a leader until his appointment is ratified in the minds and hearts of his men.

Anonymous, *The Infantry Journal*, 1948

A leader is a man who has the ability to get other people to do what they don't want to do, and like it.

President Harry S. Truman, *Memoirs*, 1954

I have been very lucky in my service. In getting on for forty years I have commanded everything from a squad of six men to an Army Group of a million and quarter, and, believe me, whether you command ten men or ten million, the essentials of leadership are the same. Leadership is that mixture of example, persuasion and compulsion which makes men do what you want them to do. If I were asked to define leadership, I should say it is the Projection of Personality. It is the most intensely personal thing in the world, because it is just plain you.

Field Marshal Viscount Slim, *Courage and Other Broadcasts*, 1957

Herd-psychology is a slightly different problem. Any schoolmaster can deal with an individual, few of them can manage a rebellious class. The test of statesmanship and leadership is ability to gauge and master mass- and mob-psychology. The hallmark of leadership is to comprehend the purpose of a crowd, its ingredients, its strength, its ability to achieve its object, its cohesive cement, its impetus and above all to search out its leadership.

Colonel Richard Meinertzhagen, *Army Diary 1899–1925*, 1960

Leadership can be taught, but not character. The one arises from the other. A successful leader of men must have character, ability and be prepared to take unlimited responsibility. Responsibility can only be learned by taking responsibility; you cannot learn the piano without playing on one. Leadership is the practical application of character. It implies the ability to command to make obedience proud and free. One of the questions demanded by the War Office before 1914 of officers recommended for the Staff College was: Does this officer give commands in a manner likely to secure cheerful obedience? My Commanding Officer – Du Maurier, whose habits were slightly unorthodox – replied: 'I do not know if this officer's orders would be cheerfully obeyed, but I can guarantee that they would be carried out.' The War Office were satisfied.

Colonel Richard Meinertzhagen, *Army Diary 1899–1925*, 1960

The good military leader will dominate the events which surround him; once he lets events get the better of him he will lose the confidence of his men, and when that happens he ceases to be of value as a leader.

Field Marshal Viscount Montgomery of Alamein, *The Memoirs of Field Marshal Montgomery*, 1958

When all is said and done the leader must exercise an effective influence, and the degree to which he can do this will depend on the personality of the man – the 'incandescence' of which he is capable, the flame which burns within him, the magnetism which will draw the hearts of other men toward him. What I personally would want to know about a leader is:

Where is he going?
Will he go all out?
Has he the talents and equipment, including knowledge, experience and courage? Will he take decisions, accepting full responsibility for them, and take risks where necessary?
Will he then delegate and decentralize, having first created an organization in which there are definite focal points of decision so that the master plan can be implemented smoothly and quickly?

Field Marshal Viscount Montgomery of Alamein, *The Memoirs of Field Marshal Montgomery*, 1958

Leadership and learning are indispensable to each other.

President John F. Kennedy, 22 November 1963, remarks prepared for delivery at the Trade Mart in Dallas

---

The behaviour of a leader *in battle* is a distinct form of social behavior which has held a special place in the history of Western civilization and, indeed, in most other cultures as well. The role of Battle Leader possesses a special mystique and attendant value system. Combat officers are not viewed as merely another component of the larger societal techno-structure that can be created cooky-cutter fashion from a training base little different from the Harvard Business School. The values, goals, and modes of action required of the combat officer are qualitively different from those associated with and expected from the commercial manager.

Lieutenant Colonel Richard Gabriel, *Crisis in Command*, 1978

. . . I hold that leadership is not a science, but an art. It conceives an ideal, states it as an objective, and then seeks actively and earnestly to attain it, everlastingly preserving, because the records of war are full of successes coming to those leaders who stuck it out just a little longer than their opponents.

General Mathew B. Ridgway, 'Leadership', *Military Review*, 9/1966

. . . one of my cardinal rules of battle leadership – or leadership in any field – is to be yourself, to strive to apply the basic principles of the art of war, and to seek to accomplish your assigned missions by your own methods and in your own way.

General Mathew B. Ridgway, *The Korean War*, 1967

Leadership must be based on goodwill. Goodwill does not mean posturing and, least of all, pandering to the mob. It means obvious and wholehearted commitment to helping followers. We are tired of leaders we fear, tired of leaders we love, and most tired of leaders who let us take liberties with them. What we need for leaders are men of the heart who are so helpful that they, in effect, do away with the need of their jobs. But leaders like that are never out of a job, never out of followers. Strange as it sounds, great leaders gain authority by giving it away.

Admiral James B. Stockdale, 'Machiavelli and Moral Leadership', in *Military Ethics*, 1987

One final aspect of leadership is the frequent necessity to be a philosopher, able to understand and to explain the lack of a moral economy in this universe, for many people have a great deal of difficulty with the fact that virtue is not always rewarded nor is evil always punished. To handle tragedy may indeed be the mark of an educated man, for one of the principal goals of education is to prepare us for failure. To say that is not to encourage resignation to the whims of fate, but to acknowledge the need for forethought about how to cope with undeserved reverses. It's important that our leadership steel themselves against the natural reaction of lashing out or withdrawing when it happens. The test of character is not 'hanging in there' when the light at the end of the tunnel is expected but performance of duty and persistence of example when the situation rules out the possibility of the light ever coming.

Admiral James B. Stockdale, *Naval War College Review*, July–August 1979

There are three types of leader: Those who make things happen; those that watch things happen; and those who wonder what happened!

American Military Saying

LESSONS LEARNED

Lesson No. 1 from Fuller: 'To anticipate strategy, imagine.'
Lesson No. 2 from Martel: 'Men, not weapons, will shape the future, so stick with fundamentals.'

Brigadier General S. L. A. Marshall, *Bringing Up the Rear*, 1978

LINES OF COMMUNICATION

Co-equal with the security of flanks, the maintenance and full use of the line of communications to the rear are of major concern to the commander. It is his responsibility that the incoming supply is equal to the needs of his deployments and that the supporting arms and fires which have been promised him keep their engagements. Or if they do not, he must raise hell about it

Brigadier General S. L. A. Marshall, *Men Against Fire*, 1947

Roads that lead from an army's position back to the main sources of food and replacements, and that are apt to be the ones the army chooses in the event of a retreat, have two purposes. In the first instance they are *lines of communication* serving to maintain an army, and in the second they are *lines of retreat.*

Major General Carl von Clausewitz, *On War*, v, 1832, tr. Howard and Paret

LOGISTICS

Without supplies neither a general nor a soldier is good for anything.

Clearchus, 401 BC, speech to the Ten Thousand, quoted in Xenophon, *The Persian Expedition*, tr. Warner

---

The commander who fails to provide his army with necessary food and other supplies is making arrangements for his own defeat, even with no enemy present.

The Emperor Maurice, *The Strategikon*, c. AD 600

An army cannot preserve good order unless its soldiers have meat in their bellies, coats on their backs and shoes on their feet.

The Duke of Marlborough, 1703, letter to Colonel Cadogan

When the Duke of Cumberland has weakened his army sufficiently, I shall teach him that a general's first duty is to provide for its welfare.

Field Marshal Maurice Comte de Saxe, 1747, before Laufeld

What I want to avoid is that my supplies should command me.

Field Marshal François Comte de Guibert (1744–1790), *Essai Général de la Tactique*, 1770

It is very necessary to attend to all this detail and to trace a biscuit from Lisbon into a man's mouth on the frontier and to provide for its removal from place to place by land or by water, or no military operations can be carried out.

The Duke of Wellington, 1811

Understand that the foundation of an army is the belly. It is necessary to procure nourishment for the soldier wherever you assemble him and wherever you wish to conduct him. This is the primary duty of a general.

Frederick the Great, *Instructions to His Generals*, 1747

Without supplies no army is brave.

Frederick the Great, *Instructions to his Generals*, 1747

What makes the general's task so difficult is the necessity of feeding so many men and animals. If he allows himself to be guided by the supply officers he will never move and his expedition will fail.

Napoleon, *Maxims of War*, 1831

The whole of military activity must relate directly or indirectly to the engagement. The end for which a soldier is recruited, clothed, armed, and trained, the whole object of his sleeping, eating, drinking, and marching *is simply that he should fight at the right place and the right time*.

Major General Carl von Clausewitz, *On War*, 1832, tr. Howard and Paret

---

The 'feeding' of an army is a matter of the most vital importance, and demands the earliest attention of the general intrusted with a campaign. To be strong, healthy, and capable of the largest measure of physical effort, the soldier needs about three pounds gross of food per day, and the horse or mule about twenty pounds. When a general first estimates the quantity of food and forage needed for any army of fifty to one hundred thousand men, he is apt to be dismayed, and here a good staff is indispensable, though the general cannot throw off on them the responsibility. He must give the subject his personal attention, for the army reposes in him alone, and should never doubt the fact that their existence overrides in importance all other considerations. Once satisfied of this, and that all has been done that can be, the soldiers are always willing to bear the largest measure of privation.

General of the Army William T. Sherman, *Memoirs of General W. T. Sherman*, 1875

One of the most effective ways of impeding the march of an army is by cutting off its supplies; and this is just as legitimate as to attack in line of battle.

Colonel John S. Mosby, *Mosby's War Reminiscences*, 1887

Victory is the beautiful, bright-coloured flower. Transport is the stem without which it never could have blossomed.

Sir Winston Churchill, *The River War*, 1899

I believe that the task of bringing the force to the fighting point, properly equipped and well-formed in all that it needs is at least as important as the capable leading of the force in the fight itself . . . In fact it is indispensable and the combat between hostile forces is more in the preparation than the fight.

General Sir John Monash (1865–1931), quoted in Horner, *The Commanders*, 1984

The more I see of war, the less I think that general principles of strategy count as compared with administrative problems and the gaining of intelligence. The main principles of stragegy, e.g., to attack the other fellow in the flank or rear in preference to the front, to surprise him by any means in one's power and to attack his morale before you attack him physically are really things that every savage schoolboy knows. But it is often outside the power of the general to act as he would have liked owing to lack of adequate resources and I think that military history very seldom brings this out, in fact, it is almost impossible that it should do so without a detailed study which is often unavailable. For instance, if Hannibal had had another twenty elephants, it might have altered his whole strategy against Italy.

Field Marshal Earl Wavell, 1942, letter to Liddell Hart

. . . the crux of generalship – superior even to tactical skill.

Field Marshal Earl Wavell (1883–1950)

Mobility is the true test of a supply system.

Captain Sir Basil Liddell Hart, *Thoughts on War,* 1944

An adequate supply system and stocks of weapons, petrol and ammunition are essential conditions for any army to be able to stand successfully the strain of battle. Before the fighting proper, the battle is fought and decided by the Quartermasters.

Field Marshal Erwin Rommel, quoted in Wavell, *Soldiers and Soldiering,* 1953

Many generals have failed in war because they neglected to ensure that what they wanted to achieve operationally was commensurate with their administrative resources; and some have failed because they over-insured in this respect. The lesson is, there must always be a nice balance between the two requirements.

Field Marshal Viscount Montgomery of Alamein, *The Memoirs of Field Marshal Montgomery,* 1958

Contemporary armies possess colossal vitality. Even the achievement of the utter defeat of the opponent at a definite moment does not assure the final victory so long as there is, behind the defeated units, a rear which is economically and morally strong. Given the presence of time and space which permit a new mobilization of human and material resources which are needed for the restoration of the fighting capacity of the army, such defeated units may re-establish the front and carry the struggle forward with hope of success.
Mikhail V. Frunze (1885–1925)

LOOTING/PILLAGE/PLUNDER

Then his majesty prevailed against them at the head of his army, and when they saw his majesty prevailing against them they fled headlong to Megiddo in fear, abandoning their horses and their chariots of gold and silver . . . Now, if only the army of his majesty had not given their heart to plundering the things of the enemy, they would have captured Megiddo at this moment when the wretched foes of Kadesh and the wretched foe of this city were hauled up in haste to bring them into this city.

Pharaoh Thutmose III, 1439 BC, at the Battle of Megiddo, quoted in Breasted, *Ancient Records of Egypt,* 1906

. . . it is sometimes to your interest to give the enemy rein, and to allow him to lay waste as much of the land as he wishes, where, while plundering and laden with spoil, he will easily suffer punishment at your hands. For in this way all that has been taken would be recovered, and those who had done the damage would receive their just deserts.

Aeneas Tacticus, *On the Defence of Fortified Positions,* 356 BC (Note: Truly there is nothing new under the sun! Aeneas could have been speaking of the slaughter of the Iraqi garrison of Kuwait City as it fled, laden with Kuwait loot, along the highway to Basra in February 1991.)

If the enemy is put to flight, our soldiers must be restrained from plundering. Otherwise while they are scattered about doing this, the enemy army might reform and attack them.

The Emperor Maurice, *The Strategikon,* c. AD 600

God be my witness, you are yourselves the destroyers, wasters and spoilers of your Fatherland. My heart is sickened when I look at you.

Gustavus II Adolphus, King of Sweden, September 1632, to the officers of his German Protestant allies who had stolen cattle during his invasion of Bavaria

I fear the Turkish Army less than their camp.

Field Marshal Prince Eugene, quoted in Chandler, *Atlas of Military Strategy*, 1980

Fire authorizes looting, which the soldier permits himself in order to save the rest from the remnants of the fire.

Napoleon, 20 September 1812, ed., Herold, *The Mind of Napoleon*, 1955

Nothing will disorganize an army more or ruin it more completely than pillage.

Napoleon, *Maxims of War*, 1831

There was only one serious feature. The success had not been as complete as it might have been, because a good division, instead of pressing on, stopped to go through an enemy supply depot.

General Erich Ludendorff, of the breakdown in discipline among hungry German troops during the 1918 *Friedenstürm*, *My War Memories 1914–1918*, 1919

War looses violence and disorder; it inflames passions and makes it relatively easy for the individual to get away with unlawful actions. But it does not lessen the gravity of the offense or make it less necessary that consituted authority put him down. The main safeguard against lawlessness and hooliganism in any armed body is the integrity of the officers. When men know that their commander is absolutely opposed to such excesses, and will take forceful action to repress any breach of discipline, they will conform. But when an officer winks at any depredation by his men, it is no different than if he had committed the act.

Brigadier General S. L. A. Marshall, *The Armed Forces Officer*, 1950

I was afraid that our wonderful captures of the last few weeks would tempt some of our Europeans to help themselves to things improperly, and I took advantage of the occasion to point out the evils of such behaviour. It must not be forgotten that war booty belongs to the State, and that the individual soldier has to notify superiors if he happens to want any particular object he has captured. An estimate is then made of the value of the object and he has to pay the amount. It was important for me to maintain the *morale* of our troops unconditionally if I was to be able to appeal to their sense of honour and make calls on their endurance.

Major General Paul von Lettow-Vorbeck, *East African Campaigns*, 1957

. . . theft is a crime no matter where or from whom. It is often from small beginnings that the rot sets in: a soldier 'lifts' a few oranges from an orchard and gets away with it; an officer, held in the highest esteem, takes a revolver as booty from the enemy, carries off a few 'finds' from an historic site, or in supposedly innocent mischief sends his wife a souvenir 'from a house in occupied territory' or 'found on an infiltrator'. In this way the dam of army morals is burst asunder and the officer is powerless to prevent the spread of corruption to which he has himself succumbed. No distinction can be drawn here between the penny and the pound; petty or large-scale, crime can be called by no other name. No commander can demand discipline and honesty from his men if he himself does not serve as an example. And plainly the higher the rank the graver the crime – and the more rigorous punishment there must be.

General Yigal Allon, *The Making of Israel's Army*, 1960

LOYALTY

In all the military works it is written: To train samurai to be loyal separate them when young, or treat them according to their character. But it is no use to train them according to any fixed plan, they must be educated by benevolence. If the superior loves benevolence then the inferior will love his duty.

Tokugawa Ieyasu (1543–1616), quoted in Sadler, *The Maker of Modern Japan: The Life of Tokugawa Ieyasu*

Those who would be military leaders must have loyal hearts, eyes and ears, claws and fangs. Without people loyal to them, they are like someone walking at night, not knowing where to step. Without eyes and ears, they are as though in the dark, not knowing where to proceed. Without claws and fangs, they are like hungry men eating poisoned food, inevitably they die . . . Therefore good generals always have intelligent and learned associates for their advisors, thoughtful and careful associates for their eyes and ears, brave and formidable associates for their claws and fangs.

Zhuge Liang (AD 180–234), *The Way of the General*

Loyalty is the marrow of honour.

Field Marshal Paul von Hindenburg, *Out of My Life*, 1920

A good officer ought to recognize that there is a higher loyalty than that to his immediate superiors – loyalty to the army and the nation. In action his first aim should be to prove a faithful subordinate, but in reflection he should maintain his critical sense. For that 'criticism is the life-blood of science' is a greater truth than the idea, prevalent among weak-minded superiors, that it is a breach of discipline. (December 1926)

Captain Sir Basil Liddell Hart, *Thoughts on War*, 1944

Man has two supreme loyalties – to country and to family. With most the second, being more personal, is the stronger. So long as their families are safe they will defend their country, believing that by their sacrifice they are safeguarding their families also. But even the bonds of patriotism, discipline, and comradeship are loosened when the family is threatened. The soldier feels instinctively that if he was at home he could at least fight for the immediate protection of his family, work to gain food for it, and, at worst, die with it. But when the enemy is closer than he is, the danger and his fears are magnified by his remoteness. Every letter, every rumour is a strain on his nerves and on his sense of duty.

Hence the deadliest form of strategy in a war between nations is that which sets the two loyalties in opposition and so imposes a breaking strain on the will of the soldier. This 'tug of war' in the soldier's mind can be created by giving a deeper sense and wider scope to the military idea of the 'rear attack' – by developing it into an attack against the rear of a people, not merely of an army. (April 1929)

Captain Sir Basil Liddell Hart, *Thoughts on War*, 1944

Those who are naturally loyal say little about it, and are ready to assume it in others. In contrast, the type of soldier who is always dwelling on the importance of 'loyalty' usually means loyalty to his own interests. (September 1932)

Captain Sir Basil Liddell Hart, *Thoughts on War*, 1944

An officer . . . should make it a cardinal principle of life that by no act of commission or omission on his part will he permit his immediate superior to make a mistake.

General Malin Craig, 12 June 1937, address to the graduating class of West Point

---

We learn from history that those who are disloyal to their own superiors are most prone to preach loyalty to their subordinates. Loyalty is a noble quality, so long as it is not blind and does not exclude the higher loyalty to truth and decency. But the word is much abused. For 'loyalty', analysed, is too often a polite word for what would be more accurately described as 'a conspiracy for mutual inefficiency'. In this sense it is essentially selfish . . . 'loyalty' is not a quality we can isolate – so far as it is real, and of intrinsic value, it is implicit in the possession of other virtues.

Captain Sir Basil Liddell Hart, *Through the Fog of War*, 1938

Loyalty is the big thing, the greatest battle asset of all. But no man ever wins the loyalty of troops by preaching loyalty. It is given him by them as he proves his possession of the other virtues. The doctrine of a blind loyalty to leadership is selfish and futile military dogma except in so far as it is ennobled by a higher loyalty in all ranks to truth and decency.

Brigadier General S. L. A. Marshall, *Men Against Fire*, 1947

There is a great deal of talk about loyalty from the bottom to the top. Loyalty from the top down is even more necessary and much less prevalent.

General George S. Patton, Jr., *War As I Knew It*, 1947

The loyalty of men simply cannot be commanded when they become embittered by selfish action.

Brigadier General S. L. A. Marshall, *The Armed Forces Officer*, 1950

Loyalty is developed through the unifying of action. The more decisive the action becomes, the greater the bond.

Brigadier General S. L. A. Marshall, *The Armed Forces Officer*, 1950

The very touchstone of loyalty is that just demands will be put upon it.

Brigadier General S. L. A. Marshall, *The Armed Forces Officer*, 1950

I also learnt that the discipline demanded from the soldier must become loyalty in the officer.

Field Marshal Viscount Montgomery of Alamein, *The Memoirs of Field Marshal Montgomery*, 1958

LUCK

. . . one can see that a general should be skillful and lucky and that no one should believe so fully in his star that he abandons himself to it blindly. If you are lucky and trust in luck alone, even your success reduces you to the defensive; if your are unlucky you are already there.

Frederick the Great, *Instructions to his Generals*, 1747

The state benefits more from a lucky general than from a brave one. The first achieves his results with little effort, whereas the other does so at some risk.

The Emperor Maurice, *The Strategikon*, c. AD 600

I base my calculation on the expectation that luck will be against me.

Napoleon (1769–1821)

The affairs of war, like the destiny of battles, as well of empires, hang upon a spider's thread.

Napoleon, *Political Aphorisms*, 1848

Luck in the long run is given only to the efficient.

Field Marshal Helmuth Graf von Moltke (1800–1891)

Luck is like a sum of gold, to be spent.

Field Marshal Viscount Allenby of Meggido (1861–1936)

There *is* such thing as luck, and as soldiers you have to believe in it.

Brigadier General S. L. A. Marshall, quoted in Fitton, ed., *Leadership*, 1990

That's the way it is in war. You win or lose, live or die – and the difference is just an eyelash.

General of the Army Douglas MacArthur, *Reminiscences*, 1964

Though it is no doubt true that luck in the long run pans out evenly, this is not true of a short campaign, still less of a battle. It can, in fact, play a disconcertingly prominent part in the issue.

Colonel Alfred H. Burne, *The Art of War on Land*, 1947

MAINTENANCE

It has been well said that Nelson took more care of his topgallant masts, in ordinary cruising, than he did of his whole fleet when the enemy was to be checked or beaten.

Rear Admiral Alfred Thayer Mahan, *Lessons of the War with Spain*, 1899

. . . vehicles are maintained properly by tools, elbow grease, and dirty hands, not by pencils and forms.

General Lesley J. McNair, in Kahn, *McNair, Educator of an Army*

We built our equipment to military specs that are petty tough. Sure, we have to change filters on engines. Sure, there will be a certain amount of corruption on helicopter blades. Yes, we will have to change engines on aircraft more often, but we know how to do all that. We know how to maintain our equipment. I absolutely do not expect to see huge breakdowns of equipment all over the kingdom.

My aide-de-camp said it is like flossing your teeth. If you forget it once, your teeth are not going to fall out. If we are smart about it, we are not going to have that sort of problem.

General H. Norman Schwarzkopf, 13 September 1990, interview with Richard Pyle, *Schwarzkopf: In His Own Words*, 1991

MANOEUVRE
(Principle of War)

The enemy should see me marching when he believes me to be fettered by calculations of subsistence; this new kind of war must astonish him, must give him no chance to breathe anywhere, and make plain at his cost this constant truth – that there is scarcely any position tenable in face of a well-constituted, temperate, patient and manoeuvring army.

Field Marshal François Duc de Guibert, *Essai général de tactique*, 1770

Nothing is more difficult than the art of manoeuvre. What is difficult about manoeuvre is to make the devious route the most direct and to turn misfortune to advantage.

Sun Tzu, *The Art of War*, c. 500 BC, tr. Griffith

Both advantage and danger are inherent in manoeuvre.

Sun Tzu, *The Art of War*, c. 500 BC, tr. Griffith

. . . for as a ship, if you deprive it of its steersman, falls with all its crew into the hands of the enemy; so, with an army in war, if you outwit or out-manoeuvre its general, the whole will often fall into your hands.

Polybius, *The Histories*, c. 125 BC

I have destroyed the enemy merely by marches.

Napoleon, 1805, after the Austrian Campaign

[victory/winning battles] . . . more by the movement of troops than by fighting.

General of the Army William T. Sherman (1820–1891)

Manoeuvres are threats; he who appears most threatening wins.

Colonel Charles Ardnant du Picq, *Battle Studies*, 1880

Battles are won by slaughter and manoeuvre. The greater the general, the more he contributes in manoeuvre, the less he demands in slaughter.

Sir Winston S. Churchill, *The World Crisis*, II, 1923

The fundamental characterization of a strategy of manoeuvre is not the formal offensive but initiative and action.

Leon Trotsky, 1921

Nearly all the battles which are regarded as masterpieces of the military art, from which have been derived the foundation of states and the fame of commanders, have been battles of manoeuvre.

Sir Winston S. Churchill, *The World Crisis*, II, 1923

MARCH TO THE SOUND OF THE GUNS

The safest way of achieving victory is to seek it among the enemy's battalions.

Field Marshal Prince Aleksandr V. Suvorov (1729–1800), quoted in Ossipov, 1945

March to the sound of the guns.

The Duke of York, 1793, to his commanders during the Dunkirk Campaign

---

The drawback to 'marching to the sound of the guns', a type of subordinate initiative that was so much praised in and after the 1870 war, is that it may serve to nullify the higher commander's strategical plan. That particularly applies where a force is being used for manoeuvre. For it may be led to swing in against the enemy's head, instead of continuing its circuit to strike the enemy's tail, and *cut off* his head. Thereby he may be given the chance to wiggle free. (September 1928)

Captain Sir Basil Liddell Hart, *Thoughts on War*, 1944

Marshal Grouchy had only left his camp at Gemlouz at ten in the morning, and was halfway to Wavres between noon and one o'clock. He heard the dreadful cannonade of Waterloo. No experienced man could have mistaken it: it was the sound of several hundred guns, and from that moment two armies were hurling death at each other. General Excelmans, commanding the cavalry, was profoundly moved by it. He went up to the Marshal and said to him: 'The Emperor is at grips with the English Army; there can be no doubt about it, such a furious fire can be no skirmish. Monsieur le Maréchal, we must march towards the sound of the guns. I am an old soldier of the Army of Italy; I have heard General Bonaparte preach this principle a hundred times. If we turn to the left we shall be on the battlefield in two hours.'

Napoleon, *The Waterloo Campaigns*, ed., Somerset de Chair, 1957

MARCHING

An army is exposed to more danger on marches than in battles. In an engagement the men are properly armed, they see their enemies before them and are prepared to fight. But on a march the soldier is less on his guard, has not his arms always ready and is thrown into disorder by a sudden attack or ambuscade. A general, therefore, cannot be too careful and diligent in taking necessary precautions to prevent a surprise on the march.

Flavius Vegetius Renatus, *Military Institutions of the Romans*, c. AD 378

We march straight on; we march to victory.

Harold Godwinson, King of England, 1066, refusing William of Normandy's offer of parly before the Battle of Hastings

---

MAXIM 9. The strength of an army, like power in mechanics, is reckoned by multiplying the mass by the rapidity. A rapid march increases the morale of an army, and increases its means of victory. Press on!

Napoleon, *Maxims of War*, 1831

One of the greatest difficulties in war is to have the men inured to marching . .. The rapidity of a march, or rather skillful marches, almost always determine the success of a war . . . It is this power of marching which constitutes the strength of infantry; and enterprises which seem to present the greatest difficulties, become comparatively easy by the advantages accruing from rapid marches.

Marshal of France Michel Ney, Duc d'Elchingen, Prince de la Moskova, *Memoirs of Marshal Ney*, 1834

The hardships of forced marches are more often painful than the dangers of battle . . . I would rather lose one man marching than five in fighting.

Lieutenant General Thomas 'Stonewall' Jackson, 1862, quoted in Henderson, *Stonewall Jackson*, 1898

MARKSMANSHIP

In archery we have three goals: to shoot accurately, to shoot powerfully, to shoot rapidly.

Anonymous Byzantine general, *On Strategy* (*Peri Strategias*) c. AD 527–65, in Dennis, *Three Byzantine Military Treatises*, 1985

Teach the soldier how to load his musket properly, whether with cartridge or loose powder, with or without the bayonet fixed; how he should give fire under the different circumstances he will encounter; teach him never to fire without an order, and never to do so without aiming, thus avoiding waste of fire to no purpose.

Marshal de Puysegur, *Art de la Guerre*, 1754

It is not sufficient that the soldier must shoot, he must shoot well.

Napoleon (1769–1821)

Good marksmanship is always the most important thing for the infantry.

General Gerhard von Scharnhorst, 1812, memorandum

You don't hurt 'em if you don't hit 'em.

Lieutenant General Lewis 'Chesty' Puller, quoted in Burke, *Marine*, 1962

MASS
(Principle of War)

Troops thrown against the enemy as a grindstone against eggs is an example of a solid acting upon a void.

Sun Tzu, *The Art of War*, c. 500 BC, tr. Griffith

Use the most solid to attack the most empty.

The 'Martial' Emperor Ts'ao Ts'ao, AD 155–220, commentary to Sun Tzu, *The Art of War*, c. 500 BC

One should not be overly fond of famous swords and daggers. For even if one has a sword valued at 10,000 cash, he will not overcome 100 men carrying spears valued at 100 cash.

Asakura Toshikage (1428–1481), *The 17 Articles of Asakura Toshikage*, in Wilson, *Ideals of the Samurai*, 1982

Move upon the enemy in one mass on one line so that when brought to battle you shall outnumber him, and from such a direction that you compromise him.

Napoleon (1769–1821)

Not by rambling operations, or naval duels, are wars decided, but by force massed and handled in skillful combinations.

Rear Admiral Alfred Thayer Mahan (1840–1914)

The art of making the weight of all one's forces successively bear on the resistances one may meet.

Marshal of France Ferdinand Foch, *Principles of War*, 1913

Do not let my opponents castigate me with the blather that Waterloo was won on the playing fields of Eton, for the fact remains geographically, historically and tactically, whether the great Duke uttered such nonsense or not, that it was won on fields in Belgium by carrying out a fundamental principle of war, the principle of mass; in other words by marching on to those fields three Englishmen, Germans or Belgians to every two Frenchmen.

Major General J. F. C. Fuller, 'Principles of War with Reference to the Campaigns of 1914–1915', *Royal United Services Institution Journal*, Vol. LXI: 1, 1916

The fifth act of the great drama in Flanders opened on the 22nd October. Enormous masses of ammunition, such as the human mind had never imagined before the war, were hurled upon the bodies of men who passed a miserable existence scattered about in mud-filled shell-holes. The horror of the shell-hole area of Verdun was surpassed. It was no longer life at all. It was more unspeakable suffering. And through this world of mud the attackers dragged themselves, slowly, but steadily, and in dense masses. Caught in the advanced zone by our hail of fire they often collapsed, and the lonely man in the shell-hole breathed again. Then the mass came on again. Rifle and machine-gun jammed in the mud. Man fought against man, and only too often the mass was successful.

General Erich Ludendorff, *My War Memoirs 1914–1918*, 1919

The concentration of all means on one . . . operation may yield a big economy of force. An enemy front capable of enduring dozens of small strikes may be broken by one big strike. In certain conditions, a certain mass of operation is necessary in order to obtain even minimal results.

General A. A. Svechin, 1927

The most common Russian form of combat was the use of mass. Human mass and mass of material were generally used unintelligently and without variation, but *they were always effective.*

The Russians repeated the same tactics again and again: employment of masses, and narrow division sectors held by large complements replenished time after time. Therefore, also the mass attacks. In the twinkling of an eye the terrain in front of our line teemed with Russian soldiers. They seemed to grow out of the earth, and nothing would stop their advance for a while. Gaps closed automatically, and the mass surged on until the supply of men was used up and the wave, substantially thinned, receded again. How often we witnessed this typical picture of a Russian attack! *It is impressive and astounding*, on the other hand, *how frequently the mass failed to recede, but rolled on and on, nothing able to stop it.* Repulsing such an attack was certainly dependent on the strength of our own forces and means for defense; primarily, however, it was a question of nerves. Only seasoned soldiers mastered the fear which instinctively gripped everyone upon the onslaught of such masses.

High German military source, quoted in Garthoff, *Soviet Military Doctrine*, 1953

The theory of human mass dominated the military mind from Waterloo to the World War. This monster was the child of the French Revolution by Napoleon. The midwife who brought it into the military world was the Prussian philosopher of war, Clausewitz, cloudily profound. He unfortunately died while his own thought was still fermenting – leaving his papers in sealed packets, with the significant note: 'Should the work be interrupted by my death, then what is found can only be called a mass of formless conceptions . . . open to endless misconceptions.' So they proved.

Captain Sir Basil Liddell Hart, *Thoughts on War*, 1944

We must not replace crushing strikes against the enemy with pinpricks . . . It is necessary to prepare an operation which will be like an earthquake.

Marshal of the Soviet Union Georgi K. Zhukov, 12 May 1944

MATERIAL

When confidence is placed in superiority of material means, valuable as they are against an enemy at a distance, it may be betrayed by the actions of the enemy. If he closes with you in spite of your superiority in means of destruction, the morale of the enemy mounts with your loss of confidence.

Colonel Charles Ardnant du Picq, *Battle Studies*, 1880

In our day wars are not won by mere enthusiasm, but by technical superiority.

V. I. Lenin, 1918, speech

We shall not discuss here the materialistic argument, except to say that if it were entirely true, savages and badly-equipped tribesmen would never have completely beaten well-armed civilized troops. Yet they have done so on frequent occasions. Witness the First Afghan War, the Zulu Wars, the American–Indian Wars, and a host of minor actions. Material only wins hands down when the *morale* of the side possessing it is at least fairly comparable with that of its opponents. Otherwise Byzantium with its 'Greek Fire' would have ruled the world.

Major Clough Williams-Ellis, *The Tank Corps*, 1919

A fighting revolutionary spirit and class self-awareness are the crucial factors in a revolutionary war, but revolutionary spirit alone without the necessary equipment cannot be victorious in modern war.

Marshal of the Soviet Union Mikhail N. Tukhachevskiy, quoted in *Voyenno-istoricheskiy zhurnal*, 2/1983

To-day, in the army we are faced with the problems of motorization and mechanization, just as the navy was seventy-odd years ago. Some think these changes good, and others bad; but their possible virtues and vices are insignificant problems if we lose sight of the greater problem which is this: *The more mechanical become the weapons with which we fight, the less mechanical must be the spirit which controls them.*

Major General J. F. C. Fuller, *Generalship: Its Diseases and Their Cure*, 1933

MAXIMS

The faculty of deciding at once and with certainty belongs only to him who, by his own experience, has tested the truth of the known maxims and possesses the manner of applying them; to him alone, in a word, who finds beforehand, in his positive acquirements, the conviction of the accuracy of his judgments.

The Archduke Charles of Austria (1771–1847)

When the application of a rule and the consequent manoeuvre have procured victory a hundred times for skillful generals, shall their occasional failure be a sufficient reason for entirely denying their value and for distrusting the effect of the study of the art? Shall a theory be pronounced absurd because it has only three-fourths of the whole number of chances in its favour?

Lieutenant General Antoine-Henri Baron de Jomini, *Summary of the Art of War*, 1838

Nothing so comforts the military mind as the maxim of a great but dead general.

Barbara W. Tuchman, *The Guns of August*, 1962

Maxims of war . . . are not so much positive rules as they are the development and applications of a few general principles . . . the maxim rooting itself in a principle, formulates a rule generally correct under the conditions; but the teacher must admit each case has its own features – like the endless variety of the one human face – which modify the application of the rule, and may even make it at times wholly inapplicable. It is for the skill of the artist in war rightly to apply the principles and rules in each case.

Rear Admiral Alfred Thayer Mahan, *Naval Strategy*, 1911

See also: QUOTATIONS

MEDICAL CORPS

War is the only proper school of the surgeon.

Hippocrates, *Wounds of the Head*, c. 415 BC

. . . eight or ten of the less-skilled soldiers in each tagma should be assigned as medical corpsmen to each bandon, especially in the first battle line. They should be alert, quick, lightly clothed, and without weapons. Their duty is to follow about a hundred feet to the rear of their own tagma, to pick up and give aid to anyone seriously wounded in the battle, or who has fallen off his horse, or is otherwise out of action, so they may not be trampled by the second line. For each person so rescued the corpsman should receive from the treasury one nomisma over above his pay . . .

The Emperor Maurice, *The Stragegikon*, c. AD 600

You medical people will have more lives to answer for in the other world than even we do.

Napoleon, quoted in O'Meara, *Napoleon in Exile*, 1822

You Gentlemen of England who sit at home in all the well-earned satisfaction of your successful cases, can have little idea from reading the newspapers of the Horror and Misery of operating upon these dying, exhausted men.

Florence Nightingale, 1854, letter from Scutari to Dr. Brown during the Crimean War

---

A corps of Medical officers was not established solely for the purpose of attending the wounded and sick . . . the labors of Medical officers cover a more extended field. The leading idea, which should be constantly kept in view, is to strengthen the hands of the Commanding General by keeping his army in the most vigorous health, thus rendering it, in the highest degree, efficient for enduring fatigue and privation, and for fighting. In this view, the duties of such a corps are of vital importance to the success of an army, and commanders seldom appreciate the full effect of their proper fulfilment [sic].

Major Jonathan Letterman (1824–1872), Medical Director of the Army of the Potomac

---

If the facilities for washing were as great those for drink, our Indian Army would be the cleanest body of men in the world.

Florence Nightingale, 1863, comments as a member of the Royal Commission on the sanitary state of the Indian Army

---

Hygiene . . . has such bearing upon the efficiency of armed forces that its place in warfare cannot be denied. As to its usefulness to line officers, I will venture to quote words of my own: 'The responsibility for the health of crews rests ultimately with the commanding officers; who, however they be guided ordinarily by the opinion of the surgeon, must be able on occasion to overrule intelligently the professional bias of the latter.' A doctor's business is to save life; the admiral's or captain's to risk it, when necessary and possible to attain a given end.

Rear Admiral Alfred Thayer Mahan, *Naval Administration and Warfare*, 1908

MEMOIRS/DIARIES

What the layman gets to know of the course of military events is usually nondescript. One action resembles another, and from a mere recital of events it would be impossible to guess what obstacles were faced and overcome. Only now and then, in the memoirs of generals or of their confidants, or as the result of close historical study, are some of the countless threads of the tapestry revealed. Most of the arguments and clashes of opinion that precede a major operation are deliberately concealed because they touch political interests, or they are simply forgotten, being considered as scaffolding to be demolished when the building is complete.

Major General Carl von Clausewitz, *On War*, 1832, i, tr. Howard and Paret

---

I have often needed a diary, an experience case record for reference. It is late to start 38½ years old, but I shall have some spare moments in the next ten days between here and Manila and I set my memory to work. Many things worthy of remembrance and reference have happened to me, especially in the last two years. To write them is to learn my lessons of experience anew. It will do me good. To live them over in memory will be like the review of a hard-studied study. It ought to polish the fixed and fix some floating principles in me, and help arrange knowledge and experience for convenient use.

Lieutenant General Robert Lee Bullard, 26 November 1899, diary entry

---

. . . this book should make a contribution towards perpetuating those experiences of the bitter war years; experiences often gained at the cost of great deprivations and bitter sacrifices.

Field Marshal Erwin Rommel, *Infantry Attacks*, 1937

. . . any memoirs of my life, distinct and unconnected with the general history of the war, would rather hurt my feelings than tickle my pride whilst I lived. I had rather glide gently down the stream of life, leaving it to posterity to think and say what they please of me, than by any act of mind to have vanity or ostentation imputed to me . . . I do not think vanity is a trait of my character.

General George Washington, 25 March 1784, *Writings of George Washington*, XXVII, 1931–1944

The last word in the Foreword to this book is 'truth'. I have tried to write the truth. I suppose everyone claims that about his memoirs! Most official accounts of past wars are deceptively well written, and seem to omit many important matters – in particular, anything which might indicate that any of our commanders ever made the slightest mistake. They are therefore useless as a source of instruction. They remind me of the French general's reply to a British protest in 1918, when the former directed the British to take over a sector from the French which had already been overrun by the Germans forty-eight hours previously. The French general said: 'Mais, mon ami, ca c'est pour l'histoire.'

Field Marshal Viscount Montgomery of Alamein, *The Memoirs of Field Marshal Montgomery*, 1958

I hear my generals are selling themselves dearly.

Sir Winston S. Churchill, of the flood of memoirs after World War II, quoted in Liddell Hart, *The Sword and the Pen*, 1976

A general who has taken part in a campaign is by no means best fitted to write its history. That, if it is to be complete and unbiased, should be the work of someone less personally involved. Yet such a general might write something of value. He might, as honestly as he could, tell of the problems he faced, why he took the decisions he did, what helped, what hindered, the luck he had, and the mistakes he made. He might, by showing how one man attempted the art of command, be of use to those who later may themselves have to exercise it. He might even give, to those who have not experienced it, some impression of what if feels like to shoulder a commander's responsibilities in war. These things I have tried to do in this book.

Field Marshal Viscount Slim, *Defeat Into Victory*, 1963

I mean to record the course of events as I saw them. I shall be as objective as I feel possible to be, but I have no intention whatever of departing, for any reason, from my own honest opinion as to events and personalities. So often, people make great play about being completely unprejudiced. Frankly, I am completely prejudiced, and I accept as a guide and as warning, Goethe's saying:

'I can promise to be upright but not to be unprejudiced.'

Air Marshal Lord Tedder, *With Prejudice: The War Memoirs of Marshal of the Royal Air Force Lord Tedder,* G.C.B., 1966

. . . the worst thing that anybody can do . . . is to keep an exact diary, because then he will put down every little resentment he had against Bill Smith and Joe Doakes. He will note everything that annoys him. Finally, his editors get hold of this and they say, 'Oh, this is a gold mine.' They take the things that he said and even emphasize them – the frictional type of entry – instead of those things which show that he thought things were going pretty well. I despise daily biographies as showing real history. I don't believe they do.

General of the Army Dwight D. Eisenhower, *General Eisenhower on the Military Churchill*, 1970

A major drawback of certain memoirs which I have read is the limited viewpoint in describing the combat events by the army commanders and even the front commanders . . . At times, the strange impression is created that seemingly an experienced and educated military man, in fighting in the war within his demarcation lines and having adjacent units to the right and left, forgets that not only his failures but also his successes are tied to their actions. He forgets that to the right and to the left of him units are fighting from the very same Soviet Army and to these proper due must be paid just as to his own units. He forgets that this is all the same army and not some other and that the Germans are fighting not specifically against his army or front but against the Soviet Army as a whole, against all the armies and all the fronts.

Marshal of the Soviet Union Georgi K. Zhukov, quoted in *Voyenno istoricheskiy zhurnal*, 12/1987

MENTOR

To act in concert with a great man is the first of blessings.

Marquis de Lafayette, 1778, letter to George Washington

I would never have believed this preference could have happened to me, that I could come out before the others in intellectual qualities, but he [Gerhard von Scharnhorst] convinced me he had thought me the most able in the school, and when I believed myself capable of this, I went ahead with confidence.

Major General Carl von Clausewitz, quoted in Parkinson, *Clausewitz*, 1971

Nothing could be more stimulating to a man who has consecrated himself to the search for the essential truth of war than this task of instructing those who will be called upon to apply it on the battlefield, where the lives of their soldiers and the fate of their country are at stake. But if, when found, he expects to instil that truth into minds, open indeed, but which the influences of practical military life have rendered somewhat sceptical toward school studies, the search must never be relaxed until at last he has seized it and made it his very own.

Marshal of France Ferdinand Foch, *The Memoirs of Marshal Foch*, 1931

By the way, I have a favor to ask. When you are handing out diplomas at West Point you will see a cadet named Clarence Gooding, about 120 in standing. I would greatly appreciate your making some personal comment to this young fellow when you hand him his diploma. He was my office orderly at Benning – for only a few months. Having just enlisted, I was greatly impressed with his efficiency, bearing and general intelligence. I found out that he had been trying for West Point and had had no success. Finally he enlisted for this purpose and was refused permission to join the candidates class of men being coached. Without his knowledge, I wrote to his congressman in Texas and got the promise of an appointment if a vacancy occurred. The first alternate dropped out at the eleventh hour and Gooding received his appointment. The principal failed and Gooding passed. He is a fine boy and would be electrified by a personal word from you.

General of the Army George C. Marshall, 2 June 1936, letter to General John J. Pershing, *Papers of George C. Marshall: The Soldierly Spirit*, 1981

The battalion returned to England in 1913 and an officer of our 2nd Battalion was posted to it who had just completed the two-year course at the Staff College at Camberley. His name was Captain Lefroy. He was a bachelor and I used to have long talks with him about the Army and what was wrong with it, and especially how one could get to real grips with the military art. He was interested at once, and helped me tremendously with advice about what books to read and how to study. I think it was Lefroy who first showed me the path to tread and encouraged my youthful ambition. He was killed later in the 1914–18 war and was a great loss to me and to the Army . . . All this goes to show how important it is for a young officer to come in contact with the best type of officer and the right influences early in his military career . . . In my case, the ambition was there, and the urge to master my profession. But it required advice and encouragement from the right people to set me on the road, and once that was forthcoming it was plainer sailing.

Field Marshal Viscount Montgomery, *The Memoirs of Field Marshall Montgomery*, 1958

Now relations with Uborevich were good. I felt that he was working with me. He took a look at me, gave me various assignments, and forced me to report. Then he assigned me . . . a report on the actions of the French Cavalry during the battle on the Po River during World War I . . . For me this report was an unusual and difficult undertaking. Particularly as I, the division commander, was to give this report in the presence of all the commanders of the district combat arms and corps commanders. But I prepared for the report and was lost only at first. I hung up all the maps and then stopped by them; I had to start but stood there in silence. But Uborevich was able to help me at that moment and by his question called me into the conversation. Subsequently, everything went normally and he judged my report as good . . . I repeat, I felt him working patiently with me.

Marshal of the Soviet Union Georgi K. Zhukov, quoted in *Voyenno istoricheskiy zhurnal*, 12/1987

MERCENARIES

The Carthaginians were in the habit of forming their armies of mercenaries drawn together from the different countries; if they did so for the purpose of preventing conspiracies, and of making the soldiers more completely under the control of their generals, they may seem perhaps, in this respect, not to have acted foolishly, for troops of this sort cannot easily unite together in factious counsels. But when we take another view of the question, the wisdom of the proceeding may be doubted, if we consider the difficulty there is to instruct, soften, and subdue the minds of an army so brought together when rage has seized them, and when hatred and resentment have taken root among them, and sedition is actually begun. In such circumstances, they are no longer men, but beasts of prey. Their fury cannot be restricted within the ordinary bounds of human wickedness or violence, but breaks out into deeds the most terrible and monstrous that are to be found in nature.

Polybius, *The Histories*, c. 125 BC

If any one supports his state by the arms of mercenaries, he will never stand firm or sure, as they are disunited, ambitious, without discipline, faithless, bold amongst friends, cowardly amongst enemies, they have no fear of God, and keep no faith with men. Ruin is only deferred as long as the assault is postponed; in peace you are despoiled by them, and in war by the enemy. The cause of this is that they have no love or other motive to keep them in the field beyond a trifling wage, which is not enough to make them ready to die for you. They are quite willing to be your soldiers so long as you do not make war.

Niccolo Machiavelli, *The Prince*, 1513

Let us therefore animate and encourage each other, and show the whole world that a Freeman contending for *Liberty* on his own ground is superior to any slavish mercenary on earth.

General George Washington, 2 July 1776, General Order to the Continental Army

---

The Greeks in the service of the Great King were not enthusiastic in his cause. The Swiss in French, Spanish, and Italian service were not enthusiastic in their causes. The troops of Frederick the great, mostly foreigners, were not enthusiastic in his cause. A good general, good training, and good discipline make good troops independently of the cause in which they fight. It is true, however, that fanaticism, love of fatherland, and national glory can inspire fresh troops to good advantage.

Napoleon, ed. Herold, *The Mind of Napoleon*, 1955

Better to lose one's life for the Fatherland, than hang a foreign tassel to one's sword.

Rhigas' immortal war song, quoted in Kolokotrones, *Theodoros Kolokotrones*, 1892

MILITARY EDUCATION

The Lacadaemonians made war their chief study. They are affirmed to be the first who reasoned on the events of battles and committed their observations thereon to writing with such success as to reduce the military art, before considered totally dependent on courage or fortune, to certain rules and fixed principles.

Flavius Vegetius Renatus, *Military Institutions of the Romans*, c. AD 378

The general should be ignorant of none of the situations likely to occur in war. Who can attempt to accomplish what he does not understand? Who is able to furnish assistance in situations whose dangers he does not understand?

The Emperor Maurice, *The Strategikon*, c. AD 600

We have acquired this knowledge not simply from hearing about it but also from having been taught by a certain amount of experience. For one thing, the men who instructed and trained us in this method were the very ones . . . who invented it. Then, on our own, we have put it into practice and, as best we could, almost made it a part of us.

The Emperor Nikephorus II Phokus, *Skirmishing*, c. AD 969, in Dennis, *Three Byzantine Military Treatises*, 1985

None other than a Gentleman, as well as seaman, both in Theory and practice is qualified to support the Character of a Commissioned Officer in the Navy, nor is any Man fit to Command a Ship of War, who is not also capable of communicating his Ideas on Paper in Language that becomes his Rank.

Admiral John Paul Jones (1747–1792)

War is not an affair of chance. A great deal of knowledge, study, and meditation is necessary to conduct it well.

Frederick the Great, *Instructions to His Generals*, 1747

Every art has its rules and maxims. One must study them: theory facilitates practice. The lifetime of one man is not long enough to enable him to acquire perfect knowledge and experience. Theory helps to supplement it, it provides a youth with premature experience and makes him skillful also through the mistakes of others. In the profession of war the rules of the art never are violated without drawing punishment from the enemy, who is delighted to find us at fault. An officer can spare himself many mistakes by improving himself. We even venture to say that he must do it, because the mistakes that he commits through ignorance cover him with shame, and even in praising his courage one cannot refrain from blaming his stupidity.

What an incentive to work hard! What reasons to travel the thorny road to glory! And what loftier and nobler compensation is there than to have one's name immortalized for his pains, and by his work.

Frederick the Great, 'Avant-propos', 1771, *Oeuvres,* Vol. XXIX, 1846–1856 in ed., Luvaas, *Frederick the Great on the Art of War,* 1966

Gneisenau, if I had only learned something, what might not have been made of me! . . . But I put off everything I should have learned. Instead of studying, I have given myself to gambling, drink, and women; I have hunted, and perpetrated all sorts of foolish pranks. That's why I know nothing now. Yes, the other way I would have become a different kind of fellow, believe me.

Field Marshal Prince Gebhardt von Blücher, to Gneisenau, his chief of staff, during the 1813 Campaign

---

Correct theories, founded upon right principles, sustained by actual events of wars, and added to accurate military history, will form a true school of instruction for generals. If these means do not produce great men, they will at least produce generals of sufficient skill to take rank next after the natural masters of the art of war.

Lieutenant General Antoine-Henri Baron de Jomini, *Summary of the Art of War,* 1838

It is necessary that the study of the military sciences should be encouraged and rewarded as well as courage and zeal. The military corps should be esteemed and honoured; this is the only way of securing for the army men of merit and genius.

Lieutenant General Antoine-Henri Baron de Jomini, *Summary of the Art of War,* 1838

An ignorant officer is a murderer. All brave men confide in the knowledge he is supposed to possess; and when the death-trial comes their generous blood flows in vain. Merciful God! How can an ignorant man charge himself with so much bloodshed? I have studied war long, earnestly and deeply, but yet I tremble at my own derelictions.

General Sir William Napier (1782–1853)

. . . the close of the siege of Vicksburg found us with an army unsurpassed, in proportion to its numbers, taken as a whole of officers and men. A military education was acquired which no other school could have given. Men who thought a company was quite enough for them to command properly at the beginning, would have made good regimental or brigade commanders; most of brigade commanders were equal to the command of a division, and one, Ransom, would have been equal to the command of a corps at least. Logan and Crocker ended the campaign fitted to command independent armies.

General of the Army Ulysses S. Grant, *Personal Memoirs of U.S. Grant,* 1885

It is evident that a theoretical knowledge is not sufficient to that end; there must be a free, practical, artistic development of the qualities of mind and character, resting of course on a previous military education and guided by experience, whether it be the experience of military history or actual experience in war.

Field Marshal Helmuth Graf von Moltke, quoted in Foch, *Principles of War,* 1913

Not to promote war but to preserve peace.

Elihu Root, 1902, motto of the US Army War College, Carlisle Barracks, Pennsylvania

---

Besides the care of the men and the education of the non-commissioned officers – an education which was also calculated to assist in their future callings – I attached the greatest importance to increasing the efficiency of the Officer Corps and the training of the younger officers. While the personnel of the Regular Officers' Corps is always the same, the officers of the Reserve, the N.C.O.s and men change continually. So the Officers' Corps is the mainspring of the Army. The officers must therefore be thoroughly conversant with the Army's great deeds and possess a comprehensive knowledge of their coutry's history, and it is expected of all men who have to lead others.

General Erich Ludendorff, *My War Memories 1914–1918*, 1919

As I perceived that for increasing the power of rapid judgment in questions of tactics, it would be useful to make the head alert by means of such a game as chess, I commenced recently to study the Englishman Jane's book on chess and although I have not yet been able to attain the results expected, still, if I can get new ideas and make some improvements, it cannot be said to have been a waste of time. If we exploit the idea of this game, we shall surely reach points which will be of use to the study of the art of war.

Admiral Marquis Togo Heihachiro, *c.* 1898

I hope the officers of her Majesty's army may never degenerate into bookworms. There is happily at present no tendency in that direction, for I am glad to say that this generation is as fond of danger, adventure, and all manly out-of-door sports as its forefathers were. At the same time, all now recognize that the officer who has not studied war as an applied science, and who is ignorant of modern military history, is of little use beyond the rank of captain.

Field Marshal Viscount Wolseley, preface to English edition of von Der Goltz, *The Conduct of War*, 1897

Special emphasis must be given to ethical training, for leadership through moral inspiration indispensable to high commanders and important staff officers. Great attention, therefore, must be directed to incalculating, besides the necessary military knowledge and techniques, military morality and martial spirit.

Generalissimo Chiang Kai-shek, 1930, description of the Army Staff College

There can be no better school for a commander of a regiment, or a brigade, or a division that the work of educating the platoon commanders . . . the best training of all is received by a teacher in teaching his pupils.

Leon Trotsky, 1922, *Military Writings of Leon Trotsky*, 1969

Though the military art is essentially a practical one, the opportunities of practising it are rare. Even the largest-scale peace manoeuvres are only a feeble shadow of the real thing. So that a soldier desirous of acquiring skill in handling troops is forced to theoretical study of Great Captains.

Field Marshal Earl Wavell, *c.* 1930, lecture to officers at Aldershot

---

For my strategy, I could find no teachers in the field: but behind me there were some years of military reading, and even in the little that I have written about it, you may be able to trace the allusions and quotations, the conscious analogies.

Colonel T. E. Lawrence (1888–1953)

Wisdom begins at the point of understanding that there is nothing shameful about ignorance; it is shameful only when a man would rather remain in that state than cultivate other men's knowledge.

Brigadier General S. L. A. Marshall, *The Armed Forces Officer*, 1950

However competent an officer may be, and however thorough his earlier training, he is always a risky investment unil he pays off in combat.

General of the Army Omar Bradley, *A Soldier's Story*, 1951

Fundamental to the training of senior officers will be the most comprehensive instruction possible in technical and organizational matters. The object of this instruction will be to induce a certain independence of mind, so that particular value will need to be laid on teaching officers to think critically on questions of basic principle. Respect for the opinion of this or that great soldier must never be allowed to go so far that nobody dares to discuss it. A sure sense of reality must be aroused. Given a well-founded knowledge of basic principles, any man of reasonably cool and logical mind can work out most of the principles for himself, provided he is not inhibited in his thinking.

Field Marshal Erwin Rommel, *The Rommel Papers,* 1953

It is sometimes thought that when an officer is promoted to the next higher command, he needs no teaching in how to handle it. This is a great mistake. There is a tremendous difference between a brigade and a division, between a division and a corps; when an officer got promotion, he needed help and advice in his new job and it was up to me to see that he got it.

Field Marshal Viscount Montgomery of Alamein, *The Memoirs of Field Marshal Montgomery,* 1958

Officers, particularly those in positions of command, must at all times be urged to expand the scope of their knowledge; nothing has a more damaging effect on the quality of the army than a hard core of commanders whose minds are narrow and inflexible. Educated, enlightened commanders will produce soldiers equally enlightened and thirsty for knowledge, soldiers who will understand why they were recruited and for what they must fight.

General Yigal Allon, *The Making of Israel's Army,* 1960

When field grade officers complain of the weakness of company-grade officers and noncommissioned officers, and in the next breath they say they do not have time for unit schools, then something is amiss in their own education.

Lieutenant General Arthur S. Collins, Jr., *Common Sense Training,* 1978

I was always being asked by the Navy brass what a destroyer skipper needs to know about Immanuel Kant; a liberally educated person meets new ideas with curiosity and fascination. An illiberally educated person meets new ideas with fear.

Admiral James B. Stockdale, 1 September 1980, *Newsweek*

MILITARY INTELLIGENCE

To the king, my lord, your servant Sennacherib: Greeting to the king my lord. All is well with Assyria, all is well with the temples, all is well with every fortress of the king. May the king be of very good cheer.

The Ukkai have sent me the following message, 'When the king of Armenia went to the land of the Cimmerians, his army perished entirely; as to him, his prefects, we brought them to his remaining forces . . . Ashur-risua has sent word as follows: "The information about Armenia that I sent was correct. A dreadful slaughter occurred among them. Now the land is at rest. Each one of his nobles has gone into his own country. Kakkadanu, his commander in chief, was captured. The king of Armenia is in the country of Uazaun."' Nbuli'u, the prefect of Halsu, has written me as follows: "When I wrote to the garrisons of the fortressees along the border inquiring news of the king of Armenia, they reported: 'When he came to the land of the Cimmerians his army perished entirely. Three of his nobles, together with their troops, were slain, but he escaped and entered his own land; his camp; however, has not yet arrived.'

Assyrian intelligence collection report to King Sargon (722–705 BC), Pfeiffer, *State Letters of Assyria,* 1935

Nothing is more worthy of the attention of a good general than the endeavour to penetrate the designs of the enemy.

Niccolo Machiavelli, *Discourses,* 1517

It is essential to know the character of the enemy and of their principal officers – whether they be rash or cautious, enterprising or timid, whether they fight on principle or from chance.

Flavius Vegetius Renatus, *Military Institutions of the Romans*, c. AD 378

The purpose of this chapter is to enable those who intend to wage war against these peoples to prepare themselves properly. For all nations do not fight in a single formation or in the same manner, and one cannot deal with them all in the same way. Some, whose boldness is unlimited, are led by an impulsive spirit, while others use good judgment and order in attacking their enemies.

The Emperor Maurice, *The Strategikon*, c. AD 600

Therefore, determine the enemy's plans and you will know which strategy will be successful and which will not . . .

Agitate him and ascertain the pattern of his movements.

Determine his dispositions and so ascertain the field of battle.

Probe him and learn where his strength is abundant and where deficient.

Sun Tzu, *The Art of War*, c. 500 BC, tr. Griffith

. . . that no war can be conducted successfully without early and good intelligence, and that such advices cannot be had but at very great expense.

The Duke of Marlborough (1650–1722)

In general it is necessary to pay spies well and not be miserly in that respect. A man who risks being hanged in your service merits being well paid.

Frederick the Great, *Instructions to His Generals*, 1747

The necessity of procuring good intelligence is apparent and need not be further urged. All that remains for me to add, is, that you keep all the whole matter as secret as possible. For upon Secrecy, Success depends in most Enterprises of the kind, and for want of it, they are generally defeated, however well-planned and promising a favourable issue.

General George Washington, 26 July 1777, letter to Colonel Elias Dayton

---

If you know the enemy's plans beforehand you will always be more than a match for him, even with inferior numbers. All generals who command armies try to procure this advantage but there are few who succeed in it.

Frederick the Great, *Instruction militaire du Roi de Prusse pour ses généraux*, 1761, in ed. Luvaas, *Frederick the Great on the Art of War*, 1966

. . . the most difficult thing is to discern the enemy's plans, and to detect the truth in all reports one receives: the remainder requires only common sense.

Napoleon, in Watson, *By Command of the Emperor, A Life of Marshal Berthier*, 1957

To guess at the intention of the enemy; to divine his opinion of yourself; to hide from both your intentions and opinion; to mislead him by feigned manoeuvres; to invoke ruses, as well as digested schemes, so as to fight under the best conditions – this is and always was the art of war.

Napoleon, quoted in Fuller, *Memoirs of an Unconventional Soldier*, 1936

The whole art of war consists in getting at what lies on the other side of the hill, or, in other words, what we do not know from what we do know.

The Duke of Wellington (1769–1852)

By 'intelligence' we mean every sort of information about the enemy and his country – the basis, in short, of our own plans and operations. If we consider the actual basis of this information, how unreliable and transient it is, we soon realize that war is a flimsy structure that can easily collapse and bury us in its ruins. The textbooks agree, of course, that we should only believe reliable intelligence, and should never cease to be suspicious, but what is the use of such feeble maxims? They belong to that wisdom which for want of anything better scribblers of systems and compendia resort to when they run out of ideas.

Major General Carl von Clausewitz, *On War*, i, 1832, tr. Howard and Paret

---

. . . most intelligence is false, and the effect of fear is to multiply lies and inaccuracies. As a rule most men would rather believe bad news than good, and rather tend to exaggerate the bad news. The dangers that are reported may soon, like waves, subside; but like waves they keep recurring, without apparent reason.

Major General Carl von Clausewitz, *On War*, i, 1832, tr. Howard and Paret

---

How can any man say what he should do himself if he is ignorant of what his adversary is about?

Lieutenant General Antoine-Henri Baron de Jomini, *Summary of the Art of War*, 1838

Nothing should be neglected in acquiring a knowledge of the geography and military statistics of other states, so as to know their material and moral capacity for attack and defence as well as the strategic advantages of the two parties. Distinguished officers should be employed in these scientific labours and should be rewarded when they demonstrate marked ability.

Lieutenant General Antoine-Henri Baron de Jomini, *Summary of the Art of War*, 1838

In order to conquer that unknown which follows us until the very point of going into action, there is only one means, which consists in looking out until the last moment, even on the battlefield, for *information*.

Marshal of France Ferdinand Foch, *Precepts and Judgments*, 1919

A general should neglect no means of gaining information of the enemy's movements. For this purpose he should make use of reconnaissance, spies, bodies of light troops commanded by capable officers, signals, and questioning of deserters and prisoners.

Ever multiply the means of obtaining information, for no matter how imperfect and contradictory they may be the truth may often be sifted from them.

Perfect reliance should be placed on none of these means.

Lieutenant General Antoine-Henri Baron de Jomini, *Summary of the Art of War*, 1838

When I took a decision, or adopted an alternative, it was after studying every relevant – and many an irrelevant – factor. Geography, tribal structure, religion, social customs, language, appetites, standards – all were at my finger-ends. The enemy I knew almost like my own side. I risked myself among them a hundred times, to *learn*.

Colonel T. E. Lawrence, 26 June 1933, letter to Liddell Hart

No operational skill can compensate for those severe consequences which can occur out of neglect or the shortcomings of an intelligence service.

Marshal of the Soviet Union Mikhail N. Tukhachevskiy, quoted in *Voyenno istorichiskiy zhurnal*, 2/1983

In battle nothing is ever as good or bad as the first reports of excited men would have it.

Field Marshal Viscount Slim, *Unofficial History*, 1959

To lack intelligence is to be in the ring blindfolded.

General David M. Shoup, 2 January 1960, remarks to the staff of the United States Marine Corps

---

Military intelligence is not, in fact, the spectacular service of the common imagining, but a much more prosaic affair, dependent on an efficient machinery for collecting and evaluating every sort of item of information – machinery that extends from the front-line right back to Supreme Headquarters. When the mass of information has been collected, the art is to sift the wheat from the chaff, and then to lay before the commander a short clear statement.

Field Marshal Earl Alexander, *The Alexander Memoirs 1940–1945*, 1961

Combat intelligence is the term applied to information of enemy forces, strength, disposition and probable movements. With personnel now assigned to Combat Intelligence, institute rigorous, continuous examination of enemy capabilities and potentialities, thereby getting the utmost value out of information of the enemy and enabling our forces to be used with the greatest effectiveness. It is particularly important to comprehend the enemy point of view in all aspects.

Admiral of the Fleet Ernest J. King, *Fleet Admiral King, A Naval Record*, 1952

Nothing helps a fighting force more than correct information. Moreover, it should be in perfect order, and done well by capable personnel.

Che Guevara, 1963, memorandum

. . . in achieving our results [intelligence collection and analysis], neither I nor my colleagues resorted to black magic. It was application, thoroughness, expert knowledge and speed . . .

General Reinhard Gehlen, description of the operation of Foreign Armies East, the German General Staff intelligence organization specializing in Soviet forces, *The Service: The Memoirs of Richard Gehlen*, 1972

MILITARY LIFE

That variety incident to a military life gives our profession some advantage over those of a more even and consistent nature. We have all our passions and affections roused and exercised, many of which must have wanted their proper employment, had not suitable occasions obliged us to exert them. Few men are acquainted with the degree of their own courage till danger prove them, and are seldom justly informed how far the love of honour and the dread of shame are superior to the love of life . . . Constancy of temper, patience, and all the virtues necessary to make us suffer with a good grace are likewise parts of our character, and . . . frequently called in to carry us through unusual difficulties . . .

Major General Sir James Wolfe, quoted in Liddell Hart, *Great Captains Unveiled*, 1927

A sense of vocation is the greatest virtue of the military men.

Field Marshal Prince Aleksandr V. Suvorov (1729–1800), quoted in *Morskoi Sbornik*, 5/1981

If we judge the future from the past, we perceive that though there may be no war, we must not rest at ease. It appears to me that a warrior's life is continual war; there is no reason why his responsibility should grow lighter or heavier by being in time of peace or war. If war breaks out, we display our military power; and if we are at peace, we cultivate it. We must always discharge our duty.

Admiral Marquis Togo Heihachiro, 21 December 1905, farewell speech to the Japanese Fleet, quoted in Ogasawara, *Life of Admiral Togo*, 1934

I am writing my own obituary . . .

I loved the Army: it reared me, it nurtured me, and it gave me the most satisfying years of my life. Thanks to it I have lived an entire lifetime in 26 years. It is only fitting that I should die in its service . . .

And yet I deny that I died FOR anything – not my country, not my Army, not my fellow man. I LIVED for these things, and the manner in which I chose to do it involved the very real chance that I would die in the execution of my duties . . .

. . . The Army is my life, it is such a part of what I was that what happened is the logical outcome of the life I loved. I never knew what it is to be too old or too tired to do anything. I lived a full life in the Army, and it has exacted the price. It is only just.

Major John Alexander Hottell, written for the *West Point Alumni Quarterly* about one year before his death in a helicopter crash on 7 July 1970

Men have joined armed forces at different times for different reasons. I do not see many young men joining for philosophical reasons. Almost always the desire for an active life has been prominent among reasons for taking up the profession of arms, but there have usually been contributory motives. These have often been ephemeral in value, and in kind accidental rather than essential.

General Sir John Hackett, *The Profession of Arms*, 1963

MILITARY SCIENCE

War is a science replete with shadows in whose obscurity one cannot move with assured step. Routine and prejudice, the natural result of ignorance, are its foundation and support. All sciences have principles and rules. War has none. The great captains who have written of it give us none. Extreme cleverness is required merely to understand them.

Field Marshal Maurice Comte de Saxe, *My Reveries*, 1732

War is a science for those who are outstanding; an art for mediocrities; a trade for ignoramuses.

Frederick the Great (1712–1786)

. . . the science of winning . . .

Field Marshal Prince Aleksandr V. Suvorov (1729–1800)

. . . nothing is better calculated to kill natural genius and to cause error to triumph, than those pedantic theories, based upon the false idea that war is a positive science, all the operations of which can be reduced to infallible calculations.

Lieutenant General Antoine-Henri Baron de Jomini, *Summary of the Art of War*, 1838

It [strategy] is more than science, it is the translation of science into practical life, the further development of the original governing idea, in accordance with the ever-changing circumstances; it is the art of taking action under the pressure of the most trying conditions.

Field Marshal Helmuth Graf von Moltke, *On Strategy*, 1871

At the present time no one thinks to assert that there might be a military science; it is as unthinkable as is a science of poetry, of painting, of music . . .

General Mikhail I. Dragomirov (1830–1905)

An aggregate of 'military principles' does not constitute a military science, for there is no more a science of war than there is a science of locksmithing. An army leader requires the knowledge of a whole number of sciences in order to feel himself fully equipped for his *art*. But military science does not exist; there does exist a military craft which can be raised to the level of military art.

Leon Trotsky, 8 May 1922, speech to the Military Scientific Society, *Military Writings*, 1969

A study of military history brings ample confirmation of Rebecca West's *mot*: 'Before a war military science seems a real science, like astronomy, but after a war it seems more like astrology.'

Captain Sir Basil Liddell Hart, *Thoughts on War*, 1944

MILITARY SERVICE

Cease to hire your armies. Go yourselves, every man of you, and stand in the ranks and either a victory beyond all victories in its glory awaits you, or falling you shall fall greatly, and worthy of your past.

Demosthenes, *Third Phillipic*, 341 BC

However long you live and whatever you accomplish, you will find that the time you spent in the Confederate army was the most profitably spent portion of your life. Never again speak of having lost time in the army!

General Robert E. Lee, quoted in Freeman, *R. E. Lee*, Vol. IV, 1935

National preparedness means . . . first of all, the moral organization of the people, an organization which creates in the heart of every citizen a sense of his obligation for service to the nation in time of war.

Major General Leonard Wood, *Our Military History*, 1916

In the first place it is more than usually difficult to attract men of outstanding ability into the army in prolonged periods of peace. What, fundamentally, tempts such men to adopt the profession of arms, is the prospect of power. There can be little doubt that no power is comparable to that of the war leader. So long, therefore, as the probability exists of some day being able to win distinction in the field, men of high quality and distinction will never be lacking worthily to fill the higher posts in the fighting services of countries with a strong military tradition. But a generation which is convinced that it will never again be called upon to fight, is unlikely to contain many individuals of first-class quality who will feel any inclination to become regular soldiers, the more so since in a pacifist period soldiers are considered of small use and have little to look forward to in terms either of reputation or money. The strong-minded, therefore, the enterprising, and those of marked personality, are not unnaturally inclined to embark on careers which will give them power and the consideration of their fellows.

Charles de Gaulle, *The Edge of the Sword*, 1932

Our soldiers join the Army to serve the people, not the officers.

Chu Teh (1886–1976), Commander of the Peoples' Liberation Army (PLA)

It is my sole wish to serve the Emperor as his shield.
I will not spare my honour or my life.

Admiral Yamamoto Isoroku, 1941

The majority of men, so long as they are treated fairly and feel that good use is being made of their powers, will rejoice in a new sense of unity with new companions even more than they will mind the increased separation from their old associations.

Brigadier General S. L. A. Marshall, *The Armed Forces Officer*, 1950

MILITARY VIRTUE/SPIRIT

The military virtues are: bravery in the soldier, courage in the officer, valour in the general, but guided by the principles of order and discipline, dominated by vigilance and foresight.

Field Marshal Prince Aleksandr V. Suvorov (1729–1800)

War is a special kind of activity, different and separate from any other pursued by man . . . An army's military qualities are based on the individual who is steeped in the spirit and essence of this activity; who trains the capacities it demands, rouses them, and makes them his own; who applies his intelligence to every detail; who gains ease and confidence through practice; and who completely immerses his personality in the appointed task.

Major General Carl von Clausewitz, *On War*, ii, 1832, tr. Howard and Paret

---

By innuring armies to labour and fatigue, by keeping them from stagnation in garrison in times of peace, by inculating their superiority over their enemies (without depreciating the latter too much), by inspiring a love for great exploits – in a word, by exciting their enthusiasm by every means in harmony with their tone of mind, by honouring courage, punishing weakness, disgracing cowardice – we may expect to maintain a high military spirit.

Lieutenant General Antoine-Henri Baron de Jomini, *Summary of the Art of War*, 1838

. . . whoever will read history unbiased, cannot fail but be convinced that of all military virtues it is the vigour, with which a war is prosecuted, that has always contributed most to military glory and success.

Major General Carl von Clausewitz, *On War*, 1832

MILITIA

Men just dragged from the tender Scenes of domestick life; unaccustomed to the din of Arms; totally unacquainted with every kind of Military skill, which being followed by a want of confidence in themselves, when opposed to Troops regularly train'd, disciplined, and appointed, superior in knowledge, and superior in Arms, makes them timid, and ready to fly from their own shadows.

George Washington, 24 September 1776, letter to the President of Congress, *The Writings of George Washington*, Vol. 6, 1931–44

No militia will ever acquire the habits necessary to resist a regular force . . . The firmness requisite for the real business of fighting is only to be attained by a constant course of discipline and service. I have never yet been witness to a single instance that can justify a different opinion, and it is most earnestly to be wished that the liberties of America may no longer be trusted, in any material degree, to so precarious a dependence.

George Washington, 15 September 1780, letter to the President of Congress, *The Writings of George Washington*, Vol. 20, 1931–44

I would not have had a swamp in the view of the militia on any consideration; they would have made for it, and nothing could have detained them from it. And as to covering my wings, I knew my adversary, and was perfectly sure I should have nothing but downright fighting. As to retreat, it was the very thing I wished to cut off all hope of. I would have thanked Tarleton had he surrounded me with his cavalry. It would have been better than placing my own men in the rear to shoot down those who broke from the ranks. When men are forced to fight, they will sell their lives dearly . . . Had I crossed the river, one half of the militia would immediately have abandoned me.

Daniel Morgan, description of his use of militia at the Battle of Cowpens, 1781

In this peaceful service I imbibed the rudiments of the language and science of tactics, which opened a new field of observation and study . . . The discipline and evolution of a modern battalion gave me a clearer notion of the Phalanx and the Legions, and the Captain of the Hampshire grenadiers (the reader may smile) has not been useless to the historians of the Roman Empire.

Edward Gibbon (1737–1794), reminiscing on his militia service during the Seven Years War, *Autobiography*, 1962

Such noble regiments they have. Three field officers, four staff officers, ten captains, thirty lieutenants, and one private with a misery in his bowels.

Lieutenant General D. H. Hill, 24 April 1863, of the Confederate Militia

MISTAKES

By their own follies they perished, the fools.

Homer, *The Iliad*, i, *c*. 800 BC

In war there is never any chance for a second mistake.

Lamarchus (465–414 BC), quoted in Plutarch, *Apothegms*

I am more afraid of our own mistakes than of our enemies' designs.

Pericles, 432 BC, speech to the Athenians, quoted in Thucydides, *History of the Peloponnesian War*, c. 404 BC

Mistakes made in ordinary affairs can generally be remedied in a short while, but errors made in war cause lasting harm.

The Emperor Maurice, *The Strategikon*, c. AD 600

When a man has committed no faults in war, he can only have been engaged in it but a short time.

Marshal of France Vicomte de Turenne, 1645, after the Battle of Marienthal

---

I am not sorry that I went, notwithstanding what has happened. One may pick up something useful from among the most fatal errors.

Major General Sir James Wolfe, 1757, of the Rochefort Expedition

---

Nor is it enough to have studied Prince Eugene as the model of the great generals. It will be no less useful to examine the mistakes that either the ministers of the court or the generals committed for lack of judgment and knowledge in devising enterprises poorly. These examples are only too numerous: I will not probe into antiquity to point out the blunders of past eras, but will cite only the modern follies that . . . are more familiar to you.

Frederick the Great, 'Réflexions sur les projets de campagne,' *Oeuvres de Frédéric le Grand*, Vol. XXIX, 1846–56, in ed. Luvaas, *Frederick the Great on the Art of War*, 1966

The training of armies is primarily devoted to developing efficiency in the detailed execution of the *attack*. This concentration on tactical technique tends to obscure the psychological element. It fosters a cult of soundness, rather than surprise. It breeds commanders who are so intent not to do anything wrong, according to 'the book', that they forget the necessity of making the enemy do something wrong. The result is that their plans have not result. For, in war, it is by compelling mistakes that the scales are most often turned.

Captain Sir Basil Liddell Hart, *Strategy* 1954

Gentlemen, when the enemy is committed to a mistake, we must not interrupt him too soon.

Admiral Lord Nelson, quoted in Mahan, *The Life of Nelson*, 1897

The greatest general is he who makes the fewest mistakes.

Napoleon (1769–1821)

Nothing is easy in war. Mistakes are always paid for in casualties and troops are quick to sense any blunder made by their commanders.

General of the Army Dwight D. Eisenhower, *Crusade in Europe*, 1948

Victory in battle – save – where it is brought about by sheer weight of numbers, and omitting all question of the courage of the troops engaged – never comes solely as the result of the victor's planning. It is not only the merits of the victor that decide the issue, but also mistakes on the part of the vanquished.

Field Marshal Erwin Rommel, *The Rommel Papers*, 1953

Happily for the result of the battle – and for me – I was, like other generals before me, to be saved from the consequences of my mistakes by the resourcefulness of my subordinate commanders and the stubborn valour of my troops.

Field Marshal Viscount Slim, *Defeat Into Victory*, 1962

To inquire if and where we made mistakes is not to apoligize. War is replete with mistakes because it is full of improvisations. In war we are always doing something for the first time. It would be a miracle if what we improvised under the stress of war should be perfect.

Vice Admiral Hyman G. Rickover, April 1964, testimony before Congress

---

You must be able to underwrite the honest mistakes of your subordinates if you wish to develop their initiative and experience.

Lieutenant General Bruce C. Clarke

MOBILITY/MOVEMENT

Aptitude for war is aptitude for movement.

Napoleon, *Maxims of War,* 1831

Force does not exist for mobility, but mobility for force.

Rear Admiral Alfred Thayer Mahan, *Lessons of the War with Spain,* 1899

Success in war depends upon mobility and mobility upon time. Mobility leads to mass, to surprise and to security. Other things being equal, the most mobile side must win: this a truism in war as in horse-racing. The tank first of all is a time-saving machine, secondly a shield – it is, in fact, a mechanical horse. If in a given time we can do three times as much as the enemy and lose a third less than he does, our possibilities of success are multiplied by nine . . .

Major General J. F. C. Fuller, memorandum of 1917, *Memoirs of an Unconventional Soldier,* 1936

Movement is the safety-valve of fear; unless we wish to risk a sudden collapse of the man's nerve control, we must open the valve. (July 1921)

Captain Sir Basil Liddell Hart, *Thoughts on War,* 1944

Two things stop the movement of armies: (a) bullet and fragments of shell which destroy the motive power of men, and (b) the confusion of the conflict.

Sir Winston S. Churchill, *The World Crisis,* 1923

We need bold and free flight, we need mobility.

Mikhail Frunze, 1921

The Principle of Movement. If concentration of weapon-power be compared to a projectile and economy of force to its line of fire, then movement may be looked upon as the propellant, and as a propellant is not always in a state of explosive energy, so neither is movement. Movement is the power of endowing mass with momentum; it depends, therefore, largely on security, which, when coupled with offensive power, results in liberty of action. Movement consquently, may be potential as well as dynamic, and, if an army be compared to a machine the power of which is supplied to it by a series of accumulators, should the object of its commander be to maintain movement, he can only accomplish this by refilling one set of accumulators while the other is in the process of being exhausted. The shorter the time available to do this the more difficult will the commander's task be; consequently, one of his most important duties, throughout war, is to increase the motive power of his troops, which depend on two main factors – moral and physical endurance.

Major General J. F. C. Fuller, *The Reformation of War,* 1923

As an individual combat, mobility is not only the link between security and surprise, but their mainstring. Mobility is economic movement, i.e., movement that fulfills the law of economy of force. The means to it are the *direction* given by the commander, the *speed* with which this is carried out, and the *cooperation* by which this fulfillment is ensured without check or waste of energy – caused by friction, loss of touch, and misunderstanding of the plan of direction. (January 1926)

Captain Sir Basil Liddell Hart, *Thoughts on War,* 1944

It is worth recalling that the Mongols, although their army was entirely composed of mobile troops, found neither the Himalayas nor the far-stretching Carpathians a barrier to progress. For mobile troops there is usually a way around. (September 1927)

Captain Sir Basil Liddell Hart, *Thoughts on War*, 1944

Our supreme tactical principle is *Mobility* . . . *Mobility* is aided by *surprise*, by the *independence of the subordinate commander* within the mission of the higher unit, and what we'd call *tactics by mission* . . . Mobility means quick decisions, quick movements, surprise attacks with concentrated force; to do always what the enemy does not expect, and to constantly change both the means and methods and to do the most improbable thing whenever the situation permits; it means to be free of all set rules and preconceived ideas. We believe that no leader who thinks or acts by stereotyped rules can ever do anything great because he is bound by such rules . . . War is not normal . . . We do not want therefore any stereotyped solutions for battle, but an understanding of the nature of war.

Captain von Bechtolsheim, 1931, Reichswehr lecturer at the US Army Cavalry School, Fort Sill, Oklahoma

---

Battle implies mobility, strategic and tactical. The army which seeks to fight another must be able to move quickly against it. On the battlefield the troops must be able to move forward in face of the enemy's fire. Once the issue is decided, the army must be able to follow up the beaten enemy, and complete the victory. Battle also implies the *immobilization of the enemy* – the paralysis of his powers of movement so that, in the first place, he may not be able to slip away, and second, that he may not be able to counter your strokes. (January 1933)

Captain Sir Basil Liddell Hart, *Thoughts on War*, 1944

MOBILIZATION

The young men shall fight; the married men shall forge weapons and transport supplies; the women will make tents and serve in the hospitals; the children will make up old linen into lint; the old men will have themselves carried into the public squares to rouse the courage of the fighting men, and to preach hatred of kings and the unity of the Republic. The public buildings shall be turned into barracks, the public squares into munitions factories; the earthen floors of cellars shall be treated with lye to extract saltpetre. All suitable firearms shall be turned over to the troops; the interior shall be policed with fowling pieces and with cold steel. All saddle horses shall be seized for the cavalry; all draft horses not employed in cultivation will draw the artillery and supply wagons.

Decree of the Committee of Public Safety, France, 23 August 1793

What does political mobilization mean? First, it means telling the army and the people about the political aim of the war. It is necessary for every soldier and civilian to see why the war must be fought and how it concerns him.

Mao Tse-tung, *On Protracted War*, May 1938

The Army used to have all the time in the world and no money; now we've got all the money and no time.

General of the Army George C. Marshall, January 1942, remark on the mobilization of World War II

Everyone will now be mobilized and all boys old enough to carry a spear will be sent to Addis Ababa. Married men will take their wives to carry food and cook. Those without wives will take any woman without a husband. Women with small babies need not go. The blind, those who cannot walk or for any reason cannot carry a spear are exempted. Anyone found at home after receipt of this order will be hanged.

Emperor Haile Selassie's mobilization order issued after the Italian invasion of Ethiopia in 1935

When a nation is without establishments and a military system, it is very difficult to organize an army.

Napoleon, *Maxims of War*, 1831

MORAL ASCENDANCY/ MORAL FORCE

Moral forces rather than numbers decide victory.

Napoleon (1769–1821)

In war the moral is to the material as three to one.

Napoleon (1769–1821)

. . . the moral elements are among the most important in war. They constitute the spirit that permeates war as a whole, and at an early stage they establish a close affinity with the will that moves and leads the whole mass of force, practically merging with it, since the will is itself a moral quantity. Unfortunately they will not yield to academic wisdom. They cannot be classified or counted. They have be seen or felt.

Major General Carl von Clausewitz, *On War*, iii, 1832, tr. Howard and Paret

---

They [moral elements] are: *the skill of the commander, the experience and courage of the troops, and patriotic spirit*. The relative value of each cannot be universally established; it is hard enough to discuss their potential, and even more difficult to weight them against each other. The wisest course is not to underrate any of them – a temptation to which human judgment, being fickle, often succumbs. It is far preferable to muster historical evidence of the unmistakable effectiveness of all three.

Major General Carl von Clausewitz, *On War*, iii, 1832, tr. Howard and Paret

---

The Army of the Tennessee had won five successive victories over the garrison of Vicksburg in the three preceding weeks. They had driven a portion of the army from Port Gibson with considerable loss, after having flanked them out of their stronghold at Grand Gulf. They had attacked another portion of the same army at Raymond, more than fifty miles farther in the interior of the State, and driven them back into Jackson with great loss in killed, wounded, captured and missing, besides loss of large and small arms; they had captured the capital of the State of Mississippi, with a large amount of materials of war and manufactures. Only a few days before, they had beaten the enemy then penned up in the town first at Champion's Hill, next at Big Black River Bridge, inflicting upon a him a loss in arms and ammunition. The Army of the Tennessee had come to believe that they could beat their antagonist under any circumstances.

General of the Army Ulysses S. Grant, *Personal Memoirs of U.S. Grant*, 1885

Combat today requires . . . a moral cohesion, a unity more binding than at any other time. If one does not wish the bonds to break, he must make them more elastic to strengthen them.

Colonel Charles Ardnant du Picq, *Battle Studies*, 1880

In battle, two moral forces, even more than two material forces, are in conflict. The stronger conquers.

Colonel Charles Ardnant du Picq, *Battle Studies*, 1880

When men become fearful in combat, the moral incentive can restore them and stimulate them to action. But when they become hopeless, it is because all moral incentive is gone. Soldiers who have ceased to hope are no longer receptive beings. They have become oblivious to all things, large and small.

Brigadier General S. L. A. Marshall, *The Armed Forces Officer*, 1950

Soldiers universally concede the general truth of Napoleon's much-quoted dictum that in war 'the moral is to the physical as three to one'. The actual arithmetical proportion may be worthless, for morale is apt to decline if the weapons are inadequate, and the strongest will is of little use if it is inside a dead body. But although the moral and physical factors are inseparable and indivisible, the saying gains its enduring value because it expresses the idea of the predominance of morale factors in all military decisions. On them constantly turns the issue of war and battle. In the history of war, they form the more constant factors, changing only in degree, whereas the physical factors are different in almost every war and every military situation.

Captain Sir Basil Liddell Hart, *Strategy*, 1954

There is no rule for the moral unifying of military forces under the pressures of the battlefield that is not equally good in the training that conditions troops to this eventuality. For the group to feel a great spiritual solidarity – for its members to be bound together by mutual confidence and the satisfactions of a rewarding comradeship – is the foundation of great enterprise. But it is more than that. Unaccompanied by a strengthening of the military virtues and a rise in the martial spirit, a friendly unity will not of itself point men directly toward the main object in training, or enable them to dispose themselves efficiently toward each other on entering battle.

Brigadier General S. L. A. Marshall, *The Officer as a Leader*, 1966

The powerful moral effect of our assuming the offensive, when nothing but surrender had been expected, seemed to bewilder them. Before they could recover from the shock of their surprise, Captain Flint, the leader, had fallen dead in their sight. Before the impetuous onset of my men they now broke and fled. No time was given them to re-form and rally.

Colonel John S. Mosby, *Mosby's War Reminiscences*, 1887

They claim that to fire human grapeshot at the enemy without preparation, gives us moral ascendancy. But the thousands of dead Frenchmen, lying in front of the German trenches, are instead those who are giving moral ascendancy to the enemy. If this waste of human material keeps on, the day is not far off when the offensive capacity of our army, already seriously weakened, will be entirely destroyed.

Lieutenant Abel Ferry, 1916, memorandum to the French Cabinet from Ferry, a Deputy of the French Assembly and a reserve officer in the field

---

The most solid morale qualities melt away under the effect of modern arms.

Marshal of France Ferdinand Foch, *Precepts and Judgments*, 1919

Moral forces may take a back seat at Committees of Imperial Defence or in War Offices; at the front they are where Joab put Uriah.

General Sir Ian Hamilton, *The Soul and Body of an Army*, 1921

Moral force is, unhappily, no substitute for armed force, but it is a very great reinforcement

Sir Winston Churchill, 21 December 1937, speech in the House of Commons

---

The thing in any organization is the creation of a soul . . .

General George S. Patton, July 1941, *The Patton Papers*, Vol. II, 1974

MORAL COURAGE

As to moral courage, I have rarely met with two-o'clock-in-the-morning courage; I mean instantaneous courage.

Napoleon, 4–5 December 1815

Words fail to describe the demands that are made of a soldier in battle. To go 'over the top' under enemy fire is, indeed, an heroic act, but it is by no means the most difficult. How much resolution and readiness to shoulder responsibility is required of a man who either has to lead or send others to certain death. Those are acts the appalling nature of which no one can imagine who has not himself had to perform them.

General Erich Ludendorff, *My War Memories 1914–1918*, 1919

For independent action in war a moral courage is needed in order to execute decisively and energetically correct and necessary knowns, without allowing oneself to err through fear of responsibility.

General Wilhelm von Blume, late 19th century German military theoretician

---

I felt it my duty to take up the cudgels. The country's safety was at stake, and I said so bluntly. The President turned the full vials of his sarcasm upon me. He was a scorcher when aroused. The tension began to boil over. For the third and last time in my life that paralyzing nausea began to creep over me. In my emotional exhaustion I spoke recklessly and said something to the general effect that when we lost the next war, and an American boy, lying in the mud with an enemy bayonet through his belly and an enemy foot on his dying throat, spat out his last curse, I wanted the name not be MacArthur, but Roosevelt. The President grew livid. 'You must not talk that way to the President!' he roared. He was right, of course, right, and I knew it almost before the words had left my mouth. I said that I was sorry and apologized. But I felt my Army career was at an end. I told him he had my resignation as Chief of Staff. As I reached the door his voice came with that cool detachment which so reflected his extraordinary self-control, 'Don't be foolish, Douglas; you and the budget must get together on this.'

Dern had shortly reached my side and I could hear his gleeful tones, 'You've saved the Army.' But I just vomited on the steps of the White House.

General of the Army Douglas MacArthur, 1933, MacArthur's stand to save the National Guard from a mortal budget slash by Roosevelt, *Reminiscences*, 1964

Determined leadership is vital, and nowhere is this more important than in the higher ranks. Other things being equal the battle will be a contest between opposing wills.

Generals who become depressed when things are not going well, and who lack the drive to get things done, and the moral courage and resolution to see their plan through to the end, are useless in battle. They are, in fact, worse than useless – they are a menace – since any lack of moral courage, or any sign of wavering or hesitation, has very quick repercussions down below.

Field Marshal Viscount Montgomery of Alamein, 10 November 1942

Now these two types of courage, physical and moral, are very distinct. I have known many men who had marked physical courage, but lacked moral courage. Some of them were in high places, but they failed to be great in themselves because they lacked it. On the other hand, I have seen men who undoubtedly possessed moral courage very cautious about taking physical risks. But I have never met a man with moral courage who would not, when it was really necessary, face bodily danger. Moral courage is a higher and a rarer virtue than physical courage.

Field Marshal Viscount Slim, *Courage and Other Broadcasts*, 1957

To teach moral courage is another matter – and it has to be taught because so few, if any, have it naturally. The young can learn it from their parents, in their homes, from school and university, from religion, from other early influences, but to inculcate it in a grown-up who lacks it requires not so much teaching as some striking emotional experience – something that suddenly bursts upon him, something in the nature of a vision. That happens rarely, and that is why you will find that most men with moral courage learnt it by precept and example in their youth.

Field Marshal Viscount Slim, *Courage and Other Broadcasts*, 1957

There are two kinds of courage, physical and moral, and he who would be a true leader must have both. Both are products of the character-forming process, of the development of self-control, self-discipline, physical endurance, of knowledge of one's job and, therefore, of confidence. These qualities minimize fear and maximize sound judgment under pressure – with some of that indispensable stuff called luck – often brings success from seemingly hopeless situations . . . both kinds of courage bespeak an untroubled conscience, a mind at peace with God . . .

General Mathew B. Ridgway, 'Leadership', *Military Review*, 9/1966

It has long seemed to me that the hard decisions are not the ones you make in the heat of battle. Far harder to make are those involved in speaking your mind about some harebrained scheme which proposes to commit troops to action under conditions where failure seems almost certain, and the only results will be the needless sacrifice of priceless lives. When all is said and done, the most precious asset any nation has is its youth, and for a battle commander ever to condone the unnecessary sacrifice of his men is inexusable. In any action you must balance the inevitable cost in lives against the objectives you seek to attain. Unless the results to be expected can reasonably justify the estimated loss of life the action involves, then for my part I want none of it.

General Mathew B. Ridgway, 'Leadership', *Military Review*, 9/1966

MORALE

You are all aware that it is not numbers or strength that bring the victories in war. No, it is when one side goes against the enemy with the gods' gift of a stronger morale that their adversaries, as a rule, cannot withstand them.

Xenophon (*c.* 431 BC–*c.* 352 BC), *The Persian Expedition*, tr. Warner

---

Alexander had been hurt by a sword-thrust in the thigh, but this did not prevent him from visiting the wounded on the day after the battle, when he also gave a splendid military funeral to the dead in the presence of the whole army paraded in full war equipment. At the ceremony he spoke in praise of every man who by his own observation or from reliable report he knew had distinguished himself in the fighting, and marked his approval in each case by a suitable reward.

Flavius Arrianius Xenophon (Arrian), *The Campaigns of Alexander the Great*, c. AD 150

Defeated troops should not be allowed to fall into despair, but they should be dealt with by stirring up hope and by various other means.

The Emperor Maurice, *The Strategikon*, c. AD 600

Now an army may be robbed of its spirit and its commander deprived of his courage.

During the early morning spirits are keen, during the day they flag, and in the evening thoughts turn toward home.

And therefore those skilled in war avoid the enemy when his spirit is sluggish and his soldiers homesick. This is control of the morale factor.

Sun Tzu, *The Art of War*, c. 500 BC, tr. Griffith

He [Chevalier Folard] supposes all men to be brave at all times and does not realize that the courage of the troops must be reborn daily, that nothing is so variable, and that the true skill of the general consists of knowing how to guarantee it.

Field Marshal Maurice Comte de Saxe, *My Reveries*, 1732

MAXIM 6. At the commencement of a campaign, to advance or not to advance is a matter for grave consideration, but when once the offensive has been assumed, it must be sustained to the last extremity. However skillful the manoeuvres, a retreat will always weaken the *morale* of an army, because in losing the chance of success, these last are transferred to the enemy. Besides, retreats cost always more men and *matériel* than the most bloody engagements, with the difference, that in a battle the enemy's loss is nearly equal to your own, whereas in a retreat the loss is on your side only.

Napoleon, *Military Maxims of Napoleon*, 1831, tr. D'Aguilar

In war, everything depends on morale; and morale and public opinion comprise the better part of reality.

Napoleon (1769–1821)

Morale makes up three quarters of the game; the relative balance of man-power accounts only for the remaining quarter.

Napoleon (1769–1821)

One fights well when the heart is light.

Napoleon, 17 February 1816, St. Helena, to General Gourgand

My lord, I never saw better horses, better clothes, finer belts and accoutrements; but money, which you don't want in England, will buy fine clothes and horses, but it can't buy the lively air, I see in everyone of these troopers.

Field Marshal Prince Eugene, June 1740, quoted in Chandler, *The Art of War in the Age of Marlborough*, 1976

A battle is lost less through the loss of men than by discouragement.

Frederick the Great, *Instructions to His Generals*, 1747

No system of tactics can lead to victory when the *morale* of an army is bad, and even when it is excellent the victory may depend upon some occurrence like the rupture of the bridges over the Danube at Essling.

Lieutenant General Antoine-Henri Baron de Jomini, *Summary of the Art of War*, 1838

Physical casualties are not the only losses incurred by both sides in the course of the engagement: their morale strength is also shaken, broken and ruined. In deciding whether or not to continue the engagement it is not enough to consider the loss of men, horses and guns; one also has to weight the loss of order, courage, confidence, cohesion, and plan. The decision rests chiefly on the state of morale, which, in cases where the victor has lost as much as the vanquished, has been the single decisive factor.

Major General Carl von Clausewitz, *On War*, iii, 1832, tr. Howard and Paret

---

The morale of an army and its chief officers has an influence upon the fate of a war. This seems to be due to a certain physical effect produced by the moral cause. For example, the impetuous attack upon a hostile line by twenty thousand brave men whose feelings are thoroughly enlisted in their cause will produce a much more powerful effect than the attack of forty thousand demoralized or apathetic men upon the same point.

Lieutenant General Antoine-Henri Baron de Jomini, *Summary of the Art of War*, 1838

Lee's army is really whipped. The prisoners we now take show it, and the action of his army shows it unmistakably. A battle with them outside of entrenchments cannot be had. Our men feel that they have gained the *morale* over the enemy, and attack him with confidence. I may be mistaken, but I feel that our success over Lee's army is already assured. The promptness and rapidity with which you have forwarded reinforcements has contributed largely to the feeling of confidence inspired in our men, and to break down that of the enemy.

General of the Army Ulysses S. Grant, 24 May 1864, *Personal Memoirs of U.S. Grant*, 1885

The art of war is subject to many modifications by industrial and scientific progress. But one thing does not change, the heart of man. In the last analysis, success in battle is a matter of morale. In all matters which pertain to an army, organization, discipline and tactics, the human heart in the supreme moment of battle is the basic factor.

Colonel Charles Ardnant du Picq, *Battle Studies*, 1880

Note the army organizations and tactical formations on paper are always determined from the mechanical point of view, neglecting the essential coefficient, that of morale. They are almost always wrong.

Colonel Charles Ardnant du Picq, *Battle Studies*, 1880

To General Swinton, too, is due the implanting into all ranks of the fundamental idea of the Tank as a weapon for saving the lives of infantry. This idea was indeed the foundation of the morale of the Tank Corps, for it spread from the fighting personnel to the depots and workshops, and even to the factories.

More than anything else, it was this sentiment which kept men ploughing through the mud of 1917, in the dark days when often the chance of reaching an objective had fallen to ten per cent.; which kept workshops in full swing all around the clock on ten and eleven hour shifts for weeks and, once, for months on end; which, finally, secured from the factories an intensive and remarkable output.

Major General Hugh Elles, Preface to Ellis, *The Tank Corps*, 1919

First of all you must build the morale of your own troops. Then you must look to the morale of your civilian population. Then, and only then, when these are in good repair should you concern yourself with the enemy's morale. And the best way to destroy the enemy's morale is to kill him in large numbers. There's nothing more demoralizing than that.

Leon Trotsky, quoted in *RUSI*, 9/1984

As a rule destroying the enemy by breaking his morale is an impossible task, since any such demoralization would turn largely on the social condition of his country. Demoralization of the remnants of an enemy army is a consequence of the destruction of his crucial main forces. It is thus a factor not just to be taken account of, but to be exploited.

Marshal of the Soviet Union Mikhail N. Tukhachevskiy, 1924, quoted in Simpkin, *Deep Battle*, 1987

The unfailing formula for production of morale is patriotism, self-respect, discipline, and self-confidence within a military unit, joined with fair treatment and merited appreciation from without. It cannot be produced by pampering or coddling an army, and is not necessarily destroyed by hardship, danger, or even calamity. Though it can survive and develop in adversity that comes as an inescapable indicent of service, it will quickly wither of indifference or injustice on the part of their government or of ignorance, personal ambition, or ineptitude on the part of military leaders.

General of the Army Douglas MacArthur, *Annual Report of the Chief of Staff, U.S. Army for the Fiscal Year Ending 30 June 1933*

Morale, only morale, individual morale as a foundation under training and discipline, will bring victory.

Field Marshal Viscount Slim, June 1941, to the officers, 10th Indian Infantry Division

---

Morale is a state of mind. It is steadfastness and courage and hope. It is confidence and zeal and loyalty. It is elan, esprit de corps and determination. It is staying power, the spirit which endures to the end – the will to win. With it all things are possible, without it everything else, planning, preparation, production, count for naught.

General of the Army George C. Marshall 15 June 1941, address at Trinity College, Hartford, Connecticut

---

Machines are nothing without men. Men are nothing without morale.

Admiral of the Fleet Ernest J. King, 19 June 1942, graduation address to the U.S. Naval Academy

Morale is when your hands and feet keep working when your head says it can't be done.

Anonymous U.S. Navy seaman in World War II

Morale is the big thing in war. We must raise the morale of our soldiery to the highest pitch; they must be made enthusiastic, and must enter this battle with their tails high in the air and with the will to win.

Field Marshal Viscount Montgomery of Alamein, 14 September 1942, instruction for Operation 'Lightfoot', the Battle of El Alamein, quoted in de Guingand, *Operation Victory*, 1947

. . . To win battles you require good Commanders in the senior ranks, and good senior staff officers; all of these must know their stuff.

You also require an Army in which the morale of the troops is right on the top line. The troops must have confidence in their Commanders and must have the light of battle in their eyes; if this is not so you can achieve nothing.

Field Marshal Viscount Montgomery of Alamein, 10 November 1942, quoted in Hamilton, *Master of the Battlefield*, 1983

Morale is the thinking of an army. It is the whole complex body of an army's thought: The way it feels about the soil and about the people from which it springs. The way that it feels about their cause and their politics as compared with other causes and politics. The way that if feels about its friends and allies, as well as its enemies. About its commanders and goldbricks. About food and shelter. Duty and leisure. Payday and sex. Militarism and civilianism. Freedom and slavery. Work and want. Weapons and comradeship. Bunk fatigue and drill. Discipline and disorder. Life and death. God and the devil.

Brigadier General S. L. A. Marshall, *Men Against Fire*, 1947

To those who have known the firing line, it would scarcely be necessary to point out that morale in combat is never a steady current of force but a rapidly oscillating wave whose variations are both immeasurable and unpredictable.

It is this respect chiefly – the rapidity and capriousness of its variations – that the morale problem in the zone of fire differs from rear area soldiering.

Brigadier General S. L. A. Marshall, *Men Against Fire*, 1947

Do not place military cemeteries where they can be seen by replacements marching to the front. This has a very bad effect on morale, even if it adds to the pride of the Graves Registration Service.

General George S. Patton, *War As I Knew It*, 1947

Very many factors go into the building-up of sound morale in an army, but one of the greatest is that men be fully employed at useful and interesting work.

Sir Winston S. Churchill, *The Gathering Storm*, 1948

It is not enough to fight. It is the spirit which we bring to the fight that decides the issue. It is morale that wins the victory.

General of the Army George C. Marshall, *Military Review*, 10/1948

Morale is the greatest single factor in successful war . . . In any long and bitter campaign morale will suffer unless all ranks thoroughly believe that their commanders are concerned first and always with the welfare of the troops who do the fighting.

General of the Army Dwight D. Eisenhower, *Crusade in Europe*, 1948

Loss of hope, rather than loss of life, is the factor that really decides wars, battles, and even the smallest combats. The all-time experience of warfare shows that when men reach the point where they see, or feel, that further effort and sacrifice can do no more than delay the end they commonly lose the will to spin it out, and bow to the inevitable.

Captain Sir Basil Liddell Hart, *Defence of the West*, 1950

To grow in knowledge of how to win a loyal and willing response from military forces, there must first be understanding of the springs of human action, what they are, and how they may be directed toward constructive ends. This done, the course which makes for the perfecting of forces during peacetime training need only be extended to harden them for the risk and stress of war.

Brigadier General S. L. A. Marshall, *The Armed Forces Officer*, 1950

Diligence in the care of men, administration of all organizational affairs according to a standard of resolute justice, military bearing in one's self, and finally, an understanding of the simple facts that men in a fighting establishment wish to think of themselves in that light and that all military information is nourishing to their spirits and their lives, are the four fundamentals by which the commander builds an all-sufficing morale in those in his charge.

Brigadier General S. L. A. Marshall, *The Armed Forces Officer*, 1950

[Spiritual foundations] Spiritual first, because no other foundation of conduct will stand firm under real strain. I use the word 'spiritual' not necessarily in the meaning of religious belief – though religion is and always has been one of the greatest foundations of morale – but in the sense of faith in a cause. A man must believe that what his organization is working for is worthy of all the labour and sacrifice he may be called upon to give – that it has a great and vitally necessary object – a noble object if you like. Next, he must feel intensely that he is part, even if only a small part, of that organization, and that what he is and what he does really counts.

[Mental Foundations] First, I think, the intelligent man must be convinced that the object of the organization is really attainable – difficult, perhaps, but not impossible. Second, he must feel that he belongs to an efficient show, one in which his efforts – perhaps his life – are not likely to be wasted. As the third mental foundation, he should, whenever possible, know why he is asked to do certain things.

[Material Foundations] . . . the material foundations of morale – where the man lives, the conditions under which he works, the tools he uses and what he gets for his labour. We are rightly increasingly concerned with these things, yet more dissatisfaction arises basically through neglect of the spiritual and intellectual foundations than in failure of the material. Often the material hardship is the irritant that inflames a deeper discontent. The specified grievance, although it may be justified, is not the real cause of the trouble. It is not that conditions are bad that upsets men so much as delay or failure to recognize that they are and to take steps to improve them. For example, few things are more discouraging than to ask men to work with obsolete or worn-out equipment – yet if it can be shown that there is a good reason why it cannot be replaced at the moment and that at the same time everyone above them is going all out to get the better tools, men will come to take a fierce pride in overcoming their difficulties.

Field Marshal Viscount Slim, *Courage and Other Broadcasts*, 1957

The morale of the soldier is the greatest single factor in war and the best way to achieve a high morale in war-time is by success in battle. The good general is the one who wins his battle with the fewest possible casualties; but morale will remain high even after considerable casualties, provided the battle has been won and the men know it was not wastefully conducted, and that every care has been taken of the wounded, and the killed have been collected and reverently buried.

Field Marshal Viscount Montgomery of Alamein, *The Memoirs of Field Marshal Montgomery*, 1958

You must watch your own morale carefully. A battle is, in effect, a contest between two wills – your own and that of the enemy general. If your heart begins to fail you when the issue hangs in the balance, your opponent will probably win.

Field Marshal Viscount Montgomery of Alamein, *The Memoirs of Field Marshal Montgomery*, 1958

Morale is a state of mind. It is that intangible force which will move a whole group of men to give their last ounce to achieve something, without counting the cost to themselves; that makes them feel they are part of something greater than themselves.

Field Marshal Viscount Slim, *Serve to Lead*, 1959

The Pentagon Whiz Kids are, I think, conscientious, patriotic people who are experts at calculating odds, figuring cost-effectiveness and squeezing the last cent out of contract negotiations. But they are heavy-handed butchers in dealing with that delicate, vital thing call 'morale'. This is the stuff that makes ships like the *Enterprise*, puts flags on top of Iwo Jima and wins wars. But I doubt if Mr. McNamara and his crew have any morale settings on their computers.

Admiral Daniel V. Gallery, *Eight Bells and All's Well*, 1965

. . . Good morale is built when men have confidence in the leaders and their comrades, when they understand and value the tasks that are set before them. Good morale depends, primarily, upon the exercise of effective leadership. And the key to effective leadership is painstaking and persistent attention to communication: learning the concerns and needs of those who work with you and for you; clearly explaining what should be done and why; conveying to others your own confidence that the task at hand can be done and done well.

Major General Ernest N. Harmon, *Combat Commander*, 1970

Morale does not develop suddenly or of its own accord. It is the result of example, of the actions and superior spirit of individuals among an élite coterie of people. Moral virtues are numbered among idealistic concepts. A man without ideas has no incentive. But those ideas can be developed only in association with known reality, with which the political and military leadership has to reckon.

Major General F. W. von Mellenthin, *German Generals of World War II*, 1977

The morale of soldiers comes from three things: a feeling that they have an important job to do, a feeling that they are trained to do it well, and a feeling that their good work is appreciated and recognized.

Lieutenant General Bruce C. Clarke, *Soldiers*, 3/1985

MOTIVATION

Because [a good general] regards his men as infants they will march with him into the deepest valleys. He treats them as his own beloved sons and they will die with him.

Sun Tzu, *The Art of War*, c. 500 BC, tr. Griffith

Men who think that their officer recognizes them are keener to be seen doing something honorable and more desirous of avoiding disgrace.

Xenophon (381 BC–c. 352 BC)

With a joyous spirit and a willing and exultant heart they [soldiers] will choose to brave dangers on behalf of our holy emperors and all the Christian people. But what is more important than all else and more basic, what arouses their enthusiasm, increases their courage, and incites them to dare what nobody else would dare is the fact that their own households and those of the soldiers serving them and everyone about them possess complete freedom. This has provided them security and protection from the beginning and from antiquity. You will find that this has been legislated by the holy and blessed emperors of old and is written down in the tactical books. In addition to freedom, though, they should enjoy proper respect and not be despised and dishonoured.

Emperor Nikephorus II Phokas, *Skirmishing*, c. AD 969, in Dennis, *Three Byzantine Military Treatises*, 1985

The two armies were now almost within striking distance. Alexander rode from one end of his line to the other with words of encouragement for all, addressing by name, with proper mention of rank and distinctions, not the officers of highest rank only but the captains of squadrons and companies; even the mercenaries were not forgotten, where any distinction or act of courage called for the mention of a name, and from every throat came the answering shout: 'Wait no longer – forward to the assault!'

Flavius Arrianus Xenophon (Arrian), Alexander the Great at the Battle of Issus, 333 BC, *The Campaigns of Alexander the Great*, c. AD 150

I had rather have a plain russet-coated captain that knows what he fights for, and loves what he knows, than that which you call a gentleman and is nothing more. I honour a gentleman that is so indeed.

Oliver Cromwell, September 1643, letter to Sir William Springe

---

Three things prompt Men to a regular discharge of their Duty in time of Action: natural bravery, hope of reward, and fear of punishment . . . The first two . . . are common to the untutor'd, and the Disciplin'd Soldiers: but the latter most obviously distinguishes the one from the other.

General George Washington, 9 February 1776, letter to the President of Congress, *The Writings of George Washington*, Vol. 4, 1931–44

I will tell you never to employ with your soldiers heartless words, dishonourable oaths, and never use profanity or degrading language in their presence. The colonel who uses these expression with his soldiers degrades himself, and if he is speaking to his officers, he commits an extremely evident blunder. Never forget the officers of your regiment are men, French, your equals, and that you must, consequently, employ a tone and language when giving them orders which is suitable to those persons in whom honour is the motivating factor.

Marshal of France Charles-Louis de Belle Isle, Duc de Gisors (1684–1761), letter to his son upon assumption of command of a French regiment, in Fitton, *Leadership*, 1990

The cavalry regiment that does not on this instant, on orders given, dash full plunge into the enemy, I will directly after the battle, unhorse and make it a garrison regiment. The infantry battalion which, meet with what it may, shows the least sign of hesitancy, loses its colours and its sabres, and I cut the trimmings from its uniforms! Now, good night, gentlemen: shortly we have either beaten the enemy, or we never see one another again.

Frederick the Great, to his officers before the Battle of Leuten, 4 December 1757

As for the soldiers, all you can do is to imbue them with *esprit de corps*, in other words the conviction that their regiment is better than all the other regiments in the universe. It sometimes happens that the officers must lead them into considerable danger, and since ambition can exert no influence on the men, they must be made to fear their officers more than the perils to which they are exposed, otherwise nobody would be able to make them attack into a storm of missiles thundering from three hundred cannon. Good will can never motivate common men in dangers of this order – we must resort to force.

Frederick the Great, *Testament Politique*, 1768

Instead of the lash, I would lead them by the stimulus of honour. I would instill a degree of emulation into their minds. I would promote every deserving soldier, as I did in France . . .

What might not be expected of the English army if every soldier hoped to be made a general provided he showed ability? [General] Bingham says, however, that most of your soldiers are brutes and must be driven by the stick. But surely the English soldiers must be possessed of sentiments sufficient to put them at least upon a level with the soldiers of other countries, where the degrading system of the lash is not used. Whatever debases man cannot be serviceable.

Napoleon, 9 November 1816, conversation on reform of the British Army at Saint Helena, ed. Herold, *The Mind of Napoleon*, 1955

A man does not have himself killed for a half-pence a day or for a petty distinction. You must speak to the soul in order to electrify him.

Napoleon (1769–1821)

Die for the Virgin, for your mother the Empress, for the royal family. The Church will pray to God for the dead. The survivor has honour and glory!

Field Marshal Prince Aleksandr V. Suvorov, *The Science of Victory*, 1796

The greatest gift a general can have is a bad temper [under control]. A bad temper gives you a sort of divine wrath, and it is only by the use of a divine wrath that you can drive men beyond their physical ability in order to save their lives.

General George S. Patton, Jr., 4 March 1944, *The Patton Papers*, Vol. 2, 1974

A few days before the invasion of Sicily, Patton had all the general officers of the Seventh Army at his headquarters for a final briefing. He turned the conduct of the discussion of details over to members of his staff and took little personal part until the close of the day's work. Then he took the floor and gave us a moving address on the theme of the quality of the American soldiers whom we were leading into action, for most of us, our first action. He described their exploits in Tunisia, how well they had stood up against the German veterans of the Afrika Korps. He described with emotion the bravery of small units and there were tears rolling down his cheeks when he told of one company which had held its position until completely destroyed. It was pretty clear that with such men under our command, he considered that if anything went wrong it would be the fault of the generals. In a grand peroration he turned on us with a roar and, waving a menacing swaggerstick under our noses, he concluded: 'Now we'll break up, and I never want to see you bastards again unless it's at your post on the shores of Sicily.' We departed, convinced that it would be well to turn up on time and at the right place on the Sicilian beaches.

General Maxwell Taylor, *Swords and Plowshares*, 1972

Awards that motivate only the top men are of little value in raising the ability of a unit. It takes awards to motivate the lower third to do that. A unit is measured by the ability of the lower third personnel in it to carry their part of the load.

Lieutenant General Bruce C. Clarke

A soldier doesn't fight very hard for someone who is going to shoot him on a whim. That's not what leadership is all about

General H. Norman Schwarzkopf, 27 February 1991, on the motivation of the Iraqi Army given during the 'Mother of All Briefings'

I hold it to be one of the simplest truths of war that the thing which enables an infantry soldier to keep going with his weapons is the near presence or the presumed presence of a comrade. The warmth which derives from human companionship is essential to the employment of the arms with which he fights as is the finger with which he pulls a trigger or the eye with which he aligns his sights.

Brigadier General S. L. A. Marshall, *Men Against Fire*, 1947

Also worth remembering is that in any man's dark hour, a pat on the back and an earnest handclasp may work a small miracle.

Brigadier General S. L. A. Marshall, *The Armed Forces Officer*, 1950

MOUNTAIN FIGHTING

Do not climb heights in order to fight. [So much for mountain warfare.]

Sun Tzu, *The Art of War*, c. 500 BC, tr. Giles

If the narrow places through which our army had been intending to pass should be seized by the enemy, and they are well equipped and in strength . . . then we must not get close to that place and certainly not send the army recklessly and at great risk through another difficult road which is occupied by the enemy. We do not say this out of fear of the enemy, but as a warning to avoid rugged terrain which involves struggling and danger. Often enough one weak person who has the advantage of a strong position can successfully fight against many brave men. For wherever you are unable to use your hands, your horse, and your weapons, while the enemy is able to do so, then the danger is obvious.

General Nikephorus Ouranos (attributed), *Campaign Organization and Tactics*, c. AD 100

Those who wage war in mountains should never pass through defiles without first making themselves masters of the heights.

Field Marshal Maurice Comte de Saxe, My *Reveries*, 1732

Very far . . . from seeing a refuge for the defensive, in a mountainous country, when a decisive battle is sought, we should rather advise a general in such a case to avoid such a field by every possible means . . . mountains, both tactically and strategically, are in general unfavourable to the defensive . . . The mountains limit the view and prevent movements in every direction; they force a state of passivity, and make it necessary to close up every avenue of approach, which always leads more or less to a war of cordons. We should therefore, if possible, avoid mountains with the principal mass of our force, and leave them on one side, or keep them before or behind us.

Major General Carl von Clausewitz, *On War*, vi, 1832

The chief characteristic of mountain defence is its complete passivity; in this light, the tendency towards the defence of mountains was very natural before armies attained to their present capability of movement . . . The offensive had thus again gained a preponderance through the greater mobility of troops; and it was only through the same means that the defence could seek for help. But mountainous ground by its nature is opposed to mobility . . .

Major General Carl von Clausewitz, *On War*, vi, 1832

In regard to mountain warfare in general, everything depends on the skill of our subordinate officers and still more on the morale of our soldiers. Here it is not a question of skillful manoeuvring, but of warlike spirit and wholehearted devotion to the cause.

Major General Carl von Clausewitz, *Principles of War*, 1812

MAXIM XIV. Among mountains, a great number of positions are always to be found very strong in themselves, and which it is dangerous to attack. The character of this mode of warfare consists in occupying camps on the flanks or in the rear of the enemy, leaving him only the alternative of abandoning his position without fighting, to take another in the rear, or to descend from it in order to attack you. In mountain warfare the assailant has always the disadvantage . . .

Napoleon, *The Military Maxims of Napoleon*, 1831, tr. D'Aguilar

Mountain operations teach us . . . that in such a country a strong and heroic will is worth more than all the precepts in the world

. . . One of the principal rules of this kind of war is, not to risk one's self in the valley without securing the heights. Shall I also say that in this kind of war, more than any other, operations should be directed upon the communications of the enemy?

Lieutenant General Antoine-Henri Baron de Jomini, *Summary of the Art of War*, 1838

With brave infantry and bold commanders mountain ranges can usually be forced.

Lieutenant General Antoine-Henri Baron de Jomini, *Summary of the Art of War*, 1838

MUTINY

Where soldiers get into trouble of this nature, it is nearly always the fault of some officer who has failed in his duty.

Field Marshal Viscount Montgomery of Alamein, quoted in Ahrenfeldt, *Psychiatry in the British Army in the Second World War*, 1958

How should base mutineers bound together by no rightful obligation, and united only by thirst for pillage and community of crime, display the intrepidity or obtain the success of honourable soldiers? Courage can never exist with the consciousness of guilt and speedily forsakes the man who has forsaken duty.

Count Belisarius, April AD 536, quoted in Mahon, *The Life of Belisarius*, 1829

That long official neglect of intolerable grievance, and inexcusable supineness towards measures of progressive improvement had, as they sooner or later infallibly do, at last aroused illegal and exasperated enforcements of redress.

Jedediah Tucker, of the British naval mutinies of 1797, *Memoirs of Earl St. Vincent*, 1830

Why does Colonel Grigsby refer to me to learn how to deal with mutineers? He should shoot them where they stand.

Lieutenant General Thomas 'Stonewall' Jackson, May 1862, on receiving a report of refusal of duty by 12-months' volunteers

NECESSITY

. . . it is the prime consideration of a general to secure victory – with its achievement unavoidable wrongs can later be remedied by kindness.

Gonzalo de Cordoba, 'El Gran Capitan', (1453–1515)

You must know, then, that there are two methods of fighting, the one by law, the other by force: the first method is that of men, the second of beasts; but as the first method is often insufficient, one must have recourse to the second. It is therefore necessary for a prince to know how to use both the beast and the man.

Niccolo Machiavelli, *The Prince*, 1513

Necessity hath no law.

Oliver Cromwell, 1654, address to Parliament

Extortions which are intolerable in their nature become excusable from the necessities of war.

Cardinal Richelieu, 1633

Great men are never cruel without necessity.

Napoleon, quoted in Herold, ed., *The Mind of Napoleon*, 1955

War, like the thunderbolt, follows its laws and turns not aside even if the beautiful, the virtuous and charitable stand in its path.

General of the Army William T. Sherman, April 1864, letter to Charles A. Dana

---

Every officer of the old army remembers how, in 1861, we were hampered with the old blue army-regulations, which tied our hands, and that to do any thing positive and necessary we had to tear it all to pieces – cut red-tape, as it was called – a dangerous things for an army to do, for it was calculated to bring law and authority into contempt; but war was upon us, and overwhelming necessity overrides all law.

General of the Army William T. Sherman, *The Memoirs of General W. T. Sherman*, 1875

NEUTRALITY

A prince is further esteemed when he is a true friend or a true enemy, when, that is, he declares himself without reserve in favour of some one or against another. This policy is always more useful than remaining neutral. For if two neighbouring powers come to blows, they are either such that if one wins, you will have to fear the victor, or else not. In either of these two cases it will be better for you to declare yourself openly and make war, because in the first case if you do not declare yourself, you will fall prey to the victor, to the pleasure and satisfaction of the one who has been defeated, and you will have no reason nor anything to defend you and nobody to receive you.

Niccolo Machiavelli, *The Prince*, 1513

A neutral is bound to be hated by those who lose and despised by those who win.

Niccolo Machiavelli, 22 December 1514, letter to Francesco Vettor

---

I shall treat neutrality as equivalent to a declaration of war against me.

Gustavus II Adolphus, King of Sweden, proclamation during the Thirty Years War, 1616–1648

The most sincere neutrality is not a sufficient guard against the depradations of nations at war.

General George Washington, 7 December 1796, Farewell Address

Never break the neutrality of any port or place, but never consider as neutral any place from whence an attack is allowed to be made.

Admiral Lord Nelson, 1804, letter of instruction

That expression 'positive neutrality' is a contradiction in terms. There can be no more positive neutrality than there can be a vegetarian tiger.

V. K. Krishna Menon, Indian Defence Minister, 18 October 1960

NIGHT FIGHTING

In the daytime the combatants see more clearly; though even then only what is going on immediately around them, and that imperfectly – nothing of the battle as a whole. But a night engagement like this in which two great armies fought . . . who could be certain of anything?

Thucydides, of the night attack on Epipolae, during the Sicilian Expedition, 414 BC, *History of the Peloponnesian War*, c. 404 BC

The more darkness in night attacks hinders and impedes the sight, the more must one supply the place of actual vision by skill and care.

Scipio Africanus (236–184 BC), quoted in Liddell Hart, *A Greater Than Napoleon: Scipio Africanus*, 1926

Some authorities state that Parmenio went to Alexander's tent and advised a night attack, because the enemy would not be expecting it, and it would naturally cause alarm and confusion. Alexander and Parmenio were not alone in the tent; others were listening, and that, perhaps, was one reason for Alexander's reply: 'I will not', he said, 'demean myself by stealing victory like a thief. Alexander must defeat his enemies openly and honestly.' However, these lofty words probably indicated confidence in danger rather than vanity, and in my own opinion they were based upon perfectly sound sense: night-fighting is a tricky business; unexpected things happen to both sides – to those who have carefully planned the attack as much as to those are taken off their guard – and often enough the better men get the worst of it, while victory, contrary to everybody's expectation, goes to the weaker side. More often than not Alexander took risks in his battles; but on this occasion he felt the chances of a night attack to be too unpredictable . . .

Flavius Arrianus Xenophon (Arrian), Alexander the Great before the Battle of Arbela, 331 BC, *The Campaigns of Alexander the Great*, c. 150 AD

The average person thinks that fighting at night is a simple matter, not calling for any special concern or precision. But such is not the case. On the contrary, very careful organization is needed. Anyone planning on night operations must first find soldiers who expressly volunteer for this kind of warfare. Both commander and men should make an agreement under oath. We should swear that when the assignment has been completed, they shall receive their proper rewards. The men should profess their readiness to accept death rather than fail to carry out their night mission. We should also promise them under oath that, if any man is killed in action, his heirs shall receive the rewards he deserved.

Second, we should seek men who are familiar with the nocturnal security arrangements of the enemy.

Third, there should be no moon that night. Otherwise the enemy will observe us marching along and, while we are still a good distance off, make preparations to confront us.

Anonymous Byzantine General, *On Strategy* (*Peri Strategias*) c. AD 527–565, in Dennis *Three Byzantine Military Treatises*, 1985

In general, I believe that night attacks are only good when you are so weak that you dare not attack the enemy in daylight.

Frederick the Great, *Instructions to His Generals*, 1747

Basically, a night attack is only an intensified raid. At first glance it looks highly effective: supposedly the defender is taken unawares, while the attacker, of course, is well prepared for what is about to happen. What an uneven contest! One imagines complete confusion on one side, and on the other an attacker concerned merely to profit by it. This image explains the many schemes for night attacks put forward by those who have neither to lead them or accept responsibility for them. In practice they are very rare.

Major General Carl von Clausewitz, *On War*, iv, 1832, tr. Howard and Paret

---

No operation of war is more critical than a night march.

Sir Winston Churchill, *The River War*, 1899

I never heard met or heard of troops who can withstand a night attack from the rear.

Bernard Newman, *The Cavalry Came Through*, 1930

Darkness is a friend to the skilled infantryman.

Captain Sir Basil Liddell Hart, *Thoughts on War*, 1944

Soldiers must be taught to move and fight at night. This is becoming more and more imperative, and it does not mean to make an approach march at night. It means to conduct lethal operations in the dark. To do this, previous and very accurate daylight reconnaissance is desirable and limited objective attacks are essential. In addition to the usual reverse following such an attack, a second reserve should be at hand to move up after daylight in case the enemy counter-attacks.

General George S. Patton, Jr., *War As I Knew It*, 1947

Night operations are a boon to troops who know the terrain like the palm of their hand and a reckless risk for those who don't . . . There is no easier way to scramble a force than to deploy it at night on jumbled, unfamiliar ground. Voice recognition is about five per cent efficient in the dark. Too quickly, men drift away, units become mixed and all control is lost.

Brigadier General S. L. A. Marshall, 'The Desert: It's Different,' *Infantry*, 12/1970

NONCOMBATANTS

Don't hurt civilians, they give us food and drink; a soldier is not a foodpad.

Field Marshal Prince Aleksandr V. Suvorov, *The Science of Victory*, 1796

It will require the full exercise of the full powers of the Federal Government to restrain the fury of the noncombatants.

General Winfield Scott, April 1861, after the Confederate attack on Fort Sumter

---

Those who live at home in peace and plenty want the duello part of this war to go on: but when they have to bear the burden by loss of property or comforts, they will cry peace.

General of the Army Philip H. Sheridan, 26 November 1864, telegram to Major General Halleck from Kernstown, Viriginia

---

Arras itself was fairly clear, as it was under shell fire; nevertheless, the Hotel de Commerce, where I pulled up for a drink, was doing good business. There was a dead soldier lying outside with a waterproof sheet over him; but inside, the place was crowded with soldiers drinking, chatting, and joking. The women serving them were now practically shell-proof, anyhow as regards nerves. You cannot terrorize a people for long, especially if there is money to be made.

Major General J. F. C. Fuller, note in 1917, *Memoirs of an Unconventional Soldier*, 1936

The morale of the civilians in a city which has not yet been touched by war is seldom as high as it is among the soldiers in the frontline.

Alan Moorehead, *Gallipoli*, 1956

NON-COMMISSIONED OFFICERS

The choice of non-commissioned officers is also an object of the greatest importance: the order and discipline of a regiment depends so much upon *their* behaviour, that too much care cannot be taken in preferring none to that trust but those who by their merit and good conduct are entitled to it. Honesty, sobriety, and a remarkable attention to every point of duty, with a neatness in their dress, are indispensable requisites; a spirit to command respect and obedience from the men, an expertness in performing every part of the exercise, and an ability to teach it, are absolutely necessary; nor can a sergeant or corporal be said to be qualified who does not write and read in a tolerable manner.

Major General Friedrich Baron von Steuben, *Revolutionary Drill Manual*, 1794

Instructions for the Sergeants and Corporals. It being on the non-commissioned officers that the discipline and order of a company in a great measure depend, they cannot be too circumspect in their behaviour towards the men, by treating them with mildness, and at the same time obliging every one to do his duty. By avoiding too great familiarity with the men, they will not only gain their love and confidence, but be treated with a proper respect whereas by contrary conduct they forfeit all regard, and their authority becomes despised.

Each sergeant and corporal will be in particular manner answerable for the squad committed to his care. He must pay particular attention to their conduct in every respect; that they keep themselves and their arms always clean; that they have their effects always ready, and put where they can get them immediately, even in the dark, without confusion; and on every fine day he must oblige them to air their effects.

Major General Friedrich Baron von Steuben, *Revolutionary Drill Manual*, 1794

The backbone of the Army is the non-commissioned man!

Rudyard Kipling, *The 'Eathen*, 1896

Even to those who best understand the reasons for the regimenting of military forces, a discipline wrongfully applied is seen only as indiscipline. Invariably it will be countered in its own terms. No average rank-and-file will become insubordinate as quickly, or react as violently, as a group of senior non-commissioned officers, brought together in a body, and then mishandled by officers who are ignorant of the customs of the service and the limits of their own authority. Not only are they conscious of their rights, but they have greater respect for the state of decency and order which is the mark of a proper military establishment than for the insignia of rank. It is this firm feeling of the fitness of things, and his unbounded allegiance to an authority when it is based on character which makes the NCO and the petty officer the backbone of discipline within the United States fighting establishment.

Brigadier General S. L. A. Marshall, *The Armed Forces Officer*, 1950

We have good corporals and sergeants and some good lieutenants and captains, and those are far more important than good generals.

General of the Army William T. Sherman (1820–1891)

As a rule it is easy to find officers, but it is sometimes very hard to find non-commissioned officers.

Napoleon, ed. Herold, 1809, *The Mind of Napoleon*, 1955

My many years in the army have demonstrated that wherever confidence in NCOs is lacking and wherever they are continuously bossed by the officers, you have no real NCOs and no really combat-worthy units.

Marshal of the Soviet Union Georgi K. Zhukov, *Reminiscences and Reflections*, 1974

NUMBERS

In war, numbers alone confer no advantage. Do not advance relying on sheer military power.

Sun Tzu, *The Art of War*, c. 500 BC, tr. Griffith

The Medes attacked at the charge, and many were killed; a second charge was made, and they were cut down in their hundreds – but they would not be driven back. All the same, it was clear to everyone, and especially to Xerxes, that his army contained plenty of men, but very few soldiers. The battle lasted throughout the day. The slaughter of the Medes was ghastly and they were as last replaced by the Persians . . . These troops were called the King's Immortals, and were expected to finish off the Greeks easily. But when they came to grips they did no better than the Medes – exactly the same, in fact – for they were fighting in a confined space, and using shorter spears than the Greeks . . . The Spartans fought a masterly battle. They were battle-scarred men of long experience fighting raw recruits, and they showed it.

Herodotus, (*c.* 484 BC–420 BC), of the Battle of Thermopolyae, 480 BC, *The Persian Wars*

Victory in war does not depend entirely upon numbers or mere courage; only skill and discipline will insure it. We find that the Romans owed the conquest of the world to no other cause than continual military training, exact observance of discipline in their camps and unwearied cultivation of the arts of war . . . A handful of men, inured to war, proceeds to certain victory, while on the contrary numerous armies of raw and undisciplined troops are but masses of men dragged to slaughter.

Flavius Vegetius Renatus, *Military Institutions of the Romans*, c. AD 378

Courage and discipline are able to accomplish more than a large number of warriors. Often enough the lie of the land has been helpful in making the weaker force come out on top.

The Emperor Maurice, *The Strategikon*, c. AD 600

I hear that General Hooker has more men than he can handle. I should like to have half as many more as I have today, and I would throw him into the river . . . We have always had to put all our troops in the fighting line, and never had enough when we needed them most.

Lieutenant General Thomas 'Stonewall' Jackson, April 1863, before the Battle of Chancellorsville

---

I have never seen a company, platoon, or squad take a hill at 100 per cent strength.

Lieutenant General Arthur S. Collins, Jr., *Common Sense Training,* 1978

That's all right . . . The greater the enemy the more they will fall over one another, and the easier it will be for us to cut through. In any case, they're not numerous enough to darken the sun for us.

Field Marshal Prince Aleksandr V. Suvorov (1729–1800), 1789, before the Battle of Rymnik

OBEDIENCE

The leader must himself believe that willing obedience always beats forced obedience, and that he can get this only by really knowing what should be done. Thus he can secure obedience from his men because he can convince them that he knows best, precisely as a good doctor makes his patients obey him. Also he must be ready to suffer more hardships than he asks of his soldiers, more fatigue, greater extremities of cold and heat.

Xenophon (*c.* 431 BC–*c.* 352 BC), *Cyropaedia*

He who wishes to be obeyed must know how to command.

Niccolo Machiavelli, *Discourses* 1517

Let your character be above reproach, for that is the way to earn men's obedience.

Field Marshal Johann von Schulenburg, 1709, letter to Field Marshal Maurice Comte de Saxe

---

Obedience to the officers and subordination is so exact that no one ever questions an order, hours are observed exactly, and however little a general knows how to make himself obeyed, he is always sure to be. No one reasons about the possibility of an enterprise, and finally, its accomplishment is never despaired of.

Frederick the Great, *Instructions to His Generals,* 1747

The [Soldiers] must learn to keep their ranks, to obey words of command and signals by drum and trumpet, and to observe good order, whether they halt, advance, retreat, are upon a march, or engaged with an enemy.

Niccolo Machiavelli, *The Art of War,* 1521

I speak harshly to no one, but I will have your head off the instant you refuse to obey me.

Marshal General Vicomte de Turenne (1611–1675), quoted in Weygand, *Turenne, Marshal of France,* 1930

Willing obedience always beats forced obedience.

Major General Frederich Baron von Steuben (1730–1794)

I was little satisfied . . . you received the order to proceed to Cairo and did not comply with it. *No event that may occur should prevent a soldier from obeying,* and talent in war consists in surmounting the difficulties liable to make an operation difficult.

Napoleon, quoted in Foch, *Principles of War,* 1913

I do not believe in the proverb that in order to be able to command one must know how to obey . . . Insubordination may only be the evidence of a strong mind.

Napoleon, 1817, letter from Saint Helena

MAXIM 72. . . . every general is culpable who undertakes the execution of a plan which he considers faulty. It is his duty to represent his reasons, to insist upon a change of plan; in short, to give his resignation rather than allow himself to become the instrument of his army's ruin. Every general-in-chief who fights a battle in consequence of superior's orders, with the certainty of losing it, is equally blameable.

In this . . . case the general ought to refuse obedience, because a blind obedience is due only to a military command given by a superior present on the spot at the moment of action . . .

Napoleon, *The Military Maxims of Napoleon*, 1831, tr. D'Aguilar

---

The Emperor, for the general direction of the operations, is in no need of advice, nor does he wish to have outlined to him any operation plans. No one knows his intentions, and it is our duty to obey him. His Majesty was all the less prepared for your movements, since you have been warned repeatedly not to take any action without orders. You can judge for yourself that partial measures will merely injure the operations as a whole and that they may even prove disastrous for the entire army . . .

Marshal of France Louis Berthier, Prince de Neuchatel, Prince de Wagram, 19 January 1807; Napoleon directed Berthier to send this letter to Marshal Ney, quoted in Foucart, *Campagne de Pologne*, n.d.

Soldiers must obey in all things. They may and do laugh at foolish orders, but they nevertheless obey, not because they are blindly obedient, but because they know that to disobey is to break the backbone of their profession.

General Sir William Napier (1782–1853)

I want none of this Nelson business in my squadron about not seeing signals.

Admiral David G. Farragut, March 1863, reprimanding Lieutenant W. S. Schley for failing to obey promptly a signal during the attack on Port Hudson

---

Men must be habituated to obey or they cannot be controlled in battle and the neglect of the least important order impairs the proper influence of the officer.

General Robert E. Lee, 1865, to the Army of Northern Virginia

---

It is difficult for the non-military mind to realize how great is the moral effort of disobeying a superior, whose order on the one hand covers all responsibility and on the other entails the most serious personal and professional injury; if violated without due cause, the burden of proving rests with the junior . . . He has to show, not that he meant to do right, but that he actually did right in disobeying in the particular instance. Under no less vigorous exactions can due military subordination be maintained.

Rear Admiral Alfred Thayer Mahan, *The Life of Nelson*, 1897

The ugly truth is revealed that fear is the foundation of obedience.

Sir Winston Churchill, *The River War*, 1899

The duty of obedience is not merely military but moral. It is not an arbitrary rule, but one essential and fundamental; the expression of a principle without which military organization would go to pieces and military success impossible.

Rear Admiral Alfred Thayer Mahan, *Retrospect and Prospect*, 1902

The spirit of obedience, as distinguished from its letter, consists in faithfully forwarding the general object to which the officer's particular command is contributing.

Rear Admiral Alfred Thayer Mahan, *Retrospect and Prospect*, 1902

All through your career of Army life, you men have bitched about what you call this chicken shit drilling. That is all for one reason, instant obedience to orders and it creates alertness. I don't give a damn for a man who is not always on his toes. You men are veterans or you wouldn't be here. You are ready. A man to continue breathing, must be alert at all times. If not, sometimes, some German son of a bitch will sneak up behind him and beat him to death with a sack full of shit.

General George S. Patton, Jr., June 1944, speech to the troops

---

In war, to obey is a difficult thing. For the obedience must be in the presence of the enemy, and in spite of the enemy, in the midst of danger, of varied and unforeseen circumstances, of a menacing unknown, in spite of fatigue of many causes.

Marshal of France Ferdinand Foch, *Principles of War*, 1913

The mind of the soldier, who commands and obeys without question, is apt to be fixed, drilled, and attached to definite rules.

Field Marshal Earl Wavell, *Generals and Generalship*, 1941

The efficiency of a war administration depends mainly upon whether decisions emanating from the highest approved authority are in fact strictly, faithfully, and punctually obeyed.

Sir Winston Churchill, *Their Finest Hour*, 1949

All I want is compliance with my wishes, after reasonable discussion.

Sir Winston Churchill, to the British Chiefs of Staff, quoted in Eisenhower, *General Eisenhower on the Military Churchill*, 1970

To know how to command obedience is a very different thing from making men obey. Obedience is not the product of fear, but of understanding, and understanding is based on knowledge.

Brigadier General S. L. A. Marshall, *The Armed Forces Officer*, 1950

There are admittedly cases where a senior commander cannot reconcile it with his responsibilities to carry out an order he has been given. Then, like Seydlitz at the Battle of Zorndorf, he has to say: 'After the battle the King may dispose of my head as he will, but during the battle he will kindly allow me to make use of it.' No general can vindicate his loss of a battle by claiming that he was compelled – against his better judgment – to execute an order that led to defeat. In this case the only course open to him is that of disobedience, for which he is answering with his head. Success usually decides whether he was right or not.

Field Marshal Erich von Manstein, *Lost Victories*, 1957

The officer's profession is not a liberal profession: a soldier's cardinal virtue is obedience, in other words the very opposite of criticism – the so-called 'manic' intellectual does not make a suitable officer while, on the other hand, the one-sided education of the professional soldier . . . results in a lack of ability to make a stand against theses which are not part of his real territory.

Field Marshal Wilhelm Keitel, *The Memoirs of Field Marshal Keitel*, 1965

THE OBJECTIVE
(Principle of War)

Victory in war is the main object in war. If this is long delayed, weapons are blunted and morale depressed . . .

Sun Tzu, *The Art of War*, c. 500 BC, tr. Griffith

Many good generals exist in Europe, but they see too many things at once: I see but one thing, and that is the masses; I seek to destroy them, sure that the minor matters will fall of themselves.

Napoleon (1769–1821)

The loss of Moscow is not the loss of Russia. My first obligation is to preserve the army, to get nearer to those troops approaching as reinforcements, and by the very act of leaving

Moscow to prepare inescapable ruin for the enemy . . . I will play for time, lull Napoleon as much as possible and not disturb him in Moscow. Every device which contributes to this object is preferable to the empty pursuit of glory.

Field Marshal Prince Mikhail I. Kutuzov, 1812, quoted in Parkinson, *Fox of the North*, 1976

Whither would you urge me? The most complete and most happy victory is to baffle the force of an enemy without imparing our own and in this favourable situation we are already placed. Is it not wiser to enjoy the advantages thus easily acquired, than to hazard them in the pursuit of more? Is it not enough to have altogether disappointed the arrogant hopes with which the Persians set out for this campaign and compelled them to a speedy and shameful retreat?

Count Belisarius, AD 531, when his army demanded an attack of a retreating Persian army, quoted in Mahon, *The Life of Belisarius*, 1829

. . . the man who tries to hang on to everything ends up by holding nothing. Your essential objective must be the hostile army . . .

Frederick the Great, *General Principles of War*, 1748

Pursue one great decisive aim with force and determination – a maxim which should take first place among all causes of victory.

Major General Carl von Clausewitz, *Principles of War*, 1812

. . . the destruction of the enemy forces is always the main object . . .

Field Marshal August Wilhelm Graf von Gneisenau (1760–1831)

. . . the aim of warfare is to disarm the enemy . . . If the enemy is to be coerced you must put him in a situation that is even more unpleasant than the sacrifice you call on him to make . . .

Major General Carl von Clausewitz, *On War*, i, 1832, tr. Howard and Paret

---

Sometimes the political and military objective is the same – for example, the conquest of a province. In other cases the political object will not provide a suitable objective. In that event, another military objective must be adopted that will serve the political purpose and symbolize it in the peace negotiations.

Field Marshal Carl von Clausewitz, *On War*, i, 1832, tr. Howard and Paret

---

What do we mean by the defeat of the enemy? Simply the destruction of his forces, whether by death, injury, or any other means – either completely or enough to make him stop fighting. Leaving aside all specific purposes of any particular engagement, the complete or partial destruction of the enemy must be regarded as the sole object of all engagements.

Field Marshal Carl von Clausewitz, *On War*, i, 1832, tr. Howard and Paret

---

No one starts a war – or rather, no one in his senses ought to do so – without first being clear in his mind what he intends to achieve by that war and how he intends to conduct it. The former is its political purpose; the later its operational objective. This is the governing principle which will set its course, prescribe the scale of means and effort which is required, and make its influence felt throughout down to the smallest operational detail.

Field Marshal Carl von Clausewitz, *On War*, viii, 1832, tr. Howard and Paret

---

To do the greatest damage to our enemy with the least exposure to ourselves, is a military axiom lost sight of only by ignorance of the true ends of victory.

Dennis Hart Mahan, 1864

Other great difficulties, experienced by every general, are to measure truly the thousand-

and-one reports that come to him in the midst of conflict; to preserve a clear and well-defined purpose at every instant of time, and to cause all efforts to converge on that end.

General of the Army William T. Sherman, *Memoirs of General W. T. Sherman*, 1875

Where Lee goes, there you will also go.

General of the Army Ulysses S. Grant, 1864, instructions to Major General Meade, Commander of the Army of the Potomac

---

. . . that exclusiveness of purpose, which is the essence of strategy, and which subordinates, adjusts, all other factors and considerations to the one exclusive aim.

Rear Admiral Alfred Thayer Mahan, *Naval Strategy*, 1911

It may be that the enemy's fleet is still at sea, in which case it is the great objective, now as always.

Rear Admiral Alfred Thayer Mahan, *Naval Strategy*, 1911

In war there is but one manner of considering every question, that is the *objective manner*. War is not an art of pleasure or sport, indulged in without other reason as one might go in for painting, music, hunting or tennis, which can be taken up or stopped at will. In war everything is co-related. Every move has some reason, seeks some *object*; once that object is determined it decides the nature and importance of the means to be employed . . .

Marshal of France Ferdinand Foch, *Principles of War*, 1913

War must entail some loss, but the less this loss is the greater will be the victory; consequently the military object of a nation is not to kill and destroy, but to enforce the policy of its government with the least possible loss of honour, life, and property. If the enemy can be compelled to accept the hostile policy without battle, so much the better. If he opposes it by military force, then it should never be forgotten that the strength of this force rests on the will of the government which employs it, and that, in its turn this will rest on the will of the nation which this government represents. If the will of the nation cannot directly be attacked, then must the will of the army protecting it be broken.

Major General J. F. C. Fuller, *The Reformation of War*, 1923

The Strategical Objective: Irrespective of the arm employed, the principles of strategy remain immutable, changes in weapons affecting their application only. The first of all strategical principles is 'the principle of the object', the object being 'the destruction of the enemy's field armies' – his fighting personnel.

Now, the potential fighting strength of a body of men lies in its organization; consequently, if we can destroy this organization, we shall destroy its fighting strength and so have gained our object.

There are two ways of destroying our object.

(i) By wearing it down (dissipating it).
(ii) By rendering it inoperative (unhinging it).

In war the first comprises the killing, wounding, capturing and disarming of the enemy's soldiers – body warfare. The second, the rendering inoperative of his power of command – brain warfare. Taking a single man as an example: the first method may be compared to a succession of slight wounds which will eventually cause him to bleed to death; the second – a shot through the brain.

Major General J. F. C. Fuller, memorandum of 24 May 1918 for 'Plan 1919', *Memoirs of an Unconventional Soldier*, 1936

Here we are dealing with the elementary object of war, war as 'politics with bloodshed', as mutual slaughter by opposing armies. The object of war is specifically 'to preserve oneself and destroy the enemy' (to destroy the enemy means to disarm him or 'deprive him of the power to resist', and does not mean to destroy every member of his forces physically.

Mao Tse-tung, *On Protracted War*, May 1938

The true military objective is a mental rather than a physical objective – the paralysis of the opposing command, not the bodies of the actual soldiers. For an army without orders, without co-ordination, without supplies, easily becomes a panic-stricken and famine-stricken mob, incapable of effective action. (March 1925)

Captain Sir Basil Liddell Hart, *Thoughts on War*, 1944

The military object may be expressed in the one word 'conquest', which presupposes victory in one form or another, and by conquest I understand that condition of success which will admit of a government imposing its will on the enemy's nation, and so attaining the execution of its policy. Conquest may also be considered as the grand strategical military idea, and victory the grand tactical means.

Major General J. F. C. Fuller, *The Foundations of the Science of War*, 1926

. . . the danger of carrying us away from the objective . . . In an operation, there must be nothing superflous; it must be the incarnation of aim-directedness. The form of the operation . . . must recall . . . not the rococo . . . but a Greek temple.

General A. A. Svechin, 1927

History shows that the unswerving pursuit of any one objective is almost certain to be barren of result. 'Variability' of objectives, like elasticity of dispositions, is necessary to fulfill an essential principle of war – *flexibility*. (August 1929)

Captain Sir Basil Liddell Hart, *Thoughts on War*, 1944

In time of war the only value that can be affixed to any unit is the tactical value of that unit in winning the war. Even the lives of those men assigned to it become nothing more than tools to be used in the accomplishment of that mission. War has neither the time nor heart to concern itself with the individual and the dignity of man. Men must be subordinated to the effort that comes with fighting a war, and as a consequence men must die that objectives might be taken. For a commander the agony of war is not in its dangers, deprivations, or the fear of defeat but in the knowledge that with each new day men's lives must be spent to pay the costs of that day's objectives.

General of the Army Omar N. Bradley, *A Soldier's Story*, 1951

The objective . . . is unquestionably the most important of all the principles of war. It is the connecting link which, alone, can impart coherence to war . . . Without the objective, all other principles are pointless. It gives the commander the 'what'. The other principles are guides in the 'how'.

Admiral C. R. Brown, *The Principles of War*, June 1949

There is no purpose in capturing these manure-filled, waterlogged villages. The purpose of our operations is to kill or capture the German personnel and vehicles . . . so that they cannot retreat and repeat their opposition. Straight frontal attacks against villages are prohibited unless after careful study there is no other possible solution.

General George S. Patton, Jr., 9 December 1944, letter of instruction for the Third Army, *The Patton Papers*, Vol. I, 1972–74

Units must not be assigned impossible missions. That only contributes to heavy losses and a breakdown of morale. It is better to mount fewer offensive operations, and marshal forces for a decisive blow.

Marshal of the Soviet Union Georgi Zhukov, 1945, quoted in Gareyev, *M. V. Frunze, Military Theorist*, 1985

In discussing the subject of the 'objective' in war it is essential to be clear about, and keep clear in our minds, the distinction between the political and military objective. The two are different but not separate. For nations do not wage war for war's sake, but in pursuance of policy. The military objective is only the means to a political end. Hence the military objective should be governed by the political objective, subect to the basic condition that policy does not demand what is militarily – that, is practically – impossible.

Captain Sir Basil Liddell Hart, *Strategy*, 1954

The alphabet of war, Comrade General, says that victory is achieved not by taking towns but by destroying the enemy. In 1812, Napoleon forgot that. He lost Moscow – and Napoleon was no mean leader of men.

General Andrei L. Getman, 1945, quoted in Ryan, *The Last Battle*, 1966

Keep your object always in mind, while adapting your plan to circumstances. Realize that there are more ways than one of gaining an object, but take heed that every objective should bear on the object. And in considering possible objectives weigh their possibility of attainment with their service to the object attained – to wander down a side-track is bad, but to reach a dead end is worse.

Captain Sir Basil Liddell Hart, *Strategy*, 1954

'You know you never defeated us on the battlefield,' said the American colonel. The North Vietnamese colonel pondered this remark a moment. 'That may be so,' he replied, 'but it is also irrelevant.'

Colonel Harry G. Summers, Jr., conversation on 25 Apirl 1975 in Hanoi between Colonel Summers, then Chief, Negotiations Division, US Delegation, Four Party Joint Military Team and Colonel Tu, Chief, North Vietnamese (DRV) Delegation, *On Strategy*, 1981

One of the most frustrating aspects of the Vietnam war from the Army's point of view is that as far as logistics and tactics were concerned we succeeded in everything we set out to do. At the height of the war the Army was able to move almost a million soldiers a year in and out of Vietnam, feed them, clothe them, house them, supply them with arms and ammunition, and generally sustain them better than any Army had ever been sustained in the field. To project an Army of that size halfway around the world was a logistics and management task of enormous magnitude, and we had been more than equal to the task. On the battlefield itself, the Army was unbeatable. In engagement after engagement the forces of the Viet Cong and of the North Vietnamese Army were thrown back with terrible losses. Yet, in the end, it was North Vietnam, not the United States, that emerged victorious. How could we have succeeded so well, yet failed so miserably?

Colonel Harry G. Summers, Jr., *On Strategy*, 1981

The first principle of war is the principle of *The Objective*. It is the first principle because all else flows from it. It is the strategic equivalent of the mission statement in tactics and we must subject it to the same rigorous analysis as we do the tactical mission.

Colonel Harry G. Summers, Jr., *On Strategy*, 1981

OBJECTIVITY/REASON/TRUTH

I never suffer my mind to be so wedded to any opinions as to refuse to listen to better ones when they are suggested to me.

Henry IV (Henry of Navarre), King of France (1553–1610)

The path of nature and of truth is narrow but it is simple and direct; the devious paths are numerous and spacious; but they all lead to error and destruction.

Robert Jackson, *A Systematic View of the Formation, Discipline and Economy of Armies*, 1804

Before beginning operations you must, without any indulgence or self-deception, examine objectively every step that the enemy might undertake to thwart your plan, and consider in each conceivable case what means are open to you to fulfill your goal. The more you anticipate the difficulties in advance, the less surprised you will be should you encounter them during the campaign. Besides, you have already thought about these obstacles deliberately, and with composure you have perceived the means of avoiding them, so nothing can surprise you.

Frederick the Great, 'Réflexions sur les projects de campagne', 1775, *Oeuvres de Frédéric le Grand*, Vol. XXIX, 1846–56, in ed., Luvaas, *Frederick the Great on the Art of War*, 1966

Nothing should be concealed from the Emperor, either good or bad: to deceive him, even about things that are likely to be disagreeable to him is a crime.

Marshal of France Louis Berthier, Prince de Neuchatel, Prince de Wagram, February 1807, letter to Marshal Lannes

---

This is the real way in which the matter must be viewed, and it is to no purpose, it is even against one's better interest, to turn away from the consideration of the real nature of affairs because of the horror of its elements excites repugnance.

Major General Carl von Clausewitz, *On War*, i, 1832, tr. Howard and Paret

---

Preconceived notions, especially in war, are dangerous, because they give their own particular colour to all information that comes in; and . . . stifled any real understanding of the actual situation . . .

General Aleksei A. Brusilov, March 1915, *A Soldier's Notebook*, 1931

As the first phase of the battle developed into a period of extremely bitter fighting, with the inevitable confusion and the weather dressing the scene in rain and mud, there was a pretty continuous stream of officers, friends, or acquaintances of mine, who called to see me on their way to the front. Those whose organizations had temporarily been withdrawn from the battle for rest, recruitment, and reorganization gave very interesting and instinctive descriptions of their experiences. Naturally, very few of these officers could discuss the operations of their units in a frankly impersonal manner. One division commander was an exception to the rule and discussed quite frankly the mistakes committed by his troops. I had previously familiarized myself with what had happened and thought I understood the reasons. He confirmed my views, but, more important than that, he impressed me with his absolute honesty by this unwillingness to cover up the shortcomings of his command, in the fear that it might lead to his relief. All he desired was an opportunity to profit by the lessons of the recent experience. As a matter of fact, I happened to hear the question of his relief discussed, and in my small way I most energetically protested. This particular General commanded his division in action most successfully a few weeks later.

General of the Army George C. Marshall, *Memoirs of My Service in the World War* (written between 1923–1926), 1977

We must have freedom of mind, no prejudices, no prepossession, no fixed ideas, no opinion accepted without discussion and merely because it has always been heard or practised. There should be only one test – reason.

Marshal of France Ferdinand Foch, quoted in Liddell Hart, *Foch: The Man of Orleans*, 1931

The butterfly is the Fact – gleaming, fluttering, settling for an instant with wings fully spread to the sun, then vanishing in the shades of the forest. Whether you believe in Free Will or Predestination, all depends on the slanting glimpses you had of the colour of his wings.

Sir Winston Churchill, *My Early Life*, 1930

There is a word in our language which I believe to be the most potent word in the dictionary. So that you do not forget it, I have written it on the blackboard – it is the word WHY. Whatever I say to you, whatever you think, ask yourselves the reason why. If you do not do so, however much you may strive to learn, you will be mentally standing at ease. Remember this: your brain is not a museum for the past, or a lumber room for the present; it is a laboratory for the future, even if the future is only five minutes ahead of you; a creative centre in which new discoveries are made and progress is fashioned.

Major General J. F. C. Fuller, *Memoirs of an Unconventional Soldier*, 1936

You have come here to acquire knowledge, to evolve it and to fit it to the men you will one day either command or administer. Knowledge is not only acquired in the lecture-room, but also in the mess and in your own private studies. Nothing clarifies knowledge like a free exchange of ideas; consequently, because I happen to be a Colonel and you a Captain or a Major, do not for a moment imagine that rank is a bar to free speech. If you disagree with me or the views of any of your instructors, openly state your disagreement; for we are all students, and the man who cannot change his opinions has mineralized his intellect – he is walking stone.

Major General J. F. C. Fuller, *Memoirs of an Unconventional Soldier*, 1936

Haig was an honourable man according to his lights – but his lights were dim. The consequences which have made 'Passchendaele' a name of ill omen may be traced to the combined effect of his tendency to deceive himself; his tendency, therefore, to encourage his subordinates to deceive him; and their 'loyal' tendency to tell a superior what was likely to coincide with his desires. Passchendaele is an object-lesson in this kind of well-meaning, if not disinterested truthfulness.

Captain Sir Basil Liddell Hart, *Through the Fog of War*, 1938

I think the difference in my attitude today compared with when I was composing those THOUGHTS ON WAR is that then I tended to assume that people could learn to think objectively about history and war; now, I just accept that few people can reason objectively about anything. I can still strive to pursue that aim, but I see the absurdities of life more than I did. I have come to see its nonsense, but one must try to live as if it could make sense.

Captain Sir Basil Liddell Hart, quoted in *Military Review*, 9/1966

The truth is sought regardless of whether pleasant or unpleasant, or whether it supports or condemns our present organization of tactics.

General Lesley J. McNair, round-robin letter to his generals, quoted in Kahn, *McNair, Educator of an Army*, 19??

OBSTACLES

. . . Country in which there are precipitous cliffs with torrents running between, deep natural hollows, confined places, tangled thickets, quagmires and crevasses, should be left with all possible speed and not approached.

While we keep away from such places, we should get the enemy to approach them; while we face them, we should let the enemy have them on his rear.

Sun Tzu, *The Art of War, c.* 500 BC, tr. Giles

Those expert at preparing defences consider it fundamental to rely on the strength of such obstacles as mountains, rivers and foothills. They make it impossible for the enemy to know where to attack. They secretly conceal themselves as under the nine-layered ground.

Tu Yu (AD 735–812), commentator to Sun Tzu, *The Art of War, c.* 500 BC, tr. Griffith

---

Mountains can be crossed wherever goats cross, and winter freezes most rivers.

Frederick the Great, *Instructions to His Generals,* 1747

Natural hazards, however formidable, are inherently less dangerous and less uncertain than fighting hazards. All conditions are more calculable, all obstacles more surmountable, than those of human resistance. By reasoned calculation and preparation they can be overcome almost to timetable. While Napoleon was able to cross the Alps in 1800 'according to plan', the little fort of Bard could interfere so seriously with the movement of his army as to endanger his whole plan.

Captain Sir Basil Liddell Hart, *Strategy*, 1954

An obstacle loses 50 per cent of its value if you stand back from it, allowing the enemy to reconnoitre the approaches and subsequently to cross without interference.

Field Marshal Viscount Montgomery of Alamein, *A History of Warfare*, 1968

THE OFFENSIVE
(Principle of War)

After God, we should place our hopes of safety in our weapons, not in our fortifications alone.

The Emperor Maurice, *The Strategikon*, c. AD 600

. . . I should say that in general the first of two army commanders who adopts an offensive attitude almost always reduces his rival to the defensive and makes him proceed in consonance with the movement of the former.

Frederick the Great, *Instructions to His Generals*, 1747

If you have resolved to attack an entrenched enemy, do it at once and do not allow him time to perfect his works.

Frederick the Great, *Instructions to His Generals*, 1747

Hit first! Hit hard! Keep on hitting!

Admiral Sir John Fisher, *Memories*, 1919

Every offensive operation offer advantages in the first half of its duration . . . the offence gradually loses the advantages deriving from surprise and preparation . . . one must not allow an offensive to drag on until its dying breath . . .

General A. A. Svechin, 1927, quoted in Leites, *The Soviet Style in War*, 1982

It is this persistent widening and intensifying of the offensive – this pushing vigorously forward on carefully chosen objectives without excessive regard to alignment or close touch – that will give us the best results with the smallest losses . . .

Marshal of France Ferdinand Foch, 26 August 1918, letter to Field Marshal Haig, quoted in Liddell Hart, *Foch: Man of Orleans*, 1931

Pushing off smartly is the road to success.

Major General Sir James Wolfe, 1758, letter to a friend

An offensive, daring kind of war will awe the Indians and ruin the French. Blockhouses and a trembling defensive encourage the meanest scoundrel to attack us.

Major General Sir James Wolfe, 1758, letter to General Jeffery Lord Amherst regarding the upcoming campaign in North America

---

In short, I think like Frederick, one should always be the first to attack.

Napoleon (1769–1821)

Make war offensively; it is the sole means to become a great captain and to fathom the secrets of the art.

Napoleon, (1769–1821), *Correspondance*, Vol. XXXI, 1858–1870

Carrying the war to the enemy and the destruction of his armed forces and his will to fight through the strategic offensive is the classic way wars are fought and won, hence the second principle of war is labeled *The Offensive*.

Harry G. Summers, *On Strategy*, 1981

When once the offensive has been assumed, it must be sustained to the last extremity.

Napoleon, *Maxims of War*, 1831

To move swiftly, strike vigorously, and secure the fruits of victory is the secret of successful war.

Lieutenant General Thomas 'Stonewall' Jackson (1824–1863)

In any fight, it's the first blow that counts; and if you keep it up hot enough, you can whip 'em as fast as they come up.

Lieutenant General Nathan Bedford Forrest (1821–1877)

War, once declared, must be waged offensively, aggressively. The enemy must not be fended off, but smitten down. You may then spare him every exaction, relinquish every gain; but till down he must be struck incessantly and remorselessly.

Rear Admiral Alfred Thayer Mahan, *The Interest of America in Sea Power*, 1896

The *offensive* alone embodies the positive qualities of combat which seek and promise decisive results. Attack seeks to strike the enemy, to destroy him on the field of battle. The attack has the advantage of the initiative. It forces action on the opponent and requires him to conform his activities to those of the attacker. It also has the factor of morale on its side.

R. Ernest Dupuy and Trevor N. Dupuy, *The Military Heritage of America*, 1984

The offensive is the basic form of combat action. Only by a resolute offensive conducted at high tempo and to a great depth is total destruction of the enemy obtained.

Colonel General V. G. Reznichenko, *Tactics*, 1984

Since I joined the Marines, I have advocated aggressiveness in the field and constant offensive action. Hit quickly, hit hard and keep right on hitting. Give the enemy no rest, no opportunity to consolidate his forces and hit back at you. This is the shortest road to victory.

General Holland M. 'Howlin' Mad' Smith, *Coral and Brass*, 1949

In war the only sure defence is offense, and the efficiency of the offense depends on the warlike souls of those conducting it.

General George S. Patton, Jr., *War As I Knew It*, 1947

The *offensive* alone embodies the positive qualities of combat which seek and promise decisive results. Attack seeks to strike the enemy, to destroy him on the field of battle. The attack has the advantage of the initiative. It forces action on the opponent and requires him to conform his activities to those of the attacker. It also has the factor of morale on its side.

R. Ernest Dupuy and Trevor N. Dupuy, *The Military Heritage of America*, 1984

The offensive is the basic form of combat action. Only by a resolute offensive conducted at high tempo and to a great depth is total destruction of the enemy obtained.

Colonel General V. G. Reznichenko, *Tactics*, 1984

OFFENCE AND DEFENCE

Invincibility lies in the defence; the possibility of victory in the attack.

One defends when his strength is inadequate; he attacks when it is abundant.

The experts in defence conceal themselves as under the ninefold earth; those skilled in attack move as from above the ninefold heavens. Thus they are capable of both protecting themselves and of gaining a complete victory.

Sun Tzu, *The Art of War*, c. 500 BC, tr. Griffith

(Commentary on the above passage) Those expert at preparing defences consider it fundamental to rely on the strength of such obstacles as mountains, rivers and foothills. They make it impossible for the enemy to know where to attack. They secretly conceal themselves as under the nine-layered ground.

Those expert in attack consider it fundamental to rely on the seasons and the advantages of the ground; they use inundations and fire according to the situation. They make it impossible for an enemy to know where to prepare. They release the attack like a lightning bolt from above the nine-layered heavens.

Tu Yu, Tang Dynasty (AD 618–905) commentator on Sun Tzu, *The Art of War*, c. 500 BC, tr. Griffith

---

Projects of absolute defence are not practicable because while seeking to place yourself in strong camps the enemy will envelop you, deprive you or your supplies from the rear and oblige you to lose ground, thus disheartening your troops. Hence, I prefer to this conduct the temerity of the offensive with the hazard of losing the battle since this will not be more fatal than retreat and timid defence. In the one case you lose ground by withdrawing and soldiers by desertion and you have no hope; in the other you do not risk more and, if you are fortunate, you can hope for the most brilliant success.

Frederick the Great, *Instructions to His Generals*, 1747

One of the falsest notions in war is to remain on the defensive and let the enemy act offensively. In the long run it is inevitable that the party which stays on the defensive will lose.

Frederick the Great, letter 8 January 1779, *Politische Correspondenz Friedrichs des Grossen*, 1879–1939

MAXIM 19. The transition from the defensive to the offensive is one of the most delicate operations in war.

Napoleon, *Military Maxims of Napoleon*, 1831, tr. D'Aguilar

If the defensive is the stronger form of conducting war, but has a negative object, it follows of itself that we must only make use of it so long as our weakness compels us to do so, and that we must give up that form as soon as we feel strong enough to aim at the positive object. Now as the state of our circumstances is usually improved in the event of our gaining a victory through the assistance of the defensive, it is therefore, also, the natural course of war to begin with the defensive and end with the offensive.

Major General Carl von Clausewitz, *On War*, 1832, tr. Howard and Paret

---

A sudden powerful transition to the offensive – the flashing sword of vengeance – is the greatest moment for the defence. If it is not in the commander's mind from the start, or rather if it is not an integral part of his idea of defence, he will never be persuaded of the superiority of the defensive from; all he will see is how much of the enemy's resources he can destroy or capture. But these things do not depend on the way in which the knot is tied, but on the way in which it is untied. Moreover, it is a crude error to equate attack with the idea of assault alone, and therefore, to conceive of defence as merely misery and confusion.

Major General Carl von Clausewitz, *On War*, vi, 1832, tr. Howard and Paret

---

The fact cannot be concealed . . . The best thing for an army on the defensive is to know *how* to take the offensive at a proper time, and *to take it*.

Lieutenant General Antoine-Henri Baron de Jomini, *Summary of the Art of War*, 1838

. . . Every army which maintains a strictly defensive attitude must, if attacked, be at last driven from its position; but if it takes advantage of the benefits of the defensive system and holds itself ready to take the offensive when occasion offers, it may hope for the greatest success.

Lieutenant General Antoine-Henri Baron de Jomini, *Summary of the Art of War*, 1838

The offensive knows what it wants . . . the defensive is in a state of uncertainty.

Field Marshal Helmuth Graf von Moltke, *Instructions for the Commanders of Large Formations*, 1869

A clever military leader will succeed in many cases in choosing defensive positions of such an offensive nature from the strategic point of view that the enemy is compelled to attack us in them.

Field Marshal Helmuth Graf von Moltke (1800–1891)

In war, the defensive exists mainly that the offensive may act more freely.

Rear Admiral Alfred Thayer Mahan, *Naval Strategy*, 1911

[Clausewitz, *On War*] 'When we say that defence is a stronger form of war, that is, that it requires a smaller force, if soundly designed, we are speaking, of course, only of one certain line of operations. If we do not know the general line of operation on which the enemy intends to attack, and so cannot mass our force upon it, then defence is weak, because we are compelled to distribute our force so as to be strong enough to stop the enemy on any line of operations he may adopt.'

Manifestly, however, a force capable of being strong enough on several lines of operation to stop an enemy possesses a superiority that should take the offensive.

Rear Admiral Alfred Thayer Mahan, *Naval Strategy*, 1911

A purely defensive battle is a duel in which one of the fighters does nothing but *parry*. Nobody would admit that, by so doing, he could succeed in defeating his enemy. On the contrary, he would sooner or later expose himself, in spite of the greatest possible skill, to being touched, to being overcome by one of his enemy's thrusts, even if that enemy were the weaker party.

Hence the conclusion that the *offensive* form alone, be it resorted to at once or only after the *defensive*, can lead to results, and must therefore *always* be adopted – at least in the end.

Marshal of France Ferdinand Foch, *Precepts and Judgments*, 1919

As they were depressed by defence their spirits rose in the offensive. The interests of the Army were best served by the offensive; in defence it was bound gradually to succumb to the ever increasing hostile superiority in men and material. This feeling was shared by everybody. In the West the Army pined for the offensive, and after Russia's collapse expected it with the most intense relief. Such was the feeling of the troops about attack and defence. It amounted to a definite conviction which obsessed them utterly that nothing but an offensive could win the war.

General Erich Ludendorff, *My War Memories 1914–1918*, 1919

Where the problem of attack is difficult of solution it is worth considering the possibilities of an alternative form of action which throws the burden of that problem on the enemy, yet keeps the initiative. The alternative is what I would term the 'baited offensive': the combination of offensive strategy with defensive tactics. Throughout history it has proved one of the most effective of moves, and its advantages have increased as modern weapons have handicapped other types of move. By rapidity of advance and mobility of manoeuvre, you may be able to seize points which the enemy, sensitive to the threat, will be constrained to attack. Thus you will invite him to a repulse which in turn may be exploited by a riposte. Such a counterstroke, against an exhausted attacker, is much less difficult than the attack on a defended position. The opportunity for it may also be created by a calculated withdrawal – what one may call the 'luring defensive'. Here is another gambit of future warfare. (November 1935)

Captain Sir Basil Liddell Hart, *Thoughts on War*, 1944

The influence of growing firepower on tactical defence is evident . . . The defensive is able more than before to carry out its original mission, which is to break the strength of the attacker, to parry his blows, to weaken him, to bleed him, so as to reverse the relation of forces and lead finally to the offensive, which is the only decisive form of warfare.

Field Marshal Ritter Wilhelm von Leeb, *Defence*, 1938

. . . the history of wars convincingly testifies . . . to the constant contradiction between the means of attack and defence. The appearance of new means of attack has always (inevitably) led to the creation of corresponding means of counteraction, and this in the final analysis has led to the development of new methods for conducting engagements, battles, operations (and war in general).

Marshal of the Soviet Union Nikolai V. Ogarkov, quoted in *Kommunist*, 1978

. . . the *defense* has the advantage of being able to choose its own ground, and often, where there is plenty of room at its disposal, of even prescribing the time when its adversary must attack. The defensive waits upon the attacker. Its object is to conserve resources and to break down the attack of the opponent. If it wishes to avoid a decision, it moves away from the enemy, as far as time and space permit; it will drag the action out. As an attack proceeds, the attacker's resources diminish; he becomes weaker and wears himself out. He gets further and further from his base of operations, while for the defender this is not the case. The further the attacker advances, the more difficult his position becomes until he has gained the decision. Time in duration handicaps an attacker but is a gain for a defender.

Colonel R. Ernest Dupuy and Colonel Trevor N. Dupuy, *The Military Heritage of America*, 1984.

THE OFFICER

Truly the only good officers are the impoverished gentlemen who have nothing but their sword and their cape; but it is essential that they should be able to live on their pay.

Field Marshal Maurice Comte de Saxe, *My Reveries*, 1732

. . . But it is not enough to cherish them at the moment when you have need of them and when their actions wring cheers from you. It is necessary that in time of peace they enjoy the reputation that they have so justly earned and that they be signally esteemed as men who have shed their blood solely for the honour and the well being of the state.

Frederick the Great, *Political Testament*, 1752

War must be carried on systematically, and to do it, you must have good Officers, there are, in my Judgment, no other possible means to obtain them but by establishing your Army upon a permanent basis; and giving your officers good pay; this will include Gentlemen, and Men of Character to engage; and till the bulk of your Officers are composed of such persons as are actuated by Principles of Honor, and spirit of Enterprize, you have little to expect from them.

General George Washington, 24 September 1776, letter to the President of Congress

Be frank with your friends, temperate in your requirements and disinterested in conduct; bear an ardent zeal for the service of your Soveriegn; love true fame, distinguish ambition from pride and vain glory; learn early to forgive the faults of others, and never forgive your own . . .

Field Marshal Prince Aleksandr V. Suvorov (1729–1800), advice to a young officer cadet, quoted in Blease, *Suvorof*

As you from this day start the world as a man, I trust that your future conduct in life will prove you both an officer and a gentleman. Recollect that you must be a seaman to be an officer; and also that you cannot be a good officer without being a gentleman.

Admiral Lord Nelson (1758–1805) (attributed), advice to a young man upon his appointment as midshipman

My brave officers . . . Such a gallant set of fellows! My heart swells at the thought of them.

Admiral Lord Nelson (1785–1805)

The officers should feel the conviction that resignation, bravery, and faithful attention to duty are virtues without which no glory is possible and no army is respectable, and that firmness amid reverses is more honourable than enthusiasm in success – since courage alone is necessary to storm a position whereas it requires heroism to make a difficult retreat before a victorious and enterprising enemy, always opposing to him a firm and unbroken front.

Lieutenant General Antoine-Henri Baron de Jomini, *Summary of the Art of War,* 1838

The more helpless the position in which an officer finds his men, the more it is his bounden duty to stay and share their fortune whether for good or ill.

Field Marshal Viscount Wolseley, quoted in Morris, *Washing of the Spears*, 1968

In every military system which has triumphed in modern war the officers have been recognized as the brain of the army, and to prepare them for their trust, governments have spared no pains to give them special education and training.

Colonel Emory Upton, *The Military Policy of the United States,* 1904

Officers used to serve, not for their starvation pay, but for love of their country and on the off-chance of being able to defend it with their lives . . . They deliberately, indeed joyously, faced up to a life of adventure, roving action, exile, and poverty because it satisfied and reposed their souls.

General Sir Ian Hamilton, *The Soul and Body of an Army*, 1921

The foundation of our National Defense system is the Regular Army, and the foundation of the Regular Army is the officer. He is the soul of the system. If you have to cut everything out of the National Defense act the last element would be the officer corps. If you had to discharge every soldier, if you had to do away with everything else, I would still professionally advise you to keep those 12,000 officers. They are the mainspring of the whole mechanism; each one of them would be worth a thousand men at the beginning of a war. They are the only ones who can take this heterogeneous mass and make it a homogeneous group.

General of the Army Douglas MacArthur, 1933, statement before Congress defending the officer corps from proposed deep cuts, *Reminiscences*, 1964

I divide my officers into four classes as follows: the clever, the industrious, the lazy, and the stupid. Each officer possess at least two of these qualities. Those who are clever and industrious I appoint to the General Staff. Use can under certain circumstances be made of those who are stupid and lazy. The man who is clever and lazy qualifies for the highest leadership posts. He has the requisite nerves and the mental clarity of difficult decisions. But whoever is stupid and industrious must be got rid of, for he is too dangerous.

Attributed to General Kurt von Hammerstein Equord, *c.* 1933

We, as officers of the army, are not only members of the oldest of honorable professions, but are also the modern representatives of the demi-gods and heroes of antiquity. Back of us stretches a line of men whose acts of valor, of self-sacrifice and of service have been the theme of song and story since long before recorded history began . . . In the days of chivalry . . . knights-officers were noted as well for courtesy and gentleness of behaviour as for death-defying courage . . . From their acts of courtesy and benevolence was derived the word, now pronounced as one, Gentle Man . . . Let us be gentle. This is, courteous and considerate of the rights of others. Let us be Men. That is, fearless and untiring in doing our duty . . . Our calling is most ancient and like all other old things it has amassed through the ages certain customs and traditions which decorate and ennoble it, which render beautiful the otherwise prosaic occupation of professional men-at-arms: killers.

General George S. Patton, Jr. (1885–1945), quoted in Fitton, ed., *Leadership*, 1990

The military officer is considered a gentleman, not because Congress wills it, nor because it has been the custom of the people at all times to afford him that courtesy, but specifically because nothing less than a gentleman is truly suited for his particular set of responsibilities.

Brigadier General S. L. A. Marshall, *The Armed Forces Officer*, 1950

The badge of rank which an officer wears on his coat is really a symbol of servitude – servitude to his men.

General Maxwell Taylor, quoted in *Army Information Digest*, 1953

Without a sense of special calling, it is unlikely that an officer corps can develop leadership within itself. Leadership requires a special sense of sacrifice, a sense of risk and a devotion in courage that can never truly develop when the climate of the organization equates these demands with the values of the marketplace.

Paul Savage and Lieutenant Colonel Richard Gabriel, 'The Environment of Military Leadership', *Military Review*, 7/1980

OFFICERS AND MEN

Pay heed to nourishing the troops; do not unnecessarily fatigue them. Unite them in spirit; conserve their strength. Make unfathomable plans for the movement of the army.

Thus, such troops need no encouragement to be vigilant. Without extorting their support the general obtains it; without inviting their affection he gains it; without demanding their trust he wins it.

Sun Tzu, *The Art of War*, c. 500 BC, tr. Griffith

. . . for if men have a spontaneous and natural love for their general, they are quick to obey his commands, they do not distrust him, and they cooperate with him in case of danger.

Onasander, *The General*, AD 58

I command a great army. It is founded on trust. The minds of those who have been given permission to depart are already set on home like arrows. Their wives and children are counting the hours that they must wait for them. Although the battle before us may be fraught with danger, that trust cannot be broken.

Zhuge Liang (AD 180–234) when urged by his officers at the approach of an enemy army to revoke the leave promised to a number of his troops. He personally urged the troops to go, but they refused, and the enemy suffered a decisive defeat, quoted in Bloodworth, *The Chinese Machiavelli*, 1976

Nothing does so much honour to the abilities and application of the Tribune as the appearance and discipline of the soldiers, when their apparel is neat and clean, their arms bright and in good order and when they perform their exercises and evolutions with dexterity.

Flavius Vegetius Renatus, *Military Institutions of the Romans*, c. AD 378

The general who is overly harsh with his subordinates and the one who is too indulgent are both unfit for command. Fear leads to great hatred, and giving in too much results in being despised. It is safe to take the middle course.

The Emperor Maurice, *The Strategikon*, c. AD 600

The general's way of life should be plain and simple like that of his soldiers; he should display a fatherly affection toward them; he should give orders in a mild manner; and he should always make sure to give advice and to discuss essential matters with them in person. His concerns ought to be with their safety, their food, and their regular pay. Without these it is impossible to maintain discipline in an army.

The Emperor Maurice, *The Strategikon*, c. AD 600

What makes soldiers in battle prefer to charge ahead rather than retreat even for survival is the benevolence of the military leadership. When soldiers know their leaders care for them as they care for their own children, then the soldiers love their leaders as they do their own fathers. This makes them willing to die in battle, to requite the benevolence of their leaders.

Liu Ji (1310–1375), *Lessons of War*

When asked one day how, after so many years, he could recollect the names and numbers of the units engaged in one of his early combats, Napoleon responded, 'Madam, this is a lover's recollection of his former mistresses.'

Napoleon, quoted in Count de Las Cases, *Memoirs of the Life, Exile and Conversations of the Emperor Napoleon*, 1836

The Duke never claimed for one moment credit to himself where he did not feel that it was thoroughly deserved. Someone saying to him 'How do you account, Duke, for your having so persistently beaten the French Marshals?' The Duke replied simply, 'Well, the fact is their soldiers got them into scrapes: mine always got me out.'

The Duke of Wellington, quoted in Fraser, *Words of Wellington*, n.d.

The one mode or the other of dealing with subordinates springs from a corresponding spirit in the breast of the commander. He who feels the respect which is due to others cannot fail to inspire in them regard for himself, while he who feels, and hence manifests, disrespect toward others, especially his inferiors, cannot fail to inspire hatred against himself.

Major General John M. Schofield, speech to the Corps of Cadets at West Point

---

For an officer to be overbearing and insulting in the treatment of enlisted men is the act of a coward. He ties the man to a tree with ropes of discipline and then strikes him in the face, knowing full well that the man cannot strike back.

Major C. A. Bach, 1917, farewell instructions to graduating student officers at Fort Sheridan, Wyoming

---

The relationship between officers and men should in no sense be that of superior and inferior nor that of master and servant, but rather that of teacher and scholar. In fact, it should partake of the nature of the relation between father and son, to the extent that officers, especially commanding officers, are responsible for the physical, mental, and moral welfare, as well as the discipline and military training of the young men under their command.

Major General John A. Lejeune, *Marine Corps Manual*, 1920

But there the fundamental condition ensuring military might is lacking: there is no link between the upper and lower strata of society. The existence of a deep gulf dividing the command class from the exploited essentially reduces to nought all efforts to create a battleworthy military force.

Mikhail Frunze (1885–1925), quoted in Gareyev, *Frunze, Military Theorist*, 1985

Sympathy and understanding are vital faculties in a commander – because they evoke a similar response in subordinates, and thus create the basis of co-operation. Arbitrary control commonly breaks down unless it rests on the co-operation of minds attuned to the same key. Without this the first is but voice control. (May 1929)

Captain Sir Basil Liddell Hart, *Thoughts on War*, 1944

Many people think that it is wrong methods that make for strained relations between officers and men and between the army and the people, but I always tell them that it is a question of basic attitude . . . of having respect for the soldiers and the people . . .

Mao Tse-tung, *On Protracted War*, May 1938

When the last bugle is sounded, I want to stand up with my soldiers.

General of the Armies John J. Pershing, quoted in Hale and Turner, *The Yanks are Coming*, 1983

Efficiency in a general, his soldiers have a right to expect; geniality they are usually right to suspect.

Field Marshal Earl Wavell, *Generals and Generalship*, 1941

The relationship between a general and his troops is very much like that between the rider and his horse. The horse must be controlled and disciplined, and yet encouraged: he should 'be cared for in the stables as if he was worth £500 and ridden in the field as if he were worth half-a-crown'.

Field Marshal Earl Wavell, *Generals and Generalship*, 1941

All officers, and particularly General officers, must be vitally interested in everything that interests the soldier. Usually you will gain a great deal of knowledge by being interested, but, even if you do not, the fact that you appear interested has a very high morale influence on the soldier.

General George S. Patton, Jr., *War As I Knew It*, 1947

Officers are responsible, not only for the conduct of their men in battle, but also for their health and contentment when not fighting. An officer must be the last man to take shelter from fire, and the first to move forward. Similarly, he must be the last man to look after his own comfort at the close of a march. He must see that his men are cared for. The officer must constantly interest himself in the rations of the men. He should know his men so well that any sign of sickness or nervous strain will be apparent to him, and he can take such action as may be necessary.

General George S. Patton, Jr., *War As I Knew It*, 1947

There is among the mass of individuals who carry rifles in a war an amount of ingenuity and efficiency. If men can talk naturally to their officers, the product of their resourcefulness becomes available to all.

General of the Armies Dwight D. Eisenhower, quoted in Marshall, *The Armed Forces Officer*, 1950

Nothing more radical is suggested here than that the leader who would make certain of the fundamental soundness of his operation cannot do better than concentrate his attention on his men. There is no other worthwhile road. They dupe only themselves who believe that there is a brand of military efficiency which consists in moving smartly, expediting papers, and achieving perfection in formations, while at the same time slighting or ignoring the human nature of those whom they command. The art of leading, in operations large or small, is the art of dealing with humanity, of working diligently on behalf of men, of being sympathetic with them, but equally, of insisting that they make a square facing toward their own problems. These are the real bases of a commander's calculations. Yet how often do we hear an executive praised as an 'efficient administrator' simply because he can keep a desk cleared, even though he is despised by everyone in the lower echelons and cannot command a fraction of their loyalty.

Brigadier General S. L. A. Marshall, *The Armed Forces Officer*, 1950

It is never a waste of time for the commander to talk to his people about their problems; more times than not, the problem will seem small to him, but so longs it looms large to the man, it cannot be dismissed with a wave of the hand. He will grow in esteem of his men as he treats their affairs with respect.

Brigadier General S. L. A. Marshall, *The Armed Forces Officer*, 1950

The habit of talking down to troops is one of the worst vices that can afflict an officer.

Brigadier General S. L. A. Marshall, *The Armed Forces Officer*, 1950

The identity of the officer with the gentleman should persist in his relations with men of all degree. In the routine of daily direction and disposition, and even in moments of exhortation, he had best bring courtesy to firmness. The finest officers that one has known are not occasional gentlemen, but in every circumstance: in commissioned company and, more importantly in contact with those who have no recourse against arrogance.

Brigadier General S. L. A. Marshall, *The Armed Forces Officer*, 1950

The commander must have contact with his men. He must be capable of feeling and thinking with them. The soldier must have confidence in him. There is one cardinal principle which must always be remembered: one must never make a show of false emotions to one's men. The ordinary soldier has a surprisingly good nose for what is true and what is false.

Field Marshal Erwin Rommel, *The Rommel Papers*, 1953

He [the soldier] will put up with this so long as he knows you are living in relatively much the same way; and he likes to see the C-in-C regularly in the forward area, and be spoken to and noticed. He must know that you really care for him and will look after his interests, and that you will give him all the pleasures you can in the midst of his discomforts.

Field Marshal Viscount Montgomery of Alamein, *The Memoirs of Field Marshal Montgomery*, 1958

The first thing a young officer must do when he joins the Army is to fight a battle, and that battle is for the hearts of his men. If he wins that battle and subsequent similar ones, his men will follow him anywhere; if he loses it, he will never do any real good.

Field Marshal Viscount Montgomery of Alamein, *The Memoirs of Field Marshal Montgomery*, 1958

The men have a highly developed collective sense. They are almost never mistaken about their commander, and are the people best qualified to judge him. It is possible to mislead them by propaganda and publicity stunts. It is possible to create a cult of admiration for some military figure who rarely comes in contact with them. But it is quite impossible to deceive them as the character and qualities of their direct commanders with whom they work day by day.

General Yigal Allon, *The Making of Israel's Army*, 1960

Soldiers who know that their commander values them and has implicit faith in them, that he will spare no effort on their behalf and will evade no risk which they must take, that he has the ability to plan and conduct a battle well and that he at no time takes unfair advantage of his rank, will understand the limitations he imposes upon them and will reconcile them to the rigorous discipline he expects of them. They will go into battle knowing that it is unavoidable and that everything possible has been done to ensure its success. They will give of their best in executing their mission.

General Yigal Allon, *The Making of Israel's Army*, 1960

The basis of all good leadership is the relationship built up between a commander and his troops. Napoleon's greatest and, as it turned out, almost his only asset on his return from Elba was the devotion of his army. His physical condition might have deteriorated, his military genius might have waned, but so long as his soldiers were prepared to die for him the French Army was still a formidable fighting machine.

Lieutenant General Sir Brian Horrocks, *Escape to Action*, 1961

I can answer questions and hold a detailed dialogue with a group of soldiers or even with a lone guard in a lookout post where there is a mutual interest in the subjects: what is happening on the enemy side? What are the defects? How can they be rectified? I am also interested in the personal lives of the men. Where do they come from? How do their families live? Do they rest or work when they are on leave? Do they support their parents?

I always found that the troops answered freely and were never afraid to criticize or complain. But these were more in the nature of question and answer than true dialogue. In this respect, I suppose there was no basic difference between my talk with an anonymous private in a Golan Heights outpost and my discussions with the top officials of the Ministry of Defence.

General Moshe Dayan, *The Story of My Life*, 1976

I don't like being separated in comparative luxury from troops who are living hard.

Field Marshal Sir Claude Auchinleck, 1976, interview with David Dimbleby

---

. . . the commander stakes his reputation, his career, all that he is and may be in his profession, on the issue of a fight in which his instinct must derive so much that is unknown and his science weigh all that is imponderable: above all, he stakes the coin he values most – the trust that his officers and men have in him.

Lieutenant General Sir Francis Tucker, quoted in Horner, *The Commanders*, 1984

Soldiers will go all out for an officer who does not waste their time through poor management when he has a tough job to do.

General Bruce C. Clarke (1901–1988)

Enlisted people with a higher quality of intelligence need a different kind of leadership. You cannot run a unit just by giving orders and having the Uniform Code of Military Justice behind you.

Admiral William Crowe, *Newsweek*, 18 April 1988

A good commander is someone who can step on your boots and still leave a shine.

A group of American soldiers

OPERATIONS

No plan of operations can with any assurance look beyond the first meeting with the main enemy forces. Only layman will think that he can see in the development of a campaign anything like a consistent pursuit of a previously conceived plan, one with all its details worked out beforehand and held to right to the end. The consecutive achievements of a war are not premeditated but spontaneous, and are guided by military instinct.

Field Marshal Helmuth Graf von Moltke (1800–1891)

This approach to an operation, this likening of it to the struggle of two wills, to the action of the 'mass psyche' of one side on the other, amounts to pernicious military idealism. An operation consists in the organized struggle of each of the armies to achieve the complete destruction of the opponent's vital material strength. The operational aim cannot be the destruction of some imaginary, abstract *nervous system* of the army; it must be the destruction of the organism itself – of the force, and of an army's *real* nervous system, its communications.

Marshal of the Soviet Union Mikhail N. Tukhachevskiy, 1924, quoted in Simpkin, *Deep Battle*, 1987

An important difference between a military operation and a surgical operation is that the patient is not tied down. But it is a common fault of generalship to assume that he is. (May 1934)

Captain Sir Basil Liddell Hart, *Thoughts on War*, 1944

OPPORTUNITY

Thus, those skilled at making the enemy move do so by creating a situation to which he must conform; they entice him with something he is certain to take, and with lures of ostensible profit they await him in strength.

Therefore, a skilled commander seeks victory from the situation and does not demand it of his subordinates.

Sun Tzu, *The Art of War*, c. 500 BC, tr. Griffith

In war opportunity waits for no man.

Pericles, reply to the Spartan ultimatum, 432 BC, quoted in Thucydides, *The Peloponnesian War*, c. 404 BC, tr. Warner

To overcome the intelligent by folly is contrary to the natural order of things; to overcome the foolish by intelligence is in accord with the natural order. To overcome the intelligent by intelligence, however, is a matter of opportunity.

There are three avenue of opportunity: events, trends, and conditions. When opportunities occur through events but you are unable to respond, you are not smart. When opportunities become active through a trend and yet you cannot make plans, you are not wise. When opportunities emerge through conditions but you cannot act on them, you are not bold.

Those skilled in generalship always achieve their victories by taking advantage of opportunities.

Zhuge Liang (AD 180–234), *The Way of the General*

The art of certain victory, the mode of harmonizing with changes, is a matter of opportunity. Who but the perspicacious can deal with it? And of all avenues of seeing opportunity, none is greater than the unexpected.

Zhuge Liang (AD 180–234), *The Way of the General*

Those designs are best of which the enemy are entirely ignorant till the moment of execution. Opportunity in war is often more to be depended on than courage.

Flavius Vegetius Renatus, *Military Institutions of the Romans*, c. AD 378

He [the general] should often deliberate about his most serious problems and carry out what he has decided with as little delay and risk as possible. For opportunity is what cures problems.

The Emperor Maurice, *The Strategikon*, c. AD 600

A good general is one who utilizes his own skills to fit the opportunities he gets and the quality of the enemy.

The Emperor Maurice, *The Strategikon*, c. AD 600

In war opportunity is fleeting and cannot be put off at all.

The Emperor Maurice, *The Strategikon*, c. AD 600

Four things come back not:
The spoken word; the sped arrow;
Time past; the neglected opportunity.

Omar I Ibn al-Khattab, Caliph (AD 581–644)

My feeling is that we should confront all the enemy's forces with all the forces of Islam; for events do not turn out according to man's will and we do not know how long a life is left to us, so it is foolish to dissipate this concentration of troops without striking a tremendous blow in the Holy War.

Saladin, July 1187, his decision before the Battle of Hattin to make use of the opportunity presented by concentration, quoted by Ibn al-Athir in Gabrieli, *Arab Historians of the Crusades*, 1957

When there is a battle that can be won or a castle that can be taken, to concern oneself with the fortuitous day or direction and let time pass is extremely regrettable. There will be little value in sending a ship out in a storm, or having a single man face great numbers even if the day is propitious. For even though the day and place be 'unlucky', if one will attain the minute details of the situation, prepare his attack in secret, adapt to the circumstances, and make strategy his foundation, the victory will surely be his.

Asakura Toshikage (1428–1481), *The 17 Articles of Asakura Toshikage*, in Wilson, *Ideals of the Samurai*, 1982

Nothing is of greater importance in time of war than knowing how to make the best use of a fair opportunity when it is offered.

Niccolo Machiavelli, *The Art of War*, 1521

Not only strike while the iron is hot, but make it hot by striking.

Oliver Cromwell (1599–1658)

Thus, on the day of battle, I should want the general to do nothing. His observations will be better for it, his judgment will be more sane, and he will be in better state to profit from the situations in which the enemy finds himself during the engagement. And when he sees an occasion, he should unleash his energies, hasten to the critical point at top speed, seize the first troops available, advance them rapidly, and lead them in person. These are the strokes that decide battles and gain victories. The important thing is to see the opportunity and to know how to use it.

Field Marshal Maurice Comte de Saxe, *My Reveries*, 1732

. . . learn to profit from local circumstances.

Field Marshal Aleksandr V. Suvorov (1729–1800)

A sea-officer cannot, like a land-officer, form plans; his object is to embrace the happy moment which now and then offers, – it may be this day, not for a month, and perhaps never.

Admiral Lord Nelson, 1796, to the British Minister at Genoa

In war there is only one favourable moment. Genius seizes it.

Napoleon, *Maxims of War*, 1831

When this movement commences, I shall move out by my left, with all the force I can, holding present intrenched lines. I shall start with no distinct view, further than holding Lee's forces from following Sheridan. But I shall be along myself and will take advantage of anything that turns up.

General of the Army Ulysses S. Grant, 1865, letter to Major General Sherman from outside Petersburg, quoted in Liddell Hart, *Sherman*, 1930

All great commanders have come with increasing experience of war to realize that opportunism is essential in war; that fixed purposes and objectives are contrary to its nature, and, if rigidly maintained, lead usually to a blank wall. But most of them have tended to develop a practice of pure opportunism – to take an obvious course and exploit any chance that occurred. A few, like Sherman, have reached a more systematized practice of opportunism, seeking and finding a solution in variability, or elasticity in the choice of a line of operation leading to alternative objectives, combined with the development of formations that give them the power to vary their course and gain whichever the enemy leaves open. (April 1925)

Captain Sir Basil Liddell Hart, *Thoughts on War*, 1944

The battle that turned the tide of the war in 1914 has given rise to an endless debate among generals and historians as to who was responsible for winning the Battle of the

Marne. It is quite as apt to pose the question: 'Was the Marne won by the Allies or lost by the Germans?' The reasonable answer is that most victories in history have been won by seizing opportunities offered by the loser. (June 1931)

Captain Sir Basil Liddell Hart, *Thoughts on War*, 1944

. . . for success, two major problems must be solved – *dislocation* and *exploitation*. One precedes and one follows the actual blow – which in comparison is a simple act. You cannot hit the enemy with effect unless you have first created the opportunity; you cannot make that effect decisive unless you exploit the second opportunity that comes before he can recover.

Captain Sir Basil Liddell Hart, *Strategy*, 1954

To our men . . . the jungle was a strange, fearsome place; moving and fighting it in were a nightmare. We were too ready to classify 'impenetrable' . . . To us it appeared only as an obstacle to movement; to the Japanese it was a welcome means of concealed manoeuvre and surprise . . . The Japanese reaped the deserved reward . . . we paid the penalty.

Field Marshal Viscount Slim, *Defeat into Victory*, 1963

ORDERS

An order is a good basis for discussion.

Old Military Saying

During any military affair, no matter how important the event may be, when something is communicated by word of mouth, the least bit of vagueness will invite grievous results.

Asakura Soteki (1474–1555), *Soteki Waki*, c. 1550, in Wilson, *Ideals of the Samurai*, 1982

The orders I have given are strong, and I know not how my admiral will approve of them, for they are, in a great measure, contrary to those he gave me; but the Service requires strong and vigorous measures to bring the war to a conclusion.

Admiral Lord Nelson, July 1795, letter to Collingwood

If all is not in order, I will hang you despite my personal high regard.

Field Marshal Prince Aleksandr V. Suvorov, 1799, administrative order to his quartermaster general during the Italian Campaign

---

I find few think as I do, but to obey orders is all perfection. What would my superiors direct, did they know what is passing under my nose? To serve my King and to destroy the French I consider as the great order of all, from which little ones spring; and if one of those little ones militate against it, I go back to obey the great order.

Admiral Lord Nelson, March 1799, letter from Palermo to the Duke of Clarence

---

I shall endeavour to comply with all their Lordship's directions in such manner as to the best of my judgment, will answer their intentions in employing me here.

Admiral Lord Nelson (1758–1805)

It is not enough to give orders, they must be obeyed.

Napoleon, 11 June 1806, to Eugene, *Correspondance*, No. 10350, Vol. XII, 1858–1870

You must avoid countermanding orders: unless the soldier can see a good reason for benefit, he becomes discouraged and loses confidence.

Napoleon, 5 August 1806, to Eugene, *Correspondance*, No. 10699, Vol. XIII, 1858–1870

Give your orders so that they cannot be disobeyed.

Napoleon, 29 March 1811, to Marshal Berthier, *Correspondance*, No. 17529, Vol. XXI, 1858–1870

Men bred as soldiers have no fancy for orders that carry want of faith on their face.

Lieutenant General James A. Longstreet, comments on Burnside's orders at the Battle of Fredericksburg, 13 December 1862, *From Manassas to Appomattax*, 1896

It is the custom of military service to accept instructions of a command as orders, but when they are coupled with conditions that transfer the responsibility of battle and defeat to the subordinate, they are not orders . . .

Lieutenant General James A. Longstreet, comments on Lee's vague orders to Ewell for the second day at Gettysburg, 2 July 1863, *From Manassas to Appomattox*, 1896

In issuing orders or giving verbal instruction, Jackson's words were few and simple; but they were so clear, so comprehensive and direct, that no officer could possibly misunderstand, and none dared disobey.

Lieutenant General John B. Gordon, quoted in Henderson, *Stonewall Jackson*, 1898

Remember, gentlemen, an order that can be misunderstood will be misunderstood.

Field Helmuth Helmuth Graf von Moltke (1800–1891)

Order, counter-order, disorder.

Field Marshal Helmuth Graf von Moltke (1880–1891)

An order shall contain everything that a commander cannot do by himself, but nothing else.

Field Marshal Helmuth Graf von Moltke (1800–1891)

The higher the commander, the shorter and simpler the orders must be.

Field Marshal Helmuth Graf von Moltke (1800–1891), *Militaerische Werke*, II, 1892–1912

I just gave an order – quite a simple matter unless a man's afraid.

General Sir Ian Hamilton, *Soul and Body of an Army*, 1921

All orders will have to be as brief as possible. They should be based on a profound appreciation of possibilities and probabilities which . . . will generally lead to a series of alternatives.

Major General J. F. C. Fuller (1878–1966)

Pétain was once asked what part of an action called for the greatest effort. 'Giving orders,' he replied. There is, indeed, something irrevocable about the intervention of the human will in a sequence of events. Whether useful or not, opportune or misjudged, it is pregant with unforeseeable consequences. The mere awareness of the audacity of such a proceeding is intimidating. Even in ordinary life there are many who do not find it easy to make decisions, and the number of those prepared to take the initiative is small in comparison with that of the obedient and submissive mass. How much more agonizing must be the call to decisive action in the case of a military commander who knows that so many poor lives depend upon his decision, and that on the highest as well as on the lowest levels it will be judged solely by its results. So heavy is the responsibility in such cases that few men are capable of shouldering the whole unaided. That is why the greatest intellectual qualities are not enough. No doubt they help; so, too, does instinct, but in the last resort, the decision is a moral one.

Charles de Gaulle, *The Edge of the Sword*, 1932

In war 'thoughtless' orders often result in the useless sacrifice of men's lives. In peace, they often contribute to the sterilization of men's reason. And they inevitably make the man who transmits the order an accomplice in the crime, however unwilling. Unfortunately, few of the transmitters have the sensitiveness of perception to feel their responsibility. (July 1933)

Captain Sir Basil Liddell Hart, *Thoughts on War*, 1944

The issuance of an order is the simplest thing in the world. The important and difficult thing is to see: first, that the order is transmitted; and, second, that it is obeyed . . .

General George S. Patton, Jr., 8 July 1941, speech to the officers of the 2nd Armored Division, *The Patton Papers*, Vol. II, 1974

I give orders only when they are necessary. I expect them to be executed at once and to the letter and that no unit under my command shall make changes, still less give orders to the contrary or delay execution through unnecessary red tape.

Field Marshal Erwin Rommel, 22 April 1944, letter of instruction to subordinate commanders

---

The best way to issue orders is by word of mouth from one general to the next. Failing this, telephone conversation which should be recorded at each end. However, in order to have a confirmatory memorandum of all oral orders given, a short written order should always be made out, not necessarily at the time of issuing the order, but it should reach the junior prior to his carrying out the order; so that, if he has forgotten anything, he will be reminded of it, and, further, in order that he may be aware that his senior has taken definite responsibility for the operation ordered orally.

General George S. Patton, Jr., *War As I Knew It*, 1947

Avoid as you would perdition issuing cover-up orders, orders for the record. This simply shows lack of intestinal fortitude on the part of the officer signing the orders, and everyone who reads them realizes it at once.

General George S. Patton, Jr., *War As I Knew It*, 1947

An officer should not ask a man: 'Would you like to do such-and-such a task?' when he has already made up his mind to assign him to a certain duty. Orders, hesitantly given, are doubtfully received. But the right way to do it is to instill the idea of collaboration. There is something irresistibly appealing about such an approach as: 'I need your help. Here's what we have to do.'

Brigadier General S. L. A. Marshall, *The Armed Forces Officer* 1950

Operations orders do not win battles without the valour and endurance of the soldiers who carry them out.

Field Marshal Earl Wavell, *Soldiers and Soldiering*, 1953

Operational command in the battle must be direct and personal, by means of visits to subordinate H.Q. where orders are given verbally. A commander must train his subordinate commanders, and his own staff, to work and act on verbal orders. Those who cannot be trusted to act on clear and concise verbal orders, but want everything in writing are useless.

Field Marshal Viscount Montgomery of Alamein, *The Memoirs of Field Marshal Montgomery*, 1958

To grasp the spirit of order is not less important than to accept them cheerfully and keep faith with the contract. But the letter of an instruction does not relieve him who receives it from the obligation to exercise common sense.

Brigadier General S. L. A. Marshall, *The Officer as a Leader*, 1966

ORGANIZATION

Generally, management of the many is the same as management of the few. It is a matter of organization.

Sun Tzu, *The Art of War*, c. 500 BC

A policy to quell disorder involves minimizing offices and combining duties, getting rid of embellishment in favour of substance.

First organize directives, then organize penalties. First organize the near at hand, then organize the far removed. First organize the inner, then organize the outer. First organize the basic, then organize the derivative. First organize the strong, then organize the weak. First organize the great, then organize the small. First organize yourself, then organize others.

Zhuge Liang (180–234 AD). *The Way of the General*

The peculiar strength of the Romans always consisted in the excellent organization of their legions. They were so denominated *ab eligendo*, from the care and exactness used in the choice of the soldiers.

Flavius Vegetius Renatus, *Military Institutions of the Romans*, c. AD 378

The mass needs, and we give it, leaders . . . We add good arms. We add suitable methods of fighting . . . We also add a rational decentralization . . . We animate with passion . . . An iron discipline . . . secures the greatest unity . . . But it depends also on supervision, the mutual supervision of groups of men who know each other well. A wise organization of comrades in peace who shall be comrades in war . . . And now confidence appears . . . Then we have an army.

Colonel Charles Ardnant du Picq, *Battle Studies*, 1880

. . . battles are rare events and that, meanwhile, the Army lives, marches, works, and has its being by organization and discipline. Further, after the first general action he will be surprised, if he thinks it over, to recollect how very great a part organization and discipline, or their absence, had to say to success or failure.

General Sir Ian Hamilton, *Soul and Body of an Army*, 1921

The primary object of organization is to shield people from unexpected calls upon their powers of adaptability, judgment and decision.

General Sir Ian Hamilton, *Soul and Body of an Army*, 1921

Organization is the vehicle of force; and force is threefold in nature; it is mental, moral, and physical. How will the idea affect these spheres of force? This is primarily a question of force and its expenditure. Thus, if the idea is complex, and does not permit of it being readily grasped by the others, mistakes are likely to occur; and if its aim is beyond the moral and physical powers of the troops, should it be pushed beyond the limit of their endurance, though organization may for the time being be maintained, ultimately demoralization will set in, and a demoralized organization is one which has become so fragile that a slight blow, especially a surprise blow, will instantaneously shatter it to pieces.

Major General J. F. C. Fuller, *The Foundation of the Science of War*, 1926

PACIFISM/ANTI-WAR ACTIVISM

If one has a free choice and can live undisturbed, it is sheer folly to go to war.

Pericles, at the beginning of the Peloponnesian War, quoted in Thucydides, *The Peloponnesian War*, c. 404 BC

If you desire peace, prepare for war.

Roman military maxim

A rich man without arms must be prey to a poor soldier well armed.

Niccolo Machiavelli, *The Art of War*, 1521

Once initiated there were but few public men who would have the courage to oppose it. Experience proves that the man who obstructs a war in which his nation is engaged, no matter whether right or wrong, occupies no enviable place in life or history. Better for him, individually, to advocate 'war, pestilence, and famine', than to act as obstructionist to a war already begun. The history of the defeated rebel will be honorable hereafter, compared with that of the Northern man who aided him by conspiring against his government while protected by it. The most favorable posthumous history the stay-at-home traitor can hope for is – oblivion.

General of the Army Ulysses S. Grant, *Personal Memoirs of U.S. Grant*, 1885

The man who has formed a clear notion of the nature of war, of its necessities, requirements and consequences, to wit, the *soldier*, will take a far more serious view of the potentialities of war than the politician or the business man who coldly weights its advantages and disadvantages. After all, it is not so difficult to sacrifice one's own life, but the professional duty of risking the lives of others weighs heavily on the conscience. The soldier, having experience of war, fears it far more than the doctrinaire who, being ignorant of war, talks only of peace; for the soldier has gazed into war's bloodshot eyes, he has observed from his point of vantage the battlefields of a world war, he has had to witness the agonies of nations, his hair has turned grey over the ashes of countless burned homesteads and he has borne the responsibility for the life and death of thousands. The figure of the sabre-rattling, fire-eating general is an invention of poisoned and unscrupulous political strife, a figure welcome to stupid comic papers, a catchword personified. There is no reason why the soldier's attitude towards war should not be called 'pacifism'. It is a pacifism established on knowledge and born of a sense of responsiblity, but it is not the pacifism engendered by national abasement or by a hazy internationalism. The soldier will be the first to welcome any effort to diminish the potentialities of war, but he does not march down the street to the slogan of 'No more war!' because he knows that war and peace are decided by higher powers than princes, statesmen, parliaments, treaties and alliances – they are decided by the eternal laws which govern the growth and decay of nations. But the kind of pacifist who would deliberately make his own nation defenceless in such fateful encounters, who prefers to weaken it in alliance with a hostile neighbour rather than support his fellow-countrymen in preparation for legitimate resistance, deserves, as he always did, to be hanged to the nearest lamppost, were it only a moral one.

Colonel General Hans von Seekt, *Thoughts of a Soldier*, 1930

We are for the abolition of war, we do not want war; but war can only be abolished through war, and to get rid of the gun, we must first grasp it in our hands.

Mao Tse-tung, *Problems of War and Strategy*, 1938

The pacifists are at it again. I met a visiting fireman of great eminence who told me this was the 'LAST WAR'. I told him that such statements since 2600 BC had signed the death warrant of millions of young men. He replied with the stock *lie* – 'Oh yes but things are different now.' My God! Will they never learn?

General George S. Patton, Jr., 3 March 1944, letter to his wife, *The Patton Papers*, Vol. II, 1974

Rational pacifism must be based on a new maxim – 'If you wish for peace, understand war.'

Captain Sir Basil Liddell Hart, *Thoughts on War*, 1944

The surest way to become a pacifist is to join the infantry.

Bill Mauldin, *Up Front*, 1945

War is never prevented by running away from it.

Air Marshal Sir John Slessor, *Strategy for the West*, 1954

PANIC

Panic, brother to blood-stained Rout.

Homer, *The Iliad*, c. 800 BC

Both officers and troops must be warned against those sudden panics which often seize the bravest armies when they are not well-controlled by discipline, and when they do not recognize that the surest hope of safety lies in order. An army seized with panic is in a state of demoralization because when disorder is once introduced all concerted action on the part of individuals becomes impossible, the voice of the officers can no longer be heard. No manoeuvre for resuming the battle can be executed, and there is no course except in ignominious flight.

Lieutenant General Antoine-Henri Baron de Jomini, *Summary of the Art of War*, 1838

The worst enemy a Chief has to face in war is an alarmist.

General Sir Ian Hamilton, *Gallipoli Diary*, 1920

. . . since panic gathers volume like a snowball, I think we can take it that every large panic starts with some very minor event and that for the general purposes of control it is more important to have exact knowledge of the small cause than the large effect. . . . the trouble began because somebody was thoughtless, somebody failed to tell other men what he was doing . . . nothing is more likely to collapse a line of infantry in combat than the sight of a few of its number in full and unexplained flight. Precipitate motion in the wrong direction is an open invitation to disaster.

Brigadier General S. L. A. Marshall, *Men Against Fire*, 1947

PATRIOTISM

This empire has been acquired by men who knew their duty and had the courage to do it, who in the hour of conflict had the fear of dishonour always present to them, and who, if ever they failed in an enterprise, would not allow their virtues to be lost to their country, but freely gave their lives to her as the fairest offering which they could present at her feast.

Pericles, 431 BC, funeral oration for the Athenian dead, quoted in Thucydides, *The Peloponnesian War*, c. 404 BC, tr. Jowett

---

These are the times that try men's souls. The summer soldier and the sunshine patriot will, in this crisis, shrink from the service of their country; but he that stands it now, deserves the love and thanks of man and woman. Tyranny, like hell, is not easily conquered; yet we have this consolation with us, that the harder the conflict, the more glorious the triumph. What we obtain too cheap, we esteem too lightly; 'tis dearness only that gives everything its value. Heaven knows how to put a price upon its goods, and it would be strange indeed if so celestial an article as freedom should not be highly rated.

Thomas Paine, *The American Crisis*, 23 December 1776

Men may speculate as they will; they may talk of patriotism; they draw a few examples from ancient story, of great achievements performed by its influence; but whoever builds upon it, as a sufficient Basis for conducting a long and bloody War, will find themselves deceived in the end . . . We must take the passions of Men as Nature has given them, and whose principles as a guide which are generally the rule of Action. I do not mean to exclude altogether the Idea of Patriotism. I know it exists, and I know it has done much in the present Contest. But I will venture to assert, that a great and lasting War can never be supported on this principle alone. It must be aided by a prospect of Interest or some reward. For a time, it may, of itself push Men to Action; to bear much, to encounter difficulties; but it will not endure unassisted by Interest.

General George Washington, 21 April 1778, letter to John Bannister, *The Writings of George Washington*, Vol. 11, 1931–44

. . . it is but justice to assign great merit to the temper of those citizens whose estates were more immediately the scene of warfare. Their personal services were rendered without constraint, and the derangement of their affairs submitted to without dissatisfaction. It was the triumph of patriotism over personal considerations. And our present enjoyment of peace and freedom reward the sacrifice.

General George Washington, 1 December 1789, to the New Jersey Legislature, *The Writings of George Washington*, Vol 30, 1931–44

. . . Whatever services I have rendered to my country, in its general approbation I have received an ample reward.

General George Washington, 24 February 1797, to the Senators of Massachusetts, *The Writings of George Washington*, Vol 35 1931–44

Our country: in her intercourse with foreign nations may she always be in the right; but our country, right or wrong!

Admiral Stephen Decatur, April 1816, toast at a dinner in Norfolk, Virginia

---

Our orders home, and tidings of the attack on Fort Sumter, came by the same mail, and some time in June. The revulsion of feeling was immediate and universal, in that distant community and foreign land, as it had been two months before in the Northern States. The doughfaces were set at once, like flint. The grave and reverend seigniors, resident merchants, who had checked any belligerent utterance among us with reproachful regret that an American should be willing to fight Americans, reconverted or silenced. Every voice but one was hushed, and that voice said, 'Fight'.

Rear Admiral Alfred Thayer Mahan, recollections of the patriotic ardour felt by the American Navy personnel and residents in Montevideo upon news of the outbreak of the Civil War, *From Sail to Steam*, 1907

Patriotism is like a plant whose roots stretch down into race and place subconsciousness; a plant whose best nutrients are blood and tears; a plant which dies down in peace and flowers most brightly in war. Patriotism does not calculate, does not profiteer, does not stop to reason: in an atmosphere of danger the sap begins to stir; it lives; it takes possession of our soul.

General Sir Ian Hamilton, *The Soul and Body of an Army*, 1921

There is nothing more soulless than a religion without good works unless it be a patriotism which does not concern itself with the welfare and dignity of the individual. Only the officer who dedicates his thought and energy to his men can convert into coherent military force their inarticulate thoughts about their country; nor is any other in a position to stimulate their desire to be of service to it.

Brigadier General S. L. A. Marshall, *Men Against Fire*, 1947

The man who is willing to fight for his country is finally the full custodian of its security. If there were no willing men, no power in government could ever rally the masses of the unwilling. But if the spririt and purpose which enable such men to find themselves and to act are to be safeguarded into the future, much more will have to be required of the country than that it point its young people toward the virtues of the production line. There is something almost fatally quixotic about a nation which professes lofty ideals in its international undertakings and yet disdains to talk patriotism to its citizens, as if this were beneath their dignity.

Brigadier General S. L. A. Marshall, *Men Against Fire*, 1947

I admire men who stand up for their country in defeat, even though I am on the other side.

Sir Winston S. Churchill, *The Gathering Storm*, 1948

In the final choice, a soldier's pack is not so heavy a burden as a prisoner's chains.

General of the Army Dwight D. Eisenhower, 20 January 1953, Inaugural Address as President of the United States

---

Ask not what your country can do for you – ask what you can do for your country.

President John F. Kennedy, 20 January 1961, Inaugural Address

PAY

Pay the troops and fear no evil.

Cornelius Tacitus (*c.* AD 56–AD 120)

Stability between nations cannot be maintained without armies, nor armies without pay, nor pay without taxation.

Cornelius Tacitus (*c.* AD 56–AD 120), *Histories*

Do not disagree between yourselves. Give the soldiers money and despise everyone else.

The Emperor Septimius Severus (AD 146–AD 211), deathbed advice to his sons, quoted in Dio Cassius

---

To raise any taxes is impracticable, since the provinces are in possession of the enemy, and the long arrear of pay which our soldiers vainly claim loosens every tie of discipline and duty. A debtor is but ill able to command.

Count Belisarius (*c.* AD 505–AD 565)

Now the chief thing incumbent upon a general, in order to maintain his reputation, is to pay well and punish soundly for if he does not pay his men duly, he cannot punish them properly when they deserve it. Suppose, for instance, a soldier should be guilty of a robbery; how can you punish him for that when you give him no pay? And how can he help robbing when he has no other means of subsistence? But if you pay them well and do not punish them severely when they offend, they will soon grow insolent and licentious; then you will become despised and lose your authority; later, tumult and discord will naturally ensue in your army; and will probably end in ruin.

Niccolo Machiavelli, *The Art of War*, 1521

I think all men of experience and judgment in matters of war do know that the first and principal thing that is requisite to assemble and form an army and to keep the same in obedience with good effect is treasure to maintain, pay, and reward, with severe execution of excellent military laws.

John Smythe (*c.* 1580–1631), *Certain Discourses Military*

Without going into the different rates of pay, I shall say only that it should be ample. It is better to have a small number of well-kept and well-disciplined troops than to have a great number who are neglected in these matters. It is not the big armies that win battles; it is the good ones. Economy can be pushed only to a certain point. It has limits beyond which it degenerates into parsimony. If your pay and allowances for officers will not support them decently, then you will only have rich men who serve for debauchery or indigent wretches devoid of spirit.

Field Marshal Maurice Comte de Saxe, *My Reveries*, 1732

The soldier should not have any ready money. If he has a few coins in his pocket, he thinks himself too much of a great lord to follow his profession, and he deserts at the opening of a campaign.

Frederick the Great, *Instructions to His Generals*, 1747

There is nothing that gives a man consequence, and renders him fit for command, like a support that renders him independent of everybody but the State he serves.

General George Washington, 24 September 1776, letter to the President of Congress from the Heights of Harlem

Every means should be taken to attach the soldier to his colours. This is best accomplished by showing consideration and respect for the old soldier. His pay likewise should increase with his length of service. It is the height of injustice to give a veteran no greater advantages than a recruit.

Napoleon, *The Military Maxims of Napoleon*, 1831, tr. D'Aguilar

---

My present pay is not wholly for present work but is in great part for past services . . . What money will pay Meade for Gettysburg? What Sheridan for Winchester? What Thomas for Chickamauga?

General of the Army William T. Sherman, 1870, letter on pending legislation to cut Army officers' pay

---

The feeling of the soldier should be that, in every event, the sympathy and preference of his government is for him who fights, rather than for him who is on provost guard duty to the rear, and, like most men, he measures this by the amount of pay. Of course, the soldier must be trained to obedience, and should be 'content with his wages', but whoever has commanded an army in the field knows the difference between a willing, contented mass of men, and one that feels a cause of grievance.

General of the Army William T. Sherman, *Memoirs of General W. T. Sherman*, 1875

PEACE

Peace is an armistice in a war that is continuously going on.

Thucydides, *History of the Peloponnesian War*, c. 404 BC

For peace, with justice and honour, is the fairest and most profitable of possessions, but with disgrace and shameful cowardice it is the most infamous and harmful of all.

Polybius, *Histories*, c. 125 BC

A bad peace is even worse than war.

Cornelius Tacitus (c. AD 56–c. AD 120), *Annals*

The first blessing is peace, as is agreed upon by all men who have even a small share of reason. It follows that if any one should be a destroyer of it, he would be most responsible not only to those near him but also to his whole nation for the troubles which come. The best general, therefore, is that one who is able to bring about peace from war.

Count Belisarius, AD 530, admonition to the Persian generals not to disrupt ongoing peace negotiations, quoted in Procopius, *History of the Wars*, I, c. AD 560

Though peace be made, yet it is interest that keeps peace.

Oliver Cromwell, 4 September 1654, to Parliament

We become reconciled with our enemies because we want to improve our situation, because we are weary of war, or because we fear defeat.

François Duc de la Rochefoucauld, *Réflexions ou sentences et maximes morales*, 1665

What my enemies call a general peace is my destruction. What I call peace is merely the disarmament of my enemies. Am I not more moderate than they?

Napoleon, 1813, ed. Herold, *The Mind of Napoleon*, 1955

We desire a peace that will be honourable to both parties. And, as I understand this document, we are leaving honour behind us, for we are now not only surrendering our independence, but we are allowing every burgher to be fettered hand and foot. Where is the 'honourable peace' for us? If we conclude peace, we have to do it as men who have to live and die here. We must not agree to a peace which leaves behind in the hearts of one party a wound that will never heal.

General Louis Botha, 19 May 1902, peace conference ending the Boer War, quoted in De Wet, *Three Years War*, 1902

Peace is generally good in itself, but it is never the highest good unless is comes as the handmaid of righteousness; and it becomes a very evil thing if it serves merely as a mask of cowardice and sloth, or as an instrument to further the ends of despotism or anarchy.

President Theodore Roosevelt, 1910, speech upon award of the Nobel Peace Prize

---

I am not worried about the war; it will be difficult but we shall win it; it is after the war that worries me. Mark you, it will take years and years of patience, courage and faith.

Field Marshal Jan Christian Smuts, qouted in Tedder, *With Prejudice*, 1948

If man does find the solution for world peace it will be the most revolutionary reversal of his record we have ever known.

General of the Army George C. Marshall, 1 September 1945, Biennial Report, Chief of Staff, U.S. Army

---

The object of war is a better state of peace – even if only from your own point of view. Hence it is essential to conduct war with constant regard to the peace you desire. That applies both to aggressor nations who seek expansion and to peaceful nations who only fight for self-preservation – although their views of what is meant by a better state of peace are very different.

Captain Sir Basil Liddell Hart, *Strategy*, 1954

PEACETIME ARMIES

I think it is evident to all men of wisdom and discretion that have read diverse notable histories with consideration and judgment, as also that have well considered of this our age, that there are two things of all others that are the greatest enemies to the art and science military and have been the occasion of the great decay, and often times the utter ruin, of many great empires, kingdoms, and commonwealths . . . the first is long peace, which ensuing after great wars to divers nations that have had notable militias and exercises military in great perfection, they by enjoying long peace have so much given themselves to covetousness, effeminacies, and superfluities that they have either in great part or else utterly forgotten all orders and exercises military.

John Smythe (*c.* 1580–1631), *Certain Discourses Military*

Our long garrison life spoiled us, and effeminacy and desire for and love of pleasure, have weakened our military virtues. The entire nation must pass through the School of Misfortune, and we shall either die in the crisis, or a better condition will be created, after we have suffered bitter misery, and after our bones have decayed.

Field Marshal August Graf von Gneisenau (1760–1831), quoted in Balck, *Development of Tactics*, 1922

In times of peace we have neglected much, occupied ourselves with frivolities, flattered the people's love of shows and neglected war.

Field Marshal August Graf von Gneisenau, 1806, on the mobilization of Prussia leading to the disasters at the Battles of Auerstadt and Jena, quoted in Seeley, *Life and Times of Stein*, 1878

One should be careful not to compare this expanded and refined solidarity of a brotherhood of tempered, battle-scarred veterans with the self-esteem and vanity of regular armies which are patched together only by service-regulations and drill. Grim severity and iron discipline may be able to preserve the military virtues of a unit, but it cannot create them. These factors are valuable, but they should not be overrated. Discipline, skill, good will, a certain pride, and high morale, are the attributes of an army trained in times of peace. They command respect, but they have no strength of their own. They stand or fall together. One crack, and the whole thing goes, like a glass too quickly cooled.

Major General Carl von Clausewitz, *On War*, iii, 1832, tr. Howard and Paret

---

It must be obvious, therefore, that periods of tranquility are rich in sources of friction between soldiers and statesmen, since the latter are for ever trying to find ways of saving money, while the former are constantly urging increased expenditure. It does, of course, occasionally happen that a lesson recently learned, or an immediate threat, compels them to agree.

Charles de Gaulle, *The Edge of the Sword*, 1832

It is in peace that regulations and routine become important and that qualities of boldness and originality are cramped. It is interesting to note how little of normal peace soldiering many of our best generals had – Cromwell, Marlborough, Wellington, and his lieutenants, Graham, Hill, Craufurd.

Field Marshal Earl Wavell, *Soldiers and Soldiering*, 1953

Our traditional fault – the one against which I coped hopelessly in earlier days of the Eighth Air Force – went all the way back to Langley Field. There they had the usual Base commander and Group commander. The Base commander wanted to mow the grass; the Group commander wanted to fly his airplanes . . . Answer? They mowed the grass.

Because why? Because the Base commander made out the efficiency report on the Group commander. He got rated whether his grass was cut or not, or whether his buildings were painted. By gad, that's what he was going to do: mow grass.

General Curtis LeMay, *Mission With LeMay*, 1965

THE PEN AND THE SWORD

We pay a high price for being intelligent. Wisdom hurts.

Euripides, *Electra*, 413 BC

I would rather have written that poem, gentlemen, than take Quebec tomorrow.

Major General Sir James Wolfe, 12 September 1759, referring to Gray's *Elegy* on day before his death in victory at the Battle of Quebec

---

It is not the business of naval officers to write books.

Rear Admiral F. M. Ramsay, 1893, endorsement on the unfavourable fitness report made out on Alfred Thayer Mahan

---

It is often said that a man who writes well cannot be a good soldier; most of the great commanders, from King David, Xenophon, and Caesar to Wellington, not only wrote well, but extremely well.

Field Marshal Viscount Wolseley, *The Story of a Soldier's Life*, Vol. II, 1903

Our regulations governing the publication by soldiers of their views on military matters are veritable Lettres de Cachet, consigning the intellects of our Service to the Bastille of ignorance.

General Sir Ian Hamilton, *The Soul and Body of an Army*, 1921

The written essay or appreciation is a good test of character and grasp of the principles of war. It is apt to explode the legend of the 'strong, silent man' – who is usually silent because his mind is so hazy that he fears to commit himself to the risk of logical argument. The man who writes gives proof that at any rate he possesses some knowledge, whereas it is quite a possibility that the mind of the inarticulate one may be a military vacuum. Further, arguments that would pass muster in verbal conference are easily seen to be lacking in logic when written, for cold print is a merciless exposer of mental fog. (March 1923)

Captain Sir Basil Liddell Hart, *Thoughts on War*, 1944

Battles are won through the ability of men to express themselves in clear and unmistakable language.

Brigadier General S. L. A. Marshall, *The Armed Forces Officer*, 1950

There is one quality above all which seems to me essential for a good commander, the ability to express himself clearly, confidently, and concisely, in speech and on paper; to have the power to translate his intentions into orders and instructions which are not merely intelligible but unmistakable, and yet brief enough to waste no time. My experience of getting on for fifty years' service has shown me that it is a rare quality amongst Army Officers, to which not nearly enough attention is paid in their education. It is one which can be acquired, but seldom is, because it is seldom taught.

Field Marshal Earl Wavell, *Soldiers and Soldiering*, 1953

PERSEVERANCE

To persevere, trusting in what hopes he has, is courage in a man. The coward despairs.

Euripides, *Heracles*, c. 422 BC

The merit of the action lies in finishing it to the end.

Genghis Khan (AD 1162–1227)

There must be a beginning of any great matter, but the continuing unto the end until it be thoroughly finished yields the true glory.

Sir Francis Drake, 17 May 1587, letter to Sir Francis Walsingham

---

We fight, get beat, rise, and fight again.

Major General Nathaniel Greene, 22 June 1781, on the campaign in the Carolinas

---

If courage is the first characteristic of the soldier, perseverance is the second.

Napoleon, *Correspondance*, No. 4855, Vol. VI, 1858–1870

. . . a general in time of war is constantly bombarded by reports both true and false; by errors arising from fear or negligence or hastiness; by disobedience born of right or wrong interpretations, of ill will, of a proper mistaken sense of duty, of laziness, or of exhaustion; by accidents that nobody could have foreseen. In short, he is exposed to countless impressions, most of them disturbing, few of them encouraging. Long experience of war creates a knack of rapidly assessing these phenomena; courage and strength of character are as impervious to them as a rock to the rippling waves. If man were to yield to these pressure, he would never complete an operation. *Perseverance* in the chosen course is the essential counterweight, provided that no compelling reasons intervene to the contrary. Moreover, there is hardly a worthwhile enterprise in war whose execution does not call for infinite effort, trouble, and privation; and as man under pressure tends to give in to physical and intellectual weakness, only great strength of will can lead to the objective. It is steadfastness that will earn the admiration of the world and posterity.

Major General Carl von Clausewitz, *On War*, iii, 1832, tr. Howard and Paret

---

I propose to fight it out on this line, if it takes all summer.

General of the Army Ulysses S. Grant, 11 May 1864, dispatch from Spotsyvlania Courthouse

---

See also: DETERMINATION, DO-OR-DIE

PERSONAL PRESENCE OF THE COMMANDER

After haranguing the Tenth Legion, Caesar started for the right wing where he saw his men under great pressure. The standards of the Twelfth were huddled in one place and the soldiers so cramped that their fighting was hampered. All the centurions of the fourth cohort were cut down, the standard bearer killed and standard lost, and almost all the centurions of the other cohorts were killed or wounded . . . the enemy did not remit their pressure . . . and were pressing in from either flank. Caesar saw that the situation was critical, and there was no reserve to throw in. He snatched a shield from a soldier in the rear . . . and moved to the front line; he called upon the centurions by name, encouraged the men to advance, and directed them to open their lines out to give freer play with their swords. His coming inspired the men with hope and gave them new heart. Even in a desperate situation each man was anxious to do his utmost when his general was looking on, and the enemy's onset was somewhat slowed down.

Julius Caesar, 57 BC, when his camp was surprised by the Nervii, *The Gallic War, c.* 51 BC, tr. Hadas

---

The general should fight cautiously rather than boldly, or should keep away altogether from a a hand-to-hand fight with the enemy. For even if in battle he shows that he is not to be outdone in valour, he can aid his army far less by fighting that he can harm it if he should be killed, since the knowledge of a general is far more important than his physical strength. Even a soldier can perform a great deed by bravery, but no one except the general can by his wisdom plan a greater one.

Onasander, *The General*, AD 58

The real reason why I succeeded . . . is because I was always on the spot. I saw everything and did everything myself.

The Duke of Wellington (1769–1852)

By God! I don't think it would have been done if I had not been there.

The Duke of Wellington (1769–1852)

. . . he [the general] must know perfectly the strength and quality of each part of his own army, as well as that of his opponent, and must be where he can personally see and observe with his own eyes, and judge with his own mind. No man can properly command an army from the rear, he must be 'at its front', and when a detachment is made, the commander thereof should be informed of the object to be accomplished, and left as free as possible to execute it in his own way; and when an Army is divided into several parts, the superior should always attend to that one which he regards as the most important.

General of the Army William T. Sherman, *Memoirs of General W. T. Sherman*, 1875

. . . Some men think that modern armies may be so regulated that a general can sit in an office and play on his several columns as on the keys of a piano; this is a fearful mistake. The directing mind must be at the very head of the army – must be seen there, and the effect of his mind and personal energy must be felt by every officer and man present with it, to secure the best results. Every attempt to make war easy and safe will result in humiliation and disaster.

General of the Army William T. Sherman, *The Memoirs of General W. T. Sherman*, 1875

When the battle becomes hot, they must see their commander, know him to be near. It does not matter even if he is without initiative, incapable of giving an order. His presence creates a belief that direction exists, and that is enough.

Colonel Charles Ardnant du Picq, *Battle Studies*, 1880

Gallwitz's army surged out against the Narew on both sides of Przasnysz. For this attack I went personally to the battlefield, not with any idea of interfering with the tactics of the

Army Headquarters Staff, which I knew to be masterly, but only because I knew what oustanding importance Main Headquarters attached to the success of the break-through they had ordered at this point. I wanted to be on the spot so that in case of need I could intervene immediately if the Army Headquarters Staff needed any further help for the execution of its difficult task from the armies under my command.

Field Marshal Paul von Hindenburg, *Out of My Life*, 1920

Perhaps the most damning comment on the plan which plunged the British Army in this bath of mud and blood [Third Battle of Ypres, 1917] is contained in an incidental revelation of the remorse of one who was largely responsible for it. The highly placed officer from General Headquarters was on his first visit to the battle front – at the end of a four months' battle. Growing increasingly uneasy as the car approached the swamp-like edges of the battle area, he eventually burst into tears, crying 'Good God, did we really send men to fight in that?' To which his companion replied that the ground was far worse ahead. If the exclamation was a credit to his heart it revealed on what a foundation of delusion and inexcusable ignorance his indomitable 'offensiveness' had been based.

Captain Sir Basil Liddell Hart, speaking of Lieutenant General Sir Lancelot Kiggell, *The Real War, 1914–1918*, 1930

. . . the most rapid way to shell-shock an army is to shell-proof its generals; for once the heart of an army is severed from its head the result is paralysis. The modern system of command has in fact guillotined generalship, hence modern battles have degenerated into saurian writhings between headless monsters.

Major General J. F. C. Fuller, *Generalship: Its Diseases and Their Cure*, 1933

What troops and subordinate commanders appreciate is that a general should be constantly in personal conduct with them, and should not see everything simply through the eyes of his staff. The less time a general spends in his office and the more with his troops the better.

Field Marshal Earl Wavell, *Generals and Generalship*, 1939

The day of a commander in the field is one of constant vigilance. He must continually be getting about to see what is actually happening and be sure that his subordinates are in fact doing what is required of them in the best possible way. This getting about not only keeps the commander informed by first hand knowledge, but the confidence of the troops is raised when they constantly see him amongst themselves. No walk I ever did in New Guinea failed to pay a dividend, either by seeing or improving something or by maintaining the morale of the soldiers by having a chat with those I met.

Major General George Vasey (1895–1945), quoted in Horner, *The Commanders*, 1984

A study of the map will indicate where critical situations exist or are apt to develop, and so indicate where the commander should be.

General George S. Patton, Jr., *War As I Knew It*, 1947

In the rear areas the commander, high and low, wins the hearts of men primarily through a zealous interest in their general welfare. This is the true basis of his prestige and the qualifying test placed upon his soldierly abilities by those who serve under him. But at the front he commands their respect as it becomes proved to them that he understands their tactical problem and will do all possible to help them solve it.

Brigadier General S. L. A. Marshall, *Men Against Fire*, 1947

The place of all commanders of armour up to the divisional commanders is *on* the battlefield, and within this wherever they have the best view of the terrain and good communications with the hard core of the tanks. I was always located where I could see and hear what was going on 'in front', that is near the enemy and around myself – namely at the focal point! Nothing and nobody can replace a personal impression.

Major General Hasso von Manteuffel, *c.* 1951, unpublished manuscript, quoted in Simpkin, *Tank Warfare*, 1979

As the situation was rather confused I spent the next day at the front again. It is of the utmost importance to the commander to have a good knowledge of the battlefield and of his own and his enemy's positions on the ground. It is often not a question of which of the opposing commanders is the higher qualified mentally, or which has the greater experience, but which of them has the better grasp of the battlefield. This is particularly the case when a situation develops, the outcome of which cannot be estimated. Then the commander must go up to see for himself; reports received second-hand rarely give the information he needs for his decisions.

Field Marshal Erwin Rommel, *The Rommel Papers*, 1953

The commander must be the prime mover of the battle and the troops must always have to reckon with his appearance in personal control.

Field Marshal Erwin Rommel, *The Rommel Papers*, 1953

In moments of panic, fatigue, or disorganization, or when something out of the ordinary has to be demanded . . . the personal example of the commander works wonders, especially if he has the wit to create some sort of legend around himself.

Field Marshal Erwin Rommel, *The Rommel Papers*, 1953

I felt utterly at sea . . . My predecessor, von Rundstedt, rightly regarded himself as the heir to the tradition of the Supreme Command in World War One . . . With his finger on the pulse of things, he issued orders from his headquarters almost never visiting the front and rarely using the telephone . . . Even if my ways were different I could still understand von Rundstedt's though I could not persuade myself to adopt them . . . The laxity of discipline everywhere required personal contact with the commanders and troops . . . One had to have a glimpse behind the scenes and into men's hearts.

Field Marshal Albert Kesselring, *The Memoirs of Field Marshal Kesselring*, 1953

One of the most valuable qualities of a commander is a flair for putting himself in the right place at the vital time.

Field Marshal Viscount Slim, *Unofficial History*, 1957

As commander of a division or smaller unit, there will rarely be more than one crisis, one really critical situation facing you at any one time. The commander belongs right at the spot, not at some rear command post. He should be there before the crisis erupts, if possible. If it is not possible, then he should get there as soon as he can after it develops. Once there, then by personal observation of terrain, enemy fire, reactions, and attitudes of his own commanders on the spot – by his eyes, ears, brain, nose, and his sixth sense – he gets the best possible picture of what is happening and can best exercise his troop leadership and the full authority of his command. He can start help of every kind to his hard-pressed subordinates. He can urge commanders to provide additional fire support, artillery, air, and other infantry weapons . . .

No other means will provide the commander with what his personal perceptions can provide, if he is present at the critical time and place. He can personally intervene, if he thinks that necessary, but only to the extent that such intervention will be helpful and not interfere with his subordinates. He is in a position to make instant decisions, to defend, withdraw, attack, exploit, or pursue.

If, at this time, he is at some rear command post, he will have to rely on reports from others, and time will be lost, perhaps just those precious moments which spell the difference between success and failure. Notwithstanding the console capabilities of future television in combat, I still believe what I have said is true. In any event, keep this time factor in mind. It is the one irretrievable, inextensible, priceless element in war.

General Mathew B. Ridgway, 'Leadership', *Military Review*, 9/1966

From the practice of the first operations I concluded that those commanders failed most often who did not visit the terrain, where action was to take place, themselves only studied it on the map and issued written orders. The commanders who are to carry out combat missions must by all means know the terrain and enemy battle formations very well in order to take advantage of weak points in his dispositions and direct the main blow there.

Marshal of the Soviet Union Georgi K. Zhukov, *Reminiscences and Reflections*, 1974

THE PETER PRINCIPLE

I have seen very good colonels become very bad generals. I have known others who were great takers of villages, excellent for manoeuvres within an army, but who, outside of that, were not even able to lead a thousand men in war, who lost their heads completely and were unable to take any decision.

If such a man arrives to the command of an army, he will seek to save himself by his dispositions, because he has no other resources. In attempting to make them understood better he will confuse the spirit of his whole army with multitudinous messages. Since the least circumstance changes everything in war, he will want to change his arrangements, will throw everything in horrible confusion, and infallibly will be defeated.

Field Marshal Maurice Comte de Saxe, *My Reveries*, 1732

At my suggestion, he had been appointed to the command of the XII Army Corps, but in this post he already showed himself only a second-rate General lacking in determination. Apart from this, he attempted to do everything himself and would not trust any of his subordinates; and as he could not be everywhere at once on his vast front he left a great deal undone. In the smallish body of cavalry which he commanded for a long time, everyone knew him, loved him, and trusted him, and he was a success at his work. But with an increased number of troops and a large establishment of officers, his mistrustfulness, his air of gloom, and his silence prevented his men from loving him, trusting him, and understanding him any more than he understood them. Events proved clearly that his case was only another example of the familiar truth that every man has his limitations, which are fixed not only by his intelligence and his learning, but to a great extent by the particular bent of his personality.

General Aleksei A. Brusilov, *A Soldier's Notebook*, 1931

Every officer has his 'ceiling' in rank, beyond which he should not be allowed to rise – particularly in war-time. An officer may do well when serving under a first class superior. But how will he shape when he finds himself the boss? It is one thing to be merely an advisor, with no real responsibility; it is quite another thing when you are the top man, responsible for the final decision. A good battalion commander does not necessarily make a good brigadier, nor a good divisional general a good corps commander. The judging of a man's ceiling in the higher ranks is one of the great problems which a commander must solve, and it occupied much of my time.

Field Marshal Viscount Montgomery of Alamein, *The Memoirs of Field Marshal Montgomery*, 1958

See also: COMMAND SELECTION

PHYSICAL COURAGE

Courage in face of personal danger is also of two kinds. It may be indifference to danger, which could be due to the individual's constitution, or to his holding life cheap, or to habit. In any case, it must be regarded as a permanent *condition*. Alternatively, courage may result from such positive motives as ambition, patriotism, or enthusiasm of any kind. In that case courage is feeling, an emotion, not a permanent state.

These two kinds of courage act in different ways. The first is the more dependable; having become second nature, it will never fail. The other will often achieve more. There is more reliability in the first kind, more boldness in the second. The first leaves the mind calmer; the second tends to stimulate, but it can also be blind. *The highest form of courage is a compound of both.*

Major General Carl von Clausewitz, *On War*, i, 1832, tr. Howard and Paret

---

. . . As to physical courage, although sheer cowardice (i.e., a man thinking of his own miserable carcass when he ought to be thinking of his men) is fatal, yet, on the other hand, a reputation for not knowing fear does not help an officer in his war discipline: in getting his company to follow him as the Artillery of the Guard followed Drouot at Wagram. I noticed this first in Afghanistan in 1879 and have often since made the same observation. If a British officer wishes to make his men shy of taking a lead from him let him stand up under fire whilst they lie in their trenches as did the Russians on the 17th of July at the battle of Motienling. Our fellows are not in the least impressed by such bravado. All they say is, 'This fellow is a fool. If he cares so little for his own life, how much less will he care for ours.'

General Sir Ian Hamilton, *The Soul and Body of an Army*, 1921

When it comes to combat, something new is added. Even if they have previously looked on him as a father and believed absolutely that being with him is their best assurance of successful survival, should he then show himself to be timid and too cautious about his own safety, he will lose hold of them no less absolutely. His lieutenant, who up till then under training conditions has been regarded as a mean creature or a sniveler, but on the field suddenly reveals himself as a man of high courage, can take moral leadership of the company away from him, and do it in one day.

On the field there is no substitute for courage, no other binding influence toward unity of action. Troops will excuse almost any stupidity; excessive timidity is simply unforgivable.

Brigadier General S. L. A. Marshall, *The Armed Forces Officer*, 1950

Complete cowards are almost non-existent. Another matter for astonishment is the large number of men and women in any group who will behave in an emergency with extreme gallantry. Who they will be you cannot tell unit they are tested. I long ago gave up trying to spot potential VCs by their looks, but, from experience, I should say that those who perform individual acts of the highest physical courage are usually drawn from two categories. Either those with quick intelligence and vivid imagination or those without imagination and with minds fixed on the practical business of living. You might almost say, I suppose, those who live on their nerves and those who have not got any nerves. The one suddenly sees the crisis, his imagination flashes the opportunity and he acts. The other meets the situation without finding it so very unusual and deals with it in a matter of fact way.

Field Marshal Viscount Slim, *Courage and Other Broadcasts*, 1957

See also: BRAVERY, COURAGE, MORAL COURAGE

PLANNING/PLANS

The execution of an enterprise is never equal to the conception of it in the confident mind of its promoter; for men are safe while they are forming plans but when the time of action comes, then they lose their presence of mind and fail.

Thucydides, *The Peloponnesian War*, c. 460 BC, tr. Jowett

Plan what you have to do at night and carry out your decisions during the day. One cannot plan and act at the same time.

The Emperor Maurice, *The Strategikon*, c. AD 600

Long and careful deliberation promises great safety in war, whereas hasty and impetuous generals usually commit serious blunders.

The Emperor Maurice, *The Strategikon*, c. AD 600

One should study the possible courses in the light of the obtacles that have to be overcome, of the inconveniences or advantages that will result from the success of each branch, and, after taking account of the more likely objections, decide on the part which can lead to the greatest advantages, while employing diversions and all else that one can do to mislead the enemy and make him imagine that the main effort is coming at some other part. And in case all these diversions, countermarches or other ruses fail of their purpose – to hide the real aim – one must be ready to profit by a second or third branch of the plan without giving one's enemy time to consider it.

General Pierre de Bourcet, *Principes de la guerre de montagnes*, 1775

In war nothing is achieved except by calculation. Everything that is not soundly planned in its details yields no result.

Napoleon, 18 September 1806, to Joseph, *Correspondance*, No. 10809, Vol. XIII, 1858–1870

If I take so many precautions it is because it is my custom to leave nothing to chance.

Napoleon, 14 March 1808, to Marshal Murat, *Correspondance*, No. 13652, Vol. XVI, 1858–1870

I am used to thinking three or four months in advance about what I must do, and I calculate on the worst.

Napoleon, (1769–1821)

If I always appear prepared, it is because before entering on an undertaking, I have meditated for long and have foreseen what may occur. It is not genius which reveals to me suddenly and secretly what I should do in circumstances unexpected by others; it is thought and meditation.

Napoleon (1769–1821)

Nothing succeeds in war except in consequence of a well-prepared plan.

Napoleon (1769–1821)

Be audacious and cunning in your plans, firm and persevering in their execution, determined to find a glorious end.

Major General Carl von Clausewitz, *Principles of War*, 1812

MAXIM 2. In forming the plan of a campaign, it is requisite to foresee everything the enemy may do, and to be prepared with the necessary means to counteract it.

Plans of campaign may be modified *ad infinitum* according to circumstances, the genius of the general, the character of the troops, and the features of the country.

Napoleon, *The Military Maxims of Napoleon*, 1831, tr. D'Aguilar

---

MAXIM 79. The first principle of a general-in-chief is to calculate what he must do, to see if he has all the means to surmount the obstacles with which the enemy can oppose him, and when he has made his decision, to do everything to overcome them.

Napoleon, *The Military Maxims of Napoleon*, 1827, ed. Burnod

---

War plans cover every aspect of a war, and weave them all into a single operation that must have a single, ultimate objective in which all particular aims are reconciled.

Major General Carl von Clausewitz, *On War*, viii, 1832, tr. Howard and Paret

No operations plan will ever extend with any sort of certainty beyond the first encounter with the hostile main force. Only the laymen believes to perceive in the development of any campaign a consistent execution of a preconceived original plan that has been thought out in all its details and adhered to to the very end.

Field Marshal Helmuth Graf von Moltke, (1800–1891), quoted in von Freytag-Loringhoven, *Generalship in the World War,* 1920

Everything comes to this: to be able to recognize the changed situation and order the foreseeable course and prepare it energetically.

Field Marshal Helmuth Graf von Moltke, *Ausgewaehlte Werke*, Vol I, 1925

I never plan beyond the first battle.

Field Marshal Helmuth Graf von Moltke, *Ausgewaehlte Werke* Vol IV, 1925

[After actual operations have begun] our will soon meets the independent will of the enemy. To be sure, we can limit the enemy's will if we are ready and determined to take the initiative, but we cannot break it by any other means than tactics, in other words, through battle. The material and moral consequences of any larger encounter are, however, so far-reaching that through them a completely different situation is created, which then becomes the basis for new measures. No plan of operations can look with any certainty beyond the first meeting with the major forces of the enemy . . . The commander is compelled during the whole campaign to reach decisions on the basis of situations which cannot be predicted. All consecutive acts of war are, therefore, not executions of a premeditated plan, but spontaneous actions, directed by military tact.

Field Marshal Helmuth Graf von Motlke (1800–1891), quoted in ed., Earle, *The Makers of Modern Strategy,* 1943

It is a delusion, when one believes that one can plan an entire campaign and carry out its planned end . . . The first battle will determine a new situation through which much of the original plan becomes inapplicable.

Field Marshal Helmuth Graf von Moltke, *Militarische Werke,* vol IV, 1892–1212

Planning is everything – Plans are nothing.

Field Marshal Helmuth Graf von Moltke (1880–1891) Sign posted above the entrance to the Joint Staff, Department of Defense, the Pentagon

---

The main thing is always to have a plan; if it is not the best plan, it is at least better than no plan at all.

General Sir John Monash, 1918, letter

The stroke of genius that turns the fate of a battle? I don't believe in it. A battle is a complicated operation, that you prepare laboriously. If the enemy does this, you say to yourself I will do that. If such and such happens, these are the steps I shall take to meet it. You think out every possible development and decide on the way to deal with the situation created. One of these developments occur; you put your plan in operation, and everyone says, 'What genius . . .' whereas the credit is really due to the labour of preparation.

Marshal of France Ferdinand Foch, April 1919, interview

Once more I spent much time at the front and with the various Army Headquarters in active interchange of ideas on the tactics of offensive fighting, and on the attack itself. Many proposals and counter-proposals, many pros and cons were laid before me . . . As was my duty, I had to give the ultimate decision. The tactical principles were considered to be correct and readily accepted by the troops. They left room for individual action in all directions.

General Erich Ludendorff, *My War Memories 1914–1918*, 1919

Now, as every policy must be plastic enough to admit of fluctuations in national conditions, so must each plan be plastic enough to receive the impressions of war, that is power to change its shape without changing or cracking its substance. This plasticity is determined psychologically by the condition of mentality in the two opposing forces. There is the determination between the commanders-in-chief, and between them and their men, and ultimately, between the two forces themselves . . .

Major General J. F. C. Fuller, *The Reformation of War,* 1923

When making a plan, try to put yourself in the enemy's mind, and think what course it is least probable he will foresee and forestall. The surest way to success in war is to choose *the course of least expectation.* (May 1930)

Captain Sir Basil Liddell Hart, *Thoughts on War,* 1944

There is a close analogy between what takes place in the mind of a military commander when planning an action, and what happens to the artist at the moment of conception. The latter does not renounce the use of his intelligence. He draws from it lessons, methods, and knowledge. But his power of creation can operate only if he possesses, in addition, a certain instinctive faculty which we call inspiration, for that alone can give the direct contact with nature from which the vital spark must leap. We can say of the military art what Bacon said of the other arts: 'They are the product of man added to nature'.

Charles de Gaulle, *The Edge of the Sword,* 1932

A plan, like a tree, must have branches – if it is to bear fruit. A plan with a single aim is apt to prove a barren pole. (January 1933)

Captain Sir Basil Liddell Hart, *Thoughts on War,* 1944

Let us now discuss the question of planning. Because of the uncertainty peculiar to war, it is much more difficult to prosecute war according to plan than is the case with other activities. Yet, since 'preparedness ensures success and unpreparedness spells failure', there can be no victory in war without advance planning and preparations. There is no absolute certainty in war, and yet it is not without some degree of relative certainty. We are comparatively certain about our own situation. We are very uncertain about the enemy's, but here too there are signs for us to read, clues to follow and sequences of phenomena to ponder. These form what we call a degree of relative certainty, which provides an objective basis for planning in war.

Mao Tse-tung, *On Protracted War,* May 1938

It may be of interest to future generals to realize that one makes plans to fit circumstances and does not try to create circumstances to fit plans. That way lies danger.

General George S. Patton, Jr., 26 February 1945, diary entry, *The Patton Papers,* Vol. II, 1974

A good plan violently executed *Now* is better than a perfect plan next week.

General George S. Patton, Jr., *War As I Knew It,* 1947

The problem that faces strategical planning (and also tactical planning, though in a more restricted sense) is twofold. We must, on the one hand, strive by all means to prevent the enemy from acting on sound principles; on the other hand, a supreme planning effort must be made to enable our forces to exploit those principles, in order to facilitate the achievement of our aims and objectives. For this purpose every principle which the enemy is likely to apply must serve as a target for the ingenuity of those who plan the operation of our forces.

General Yigael Yadin, article in *The Israeli Forces' Journal,* 9/1949, quoted in Liddell Hart, *Strategy,* 1954

To be practical, any plan must take account of the enemy's power to frustrate it; the best chance of overcoming such obstruction is to have a plan that can easily be varied to fit the circumstances met; to keep such adaptability, while still keeping the initiative, the best way is to operate along a line which offers alternative objectives. For thereby you put your opponent on the horns of a dilemma, which goes far to assure the gaining of at least one objective – whichever is least guarded – and may enable you to gain one after the other.

Captain Sir Basil Liddell Hart, *Strategy*, 1954

On the operational side a C-in-C must draw up a master plan for the campaign he envisages and he must always think and plan two battles ahead – the one he is preparing to fight *and* the next one – so that success gained in one battle can be used as a spring-board for the next. He has got to read the mind of his opponent, to anticipate enemy reactions to his moves, and to take quick steps to prevent enemy interference with his own plans. He has got to be a very clear thinker and able to sort out the essentials from the mass of factors which bear on every problem.

Field Marshal Viscount Montgomery of Alamein, *The Memoirs of Field Marshal Montgomery*, 1958

The plan of operations must always be made by the commander and must not be forced on him by his staff, or by circumstances, or by the enemy. He has got to relate what is strategically desirable with that which is tactically possible with the forces at his disposal; if this is not done he is unlikely to win. What is possible, given a bit of luck? And what is definitely impossible? That is always the problem. The plan having been made, there will be much detailed work to be done before the operation is launched; this detailed work must be done by the staff. The commander himself must stand back and have time to think: his attention must be directed to ensuring that the basic foundations and corner-stones of the master plan are not broken down by the mass of detail which will necessarily occupy the attention of the staff. If all these things are to be done successfully, a good Chief of Staff is essential.

Field Marshal Viscount Montgomery of Alamein, *The Memoirs of Field Marshal Montgomery*, 1958

The commander must decide how he will fight the battle *before it begins*. He must then decide how he will use the military effort at his disposal to force the battle to swing the way he wishes it to go; he must make the enemy dance to his tune from the beginning, and never vice versa. To be able to do this, his own dispositions must be so balanced that he can utilise but need not react to the enemy's move but can continue relentlessly with his own plan. The question of 'balance' was a definite feature of my military creed.

Field Marshal Viscount Montgomery of Alamein, *The Memoirs of Field Marshal Montgomery*, 1958

But in truth, the larger the command, the more time must go into planning; the longer it will take to move troops into position, to reconnoiter, to accumulate ammunition and other supplies, and to coordinate other participating elements on the ground and in the air. To a conscientious commander, time is the most vital factor in his planning. By proper foresight and correct preliminary action, he knows he can conserve the most precious elements he controls, the lives of his men. So he thinks ahead as far as he can. He keeps his tactical plan simple. He tries to eliminate as many variable factors as he is able. He has a firsthand look at as much of the ground as circumstances render accessible to him. He checks each task in the plan with the man to whom he intends to assign it. Then – having secured in almost every instance his subordinates' wholehearted acceptance of the contemplated mission and agreement on its feasibility – only then does he issue an order.

General Mathew B. Ridgway, *The Korean War*, 1967

Naturally, in the course of a battle, one would like to fulfill the initial plan . . . – but what does it mean to plan in war? We plan alone, but we fulfill our plans, if one may do so, together with the enemy, that is, taking account of his counteraction.

Marshal of the Soviet Union Ivan S. Konev (1897–1793), 1972, quoted in Leites, *The Soviet Style in War*, 1982

I grew big with these plans. I had merely to cross a river, capture Brussels and then go on and take the port of Antwerp. And all this in the worst months of the year, December, January, February, through the countryside where snow was waist-deep and there wasn't room to deploy four tanks abreast, let alone six armoured divisions; when it didn't get light until eight in the morning and was dark again at four in the afternoon; with divisions that had just been reformed and contained chiefly raw, untried recruits; and at Christmas time.

General Sepp Dietrich, his sarcastic opinion of the German planning for the Battle of the Bulge, in Messenger, *Hitler's Gladiator*, 1987

I emphasized meticulous planning not simply because I thought it was the most effective approach, which it is, but because by taking that approach you enforce on your subordinates the same necessity. They have to learn every detail of the topography, every position, every soldier they will be facing. And once they do that, they will be able to decide rationally – not intuitively – on the steps they will have to take. They will make their decisions on the basis of knowledge. Experience had also taught me that if you lay your plans in detail before you are under the stress of fighting, the chances are much greater that you will be able to implement at least the outlines of the plans despite the contingencies of battle.

General Ariel Sharon, *Warrior*, 1989

The plan was smooth on paper, only they forgot about the ravines.

Russian military proverb

PRAISE

Praise from a friend, or censure from a foe,
Are lost on hearers that our merits know.

Homer, *The Iliad*, c. 800 BC, tr. Pope

Mankind will tolerate the praises of others as long as each hearer thinks he can do as well or nearly well himself.

Thucydides, *History of the Peloponnesian War*, c. 460 BC

The most pleasing of all sounds that of your own praise.

Xenophon (c. 431 BC–c. 352 BC), *Hiero*

When a general gives a public speech he ought also to say something in praise of the enemy. This will convince our men, even when you are praising others, that you will never deprive us of the praise we might receive from others and adorn them with our honours.

The Emperor Maurice, *The Strategikon*, c. AD 600

A refusal of praise is a desire to be praised twice.

There are reproaches which praise, and praises which defame.

Generally we praise only to be praised.

François Duc de La Rochefoucauld, *Maxims*, 1665

The desire to imitate brave actions will be aroused by praise. And these trifles will diffuse a spirit of emulation among troops which affects both officers and soldiers and in time will make them invincible.

Field Marshal Maurice Comte de Saxe, *My Reveries*, 1732

The general can even discuss the war with those of his corps commanders who are the most intelligent, and permit them to express their sentiments freely in conversation. If you find some good among what they say, you should not remark about it then, but make use of it. When this has been done, you should speak about it in the presence of many others, it was so-and-so who had this idea; praise him for it. This modesty will gain the general the friendship of thinking men, and he will more easily find persons who will speak their sentiments sincerely to him.

Frederick the Great, *Instructions to His Generals*, 1747

Praise from enemies is suspicious; it cannot flatter an honourable man unless it is given after the cessation of hostilities.

Napoleon, *The Military Maxims of Napoleon*, 1827, ed. Burnod

---

And don't forget a good word for the cooks.

Lieutenant General Sir Leslie Morshead (1889–1959), at a commanders' conference after the Battle of El Alamein, quoted in Horner, *The Commanders*, 1984

Humility must always be the portion of any man who receives acclaim earned in the blood of his followers and the sacrifices of his friends.

General of the Army Dwight D. Eisenhower, 12 June 1945, address in London

All a soldier desires to drive him forward is recognition and appreciation of his work.

General George S. Patton, Jr. (1885–1947)

To honour or otherwise enhance the prestige of junior officers or men in secondary positions of command – when this is their due – is to the army's benefit. This rule is not, however, to be applied solely to commanders. The whole unit deserves a word of appreciation or praise no less, sometimes more, than highly-placed or outstanding individuals. Yet since military activity is not an end in itself but rather an imposition which is to be thrown off as soon as conditions permit, the myth of military valour must be played down and a constant attempt made to give a rational account of the army's triumphs. In this way experience may be more realistically instructive and the soldiers educated towards the healthier and more relevant approach to their future in the army, an approach devoid of the complex superiority, self-importance and thoughtlessness so common in some armies.

General Yigal Allon, *The Making of Israel's Army*, 1960

You must get around and show interest in what your subordinates are doing even if you don't know much about the technique of their work. And when you are making these visits, try to pass out praise then due, as well as corrections and criticisms.

General of the Army Omar N. Bradley, 16 May 1967, speech to the U.S. Army Command and General Staff College

---

A general must never be chary in allotting praise where it is due. People like to be praised when they have done well. In this connection Sir Winston Churchill once told me of the reply made by the Duke of Wellington, in his last years, when a friend asked him: 'If you had your life over again, is there any way in which you could have done better?' The old Duke replied: 'Yes, I should have given more praise.'

Field Marshal Viscount Montgomery of Alamein, *A History of Warfare*, 1968

In command and leadership many qualities, attributes and techniques are required – including drive, force, judgment, perception and others. But nothing can replace the inspiration and lift that comes from commending a job well done.

Major General Aubrey 'Red' Newman, *Follow Me*, 1981

PRAYERS/RELIGION

If thou wishest the gods to be propitious to thee, thou must honour the gods.

Xenophon (c. 431 BC–c. 352 BC), *Memorabilia*

Before getting into danger, the general should worship God. When he goes into danger, then, he can with confidence pray to God as a friend.

The Emperor Maurice, *The Strategikon*, c. AD 600

Dear Lord, I pray Thee to suffer me not to see Thy Holy City, since I cannot deliver it from the hands of Thy enemies.

King Richard I, 'The Lion Heart', King of England, 1192, on seeing Jerusalem, the object of his Crusade, that he was never to conquer

---

I think that when God grants me victory over the rest of Palestine I shall divide my territories, make a will stating my wishes, then set sail on this sea for their far-off lands and pursue the Franks there, so as to free the earth of anyone who does not believe in God, or die in the attempt.

Saladin (1138–93), quoted by Baha Ad-Din in Gabarieli, *Arab Historians of the Crusades*, 1957

It will give me great pleasure to fight for my God against your gods, who are a mere nothing.

Hernan Cortes, 1521, to the Aztec priesthood, quoted in *Five Letters*, 1522–5

O Lord God, when Thou givest to they servants to endeavour any great matter, grant us also to know that it is not the beginning, but the continuing of the same until it is thoroughly finished which yieldeth the true glory.

Sir Francis Drake (c. 1540–1596)

Praying hard is fighting hard.

Gustavus II Adolphus (1594–1632), King of Sweden

You may win salvation under my command, but hardly riches.

Gustavus II Adolphus (1594–1632), King of Sweden

Oh Lord! Thou knowest how busy I must be this day: if I forget Thee, do not Thou forget me. March on, boys!

Sir Jacob Astley, 1642, before the battle of Edgehill

I can say this of Naseby that when I saw the enemy drawn up and march in gallant order towards us, and we a company of poor ignorant men, to seek how to order our battle, the General having commanded me to order all the horse, I could not, riding alone about my business, but smile out to God in praises in assurance of victory, because God would, by things that are not, bring to naught things that are. Of which I had great assurance – and God did it.

Oliver Cromwell, after the Battle of Naseby, 14 June 1644, quoted in Churchill, *A History of the English Speaking Peoples: The New World*, 1956

Truly I think he that prays and preaches best will fight best.

Oliver Cromwell, 25 December 1650, letter to Colonel Francis Hacker, quoted in Ashely, ed., *Cromwell*, 1969

Put your trust in God, but be sure to see that your powder is dry.

Oliver Cromwell (1599–1658), to his troops when they were about to cross a river

---

Oh God, let me not be disgraced in my old days. Or if Thou wilt not help me, do not help these scoundrels; but leave us to try it ourselves.

Leopold I of Anhalt-Dessau, 'The Old Dessauer', 14 December 1745, before the Battle of Kesselsdorff

---

May the Great God whom I worship grant to my Country, and for the benefit of Europe in general, a great and glorious victory; and may no misconduct in any way tarnish it; and may humanity after Victory be the predominant feature in the British Fleet. For myself individually, I commit my life to Him who made me, and may his blessing light upon my endeavours for serving my Country faithfully. To Him I resign myself and the just cause which is entrusted to me to defend. Amen. Amen. Amen.

Admiral Lord Nelson, 21 October 1805, diary prayer before the Battle of Trafalgar

---

O Lord, if Thou wilt not be for us today, we ask that Thou be not against us. Just leave it between the French and ourselves.

Lieutenant General Sir Alan Campbell of Erracht, before battle, *c.* 1809

Fear not. God has pledged His word for the liberty of Greece, and He will not retract. We have been baptized once with holy ointment; we shall again be baptized in blood for the freedom of the Fatherland!

Theodoros Kolokotrones, 1821, at the beginning of the Greek Revolution – Zeeto Ellas! *Theodoros Kolokotrones: Old Man of the Morea*, 1892

Captain, my religious belief teaches me to feel as safe in battle as in bed. God has fixed the time for my death. I do not concern myself about that, but to be always ready, no matter when it may overtake me. That is the way all men should live, and then all would be equally brave.

Lieutenant General Thomas 'Stonewall' Jackson, quoted in Henderson, *Stonewall Jackson*, 1898

Please God – let there be victory, before the Americans arrive.

Sir Douglas Earl Haig, 1917 diary entry (attributed)

Finally, knowing the vanity's of man's effort and the confusion of his purpose, let us pray that God may accept our services and direct our endeavours, so that when we shall have done all we shall see the fruits of our labours and be satisfied.

Major General Orde Wingate, February 1943, Order of the Day, Imphal

---

God of our Fathers, who by land and sea has ever led us on to victory, please continue Your inspiring guidance in this the greatest of our conflicts.

Strengthen my soul so that the weakening instinct of self-preservation, which besets all of us in battle, shall not blind me to my duty to my own manhood, to the glory of my calling, and to my responsibility to my fellow soldiers.

Grant to our armed forces that disciplined valor and mutual confidence which ensures success in war.

Let me not mourn for the men who have died fighting, but rather led me be glad that such heroes have lived.

If it be my lot to die, let me do so with courage and honor in a manner which will bring the greatest harm to the enemy, and please, oh, Lord, protect and guide those I shall leave behind.

Give us victory, Lord.

General George S. Patton, Jr. (1885–1947)

Almighty and most merciful Father, we humbly beseech Thee, of Thy great goodness, to restrain these immoderate rains with which we have to contend. Grant us fair weather for Battle. Graciously hearken to us as soldiers who call upon Thee that armed with Thy power, we may advance from victory to victory, and crush the oppression and wickedness of our enemies, and establish Thy justice among men and nations. Amen.

Patton's 'Weather Prayer', 24 December 1944, actually written at his command by Chaplain James O'Neill

---

Finally, I do not believe that today a commander can inspire great armies, or single units, or even individual men, and lead them to achieve great victories, unless he has a proper sense of religious truth. He must always keep his finger on the spiritual pulse of his armies; he must be sure that the spiritual purpose which inspires them is right and true, and is clearly expounded to one and all. Unless he does this, he can expect no lasting success. For all leadership, I believe, is based on the spiritual quality, the power to inspire others to follow; this spiritual quality may be for good, or evil. In many cases in the past this quality has been devoted towards personal ends, and was partly or wholly evil; whenever this was so, in the end it failed. Leadership which is evil, while it may temporarily succeed, always carries within itself the seeds of its own destruction.

Field Marshal Viscount Montgomery of Alamein, *The Memoirs of Field Marshall Montgomery*, 1958

PRISONERS OF WAR

When the war is concluded, I am definitely of the opinion that all animosity should be forgotten, and that all prisoners should be released.

The Duke of Wellington, 1804, letter to E. S. Warring

MAXIM 109. Prisoners of war do not belong to the power for which they have fought; they are all under the safeguard of honour and generosity of the nation that has disarmed them.

Napoleon, *The Military Maxims of Napoleon*, 1827, ed. Burnod

---

Losses incurred during the battle consist mostly of dead and wounded; after the battle, they are usually greater in terms of captured guns and prisoners. While the former are shared more or less evenly by winner and loser, the latter are not. For that reason they are usually found on one side, or at any rate in significant numbers on one side.

That is why guns and prisoners have always counted as the real trophies of victory: they are also its measure, for they are the tangible evidence of its scale. They are a better index to the degree of superior morale than any other factors, even when one relates them to the casualty figures . . .

Major General Carl von Clausewitz, *On War*, iv, 1832, tr. Howard and Paret

---

It's cheaper to feed them than to fight them.

Lieutenant General Thomas 'Stonewall' Jackson, 30 June 1862

Prisoner of war guard companies, or an equivalent organization, should be as far forward as possible in action to take over prisoners of war, because troops heated with battle are not safe custodians. Any attempt to rob or loot prisoners of war by escorts must be strictly dealt with.

General George S. Patton, Jr., *War As I Knew It*, 1947

A prisoner of war is a man who tries to kill you and fails, and then asks you not to kill him.

Sir Winston S. Churchill, 1952, quoted in the *Observer*

I saw a group of soldiers clustered around a prisoner; as I drove up, one of them was hitting the Egyptian. I court-martialled the soldier on the spot, sentencing him to thirty-five days in the stockade. It was the kind of thing that brought my anger to the boiling point. In battle you fight and you have to kill. That's the nature of it. But once a man is your prisoner you never touch him.

General Ariel Sharon, *Warrior*, 1989

PREPARATION/PREPAREDNESS

Now, it happened through this god, the lord of gods, that I was prepared and armed to trap them like wild fowl. He furnished my strength and caused my plans to prosper. I went forth, directing these marvellous things. I equipped my frontier in Zahi, prepared before them. The chiefs, the captains of infantry, the nobles, I caused to equip the harbour-mouths, like a strong wall, with warships, galleys and barges . . . They were manned completely from bow to stern with valiant warriors bearing their arms, soldiers of all the choicest of Egypt, being like lions roaring upon the mountain tops. The charioteers were warriors . . ., and all good officers . . . ready to crush the countries under their feet. I was valiant Montu, stationed before them, that they might behold the hand-to-hand fighting of my own arm. I, King Ramses III, was made a far-striding hero, conscious of his might, valiant to lead his army in the day of battle.

Ramses III, Pharaoh of Egypt, c. 1190 BC, the Northern War against Peoples of the Sea, quoted in Breasted, *Ancient Records of Egypt*, 1906

To rely on rustics and not prepare is the greatest of crimes; to be prepared beforehand for any contingency is the greatest of virtues.

Ho Yen-hsi, early commentator to Sun Tzu, *The Art of War*, c. 500 BC, tr. Griffith

---

He, therefore, who aspires to peace should prepare for war.

Flavius Vegetius Renatus, *Military Institutions of the Romans*, c. AD 378

The general should not go to sleep before reflecting on what he should have done that he might have neglected and on what he has to do the next day.

The Emperor Maurice, *The Strategikon*, c. AD 600

Sages are very careful not to forget about danger when secure, not to forget about chaos in times of order. Even when there is peace in the land, it will not do to abandon the military altogether. If you lack foresight, you will be defenceless. It is necessary to develop cultured qualities internally while organizing military preparedness externally. Be considerate and gentle with foreigners, beware of the unexpected. Routine military exercises in each of the four seasons is the way to show that the nation is not oblivious to warfare. Not forgetting about warfare means teaching the people not to give up the practice of martial arts.

The rule is 'Even if the land is at peace, to forget about warfare leads to collapse.'

Liu Ji (1310–1375), *Lessons of War*

There cannot be good laws where there are not good arms.

Niccolo Machiavelli, *The Prince*, 1513

The man who is prepared has his battle half fought.

Miguel de Cervantes, *Don Quixote*, 1615

Walled towns, stored arsenals and armouries, goodly races of horse, chariots of war, elephants, ordnance, artillery, and the like; all this is but a sheep in a lion's skin, except the breed and disposition of the people be stout and warlike. Nay, number itself in armies importeth not much, where the people is of weak courage; for as Virgil sayeth: 'It never troubles the wolf how many the sheep are.'

Francis Bacon (1561–1621), *Of the Greatness of Kingdoms and Estates*

The arts of peace and the arts of war are like two wheels of a cart which, lacking one, will have difficulty in standing. Naturally, the arts of peace are used during times of tranquillity and those of war during times of confusion, but it is most essential to not forget the military during peaceful times nor to disregard scholastics during time of war. When the master of a province feels that the world is in peace and forgets the arts of war, first, military tactics will fall into disuse, the warriors of his clan will naturally become effeminate and lose interest in martial ways, the martial arts will be neglected, the variety of weapons will be insufficient, weapons handed down through the generations will become rusty and rot, and there will be nothing of any use during times of emergency. If the Way of the Warrior is thus neglected, ordinary military tactics will not be established; if a military situation were to suddenly arise there would be panic and confusion, consultation would be unprepared for, and the establishment of strategy would be difficult. When one has been born into the house of military commander, he should not forget the arts of war even for a moment.

Kuroda Nagamasa (1568–1623), *Notes on Regulations*, 1622, in Wilson, *Ideals of the Samurai*, 1982

Fritz, pay close attention to what I am going to say to you. Always keep up a good and strong army – you won't have a better friend and you can't survive without it. Our Neighbours want nothing more than to bring about our ruin – I am aware of their intentions, and you will come to know them as well. Believe me, don't let wishful thinking run away with you – stick to what is real. Always put your trust in a good army and in hard cash – they are the things which keep rulers in peace and security.

Frederick William I, 1731, advice to his son, the future Frederick the Great

---

After a fatal procrastination, not only vigorous measures but of preparations for such, we took a step as decisive as the passage of the Rubicon, and now find ourselves plunged at once in most serious war without a single requisition, gunpowder excepted, for carrying it on.

Lieutenant General Sir John Burgoyne, April 1775, letter from Boston after the Battle of Lexington

---

I am persuaded and as fully convinced, as I am of any one fact that has happened that our Liberties must of necessity be greatly hazarded, if not entirely lost, if their defence is left to any but a permanent standing Army, I mean one to exist during the War.

General George Washington, 2 September 1776, letter to the President of Congress, *The Writings of George Washington*, Vol. 6, 1931–44

There is nothing so likely to produce peace as to be well prepared to meet an enemy.

General George Washington, 29 January 1780, letter to Elbridge Gerry

---

Had we kept a permanent Army on foot, the enemy would have had nothing to hope for, and would, in all probability, have listened to terms long since.

General George Washington, 10 August 1780, letter to the President of Congress, *The Writings of George Washington*, Vol. 10, 1931–44

That government is a murderer of its citizens which sends them to the field uninformed and untaught, where they are to meet men of the same age and strength mechanized by education and discipline for battle.

Major General Henry 'Light Horse Harry' Lee (1756–1818)

You have not done anything in as much as you still have something to do.

Napoleon (1769–1821)

The country must have a large and efficient army, one capable of meeting the enemy abroad, or they must expect to meet him at home.

The Duke of Wellington, 28 January 1811

Victory smiles in general upon those only who know how to command it by good preparations. It is seldom the effect of chance or of unexpected good luck, but the fruit and recompense culled by the experienced soldier, whose discernment is supported by the resolution and boldness of his undertakings.

Marshal of France Michel Ney, Duc d'Elchingen, Prince de la Moskova, *Memoirs of Marshal Ney*, 1834

It is particularly necessary to watch over the preservation of armies in the interval of a long peace, when they are most likely to degenerate. It is important to foster the military spirit in the armies and to exercise them in great manoeuvres which, though but faintly resembling those of actual war, still are of decided advantage in preparing them for war.

Lieutenant General Atoine-Henri Baron de Jomini, *Summary of the Arts of War*, 1838

What our sword has won in half a year, our sword must guard for half a century.

Field Marshal Helmuth Graf von Moltke (1800–1891), after the Franco–Prussian War, 1871

The real objective of having an army is to provide for war.

Secretary of War Elihu Root, December 1899, *Annual Report of the Secretary of War*

The nation that will insist on drawing a broad line of demarkation between the fighting man and the thinking man is liable to find its fighting done by fools and its thinking done by cowards.

Lieutenant General Sir William Butler (1838–1910)

It cannot be too often repeated that in modern war, and especially in modern naval war, the chief factor in achieving triumph is what has been done in the way of thorough preparation and training before the beginning of war.

President Theodore Roosevelt, June 1902, graduation address at the U.S. Naval Academy

The enemy opened fire at 2.08 and our First Division bore it for a few minutes and replied at about 2.11. The number of enemy shells fired during these few minutes exceeded 300 and the *Mikasa* was damaged and had casualties before she had fired a shot. About a half hour later the enemy's battle formation was entirely out of order, so that the fate of our empire was really settled within this first half an hour. The *Mikasa* and the eleven others of the main force had taken years of labour to design and build, and yet they were used for only half an hour of decisive battle. We, too, studied the art of war and trained ourselves in it, but it was put to use for only that short period. Though the decisive battle took such a short time, it required ten years of preparation.

Admiral Marquis Togo Heihachiro (1846–1934), speaking of the Battle of Tsushima and his flagship, the *Mikasa*, quoted in Warner, *The Tide at Sunrise*, 1974

DEFENSIVE POLICY of the U.S. – Does not alter PACIFIC POLICY of the U.S. Be as pacific as you please but do not let the other fellow catch you unprepared.

Lieutenant General Robert L. Bullard, 1911, journal entry

Much was attempted and much was done to supplement the lack of opportunity by demonstration, lectures, attachments. But by reason of the incomplete military education of our hastily-trained troops it was necessary to limit manoeuvre and tactics on the battlefield to the simplest elements. Anything in the nature of finesse had to be avoided. Skilful use of ground and mutual fire support were things hoped for more often than achieved.

It was a question of bulk production against time, but the results obtained only prove how much more could be achieved with the same material had conditions of training been those of peace time with its long service and rigorous and plentiful supervision.

Major General Hugh Elles, preface to Elles, *The Tank Corps*, 1919

So long as a country is not immediately threatened, public opinion will be strongly opposed to increasing the burden of armaments, and to accepting the need for additional manpower. Recruiting, whether in the form of conscription or the press gang, has always been regarded as an attack on personal liberty which arouses violent feelings in the populace. Only too often, the money spent on an army which is not engaged in actual combat is looked upon as sheer waste. The very idea of military discipline causes a muttering of discontent as being a sin against the spirit of independence. How can a government which cannot do without popular vote ignore such feelings? One of its duties is to produce a budget, and in this the army estimates make an ugly hole, besides which, rumours are sure to be disseminated of the anxiety felt abroad (as it always is) by the news of increased expenditure on armaments, though about these the government must keep silent. At the same time, as the guide and father of the people, it must, in the interests of good sense, proclaim its pacific intentions, and seem to be under no compulsion to forge the weapons of war. History contains no example of a conqueror who did not, in good faith, loudly declare that he wanted peace.

Naturally enough, military problems are anathema to the men in power. Nobody, whether a spendthrift or a miser, likes getting bills. Although, in the last analysis, armaments are the consequences of policy, governments are afraid to impose them until immediate danger makes them obviously essential. On the contrary, when a long period of peace is promised, they are the first to put ships out of commission and disband regiments.

Charles de Gaulle, *The Edge of the Sword*, 1932

To maintain in peace a needlessly elaborate military establishment entails economic waste. But there can been no compromise with minimum requirements. Second best is to be defeated, and military defeat carries with it national disaster – political, economic, social, and spiritual disaster.

General of the Army Douglas MacArthur, *Annual Report of the Chief of Staff, June 30, 1935*

Without preparedness superiority is not real superiority and there can be no initiative either. Having grasped this point, a force which is inferior but prepared can often defeat a superior enemy by surprise attack.

Mao Tse-tung, *On Protracted War*, May 1938

Virtuous motives, trammelled by inertia and timidity, are no match for armed and resolute wickedness. A sincere love of peace is no excuse for muddling hundreds of millions of humble folk into total war. The cheers of weak, well-meaning assemblies soon cease to echo, and their votes soon cease to count. Doom marches on.

Sir Winston S. Churchill, *The Gathering Storm*, 1948

No foreign policy can have validity if there is no adequate force behind it and no national readiness to make the necessary sacrifices to produce that force.

Sir Winston S. Churchill, *The Gathering Storm*, 1948

The lack of manpower became more and more desperate during the war, and the almost untrained troops who were brought up to fill the gaps in the ranks suffered disproportionate heavy losses. The economies achieved in time of peace had now to be paid for in blood. The officer cadres of peacetime were insufficient and largely used up after years of exhausting work. Both officers and non-commissioned officers of the reserve carried out their duties with the greatest devotion, but they could naturally not take the place of regular cadres. The war trained them, but the price was too high.

Field Marshal Carl Gustav Baron von Mannerheim, *The Memories of Marshal Mannerheim*, 1953

Generals have often been reproached with preparing for the last war instead of for the next – an easy gibe when their fellow countrymen and their political leaders, too frequently, have prepared for no war at all. Preparation for war is an expensive, burdensome business, yet there is one important part of it that costs little – study. However changed and strange the new conditions of war may be, not only generals, but politicians and ordinary citizens, may find there is much to be learned from the past that can be applied to the future and, in their search for it, that some campaigns have more than others foreshadowed the coming modern war. I believe that ours in Burma was one of these.

Field Marshal Slim, *Defeat Into Victory*, 1963

It is customary in the democratic countries to deplore expenditures on armaments as conflicting with the requirements of social services. There is a tendency to forget that the most important social service a government can do for its people is to keep them alive and free.

Air Marshal Sir John Slessor (1897–1979)

The most difficult military problem to resolve is that of establishing a security system, as inexpensively as possible in time of peace, capable of transforming itself very rapidly into a powerful force in case of the danger of aggression.

General Andre Beaufre, *Strategy for Tomorrow*, 1974

If the United States is ever defeated in a war, The Congress and the people will not be impressed by the services' education and social rehabilitation programs.

Lieutenant General Arthur S. Colllins, Jr., *Common Sense Training*, 1978

THE PRESS

. . . four hostile newspapers were more to be feared than a thousand bayonets.

Napoleon, quoted in Liddell Hart, *The Sword and the Pen*, 1976

So you think the papers ought to say more about your husband! My brigade is not a brigde of newspaper correspondents.

Lieutenant Thomas 'Stonewall' Jackson, July 1861, letter to his wife after the Battle of First Manassas

---

I have made arrangements for the correspondents to take the field . . . and I have suggested to them that they wear a white uniform to indicate the purity of their profession.

Major General Irvin McDowell, quoted in Russell, *My Diary North and South*, 1863

I will never command an army in America if we must carry along paid spies. I will banish myself to some foreign country first.

General of the Army William T. Sherman, February 1863, letter to his wife referring to war correspondents who had been attacking him in the Press

---

Those newly invented curse to armies . . . that race of drones who are an emcumbrance to an army; they eat the rations of the fighting man, they do not work at all.

Field Marshal Viscount Wolseley, *The Soldier's Pocket Book*, 1869

Newspaper correspondents with an army, as a rule, are mischievous. They are the world's gossips, pick up and retail the camp scandal, and gradually drift to the headquarters of some general, who finds it easier to make reputation at home than with his own corps or division. They are also tempted to prophesy events and state facts which, to an enemy, reveal a purpose in time to guard against it. Moreover, they are always bound to see facts coloured by the partisan or political character of their own patrons, and thus bring army officers into the political controversies of the day, which are always mischievous and wrong. Yet, so greedy are the people at large for war news, that it is doubtful whether an army commander can exclude all reporters, without bringing down on himself a clamor that may imperil his own safety. Time and moderation must bring a just solution to this modern difficulty.

General of the Army William T. Sherman, *The Memoirs of General W. T. Sherman*, 1875

The British public likes to read sensational news, and the best war correspondent is he who can tell the most thrilling lies.

Field Marshal Earl Haig, 1898, of the Sudan Campaign, letter to his sister

---

'All the danger of war and one-half per cent of the glory': such is our motto, and that is the reason why we [the press] expect large salaries.

Sir Winston Churchill, *Ian Hamilton's March*, 1990

The printing press is the greatest weapon in the armoury of the modern commander.

Colonel T. E. Lawrence (1888–1935)

Who can ascertain the truth about a cannon shot fired in the thick of night from a mutinous ship at a Czar's palace where the last government of the possessing classes is going out like an oil-less lamp? But just the same, the historian will make no mistake if he says that on October 25th not only was the electric current shut off in the government printing plant but an important page was turned in the history of mankind.

Leon Trotsky, *History of the Russian Revolution*, 1932

It may well be that between press and officials there is an inherent, built-in conflict of interest. There is something to be said for both sides, but when the nation is at war and men's lives are at stake, there should be no ambiguity.

General William Westmoreland, *A Soldier Reports*, 1976

There is no parallel . . . between the work of a war correspondent and the obligation of the serving soldier. The latter has to stick it and has no choice about the risks he will run. With very few notable exceptions, our war correspondents will not stay with combat danger long enough to begin to understand the ordeal of troops. They are another variety of sightseer. They flit in and out of the scene, hear a few shells explode, take a quick look at frontal living conditions, ask a few trivial quesitons of the hometown boys, and then beat it back to secure billets to pound out tear-jerking pieces about the horrors of war. With few exceptions, they will not stay on the job.

Brigadier General S. L. A. Marshall, *Bringing Up the Rear*, 1979

I guess the one thing I would say to the press what I was delighted with is, in the very, very early stages of this operation, when we were over here building up and we didn't have very much on the ground, you all were giving us credit for a whole lot more over here. And as a result, that gave me quite a feeling of confidence that we might not be attacked quite as quickly as I thought we were going to be attacked.

General H. Norman Schwarzkopf, 27 February 1991, 'The Mother of All Briefings' in Riyadh as Operation 'Desert Storm' was concluding

PRINCIPLES OF WAR

. . . I have made my boundary beyond that of my fathers; I have increased that which bequeathed me. I am a king who speaks and executes; that which my ear conceives is that which comes to pass by my hand; one who is eager to possess, and powerful . . . not allowing a matter to sleep in his heart . . . attacking him who attacks, silent in a matter, or answering a matter according that which is in it; since if one is silent after an attack, it strengthens the heart of the enemy. Valiance is eagerness, cowardice is to slink back; he is truly a craven who is repelled on his own border . . .

Sesostris III, Pharaoh of Egypt (1887–1849 BC), c. 1871 BC; an early exposition of several of the principles of war: the objective, the offensive, security, quoted in Breasted, *Ancient Records of Egypt*, 1906

To make war with success the following principles should never be departed from:

To be superior to your enemy in numbers as well as in *morale*; to fight battles in order to spread terror in the country; to divide your army into as many corps as may be effected without risk, in order to undertake several objects at the same time; to treat *well* those who yield, to *ill*-treat those who resist; to secure your rear, and occupy and strengthen yourself at the outset in some post which shall serve as a Central base point for the support of your future movements; to make yourself master of the great rivers and principal passes, and to establish your line of communications by getting possession of the fortresses by laying siege to them, and of the open country by giving battle; for it is vain to expect that conquests are to be achieved without combats, although when the victory is won they will be best maintained by uniting mildness with valour.

Field Marshal Prince Raimundo Montecuccoli (1609–1680), *Memoirs*, quoted in Chandler, *The Military Maxims of Napoleon*, 1988

The art of war owns certain elements and fixed principles. We must acquire that theory, and lodge it in our heads – otherwise we will never get very far.

Frederick the Great, quoted in Duffy, *The Military Life of Frederick the Great*, 1986

Get your principles straight; the rest is a matter of detail.

Napoleon (1769–1821)

These three things you must always keep in mind: concentration of strength, activity, and a firm resolve to perish gloriously. They are the three principles of the military art which have disposed luck in my favour in all my operations. Death is nothing; but to live defeated and without glory is to die every day.

Napoleon, 1804, letter to General Lauriston, quoted in Herold, ed., *The Mind of Napoleon*, 1955

The principles of warfare are those which guided the great captains whose high deeds history has transmitted to us – Alexander, Hannibal, Caesar, Gustavus Adolphus, Turenne, Eugene of Savoy, Frederick the Great . . . The history of their eighty-three campaigns would constitute a complete treatise on the art of war; the principles that must be followed in defensive and offensive warfare would stem from it as from a common source.

Napoleon, at Saint Helena, quoted in Herold, ed., *The Mind of Napoleon*, 1955

MAXIM 112. All the great captains have done great things only by conforming to the rules and natural principles of the art; that is to say, by the wisdom of their combinations, the reasoned balance of means with consequences, and efforts with obstacles. They have succeeded only by thus conforming, whatever may have been the audacity of their enterprises and the extent of their success. They have never ceased to make war a veritable science. It is only under this title that they are our great models, and it is only in imitating them that one can hope to approach them.

Napoleon, *The Military Maxims of Napoleon*, 1827, tr. Burnod

---

Keeping your forces united, being vulnerable at no point, moving rapidly on important points – these are the principles which assure victory, and, with fear, resulting from the reputation of your arms, maintains the faithfulness of allies and the obedience of conquered peoples.

Napoleon, *Maxims of War*, 1831

The principles of war are the same as those of a seige. Fire must be concentrated at one point, and as soon as the breach is made, the equilibrium is broken and the rest is nothing.

Napoleon, *Maxims of War*, 1831

MAXIM 5. All wars should be governed by certain principles for every war should have a definite object, and be conducted according to the rules of art. War should only be undertaken with forces proportioned to the obstacles to be overcome.

Napoleon, *The Military Maxims of Napoleon*, 1831, tr. D'Aguilar

---

I repeat once more that there are no unfailing recipes in the art of war; that any number of accidents, unusual determination of the enemy, and a not always equal determination on our side can ruin the best-laid plans. I repeat that there are however general principles, abstracted from experience, which can be safely used as the basis of one's dispositions: *but that the execution of this principle must be coupled with determination and prudence,* and that evolutions designated for the level drill-field play a minimal role in broken terrain . . .

Field Marshal Johann Hans Graf Yorck von Wartenburg (1759–1830), quoted in Paret, *Yorck and the Era of Prussian Reform 1807–1815*, 1966

Only those general principles and attitudes that result from clear and deep understanding can provide a *comprehensive* guide to action. It is to these that opinions on specific problems should be anchored. The difficulty is to hold fast to these results of contemplation in the torrent of events and new opinions. Often there is a gap between principles and actual events that cannot always be bridged by a succession of logical deductions. Then a measure of self-confidence is needed, and a degree of scepticism is also salutary. Frequently nothing short of an imperative principle will suffice, which is not part of the immediate thought-process, but dominates it: that principle is in all doubtful cases *to stick to one's first impression and to refuse to change unless forced to do so by a clear conviction.* A strong faith in the overriding truth of tested principles is needed; the *vividness* of transient impressions must not make us forget that such truth as they contain is of a lesser stamp. By giving preference, in case of doubt, to our earlier convictions, by holding to them stubbornly, our actions acquire that quality of steadiness and consistency which is termed strength of character.

Major General Carl von Clausewitz, *On War*, i, 1832, tr. Howard and Paret

---

There exists a small number of fundamental principles of war, which could not be deviated from without danger, and the application of which, on the contrary, has been in almost all time crowned with success.

Lieutenant General Antoine-Henri Baron de Jomini, *Summary of the Art of War*, 1838

One great principle underlies all the operations of war – a principle which must be followed in all good combinations. It is embraced in the following maxims:

1. To throw by strategic movements the mass of an army, successively, upon the decisive points of a theatre of war, and also upon the communications of the enemy as much as possible without compromising one's own.
2. To manoeuvre to engage fractions of the hostile army with the bulk of one's forces.
3. On the battlefield, to throw the mass of the forces upon the decisive point, or upon the portion of the hostile line which it is of the first importance to overthrow.
4. To so arrange that these masses shall not only be thrown upon the decisive point, but that they shall engage at the proper times and with ample energy.

Lieutenant General Antoine-Henri Baron de Jomini, *Summary of the Art of War*, 1838

In war, as in every other art based upon settled principles, there are exceptions to all general rules. It is in discovering these cases that the talent of the general is shown.

Dennis Hart Mahan (1802–1871), 1864

The first principle of war is to concentrate superior force at the decisive point, that is, upon the field of battle. But it is exceedingly seldom that by standing still, and leaving the initiative to the enemy, that this principle can be observed, for a numerically inferior force, if it once permits its enemy to concentrate, can hardly hope for success; to strike the enemy in detail and overthrow his columns in succession. And the highest art of all is to compel him to disperse his army, and then to concentrate superior force against each fraction in turn.

Colonel George F. Henderson, *Stonewall Jackson*, 1898

The principles of the art of war are within reach of the most ordinary intelligence, but that does not mean that it is capable of applying them.

General Mikhail I. Dragomirov (1830–1905), quoted in Foch, *Principles of War*, 1913

War acknowledges principles, and even rules, but these are not so much fetters, or bars, which compel its movement aright, as guides which warn us when it is going wrong.

Rear Admiral Alfred Thayer Mahan (1840–1914)

The fundamental principles of war are neither very numerous nor in themselves very abstruse, but the application of them is difficult and cannot be made subject to rules. The correct application of principles to circumstances is the outcome of sound military knowledge, built up by study and practice until it has become instinct.

British Army *Field Service Regulations 1909*, quoted in Fuller, *The Foundations of the Science of War*, 1926

To create a formula or a general rule . . . that would be appropriate in all situations is absurd. One must be able to think on one's feet to be able to make sense of each separate case.

V. I. Lenin, quoted in Gareyev, *Frunze, Military Theorist*, 1985

There are eight principles of war, and they constitute the laws of every scientifically fought boxing match as of every battle. These principles are:

1st Principle. – The Principle of the objective
2nd Principle.– The Principle of the offensive
3rd Principle. – The Principle of security
4th Principle. – The Principle of concentration
5th Principle. – The principle of economy of force
6th Principle. – The Principle of movement
7th Principle. – The Principle of surprise
8th Principle. – The Principle of co-operation.

No one of the above eight principles is of greater value than the other. No plan of action can be considered in harmony unless all are in harmony, and none can be considered in harmony unless weighed against the conditions which govern their application. Seldom can a perfect plan be arrived at because the fog of war seldom, if ever, rises. It is, however, an undoubted fact that the general who places his trust in the principles of war, and who trusts in them the more strongly the fog of war thickens, almost inevitably beats the general who does not.

A general will seldom win without attacking, and he will seldom attack correctly unless he has chosen his objective with reference to the princples of war, and unless his attack is based on these principles. Imagination is a great detective, but imagination which is not based on the sound foundation of reason is at best a capricious leader. Even genius itself, unless it be stiffened by powerful weapons, a high morale, discipline and training, can only be likened to a marksman armed with a blunderbuss – ability wasted through insufficiency of means. Conversely, an efficient army led by an antiquated soldier may be compared to a machine-gun in the hand of an arbalister.

Major General J. F. C. Fuller, *The Reformation of War*, 1923; Fuller was the first to define the principles of war as a coherent, interactive set of concepts; these principles subsequently modified are the basis of the Principles of War used by the US and British Armies

I would give you a word of warning on the so-called principles of war, as laid down in *Field Service Regulations*. For heaven's sake, don't threat those as holy writ, like the Ten Commandments, to be learned by heart, and as having by their repetition some magic, like the incantations of savage priests. They are merely a set of common-sense maxims, like 'cut your coat according to your cloth', 'a rolling stone gathers no moss', 'honesty is the best policy', and so forth . . . Clausewitz has a different set, so has Foch, so have other military writers. They are all simply common sense, and are instinctive to the properly trained soldier.

Field Marshal Earl Wavell, *c.* 1930, officer lecture at Aldershot

---

. . . The military student does not seek to learn from history the minutiae of method and technique. In every age these are decisively influenced by the characteristics of weapons currently available and by means at hand for maneuvering, supplying and controlling combat forces. But research does bring to light those fundamental principles, and their combinations and applications, which, in the past, have been productive of success. These principles know no limitation of time. Consequently, the army extends its analytical interest to the dust-buried accounts of wars long past as well as to those still reeking with the scent of battle.

General of the Army Douglas MacArthur, *Annual Report of the Chief of Staff for Fiscal Year Ending June 30, 1935*

Adherence to one principle frequently demands violation of another. Any leader who adheres inflexibly to one set of commandments is inviting disastrous defeat from a resourceful opponent.

Admiral C. R. Brown, *Principles of War,* June 1949

Respect for the opinion of this or that great soldier must never be allowed to go so far that nobody dares to discuss it. A sure sense of reality must be aroused. Given a well-founded knowledge of basic principles, any man of reasonably cool and logical mind can work out most of the details for himself, provided he is not inhibited in his thinking.

Field Marshal Erwin Rommel, *The Rommel Papers,* 1953

Against the principle of surprise – continuous activity by the various intelligence agencies. Against the principle of maintenance-of-aim – tactical diversionary attacks and strategical, psychological and political offensives. Against the principle of economy-of-force – attacks against lines of communications and stores in the rear, thereby pinning down the enemy's forces and dispersing them. Against the principle of coordination – strike against the channels of administration. Against the principle of concentration – diversionary attacks and air activity to split up the enemy's forces. Against the principle of security – sum total of the above activities and those that follow. Against the spirit of offensive-spirit – offensive spirit. Against the principle of mobility – destruction of the lines of communication.

General Yigael Yadin, *Israeli Forces Journal*, 9/1949, reprinted in Liddell Hart, *Strategy*, 1954

The principles of war, not merely one principle, can be condensed into a single word – 'concentration'. But for truth this needs to be amplified as the 'concentration of strength against weakness'. And for any real value it needs to be explained that the concentration of strength against weakness depends on the dispersion of your opponent's strength, which in turn is produced by a distribution of your own that gives the appearance, and partial effect of dispersion. Your dispersion, his dispersion, your concentration – such is the sequence, and each is a sequel. True concentration is the fruit of calculated dispersion.

Captain Sir Basil Liddell Hart, *Strategy*, 1954

In most military text books, more particularly *Field Service Regulations*, a list of the principles of war will be found . . . There are several versions of these so-called principles, but they are no more than pegs on which to hang our tactical thoughts. There is nothing irrevocable about them; sometimes they may be discarded with impunity; but as a study of military history will show, they should only be discarded after deep consideration. They are very important guides rather than principles, and in the writer's opinion the simplest and most useful are derived from the seven tactical elements . . . aim or object, security, mobility, offensive power, economy of force, concentration of force, and surprise. Further, they are as applicable to strategy (operations in plan) as to tactics (operations in action), two terms which should never be separated by a bulkhead, because their components flow into each other and together constitute the art of war . . .

Major General J. F. C. Fuller, *The Generalship of Alexander the Great*, 1958

War remains an art and, like all arts, whatever its variation, will have its enduring principles. Many men, skilled either with sword or pen, and sometimes with both, have tried to expound those principles. I heard them once from a soldier of experience for whom I had a deep and well-founded respect. Many years ago, as a cadet hoping someday to be an officer, I was poring over the 'Principles of War', listed in the old Field Service Regulations, when the sergeant-major came upon me. He surveyed me with kindly amusement. 'Don't bother your head about all them things, me lad,' he said. 'There's only one principle of war, and that's this. Hit the other fellow as quick as you can and as hard as you can, where it hurts him the most, when he ain't lookin'!' As a recruit, I earned that great man's reproof often enough; now, as an old soldier, I would hope to receive his commendation. I think I might, for we of the Fourteenth Army held to his 'Principle of War'.

Field Marshal Viscount Slim, *Defeat Into Victory*, 1963

PROFESSION OF ARMS

The Profession of a Souldier is allowed to be lawful by the Word of God; and so Famous and Honourable amongst Men, that Emperours and Kings do account it a Great Honour to be of the Profession, and to have Experience in it . . .

General George Monck (1608–1670) *Observations Upon Military and Political Matters*, 1671

The man who devotes himself to war should regard it as a religious order into which he enters. He should have nothing, know no other home than his troop, and should hold himself honoured in his profession.

Field Marshal Maurice Comte de Saxe, *My Reveries*, 1732

No matter how clearly we see the citizen and the soldier in the same man, how strongly we conceive of war as the business of the entire nation, opposed diametrically to the pattern set by the *condottieri* of former times, the business of war will always remain individual and abstract. Consequently for as long as they practise this activity, soldiers will think of themselves as members of a kind of guild, in whose regulations, laws, and customs the spirit of war is given pride of place. And that does seem to be the case. No matter how much one may be inclined to take the most sophisticated view of war, it would be a serious mistake to underrate professional pride (*esprit de corps*) as something that may and must be present in an army to greater or lesser degree. Professional pride is the bond between the various natural forces that activate the military virtues; in the context of this professional pride they crystallize more readily.

Major General Carl von Clausewitz, *On War*, ii, tr. Howard and Paret

---

War is a special profession, however general its relation may be and even if all the male population of a country capable of bearing arms were able to practise it, war would still continue to be different and separate from any other activity which occupies the life of man.

Major General Carl von Clausewitz (1780–1831)

In no event will there be money in it; but there may always be honor and quietness of mind and worthy occupation – which are better guarantees of happiness.

Rear Admiral Alfred Thayer Mahan, *The Navy as a Career*, 1985

I do not regard and have never regarded permanent soldiering as an attractive proposition for any man who has some other profession at his command . . . if a man could command an income no larger in private practice than he could in military employment, I would recommend to him to stick to private practice every time. There is something about permanent military occupation which seems to confine a man's scope and limit his opportunities, and after he has had a few years under the circumscribed conditions of official routine, he generally finds himself wholly out of touch with civil occupation.

General Sir John Monash, 6 Jan 1911, quoted in Horner, *The Commanders*, 1984

There is scope and occupation in the army for the highest mental as well as the highest physical efficiency. Whereas delight in the trade of arms, inherited and indulged from generation to generation, drives one man to the army, another may join from the desire to devote his ability and knowledge to the most immediate form of national service; one is attracted by a profession which gives pride of place to manly strength and personal value, and another by the prospect of activity in the open, another because he is interested in the machinery of war, and yet another because he loves horses. '**Αὐτὸς γὰρ ἐφέλκεται ἄνδρα σίδηρος**' (Iron attracts man).

Colonel General Hans von Seekt, *Thoughts of a Soldier*, 1930

Men who adopt the profession of arms submit, of their own free will, to a law of perpetual constraint. Of their own accord they reject the right to live where they choose, to say what they think, to dress as they like. From the moment they become soldiers it needs but an order to settle them in this place, to move them to that, to separate them from their families and dislocate their normal lives. On the word of command they must rise, march, run, endure bad weather, go without sleep or food, be isolated in some distant post, work till they drop. They have ceased to be the masters of their fate. If they drop in their tracks, if their ashes are scattered to the four winds, that is all part and parcel of their job.

Charles de Gaulle, *The Edge of the Sword*, 1932

War is the professional soldier's time of opportunity.

Captain Sir Basil Liddell Hart, *Defence of the West*, 1956

I find I have liked all the soldiers of different race who have fought with me and most of those who have fought against me. This is not strange, for there is a freemasonry among fighting soldiers that helps them to understand one another even if they are enemies.

Field Marshal Viscount Slim, *Unofficial History*, 1959

When there is a visible enemy to fight in open combat . . . many serve, all applaud, and the tide of patriotism runs high. But when there is a long, slow struggle, with no immediate visible foe, your choice will seem hard indeed.

President John F. Kennedy, address to a graduating class of the U.S. Naval Academy

---

The function of the profession of arms is the ordered application of force in the resolution of a social problem. Harold Lasswell describes it as the management of violence, which is rather less precise. The bearing of arms among men for the purpose of fighting other men is found as far back as we can see. It has become at some times and in some places a calling resembling the priesthood in its dedication. It has never ceased to display a strong element of the vocational.

It has also become a profession, not only in the wider sense of what is professed, but in the narrower one of an occupation with a distinguished corpus of specific technical knowledge and doctrine, a more or less exclusive group coherence, a complex of institutions peculiar to itself, an educational pattern adapted to its own needs, a career structure of its own and a distinct place in the society which has brought it forth. In all these respects it has strong points of resemblance to medicine and law, as well as holy orders.

General Sir John Hacket, *The Profession of Arms*, 1963

It is well to remind these who quote the injunction in the Sermon on the Mount to 'turn the other cheek also' to him who 'shall smite thee on the right cheek', that, in the version in Saint Matthew's Gospel, this is preceded by the words, 'I say unto you that ye resist not evil'. The sad fact of life is that, if evil is not resisted, it will prevail. That is the justification for the use of force to deter, and if necessary, defeat those who turn to it to further their own ends, the justification for maintaining in the service of the community and the state, forces who are trained, skilled and well-equipped to meet that challenge when and wherever it arises. Their profession is an honourable one.

Field Marshal Sir Michael Carver, *Warfare Since 1945*, 1980

PROMOTION

I sailed with King Amenhotep I triumphant, when he ascended the river to Kush in order to extend the borders of Egypt . . . I was at the head of our army; I fought incredibly; his majesty beheld my bravery. I brought off two hands, and took them to his majesty. . . . I brought his majesty in two days to Egypt from the upper well; he [the king] presented me with gold. Then I brought away two female slaves, in addition to those which I had taken to his majesty. One [the king] appointed me as 'Warrior of the Ruler'.

Ahmose son of Ebana, Egyptian naval officer during the reign of Pharaoh Amenhotep I (1557–1501 BC), quoted in Breasted, *Ancient Records of Egypt*, 1906

In choosing their centurions the Romans look not so much for the daring or fire-eating type, but rather for men who are natural leaders and possess a stable and imperturbable temperament, not men who will open the battle and launch attacks, but those who will stand their ground even when worsted or hard-pressed, and will die in defence of their posts.

Polybius, *The Rise of the Roman Empire*, c. 125 BC

If the skill of a general is one of the surest elements of victory, it will readily be seen that the judicious selection of generals is one of the most delicate points in the science of government and one of the most essential parts of the military policy of a state. Unfortunately, this choice is influenced by so many petty passions that chance, rank, age, favour, party spirit, or jealousy will have as much to do with it as the public interest and justice.

Lieutenant General Antoine-Henri Baron de Jomini, *Summary of the Art of War*, 1838

As to rewards and promotion, it is essential to respect long service and at the same time to open a way for merit. Three-fourths of the promotions in each grade should be according to the roster, with the remaining fourth reserved for those distinguished men of merit and zeal. In time of war, however, the regular order of promotion should be suspended, or at least reduced to a third of the promotions, leaving the majority for brilliant conduct and marked services.

Lieutenant General Antoine-Henri Baron de Jomini, *Summary of the Art of War*, 1838

Differences of opinion must exist between the best of friends as to policies in war, and of judgments as to men's fitness. The officer who has the command, however, should be allowed to judge the fitness of the officers under him, unless he is very manifestly wrong.

General of the Army Ulysses S. Grant, *Personal Memoirs of U.S. Grant*, 1885

Originality never yet led to preferment.

Admiral Sir John Fisher, RN, *Memories*, 1919

We are now at war, fighting for our lives, and we cannot afford to confine Army appointments to persons who have excited no hostile comment in their career.

Sir Winston S. Churchill, 19 October 1940, note to the Chief of the Imperial General Staff

---

The value of 'tact' can be over-emphasized in selecting officers for command: positive personality will evoke a greater response than negative pleasantness.

Captain Sir Basil Liddell Hart, *Thoughts on War,* 1944

More and more does the 'System' tend to promote to *control*, men who have shown themselves efficient *cogs* in the machine . . . There are few commanders in our higher commands. And even these, since their chins usually outweigh their foreheads, are themselves outweighed by the majority – of commanders who are essentially staff officers.

Captain Sir Basil Liddell Hart, *Thoughts on War,* 1944

The flaw in all these conscientious attempts to improve the system of promotion is that they tend inevitably to stress the formal record at the expense of the human element and that they aim at an improved flow of promotion when nothing less than an eruption will avail. In fact, the effort to make the Army 'fairer' as a profession is contrary to the very nature of its object – war, which is essentially 'unfair'. The professional attitude was epitomized in a recent article by Admiral Harper, who scathingly critized Mr. Churchill's selection of an officer for a certain command on the ground that Mr. Churchill had said that his choice was based on personal impressions and conversations rather than on official records. Yet outside the closely regulated professions what employer choosing a man for an important post would not weigh personality more than record? Admittedly, the value of a choice based on personal knowledge depends on the man who makes it, and even at the best it is highly unfair to those who fail through no fault of their own to catch the selector's eye. But it is the only system, or lack of a system, under which genius can be brought rapidly to the top. The lesson of military history is that genius and not mere competence decides the fate of warring nations. It was the shrewd dictum of one who, whatever his defects, was assuredly a genius – Lord Fisher – that 'favouritism is the secret of efficiency'. In our zeal for equalitarian fairness we have almost ruled out 'favouritism', and the stalemate of 1914–1918 is a testimonial to our good intentions – which 'paved the patch to hell'.

Captain Sir Basil Liddell Hart, *Thoughts on War,* 1944

If evolution were destined to favour the rise of those who, in the tragic hours when the storm sweeps away conventions and habits, are the only ones to remain on their feet and to be, therefore, necessary, would not that be all to the good?

Charles de Gaulle, *War Memoirs: The Call to Honour 1940–1942*, 1955

The next point, still a human one, is the selection of commanders. Probably a third of my working hours were spent in the consideration of personalities. In dealing with subordinates, justice and a keen sense of fairness are essential – as also is a full measure of human consideration. I kept command appointments in my own hand, right down to and including battalion or regimental level. Merit, leadership and ability to do the job, were the sole criteria; I made it my business to know all commanders, and to insist on a high standard.

Field Marshal Viscount Montgomery of Alamein, *The Memoirs of Field Marshal Montgomery*, 1958

An extensive use of weedkiller is needed in the *senior* ranks after a war; this will enable the first-class young officers who have emerged during the war to be moved up.

Field Marshal Viscount Montgomery of Alamein, *The Memoirs of Field Marshal Montgomery*, 1958

By good fortune in the game of military snakes and ladders, I found myself a general.

Field Marshal Viscount Slim, *Unofficial History*, 1959

No written examination is a test of character, personality or leadership. Examinations are not a true test of ability and are no test at all of character or leadership. Some shine at examinations, others fail to disclose their knowledge of the subject. Brains of themselves are only of academic value. I have seen so many scholars fail in life through lack of application, character and personality . . . Entrance examinations into the Army used to be particularly harmful. How can written examinations be a test of good officership? . . . Academic knowledge without character and personality is useless. The boy with average intelligence, with drive, character and practical ability – I knew so many of them who have had brilliant careers, leaving scholars behind. Scholars seldom enter the Army, preferring professions or politics. Successful soldiers have not been scholars, the one exception being Lord Wavell. I can think of no soldier who has reached the top of the ladder in recent years whose main assets have been scholarly ability. Our leading soldiers in both World Wars reached the top through character, personality, determination and leadership. All the very necessary and arduous brainwork is done by scholarly staff officers.

Colonel Richard Meinertzhagen, *Army Diary 1899–1925*, 1960

Those of you who may regard my profession of political life with some disdain should remember that it made it possible for me to move from being an obscure lieutenant in the United States Navy to Commander-in-Chief in fourteen years with very little technical competence.

President John F. Kennedy, 12 October 1961, speech at the University of North Carolina

---

We make generals today on the basis of their ability to write a damned letter. Those kinds of men can't get us ready for war.

Lieutenant General Lewis B. 'Chesty' Puller, quoted in Davis, *Marine*, 1962

In the early days of a tremendous undertaking, such as this war, appointments and selection of individuals are based upon a number of unimportant factors – among which, as we all know, are: personal propinquity, wild guesses, school records, past acquaintanceship, and a number of others, of which few really search down into the depths of character and ability. But the stark realities of distress, privation, and discouragement will bring character and ability into their own.

General of the Army Dwight D. Eisenhower, *At Ease: Stories I Tell My Friends*, 1967

This is a long tough road we have to travel. The men that can do things are going to be sought out just as surely as the sun rises in the morning. Fake reputations, habits of glib and clever speech, and glittering surface performance are going to be discovered and kicked overboard. Solid, sound leadership, with inexhaustible nervous energy to spur on the efforts of lesser men, and ironclad determination to face discouragement, risk, and increasing work without flinching, will always characterize the man who has a sure-enough, bang-up fighting unit. Added to this he must have a darned strong tinge of imagination – I am continuously astonished by the utter lack of imaginative thinking among so many of our people that have reputations for being really good officers. Finally, the man has to be able to forget himself and his personal fortunes.

General of the Army Dwight D. Eisenhower, *At Ease: Stories I Tell My Friends*, 1967

We must choose the boldest. Only a man of complete bravery can take a plane off from a rolling deck and, without relying on any alternate airfields, head off into the blue, engage in combat, win and then, having located one's ship, bring one's aircraft back precisely on the dime of the runway . . .

Admiral of the Soviet Fleet of the Soviet Union Sergei G. Gorshkov, quoted in Ablamonov, *Admiral*, 1986

PROMPTNESS

It is essential to be cautious and take your time in making plans, and once you come to a decision, carry it out right away without any hesitation or timidity. Timidity after all is not caution, but the invention of wickedness.

The Emperor Maurice, *The Strategikon*, c. AD 600

You will have seen by what I have had occasion to delineate concerning war that promptness contributes a great deal to success in marches and even more in battles. That is why our army is drilled in such a fashion that it acts faster than others. Drill is the basis of these manoeuvres which enable us to form in the twinkling of an eye . . .

Frederick the Great, *Instructions to His Generals*, 1747

I have always been a quarter of an hour before my time, and it has made a man of me.

Attributed to Admiral Lord Nelson (1758–1805)

Promptness is the greatest of military virtues, evincing, as it does, zeal, energy and discipline. The success of arms depends more upon celerity than any one thing else.

Lieutenant General D. H. Hill, 7 September 1863, circular to his troops in Chattanooga

PROPAGANDA

His victories were published in all lands, to cause that every land together may see, to cause the glory of his conquests to appear; King Merneptah, the Bull, lord of strength, who slays his foes, beautiful upon the field of victory . . . He has penetrated the land of Temeh in his lifetime, and put eternal fear in the heart of Meshwesh. He has turned back Libya, who invaded Egypt, and great fear of Egypt is in their hearts.

Merneptah, Pharaoh of Egypt, *c.* 1220 BC, celebration of his victory over the Libyans, quoted in Breasted, *Records of Ancient Egypt*, 1906

We are here to guide public opinion, not to discuss it.

Napoleon, 1804, meeting of the Conseil d'Etat, quoted in Herold, ed., *The Mind of Napoleon*, 1955

Barere still believes that the masses must be stirred. On the contrary, they must be guided without noticing it.

Napoleon, 1804, letter to Fouche, quoted in Herold, ed., *The Mind of Napoleon*, 1955

Instead of all the stupidities with which the daily press is filled, why do you not send commissioners to visit the districts from which we have expelled the enemy and make them collect the details of the crimes that have been committed there? Nothing more powerful could be found to stir the minds than a recital of the details. What we need at this moment is real and serious things, not wit in prose and verse. My hair stands on end when I hear of the crimes committed by the enemy, and the police have not even though of obtaining a single account of these happenings . . . A picture drawn in larger strokes will not convince the people. With ink and paper you can draw any pictures you like. Only by telling the facts simply and with detail can we convince them.

Napoleon, 27 February 1814, letter to Savary, quoted in Herold, ed., *The Mind of Napoleon*, 1955

We were hypnotized by the enemy propaganda as a rabbit is by a snake. It was exceptionally clever and conceived on a great scale. It worked by strong mass-suggestion, kept in the closest touch with the military situation, and was unscrupulous as to the means it used . . . The Army found no ally in a strong propaganda directed from home. While her Army was victorious on the field of battle, Germany failed in the fight against the *morale* of the enemy peoples.

General Erich Ludendorff, *My War Memories, 1914–1918*, 1919

Propaganda, as inverted patriotism, draws nourishment from the sins of the enemy. If there are no sins, invent them! The aim is to make the enemy appear so great a monster that he forfeits the rights of a human being. He cannot bring a libel action, so there is no need to stick at trifles.

General Sir Ian Hamilton, *The Soul and Body of an Army*, 1921

The printing press is the greatest weapon in the armoury of the modern commander.

Colonel T. E. Lawrence (1888–1935)

It was more subtle than tactics, and better worth doing, because it dealt with uncontrollables, with subjects incapable of direct command. It considered the capacity for mood of our men, their complexities and mutability, and the cultivation of whatever in them promised to profit out intention. We had to arrange their minds in order of battle just as carefully and as formally as other officers would arrange their bodies. And not only our men's minds, though naturally they came first. We must also arrange the minds of the enemy, so far as we could reach them; then those other minds of the nation supporting us behind the firing line, since more than half the battle passed there in the back; then the minds of the enemy nation waiting the verdict; and of the neutrals looking on; circle beyond circle.

Colonel T. E. Lawrence, *The Seven Pillars of Wisdom*, 1926

During the great war a battle of propaganda was waged by all belligerents, though at its beginning few were prepared to wage it. Our object was to prove that the Germans were 'dirty dogs', and that it was they who had started the war. I do not suggest that our contentions were wrong, but I cannot help feeling that when the Germans retaliated the means we employed to protect our national character were not of the best. In place of maintaining our reputation for fair play we hired a pack of journalists to defend us. These people, who had spent their lives in raking filth out of the law courts, went to mud with the alacrity of eels, and, though they undoubtedly succeeded in blackening the German nation, we ourselves became somewhat piebald in these gutter attacks.

Major General J. F. C. Fuller, *The Foundations of the Science of War*, 1926

All they need do really is quietly let people know the truth. There is no need to bang the big drum. Official reports should stick to the absolute truth – once you start lying, the war's as good as lost. Information Division's outlook is all wrong. All this talk of guiding public opinion and maintaining the national morale is so much empty puff.

Admiral Yamamoto Isoroku (1884–1903), 1942, quoted in Agawa, *The Reluctant Admiral*, 1979

PRUDENCE

There are some roads not to follow; some troops not to strike; some cities not to assault; and some ground which should not be contested.

Sun Tzu, *The Art of War*, c. 500 BC, tr. Griffith

The best soldier is one who does not willingly engaged in a hazardous and highly uncertain battle and refrains from emulating those who carry out operations recklessly and are admired for their brilliant success, but one who, while keeping the enemy on the move, remains secure and always in circumstances of his own choosing.

The Emperor Maurice, *Strategikon*, c. AD 600

In war, when adversaries are orderly in their movements and are at their sharpest, it is not yet time to fight with them; it is best to fortify your positions and wait. Watch for their energy to wane after being on alert for a long time; then rise and strike them. You will not fail to win.

Liu Ji (AD 1310–1375), *Lessons of War*

Good commanders never come to an engagement unless they are compelled to by absolute necessity, or the occasion calls for it.

Niccolo Machiavelli, *The Art of War,* 1521

PSYCHOLOGICAL OPERATIONS

To seduce the enemy's soldiers from their allegiance and encourage them to surrender is of especial service, for an adversary is more hurt by desertion than by slaughter.

Flavius Vegetius Renatus, *The Military Institutions of the Romans,* c. AD 378

Deception is often helpful in warfare. An enemy soldier who deserts to us, apart from some plot, is of the greatest advantage, for the enemy is hurt by deserters more than if the same men were killed in action.

The Emperor Maurice, *The Strategikon,* c. AD 600

Heart is that by which the general masters. Now order and confusion, bravery and cowardice, are qualities dominated by the heart. Therefore the expert at controlling his enemy frustrates him and then moves against him. He aggravates him to confuse him and harasses him to make him fearful. He thus robs the enemy of his heart and of his ability to plan.

Chang Yu, *c.* 1000, late Sung Dynasty commentator to Sun Tzu, *The Art of War, c.* 500 BC, tr. Griffith

---

It is your attitude, and the suspicion that you are maturing the boldest designs against him, that imposes on your enemy.

Frederick the Great, *Instructions to His Generals*, 1747

As the excited passions of hostile people are of themselves a powerful enemy, both the general and his government should use their best efforts to ally them.

Lieutenant General Antoine-Henri Baron de Jomini, *Summary of the Art of War,* 1838

The enemy bombards our front not only with a drumfire of artillery, but also with a drumfire of printed paper. Besides bombs which kill the body, his airmen also throw down leaflets which are intended to kill the soul.

Field Marshal Paul von Hindenburg (1847–1934)

The real target in war is the mind of the enemy command, not the bodies of his troops. If we operate against his troops it is fundamentally for the effect that action will produce on the mind and will of the commander; indeed, the trend of warfare and the development of new weapons – aircraft and tanks – promise to give us increased and more direct opportunities of striking at this psychological target. (June 1926)

Captain Sir Basil Liddell Hart, *Thoughts on War,* 1944

For a strong adversary (corps) the opposition of twenty-four squadrons and twelve guns ought not to have appeared very serious, but in war the psychological factors are often decisive. An adversary who feels inferior is in reality so.

Field Marshal Carl Gustav Baron von Mannerheim, *The Memoirs of Field Marshal Mannerheim*, 1953

Lenin had a vision of fundamental truth when he said that 'the soundest strategy in war is to postpone operations until the moral disintegration of the enemy renders delivery of the mortal blow both possible and easy'. This is not always practicable, nor his methods of propaganda always fruitful. But it will bear adaptation – 'The soundest strategy in any campaign is to postpone battle and the soundest tactics to postpone attack, until the moral dislocation of the enemy renders the delivery of a decisive blow practicable'.

Captain Sir Basil Liddell Hart, *Strategy*, 1954

PURSUIT

A pursuit gives even cowards confidence.

Xenophon (*c.* 431 BC–*c.* 352 BC), *Anabasis*

He who rashly pursues a flying enemy with troops in disorder, seems bent upon throwing away that victory which he had before obtained.

Flavius Vegetius Renatus, *Military Institutions of the Romans*, c. AD 378

After gaining a victory the general who pursues the enemy with a scattered and disorganized army gives away his victory to the foe.

The Emperor Maurice, *The Strategikon*, c. AD 600

The words of the proverb: 'A bridge of gold should be made for the enemy', is followed religiously. This is false. On the contrary, the pursuit should be pushed to the limit. And the retreat which had appeared such a satisfactory solution will be turned into a route [sic]. A detachment of ten thousand men can destroy an army of one hundred thousand in flight. Nothing inspires so much terror or occasions so much damage, for everything is lost. Substantial efforts are required to restore the defeated army, and in addition you are rid of the enemy for a good time. But many generals do not worry about finishing the war so soon.

Field Marshal Maurice Comte de Saxe, *My Reveries*, 1732

Once the enemy has taken flight they can be pursued with no better weapons than air-filled bladders. But if the officer you have ordered in pursuit prides himself on the regularity of his formations and the precautions of his march . . . there is no use in having sent him. He must attack, push and pursue without cease.

Field Marshal Maurice Comte de Saxe, *My Reveries*, 1732

The enemy is pursued to the first defile; all the harm possible is to be done to him, but you should not allow yourself to become so drunk with success that you become imprudent. If the enemy is thoroughly defeated, make several marches after him and you will gain a prodigious amount of territory. But always camp in accordance with regulations!

Frederick the Great, *Instructions to His Generals*, 1747

Only pursuit destroys a running enemy.

Field Marshal Prince Aleksandr V. Suvorov (1729–1800)

Do not delay in the attack. When the foe has been split off and cut down, pursue him immediately and give him no time to assemble or form up . . . spare nothing. Without regard for difficulties, pursue the enemy day and night until he has been annihilated.

Field Marshal Prince Aleksandr V. Suvorov (1729–1800)

Pursue the last man to the Adda and throw the remains into the river.

Field Marshal Prince Aleksandr V. Suvorov, 1799, during the Italian Campaign

---

A strong pursuit, give no time for the enemy to think, take advantage of victory, uproot him, cut off his escape.

Field Marshal Prince Aleksandr V. Suvorov (1729–1800)

In cases of obstacles arising, don't be too distracted by them. Time is more valuable than anything else – one must know how to save it. Often our previous victories remained without results because of insufficient men. The falsest of rules is the conviction that after an enemy defeat everything is over, whereas the fact is that it is necessary to try for even greater success.

Field Marshal Prince Aleksandr V. Suvorov (1729–1800)

Not days nor hours, but even moments are valuable in such a situation; want of provisions cannot serve you as an excuse.

Field Marshal Prince Aleksandr V. Suvorov (1729–1800)

MAXIM 83. A general-in-chief should never allow any rest either to the conqueror or to the conquered.

Napoleon, *The Military Maxims of Napoleon*, 1827, ed. Burnod

---

Next to victory, the act of pursuit is the most important in war.

Major General Carl von Clausewitz, *Principles of War*, 1812

Pursuit of a beaten enemy begins the moment he concedes the fight and abandons his position. At this juncture victory, while assured, is usually still limited and modest in its dimensions. Little positive advantage would be gained in the normal course of events unless victory were consummated by pursuit on the first day. It is usually only then . . . that trophies [prisoners, guns, etc.] tend to be taken which will embody the victory . . .

Major General Carl von Clausewitz, *On War*, iv, 1832, tr. Howard and Paret

---

. . . to follow up the success we gain with the utmost energy. The pursuit of the enemy when defeated is the only means of gathering up the fruits of victory.

Major General Carl von Clausewitz, *On War*, 1832

A pursuit should generally be executed as boldly and actively as possible, especially when it is subsequent to a battle gained, because the demoralized army may be wholly dispersed if vigorously followed up.

Lieutenant General Antoine-Henri Baron de Jomini, *Summary of the Art of War*, 1838

When you strike the enemy and overcome him, never give up the pursuit as long as your men have strength to follow; for an enemy routed, if hotly pursued, becomes panic-stricken, and can be destroyed by half their number.

Lieutenant General Thomas 'Stonewall' Jackson (1824–1863)

Do not fail in that event to make the most of the opportunity by the most vigorous attack possible, as it may save us what we have most reason to apprehend – a slow pursuit, in which he gains strength as we lose it.

General of the Army William T. Sherman, 1864, quoted in Liddell Hart, *Sherman*, 1930

A prompt and vigorous pursuit is the only means of ensuring complete success.

General of the Army Philip Sheridan (1831–1888)

Strenuous, unrelaxing pursuit is therefore as imperative after a battle as is courage during it.

Rear Admiral Alfred Thayer Mahan, *Naval Strategy*, 1911

In pursuit you must always stretch possibilities to the limit. Troops having beaten the enemy will want to rest. They must be given as objectives, not those that you think they will reach, but the farthest they could possibly reach.

Field Marshal Viscount Allenby of Meggido, 1917, Order to XXI Corps, Philistia

---

Pursuit will hold the first place in your thoughts. It is at the moment when the victor is most exhausted that the greatest forfeit can be extracted from the vanquished.

Sir Winston Churchill, 13 December 1940, to Lord Wavell after his victory in the Libyan desert against the Italians

---

While coolness in disaster is the supreme proof of a commander's courage, energy in pursuit is the surest test of his strength of will.

Field Marshal Earl Wavell, unpublished, *Recollections*, 1946

QUALITY

The commander who has five or six thousand of our heavy cavalry and the help of God will need nothing more.

The Emperor Nicephorus II Phocas (AD 912–969

At my first going out into this engagement I saw our men beaten at every hand . . . 'Your troops', said I, 'are most of them old decayed serving-men, and tapsters, and such like fellows; and', said I, 'their troops are gentlemen's sons, younger sons and persons of quality: do you think that the spirits of such base mean fellows will ever be able to encounter gentlemen, that have honour and courage and resolution in them? . . . You must get men of spirit: and take it not ill what I say, – I know you will not, – of a spirit that is likely to go on as far as gentlemen will go: or else you will be beaten still.'

Oliver Cromwell, October 1642, to John Hampden after the Battle of Edgehill

---

It is not the big armies that win battles; it is the good ones.

Field Marshal Maurice Comte de Saxe, *My Reveries*, 1732

Not alone is the strength of the Fleet measured by the number of its fighting units, but by its efficiency, by its ability to proceed promptly where it is needed and to engage and overcome an enemy.

Admiral Richard Wainwright, USN, 1911

He who reckons solely by the visible in war is reckoning falsely. The inherent worth of the soldier is everything. It was on that that I based my confidence. What I thought to myself was this:

The Russian may invade our Fatherland, and contact with the soil of Germany may lift up his heart, but that does not make him a German soldier, and those who lead him are not German officers. The Russian soldier has fought with the greatest obedience on the battlefields of Manchuria although he had no sympathy with the political ambitions of his rulers in the Pacific. It did not seem unlikely that in a war against the Central Powers the Russian Army would have greater enthusiasm for the war aims of the Tsar's Empire. On the other hand, I considered that, taking it all round, the Russian soldier and officer would not display higher military qualities in the European theatre than they had in the Asiatic, and believed that in comparing the two forces I was entitled to credit our side with a plus on the ground of instrinsic value instead of a minus for our numerical superiority.

Field Marshal Paul von Hindenburg, *Out of My Life*, 1920

Thus, the creation of new weapons and military equipment . . . also entails corresponding transformations in the methods of conducting military operations. But this does not take place immediately upon the appearance of new means of combat, but only when they begin to be used in a quantity which inevitably causes a new qualitative state of the phenonemon. As long as the new weapons and military equipment are used in limited numbers, they are most often merely adapted to the existing methods of armed conflict or, at best, introduce in them only a few partial adjustments.

Marshal of the Soviet Union Nikolai V. Ogarkov, *History Teaches Vigilance*, 1985

QUOTATIONS

I hate quotations. Tell me what you know.

Ralph Waldo Emerson, *Journals*, 1849

It is a good thing for an uneducated man to read books of quotations.

Sir Winston S. Churchill, *My Early Life*, 1930

Catchwords and trite phrases are not the same thing as quotations, although not unrelated; for quotations also tend to have ridiculous and dangerous associations. At the same time it is undoubtedly convenient to find that someone else has already expressed the same though in a happy and generally accepted form; not to mention the fact that literary people are agreeably surprised or impressed when they find a soldier occasionally quoting Goethe or even Greek, suggesting an intellectual capacity in excess of that required for reading the drill book. That is why I sometimes make quotations myself.

Colonel General Hans von Seekt, *Thoughts of a Soldier*, 1930

There was a certain general, now dead but universally esteemed and respected in his day, who was a veritable fountain of information. Whenever he was asked to express an opinion on some military situation he would always begin by saying, 'In such a situation Frederick the Great would say,' etc., and then would follow some invariably apposite quotation. But the best quotation, the best parallel present to the mind will not relieve the soldier of the difficulty of decision.

Colonel General Hans von Seekt, *Thoughts of a Soldier*, 1930

Professional soldiers are sentimental men, for all the harsh realities of their calling. In their wallets they carry bits of philosophy, fragments of poetry, quotations from the Scriptures which in time of stress and danger speak to them with great meaning.

General Mathew B. Ridgway, *My Battles in War and Peace*, 1956

RANK

Do not look down on others because of your own elevated rank.

Zhuge Liang (AD 189–234), *The Way of the General*

There is merit without rank, but there is no rank without some merit.

François Duc de La Rochefoucauld, *Réflexions ou sentences et maximes morales*, 1665

When fortune surprises us by giving us an important position, without having led us to it by degrees, or without our being elevated to it by our hopes, it is almost impossible for us to maintain ourselves suitably in it, and appear worthy of possessing it.

François Duc de la Rochefoucauld, *Réflexions ou sentences et maximes morales*, 1665

I will fight under your command with the noblest abnegation. There are no rivalries as far as I am concerned when the independence of America is at stake. Be assured, General; come to Peru; count on my sincere cooperation. I shall be your lieutenant.

General José de San Martin, *c.* 1822, letter to Bolivar, quoted in Rojas, *San Martin*, 1957

At that time he was my senior in rank and there was no authority of law to assign a junior to command a senior of the same grade. But every boat that came up with supplies or reinforcements brought a note of encouragement from Sherman, asking me to call upon him for any assistance he could render, saying that if he could be of service at the front I might send for him and he would waive rank.

General of the Army Ulysses S. Grant, 1862, quoted in Liddell Hart, *Sherman*, 1930

I think rank of but trivial importance so that it is sufficient for the individual to exercise command.

General Robert E. Lee (1807–1870)

The barrier of rank is the highest of all barriers in the way of access to the truth.

Captain Sir Basil Liddell Hart, *Thoughts on War*, 1944

The badge of rank which an officer wears on his coat is really a symbol of servitude to his men.

General Maxwell D. Taylor, *The Field Artillery Journal*, Jan/Feb 1947

Rank should be used to serve one's subordinates. It should never be flaunted or used to get the upper hand of a subordinate in any situation save where he had already discredited himself in an unusually ugly or unseemly manner.

Brigadier General S. L. A. Marshall, *The Armed Forces Officer*, 1950

Rank is only given you in the Army to enable you to better serve those below you and those above you. Rank is not given for you to exercise your idiosyncrasies.

General Bruce C. Clarke, *Thoughts on Leadership*, n.d.

One of the problems inherent in possessing rank and owning the responsibility of command is that people don't think you're quite human. You can't get drunk even every now and then. (But there have never been too many teetotalers among the military, if we leave out Stonewall Jackson and – I suppose – Sir Galahad.)

General Curtis LeMay, *Mission With LeMay*, 1965

RASHNESS

Remember, that even intrepidity must be restrained within certain moderate limits, and, when it becomes pernicious, ceases to be honourable.

Count Belisarius (c. AD 505–AD 565) urging restraint upon his troops, quoted in Mahon, *Life of Belisarius*, 1829

Our commander ought to adapt his strategems to the disposition of the enemy general. If the latter is inclined to rashness, he may be enticed into premature and reckless action; if he is on the timid side, he may be struck down with continual surprise attacks.

The Emperor Maurice, *The Strategikon*, c. AD 600

Long and careful deliberation promises great safety in war, whereas hasty and impetuous generals usually commit serious blunders.

The Emperor Maurice, *The Strategikon*, c. AD 600

I may be accused of rashness but not of sluggishness.

Napoleon, 6 May 1796, to the Executive Direction, *Correspondance*, No. 337, Vol. I, 1858–1870

Rashness succeeds often, still more often fails.

Napoleon (1769–1821)

Rashness in war is prudence.

Admiral Sir John Fisher, *Memories*, 1919

READINESS

It is a doctrine of war not to assume the enemy will not come, but rather to rely on one's readiness to meet him; not to presume that he will not attack, but rather to make one's self invincible.

Sun Tzu, *The Art of War*, c. 500 BC, tr. Griffith

A general who desires peace must be prepared for war, for the barbarians become very nervous when they face an adversary all set to fight.

The Emperor Maurice, *The Strategikon*, c. AD 600

Orders should be given to the soldiers that they should be ready to march out on a holiday, in the rain, day or night. For this reason they should not be told the time or the day beforehand, so they may always be prepared.

The Emperor Maurice, *The Strategikon*, c. AD 600

MAXIM 7. An army should be ready every day, every night, and at all times of the day and night, to oppose all the resistance of which it is capable. With this view, the soldier should be invariably complete in arms and ammunition; the infantry should never be without its artillery, its cavalry, and its generals . . .

Napoleon, *The Military Maxims of Napoleon,* 1831, tr. D'Aguilar

---

The Gods who give the crown of victory to those who, by their training in peaceful times, are already victorious before they fight, refuse it to those who, satisfied with one victory, rest contentedly in peace. As the old saying goes, 'After a victory, tighten the strings of your helmet'.

Admiral Marquis Togo Heihachiro, 21 December 1905, farewell to the Japanese Fleet after the Battle of Tsushima, quoted in Ogasawara, *Life of Admiral Togo*, 1934

It is too late to learn the technique of warfare when military operations are already in progress, especially when the enemy is an expert in it.

General Aleksei A. Brusilov, *A Soldier's Notebook,* 1931

I am not a bit anxious about my battles. If I am anxious I don't fight them. I wait until I am ready.

Attributed to Field Marshal Viscount Montgomery of Alamein (1887–1976)

A 'ready' state of mind requires a positive, constructive approach to the Army's problems. The Army is not without problems now, nor will it ever be. Some of these problems have their roots outside the Army's gate, others are wholly our own, but we cannot allow ourselves to be overcome by them, nor can we stand idly by waiting for others to solve them. By facing these problems positively and fairly, without cant and without whitewash, we are improving our readiness and our ability to provide this country the land fighting force it needs. If we are to be ready, every member of the Army – our Army – must step forward to accept these challenges, bypass the trap of self-pity, and avoid becoming the unwitting captive of events . . . We cannot [merely] look for readiness sometime in the future, but be ready in mind, in spirit and in attitude today. A better weapon on the drawing-board, a new command and control system for the next decade, or a new concept for organizing the Army in the field may contribute to our readiness eventually, but as promises, they contribute little to our strength today. We cannot let tomorrow's promises deter us from meeting today's requirements.

General Creighton Ăbrams, quoted in *Military Review,* 4/1985

See also: PREPAREDNESS

THE REAR/REAR ATTACK/ REAR GUARDS

Most terrible, or rather most effective, of all manoeuvres, is sudden attack against the enemy's rear . . . For no hope of safety would remain for them in flight, and they would be unable to turn backwards, since the opposing army would attack, or to go forward, because of the detachment assailing their rear.

Onasander, *The General,* AD 58

Rear guards are the safety of armies and often they carry victory with them.

Frederick the Great, *Instructions to His Generals,* 1747

The risk of having to fight on two fronts, and the even greater risk of finding one's retreat cut off, tend to paralyze movement and the ability to resist, and so affect the balance between victory and defeat. What is more, in the case of defeat, they increase the losses and can raise them to their very limit – to annihilation. A threat to the rear can, therefore, make a defeat *more probable*, as well as *more decisive*.

Major General Carl von Clausewitz, *On War,* iv, 1832, tr. Howard and Paret

---

I never saw the rear of an army engaged in battle but I feared that some calamity had happened at the front – the apparent confusion, broken wagons, crippled horses, men lying about dead and maimed, parties hastening to and fro in seeming disorder, and a general apprehension of something dreadful about to ensue; all these signs, however, lessened as I neared the front, and there the contrast was complete – perfect order, men and horses full of confidence, and it was not unusual for general hilarity, laughing, and cheering. Although cannon might be firing, the musketry clattering, and the enemy's shot hitting close, there reigned a general feeling of strength and security that bore a marked contrast to the bloody signs that had drifted rapidly to the rear; therefore, for comfort and safety, I surely would rather be at the front than the rear line of battle. So also on the march, the head of the column moves on steadily, while the rear is alternately halting then rushing forward to close up the gap; and all sorts of rumors, especially the worst, float back to the rear.

General of the Army William T. Sherman, *The Memoirs of General W. T. Sherman*, 1885

It is just as legitimate to fight an enemy in the rear as in the front. The only difference is the danger.

Colonel John S. Mosby, *Mosby's War Reminiscences*, 1887

The object of the rear attack is not itself to crush the enemy, but to unhinge his morale and dispositions so that his dislocation renders the subsequent delivery of a decisive blow both practicable and easy. (April 1929)

Captain Sir Basil Liddell Hart, *Thoughts on War*, 1944

. . . give the enemy a spanking from behind. You can kill more soldiers by scaring them to death from behind with a lot of noise than by attacking them from the front.

General George S. Patton, Jr., 1940, *The Patton Papers*, Vol II, 1974

. . . a move round the enemy's front against the rear has the aim not only of avoiding resistance on its way but in its issue. In the profoundest sense, it takes the *line of least resistance*. The equivalent in the psychological sphere is the *line of least expectation*. They are the two faces of the same coin, and to appreciate this is to widen our understanding of strategy.

Captain Sir Basil Liddell Hart, *Strategy*, 1954

RECONNAISSANCE

Agitate the enemy and ascertain the pattern of his movement. Determine his dispositions and so ascertain the field of battle. Probe him and learn where his strength is abundant and where deficient.

Sun Tzu, *The Art of War*, vi, c. 500 BC, tr. Griffith

Those who do not know the conditions of mountains and forests, hazardous defiles, marshes and swamps, cannot conduct the march of an army. Those who do not use local guides are unable to obtain the advantages of the ground.

Sun Tzu, *The Art of War*, vii, c. 500 BC, tr. Griffith

Since it is good not to neglect any one of the factors which contribute to the common benefit of the army, it is necessary to have experienced and intelligent guides. We should treat them well, look out for them, and take good care of them, for without them nothing worthwhile will be accomplished. The men we call guides are not simply men who know the roads, for the lowliest peasant can do that, but men who, in addition to knowing roads are able to conduct the army through the mountain passes, who can plan ahead, and who know the proper distances for the campsites, locations which are suitable and which have plenty of water, so the camp will not find itself in dire straits. They should know the topography of the enemy's country in detail, so they can lead the army into it to plunder and take captives.

General Nikephorus Ouranos, *Campaign Organization and Tactics*, c. AD 994, in

Dennis, *Three Byzantine Military Treatises*, 1985

When it is desired to apply oneself to this essential part of war, the most detailed and exact maps of the country that can be found are taken and examined and re-examined frequently. If it is not in time of war, the places are visited, camps are chosen, roads are examined, the mayors of villages, the butchers, and the farmers are talked to. One becomes familiar with the footpaths, the depths of the woods, their nature, the depth of the rivers, the marshes that can be crossed and those which cannot . . . the road is chosen for such and such a march, the number of columns in which the march can be made estimated and all strong camping places on the route are examined.

Frederick the Great, *Instructions to His Generals*, 1747

. . . a peasant and a drover are not military men, and you will find that quite different descriptions of the same stretch of country will be given by an economist, a carter, a huntsman or a soldier.

Frederick the Great, 'Des Marches d'Armée', 1777, *Ouevres*, XXIX

Single men in the night will be more likely to ascertain facts than the best glasses in the day.

General George Washington, 10 July 1779, letter to General Anthony Wayne, *The Writings of George Washington*, Vol. 15, 1931–44

A reconnaissance? I don't need one. They're only necessary for the timid, and forewarn the enemy. If you really want to find the enemy, you'll find him without them. Bayonets, cold arms, attack, punch – these are my reconnaissances.

Field Marshal Prince Aleksandr V. Suvorov, 1799, during the Italian Campaign, quoted in Longworth, *The Art of Victory*, 1966

. . . skilfully reconnoitring defiles and fords, providing himself with trusty guides, interrogating the village priest and the chief of relays, quickly establishing relations with the inhabitants, seeking out spies, seizing letters.

Napoleon, *Maxims of War*, 1831

I have no confidence in any scout.

General Robert E. Lee (1807–1870)

We would seriously overestimate the value of our excellent modern maps, however, to imagine we can now dispense with personal reconnaissance. No matter how skilfully one may read today's maps of former battlefields, one cannot fully understand, without an examination of the ground, the reasons for events as they actually occurred. At Kolin, for example, the full difficulty of the attack could be appreciated only from the Austrian defensive position.

General der Infantrie Hugo Baron von Freytag-Loringhoven, *The Power of Personality in War*, 1911

Time spent on reconnaissance is seldom wasted.

British Army Field Service Regulations (F.S.R.) 1912

In order to conquer that unknown which follows us until the very point of going into action, there is only one means, which consists in looking out until the last moment, even on the battlefield, for *information*.

Marshal of France Ferdinand Foch, *Precepts and Judgments*, 1919

You can never do too much reconnaissance.

General George S. Patton, Jr., *War As I Knew It*, 1947

From the battalion down, the use of maps is of no value and is frequently fraught with great danger. I have never seen a good battalion commander direct his units from a map. I have seen many bad battalion commanders indulge in this pusillanimous method of command.

General George S. Patton, Jr., *War As I Knew It*, 1947

Junior officers of reconnaissance units must be very inquisitive. Their reports must be accurate and factual. Negative information is as important as positive information.

General George S. Patton, Jr., *War As I Knew It*, 1947

RECRUITS/RECRUITMENT

An army raised without proper regard to the choice of its recruits was never yet made good by the length of time.

Flavius Vegetius Renatus, *Military Institutions of the Romans*, c. AD 378

No one, I imagine, can doubt that the peasants are the most fit to carry arms for they from their infancy have been exposed to all kinds of weather and have been brought up to the hardest labour. They are able to endure the most intense heat of the sun, are unaquainted with use of baths and are strangers to the other luxuries of life. They are simple, content with little, inured to fatigue, and prepared in some measure for a military life by their continual employment in their farm work, in handling the spade, digging trenches and carrying burdens.

Flavius Vegetius Renatus, *Military Institutions of the Romans*, c. AD 378

Men just dragged from the tender Scenes of domestick life; unaccustomed to the din of Arms; totally unacquainted with every kind of Military skill . . . when opposed to Troops regularly train'd, disciplined, and appointed, superior in knowledge, and superior in Arms, makes them timid and ready to fly from their own shadows.

General George Washington, 24 September 1776, letter to the President of Congress

The country lad is generally the most humble, obedient, and easily governed; but the city-born recruit is, as a rule, the smartest, tidiest, and most easily trained.

John Menzies, *Reminiscences of an Old Soldier*, 1883

There are those who say, 'I am a farmer', or, 'I am a student'; 'I can discuss literature but not the military arts.' This is incorrect. There is no profound difference between the farmer and the soldier. You must have courage. You simply leave your farms and become soldiers . . . When you take your arms in hand, you become soldiers; when you are organized, you become military units.

Mao Tse-tung, *On Guerrilla War*, 1937

When a man puts on a soldier's uniform he leaves behind his former snug and complex life for ever. All that filled his life yesterday becomes like shadows in a dream.

Ilya Ehrenburg, *Red Star*, 1941

See also: TRAINING

REFLECTION/MEDITATION

Nowhere can man find a quieter or more untroubled retreat than in his own soul.

The Emperor Marcus Aurelius (AD 121–180), *Meditations*

The general should not go to sleep before reflecting on what he should have done that he might have neglected and on what he has to do the next day.

The Emperor Maurice, *The Strategikon*, c. AD 600

What is the good of experience if you do not reflect.

Frederick the Great (1712–1786)

If I appear to be always ready to reply to everything, it is because, before undertaking anything, I have meditated for a long time – I have foreseen what might happen. It is not a spirit which suddenly reveals to me what I have to say or do in a circumstance unexpected by others: it is reflexion, meditation.

Napoleon (1769–1821)

Generalship, at least in my case, came not by instinct, unsought, but by understanding, hard study and brain concentration. Had it come easily to me, I should not have done it as well.

Colonel T. E. Lawrence, 1932, letter to Liddell Hart

An acquaintance with the 'friction' of military movements and the actual conditions of the battlefield is invaluable. But the duration of experience matters less than the capacity for reflection. And the measure of this, in turn, counts for more than the extent of study. Given some acquaintance with actual war, and time for study, the mastery of strategy and tactics is likely to be in proportion to the capacity for reflection, analysis, and originality of thought.

Captain Sir Basil Liddell Hart, *Thoughts on War*, 1944

REGIMENTS

Our home was the regiment, and the farther we got from our native state the more we became attached to it.

Private William Watson, 3rd Louisiana Infantry, CSA, quoted in *Tenting Tonight*, 1984

The regiment is the family. The colonel, as the father, should have a personal acquaintance with every officer and man, and should instill a feeling of pride and affection for himself, so that his officers and men naturally look to him for personal advice and instruction. In war the regiment should never be subdivided, but should always be maintained entire. In peace this is impossible.

General of the Army William T. Sherman, *The Memoirs of General W. T. Sherman*, 1875

The noble courage that has its origins in love of country and sense of duty is not confined to the well-born; it is to be equally found in the uneducated private soldier. What can be finer than his love of regiment, his devotion to its reputation, and his determination to protect its honour! To him 'The Regiment' is mother, sister and mistress. That its fame may live and flourish he is prepared to risk all and to die without a murmur. What other cause calls forth greater enthusiasm? It is a high, an admirable phase of patriotism, for, to the soldier, his regiment is his country.

Field Marshal Viscount Wolseley, *The Story of a Soldier's Life*, 1903

Keep your hands off the Regiments, you iconoclastic civilians who meddle and muddle in Army matters: you are not soldiers and you do not understand them.

Field Marshal Viscount Wolseley, *The Story of a Soldier's Life*, 1903

Never forget: the Regiment is the foundation of everything.

Field Marshal Earl Wavell (1883–1950), quoted in Fergusson, *The Trumpet in the Hall*, 1970

There is a great source of strength in the regimental system itself. This sets up in the group a continuing focus of affection, trust and loyalty. It uses insignia, totems and a good deal of almost mystical paraphernalia to increase the binding grip of the whole upon its members. Some day I want to turn the ethnologists on to the study of the regimental system as a means of strengthening group resistance to stress. Their advice may well be that instead of a quaint and decorative traditional survival, we have in the British regimental system a military instrument of deadly efficiency.

General Sir John Hackett, *The Profession of Arms*, 1983

REGULARS AND VOLUNTEERS

. . . Regular Troops alone are equal to the exigencies of modern war, as well as for defence as offence, and whenever a substitute is attempted it must prove illusory and ruinous.

General George Washington, 15 September 1780, to the President of Congress, *The Writings of George Washington*, Vol. 20, 1931–44

No militia will ever acquire the habits necessary to resist a regular force.

General George Washington, 15 September 1780, to the President of Congress, *The Writings of George Washington*, Vol. 20, 1931–44

. . . The firmness requisite for the real business of fighting is only to be attained by a constant course of discipline and service. I have never yet been witness to a single instance that can justify a different opinion; and it is most earnestly to be wished the liberties of America may no longer be trusted in any material degree to so precarious a dependence.

General George Washington, 15 September 1780, to the President of Congress, *The Writings of George Washington*, Vol. 20, 1931–44

When the perfect order and discipline which are essential to regular troops are contemplated, and with what ease and precision they execute the difficult manoeuvres indispensable to the success of offensive or defensive operations, the conviction cannot be resisted that such troops will always have a decided advantage over the more numerous forces composed of uninstructed militia or undisciplined recruits.

Alexander Hamilton (1755–1804)

When defending itself against another country, a nation never lacks men, but too often, *soldiers*.

Napoleon, *Political Aphorisms*, 1848

The victories in Mexico were, in every instance, over vastly superior numbers. There were two reasons for this. Both General Scott and General Taylor had such armies as are not often got together. At the battles of Palo Alto and Resaca de la Palma, General Taylor had a small army, but it was composed exclusively of regular troops, under the best drill and discipline. Every officer, from the highest to the lowest, was educated in his profession, not at West Point necessarily, but in the camp, in garrison, and many of them in Indian wars. The rank and file were probably inferior, as material out of which to make an army, to the volunteers that participated in all the later battles of the war; but they were brave men, and then drill and discipline brought out all there was in them. A better army, man for man, probably never faced an enemy than the one commanded by General Taylor in the earliest two engagements of the Mexican war. The volunteers who followed were of better material, but without drill or disciplined at the start. They were associated with so many disciplined men and professionally educated officers, that when they went into engagements it was with a confidence they would not have felt otherwise. They became soldiers themselves almost at once . . .

General of the Army Ulysses S. Grant, *Personal Memoirs of U.S. Grant*, 1885

That when a nation relies upon a system of regulars and volunteers, or regulars and militia, the men, in the absence of compulsion, or very strong inducements, will invariably enlist in the organizations most lax in discipline.

Colonel Emory Upton, *The Military Policy of the United States*, 1904

That troops become reliable only in proportion as they are disciplined; that discipline is the fruit of long training, and cannot be attained without the existence of good corps of officers.

Colonel Emory Upton, *The Military Policy of the United States*, 1904

During the war young men of two or three years' service had to lead companies. Many succeeded, but others failed in many ways. The capacity for leadership is a gift, the result of education and tact. Zeal and courage cannot always take their place. Everything was done, at home and at the front, to secure the thorough training of company commanders, but there is no doubt that the complaints of the men as to their inexperience were, at bottom, justified. This was a very serious matter, involving the admirable relations that had hitherto existed between officers and men . . . The excellent regular officer, so often the object of attacks, was no longer available. The green grass was growing on his grave. In the short period of the war, it was impossible to train a new generation of these men, with the same high professional qualities, the same thorough knowledge, and the same sense of responsibility towards their men as had been possessed by officers trained through a long course of years. Nothing could provide a more striking justification of our whole army system than the events of this war.

General Erich Ludendorff, *My War Memories 1914–1918*, 1919

In the big war companies, 250 strong, you could find every sort of man, from every sort of calling. There were Northwesterners with straw-colored hair that looked white against their tanned skins, and delicately spoken chaps with the stamp of eastern universities on them. There were large-boned fellows from Pacific-coast lumber camps, and tall, lean Southerners who swore amazingly in gentle, drawling voices. There were husky farmers from the corn-belt, and youngsters who had sprung, as it were, to arms from the necktie counter. And there were also a number of diverse people who ran curiously to type, with drilled shoulders and a bone-deep sunburn, and a tolerant scorn of everything on earth. Their speech was flavored with navy words, and the words culled from all the folk who live on the seas and the ports where our war-ships go. In easy hours their talk ran from the Tatar Wall beyond Pekin to the Southern Islands, down under Manila; from Portsmouth Navy Yard – New Hampshire and very cold – to obscure bushwhackings in the West Indies, where Cacao chiefs, whimsically sanguinary, barefoot generals with names like Charlemagne and Christophe, waged war according to the precepts of the French Revolution and the Cult of the Snake. They drank their *eau de vie* of Haute-Marne, and reminisced on saki, and vino, and Bacardi Rum – strange drinks in strange *cantinas* at the far ends of the earth; and they spoke fondly of Milwaukee beer. Rifles were high and holy things to them, and they knew five-inch broadside guns. They talked patronizingly of the war, and were concerned about rations. They were the Leathernecks, the Old Timers: collected from the ship's guard and shore stations all over the earth to form the 4th Brigade of Marines, the two rifle regiments, detached from the navy by order of the President for service with the American regulars, regarding the service as home and war as an occupation; and they transmitted their temper and character and view-point to the high-hearted volunteer mass which filled the ranks of the Marine Brigade.

Colonel John W. Thompson, Jr., *Fix Bayonets*, 1926

After hardly three months of war the greater part of our regular, professional officers and trained men had vanished, leaving only skeleton forces which had to be hastily filled with men wretchedly instructed who were sent to me from depots; while the strength of the officers was kept up by promoting subalterns, who likewise were inadequately trained. From this period onwards the professional character of our forces disappeared, and the Army became more and more like a sort of badly trained militia. The question of N.C.O.s became a particularly acute one; we had to institute training squads so as to provide, hastily and anyhow, N.C.O.s who assuredly could not take the place of their well-trained predecessors.

General Aleksei A. Brusilov, *A Soldier's Notebook*, 1931

From the beginning I had to suppose that the attitude of the Austrian and German troops, standing armies with century-old traditions, would be one of profound scepticism as the military value of our volunteer formation with the character of a militia. I was prepared for this, and knowing well the lofty ambition of the Strzelcy, I was very much afraid that I might not only wound this ambition of theirs at the first reverse, but even worse, destroy their faith in themselves as soldiers. And a reverse was very likely in the extremely poor state of our technical armament and equipment . . . Finally, as in every new foundation, our internal relations were so little consolidated that each individual commander had to devote much time to bringing order into everyday details, to founding some sort of internal *modus vivendi* between the men. I had constantly to settle a number of routine questions of a personal character resulting from the friction of the whole military machine. Questions of seniority amongst fellow officers, questions of the sphere of each man's competence, this was the hell in which I lived at the beginning of the War. I had to defend the soldiers not only from external humiliations, but also from the internal humiliation, to which a consciousness of being less efficient than those around them, and of being unable to execute the tasks they had taken upon themselves, could not but give rise.

Marshal Joseph Pilsudski, of the founding of the Polish Army as volunteer allies of the German and Austrians in the First World War, *Memoirs of a Polish Revolutionary and Soldier*, 1931

I longed for more Regular troops with which to rebuild and expand the Army. Wars are not won by heroic militia.

Sir Winston S. Churchill, *Their Finest Hour*, 1949

REGULATIONS

As far as propriety, laws, and decrees are concerned, the army has its own code, which it ordinarily follows. If these are made identical with those used in governing a state, the officers will be bewildered.

Tu Mu, AD 803–852, commentator to Sun Tzu, *The Art of War*, c. 500 BC, tr. Griffith

---

. . . not to follow regulations is the same as a blind man not following a wall.

Peter I 'The Great', quoted in Savkin, *Basic Principles of Operational Art and Tactics*, 1972

Nobody in the British Army ever reads a regulation or an order as if it were to be a guide for his conduct, or in any other manner than as an amusing novel.

The Duke of Wellington (1769–1852)

Regulations are all very well for drill but in the hour of danger they are no more use . . . You must learn to think.

Marshal of France Ferdinand Foch (1851–1929)

Method, I explained, was laid down in the Training Manuals, which were written not for sages, but for normal men, many of whom are fools. Though they must be followed, 'do not imagine for a moment', I said, 'that they have been written to exonerate you from thinking.'

Major General J. F. C. Fuller, *Memoirs of an Unconventional Soldier*, 1936

You do not rise by the regulations, but in spite of them. Therefore in all matters of active service the subaltern must never take 'No' for an answer. He should get to the front at all costs.

Sir Winston S. Churchill, *Ian Hamilton's March*

One of the first regulations might be to think.

Major General Orlando Ward, under pseudonym MacKenzie Hill, *The Field Artillery Journal*, c. 1938

Well, ultimately, the regulations are made by men. And not everything can be anticipated in the regulations. Of course, there are exceptional instances.

Admiral of the Fleet of the Soviet Union Sergei G. Gorshkov, quoted in Ablamonov, *Admiral*, 1986

REINFORCEMENTS

. . . our best chance of establishing a panic among them, as a fresh assailant has always more terrors for an enemy than the one he is immediately engaged with.

Brasidas, the Spartan, 422 BC, speech to his army before the Battle of Amphipolis, quoted by Thucydides, *History of the Peloponnesian War*, c. 404 BC, tr. Crawley

. . . if a little help reaches you in the action itself, it determines the turn of fortune for you. The enemy is discouraged and his excited imagination sees the help as being at least twice as strong as it really is.

Frederick the Great, *Instructions to His Generals*, 1747

One always has enough troops when he knows how to use them.

Napoleon, 26 June 1806, to Joseph, *Correspondance*, No. 10416, Vol. XII, 1858–1870

Generals always make requests – it is in the nature of things. There is not a one who cannot be counted upon for that. It is quite natural that the man who is entrusted with only one task thinks only about it, and the more men he has the better guarantee he has for success.

Napoleon, 4 March 1809, to Joseph, *Correspondance*, No. 14846, Vol. XVIII, 1858–1870

A seasonable reinforcement renders the success of a battle certain, because the enemy will always imagine it stronger than it really is, and lose courage accordingly.

Napoleon, *Maxims of War*, 1831

General:

In reply to your letter of the 14th, I will state that the extent of the reinforcements you can send to the Army of Northern Virginia must necessarily depend upon the strength of the enemy in your front. The plan you propose of exchanging your full for its reduced brigades I fear will add but little to its real strength. It would increase it numerically but weaken it instrinsically by taking away tired troops under experienced officers & replacing them with fresh men & uninstructed commanders. I should therefore have more to feed but less to depend on . . .

General Robert E. Lee, 16 May 1863, to General Daniel H. Hill, commanding The Department South of the James River, *Wartime Papers of Robert E. Lee*, 1987

REINFORCE SUCCESS

The moment it becomes certain that an assault cannot succeed, suspend the offensive; but when one does succeed, push it vigorously, and if necessary pile in troops at the successful point from wherever they can be taken.

General of the Army Ulysses S. Grant, 3 June 1864, at the Battle of Cold Harbor, quoted in Porter, *Campaigning With Grant*, 1906

A commander should never renew an attack along the same line (or in the same form) after it has once failed. A mere reinforcement of weight is not sufficient change, for it is probable that the enemy also will have strengthened himself in the interval. It is even more probable that his success in repulsing you will have strengthened him morally. (May 1930)

Captain Sir Basil Liddell Hart, *Thoughts on War*, 1944

RELIEF

What am I going to do with him? He is not growing in his job, but where is a better man? Do you know one?

General of the Armies John J. 'Black Jack' Pershing, 1918, discussing a deficient officer he was about to relieve, quoted in Vandiver, *Black Jack*, 1977

There were those who had been relieved from their commands who came in to tell me of the injustice which had been done them. It was hard to talk to men of this class, because in most instances I was convinced by what they told me that their relief was justified. To discuss with an old friend the smash-up of his career is tragic and depressing at best, and more practically when he feels that he has been treated unfairly and an honorable record forever besmirched.

General of the Army George C. Marshall, *Memoirs of My Service in the World War*, written between 1923–26, published in 1976

In war you must either trust your general or sack him.

General Sir John Dill, 1941, letter to Field Marshal Earl Wavell

---

. . . there were instances in Europe where I relieved commanders for their failure to move fast enough. And it is possible that some were the victims of circumstance. For how can the blame for failure be laid fairly on a single man when there are in reality so many factors that can effect the outcome of any battle? Yet each commander must always assume total responsibility for every individual in his command. If his battalion or regimental commanders fail him in the attack, then he must relieve them or be relieved himself. Many a division commander has failed not because he lacked the capacity for command but only because he declined to be hard enough on his subordinates.

In the last analysis, however, the issue of relief resolves itself into one of mutual confidence.

General of the Army Omar N. Bradley, *A Soldier's Story*, 1951

The occasion for the relief of commanders may regrettably arise. If it does, there are three points to consider: Is your decision based on personal knowledge and observation, or on secondhand information? What will be the effect on the command concerned? Are you relieving a commander whose men think highly of him – even with affection – regardless of professional competence? And finally, have you a better man available?

Every man is entitled to go into battle with the best chance of survival your forethought as a leader can provide. What best helps you discharge this responsibility? Sharing things with your men; to be always in the toughest spots; always where the crisis is, or seems most likely to develop; always thinking of what help you can give your commanders who are executing your orders; doing your utmost to see that the best in rations, shelter, first aid, and evacuation facilities are available; being generous with praise, swift and fair with punishment when you have the facts, intolerant of demonstrated failure in leadership on which lives depend, yet making full allowances for human weaknesses and the stresses and strains of battle on individuals.

General Mathew B. Ridgway, 'Leadership', *Military Review*, 10/1966

REMINISCENCES

Chief of the sailors, Ahmose, son of Ebana, triumphant; he says: 'I will tell you, O all ye people; I will cause you to know the honours which came to me. I was presented with gold seven times in the presence of the whole land; male and female slaves likewise. I was endowed with many fields.' The fame of one valiant in achievements shall not perish in this land forever.

Ahmose, son of Ebana, Egyptian naval officer in the reign of Pharaohs Ahmose, Amenhotep I, and Thutmose I (1580–1501 BC), quoted in Breasted, *Ancient Records of Egypt*, 1906

Let me urge you not despair of success since this province seems destined to revive the arts and sciences which have seemed long since dead, as we see it has already raised poetry, painting, and sculpture – as it were – from the grave. As to myself, I cannot expect to see so happy a change at my time of life. Indeed if *Fortuna* had indulged me some years ago with a territory fit for such an undertaking, I think I should soon have convinced the world of the excellence of the ancient military discipline, for I would either have increased my own dominions with glory or, at least, not have lost them with infamy and disgrace.

Niccolo Machiavelli, *The Art of War*, 1521

I should have given Suchet an army-corps under my command, I should have sent Davout a month earlier to organize my army, and appointed Clauzel Minister of War. Or I ought to have given Soult the command of the Guard. He did not wish me to employ Ney. I should have spent the night of the 15th in Fleurus, and given Grouchey's command to Suchet, and given the former the command of all the cavalry, as I had not got Murat. The soldiers did not know each other well enough to possess the proper *esprit de corps*. The cavalry were better than the infantry. It is a pity that I did not fall at Waterloo, for that would have been a fine ending. My situation is frightful! I am like a dead man, yet full of life!

Napoleon, of the Waterloo Campaign, quoted in Kircheisen, *The Memoirs of Napoleon I*, n.d.

In conclusion, I must state that, as I have dedicated myself entirely to the mobile profession of defending my country, these last *Memoirs* may be lacking literary style. In reading these pages, the reader will soon discover that the author is a soldier, not a writer. I leave it to my readers, in their infinite wisdom, to judge my character and deeds without prejudice. I will abide by their sentence.

General Antonio Lopez de Santa Anna, *The Eagle: The Autobiography of Santa Anna*, c. 1870

To be at the head of a strong column of troops, in the execution of some task that requires brain, is the highest pleasure of war – a grim one and terrible, but which leaves on the mind and memory the strongest mark; to detect the weak point of an enemy's line; to break through with vehemence and thus lead to victory; or to discover some key-point and hold it with tenacity; or to do some other distinct act which is afterward recognized as the real cause of success. These all become matters that are never forgotten.

General of the Army William T. Sherman, *The Memoirs of General W. T. Sherman*, 1875

It was given to us to learn at the outset that life is a profound and passionate thing. And it is for us to bear the report to those who come after . . . the one and only success it is man's to command is to bring to his work a mighty heart.

Justice Oliver Wendell Holmes (1841–1935)

To conclude, I would again put on record the deep love and gratitude that I have always felt for all the troops who were in my charge. At a word from me, they faced wounds, agony, and death for Russia's sake. And all in vain! May I be pardoned, for the guilt is not mine; and I could not foresee the future.

General Aleksei A. Brusilov, *A Soldier's Notebook*, 1931

I came to the conclusion that the simplest way to get rid of the burden of longing would be to attempt to put my memories on paper, to take a pen in my hand, and by its mechanical work to bind myself more closely to a life which, though so poor in experience, was yet real.

And then came into my mind the ten years of study I had devoted, before the war, to the phenomenon of war in the world. For ten years I had burrowed into the essence of the work of leadership, the element – as Clausewitz says – of danger, the element of uncertainty, finally, as I put it, the element of eternal contradictions, which are not be solved, but are cut like the Gordian knot by the sword of decision, the sword of command. I remembered that when I went to war in August, 1914, I decided to observe its phenomena carefully, to analyse myself in order to be able to provide myself with the solution to a hundred doubts, to answer a hundred questions which had remained in my spirit and my head from the time spent over books. Now in Magdeburg I decided to try sincerely and quietly to illustrate in myself the truth of the essence of commanding weighed down as it is by the burden of dangers, uncertainties and contradictions. Every soldier struggles with them because they are the essence of war. A commander bears in addition, the weight of responsiblity for his subordinates and must feel on his cheek the stinging shame of humiliation when his work of commanding has failed and others have paid with their blood for his unsuccess.

Marshal Joseph Pilsudski, *The Memoirs of a Polish Revolutionary and Soldier*, 1931

It was seventeen years ago – those days of old have vanished, tone and tint; they have gone glimmering through the dreams of things that were. Their memory is a land where flowers of wondrous beauty and varied colors spring, watered by tears and coaxed and caressed into fuller bloom by the smiles of yesterday. Refrains no longer rise and fall from that land of used-to-be. We listen vainly, but with thirsty ear, for the witching melodies of days that are gone. Ghosts in olive drab and sky blue and German grey pass before our eyes; voices that have stolen away in the echoes from the battlefields no more ring out. The faint, far whisper of forgotten songs no longer floats through the air. Youth, strength, aspirations, struggles, triumphs, despairs, wide winds sweeping, beacons flashing across uncharted depths, movements, vividness, radiance, shadows, faint bugles sounding reveille, far drums beating the long roll, the crash of guns, the rattle of musketry – the still white crosses!

And tonight we are met to remember.

General of the Army Douglas MacArthur, 14 July 1935, address to the veterans of the Rainbow (42nd) Division, *A Soldier Speaks*, 1942

This I have always felt; as a youth and now as a man well in middle age: that truth is courage intellectualized. Thus the idea of the great man is the human coping stone of my philosophy as it was of the philosophy of that great man – Thomas Carlyle. Therefore in my study of war I have always put the great man first. As my system was founded on the organization of the human body, of necessity it follows that to breathe life into it, those who can do so must be men who at least aspire towards greatness.

Major General J. F. C. Fuller, *Memoirs of an Unconventional Soldier*, 1936

At a difficult time a prince of my royal family once sent me a small portrait of Frederick the Great on which he had inscribed these words that the great king addressed to his friend, the Marquis d'Argens, when his own defeat seemed imminent. 'Nothing can alter my inner soul: I shall pursue my own straight course and shall do what I believe to be right and honourable.' The little picture I have lost, but the King's words remain engraved on my memory and are for me a model. If, despite everything, I could not prevent the defeat of my country, I must ask my readers to believe that this was not for lack of a will to do so.

Colonel General Heinz Guderian, *Panzer Leader*, 1953

I believe it was the transparency of our aims, the love of our Fatherland, the strong sense of duty and the spirit of self-sacrifice which animated each of our few Europeans and communicated themselves, consciously or unconsciously, to our brave black soldiers that gave our operations that impetus which they possessed to the end. In addition there was a soldierly pride, a feeling of firm mutual co-operation and a spirit of enterprise without which military success is impossible in the long run. We East Africans knew only too well that our achievements could not be compared with the military deeds and devotion of those in the homeland. No people in history had ever done more.

If we East Africans received so kindly a reception in the homeland it was because everyone seemed to think that we had preserved some part of Germany's soldierly traditions, had come back home unsullied, and that the Teutonic sense of loyalty peculiar to Germans had kept its head high even under the conditions of war in the tropics.

Major General Paul von Lettow-Vorbeck, *East African Campaigns*, 1957

Well, I'd like to do it all over again. The whole thing.
And more than that – more than anything – I'd like to see once again the face of every Marine I've ever served with.

General Lewis 'Chesty' Puller, 26 June 1960, to his wife when she asked him what he would like to do after his retirement, quoted in Davis, *Marine!*, 1962

With every passing year we draw further and further away from those years of war. A new generation of people has grown up. For them war only means our reminiscences of it. And the numbers of us who took part in those historic events are dwindling fast. But I am convinced that time cannot cause the greatness to fade. Those were extraordinarily difficult, but also truly glorious years. Once a person has undergone great trials and come through victorious, then throughout his life he draws strength from this victory.

Marshal of the Soviet Union Georgi K. Zhukov, *Reminiscences and Reflections*, 1974

I would like to be rated not by what I wrote but by how I lived, for I wrote to live and not the other way around. Having tried in this accounting to bestride the various periods in my journeying as if they were so many stages along a road, largely with the hope of explaining fortune that I hardly understand myself, I find on looking back that there is little to lament and much to cheer. Good companions were always alongside to help me through trial; I needed them as they in turn needed me. My yesterdays were the best the country ever knew, or so I believe, for in living them to the hilt I helped make them.

I think I will leave it at that.

Brigadier General S. L. A. Marshall, *Bringing Up the Rear*, 1979

REORGANIZATION

Any military reorganization should conform to certain set principles:

(1) Power must go with responsibility.

(2) The average human brain finds its effective scope in handling from three to six other brains.

If a man divides the whole of his work into two branches and delegates his responsibility, freely and properly, to two experienced heads of branches he will not have enough to do. The occasions when they would have to refer to him would be too few to keep him fully occupied. If he delegates to three heads he will be kept fairly busy whilst six heads of branches will give most bosses a ten hours' day. Those data are the result of centuries of the experiences of soldiers, which are greater, where organization is in question, than those of politicians, business men or any other class of men by just so much as an Army in the field is a bigger concern than a general election, the Bank of England, the Standard Oil Company, the Steel Trusts, the Railway Combines of America, or any other part of politics or business. Of all the ways to waste there is none so vicious as that of your clever politician trying to run a business concern without having any notion of self-organization. One of them who took over Munitions for a time had so little idea of organizing his own energy that he nearly dies of overwork *through holding up the work of others*; by delegating responsibility coupled with *direct access to himself* to seventeen sub-chiefs!

Now, it will be understood why a battalion has four companies (and not seventeen); why a brigade has three or four battalions (and not seventeen).

General Sir Ian Hamilton, *the Soul and Body of an Army*, 1921

The hypothesis of every reorganization of an army is, first of all, peace on the outer borders. That means several years of internal peace and a state of political calm . . . Before these conditions are achieved a successful military reorganization cannot be accomplished. Success cannot be attained while you are in a continual state of war.

Colonel General Hans von Seekt, while a military adviser to Chiang Kai-shek during the 1930s, *Denkschrift fuer Marschall Chiang Kai-shek*, n.d.

REPLACEMENTS

I don't know what effect these men will have on the enemy, but, by God, they frighten *me*.

The Duke of Wellington, 1809, impression of replacements sent to him in Spain

---

The fighting troops are not being replaced effectively, although masses of drafts are sent to the technical and administrative services, who were originally on the most lavish scale, and who have since hardly suffered at all by the fire of the enemy. The first duty of the War Office is to keep up the rifle infantry strength.

Sir Winston Churchill, 3 May 1943, note to the Chief of the Imperial General Staff

---

Replacements are like spare parts – supplies. They must be asked for in time by the front line, and the need for them must be anticipated in the rear.

General George S. Patton, Jr., *War As I Knew It*, 1947

REPORTS

I came, I saw, I conquered.

Julius Caesar, quoted in Suetonius (c. AD 69–after AD 122), *Lives of the Caesars*

They enemy came to us. They are beaten. God be praised! I have been a little tired all day. I bid you good-night and am going to bed.

General Marshal Vicomte de Turenne, 14 June 1658, report of his victory that day in the Battle of the Dunes, quoted in Weygand, *Turenne, Marshal of France*, 1930

All reports should be written clearly, precisely, as far as possible avoiding any inaccuracy, length or beauty of expression, in order not to cloud the thought.

Field Marshal Prince Aleksandr V. Suvorov, 1799, during his campaign in Northern Italy

---

Reports are not self executive.

Florence Nightingale, 1857, marginal comment on a document

. . . as it was, nothing came of all the loss and effort, except a report which I sent over to the British headquarters in Palestine for the Staff's consumption. It was mainly written for effect, full of quaint similes and mock simplicities; and made them think me a modest amateur, doing his best after the great models; not a clown, leering after them where they with Foch, bandmaster, at their head went drumming down the old road of effusion of blood into the house of Clausewitz. Like the battle, it was a nearly-proof parody of regulation use. Headquarters loved it, and innocently, to crown the jest, offered me a decoration on the strength of it. We should have more bright breasts in the Army if each man was able without witnesses, to write his own despatch.

Colonel T. E. Lawrence, *The Seven Pillars of Wisdom*, 1926

In war nothing is ever as bad, or as good, as it is reported to Higher Headquarters. Any reports which emanate from a unit after dark – that is, where the knowledge has been obtained after dark – should be viewed with scepticism by the next higher unit. Reports by wounded men are always exaggerated and favor the enemy.

General George S. Patton, Jr., *War As I Knew It*, 1947

When you receive reports of counter-attacks, find out who sent them – that is, the size of the unit which sent them. A squad occupying a position will report an enemy section approaching it as a counter-attack, but such a counter-attack has no material effect on a division or a corps.

General George S. Patton, Jr., *War As I Knew It*, 1947

Bad news is not like fine wine – it does not improve with age.

General Creighton Abrams, quoted in *Parameters*, Summer 1987

REPUTATION

Achilles absent was still Achilles.

Homer, *The Iliad*, xxii, *c.* 800 BC

What of the two men in command? You have Alexander, they – Darius!

Alexander the Great, 333 BC, address to his army before the Battle of Issus, quoted in Arrian, *The Campaigns of Alexander the Great*, *c.* AD 150

But how many ships do you reckon my presence is worth?

Antigonas II Gonatas (*c.* 319 BC–239 BC), quoted in Plutarch

---

But what most commonly keeps an army united, is the reputation of the general, that is, of his courage and good conduct; without these, neither high birth nor any sort of authority is sufficient.

Niccolo Machiavelli, *The Art of War*, 1521

I have tamed men of iron in my day – shall I not crush men of butter?

Alvarez de Toledo, Duke of Alva (1507–1582) of his attempt to pacify the Netherlands

---

. . . it is more the nature of men to be less interested in things which relate to others than about those in which they themselves are concerned. The reputation of an organization becomes personal just as soon as it is an honour to belong to it.

Field Marshal Maurice Comte de Saxe, *My Reveries*, 1732

Had I succeeded, I should have died with the reputation of the greatest man that ever lived.

Napoleon, quoted in O'Meara, *Napoleon in Exile*, 1822

I used to say of him that his presence on the field made the difference of forty thousand men.

The Duke of Wellington, 1831, letter to Lord Stanhope

I desire that the terror of my name shall guard our frontiers more potently than chains or fortresses.

General Mikhail Yermolov, during the pacification of the Caucasus in the 1830s, quoted in Blanch, *The Sabres of Paradise*, 1960

RESERVE

Wars are paid for by the possession of reserves.

Thucydides, *Peloponnesian Wars*, *c.* 404 BC, tr. Warner

The general should also have somewhere a picked corps, stationed apart from the phalanx as military reserves, that he may have them ready to give assistance to those detachments of his force that are exhausted. These fresh troops are of not a little advantage in attacking tired men; for, besides relieving those of their own men who are worn out, they attack in their full freshness a wearied enemy.

Onasander, *The General*, AD 58

It is better to have several bodies of reserves than to extend your front too much.

Flavius Vegetius Renatus, *Military Institutions of the Romans*, *c.* AD 378

A general should always have a body of chosen troops about him, whom he can send to the support of sections of the army which are hard pressed.

The Emperor Maurice, *The Strategikon*, *c.* AD 600

When one has a good reserve, one does not fear one's enemies.

Richard I, 'The Lion Heart', King of England, 1194, after his victory at the Battle of Freteval, quoted in *Histoire de Guillaume le Maréchal*, *c.* 1220

Providence is always on the side of the last reserve.

Napoleon, *Sayings of Napoleon*

Fatigue the opponent, if possible, with few forces and conserve a decisive mass for the critical moment. Once this decisive mass has been thrown in, it must be used with the greatest audacity.

Major General Carl von Clausewitz, *Principles of War*, 1812

The great secret of battle is to have a reserve. I always had.

The Duke of Wellington (1769–1852)

MAXIM 96. A general who retains fresh troops for the day after a battle is almost always beaten. He should, if helpful, throw in his last man, because on the day after a complete success there are no more obstacles in front of him; prestige alone will insure new triumphs to the conqueror.

Napoleon, *The Military Maxims of Napoleon*, 1827, ed. Burnod

---

A reserve has two distinct purposes. One is to prolong and renew the action; the second, to counter unforeseen threats. The first purpose presupposes the value of the successive use of force . . . The case of a unit being sent to a point that is about to be overrun is clearly an instance of the second category, since the amount of resistance necessary at that point had obviously not been foreseen. A unit that is intended merely to prolong the fighting in a particular engagement and for that purpose is kept in reserve, will be available and subordinate to the commanding officer, though posted out of the reach of fire.

Major General Carl von Clausewitz, *On War*, iii, 1832, tr. Howard and Paret

---

The system of holding out a reserve as long as possible for independent action when the enemy has used his own, ought to be applied downwards. Each battalion should have its own, each regiment its own, firmly maintained.

Colonel Charles Ardnant du Picq, *Battle Studies*, 1880

In the battle line, tactics merely consist in overcoming hostile resistance by a slow and progressive wear of the enemy's resources; for that purpose, the fight is kept up everywhere. It must be supported, and such is the use made of the reserve. They must become warehouses into which one dips to replace the wear and tear as it occurs. Art consists in still having a reserve when the opponent no longer has one, so as to have the last word in a struggle in which wearing-down is the only argument employed.

In the battle of manoeuvres, the reserve is a sledgehammer planned and carefully preserved to execute the only action from which any decisive result is to be expected: the final attack. The reserve is meanwhile husbanded with the utmost caution, in order that the tool be as strong, the blow as violent as possible.

Marshal of France Ferdinand Foch, *Principles of War*, 1913

I have no more reserves. The only men I have left are the sentries at my gates. I will take them with me to where the line is broken, and the last of the English will die fighting.

Field Marshal Sir John French, 1914, message to General Foch during the Battle of Ypres, quoted in Spears, *Liaison 1914, A Narration of the Great Retreat*

The reserve is a *club*, prepared, organized, reserved, carefully maintained with a view to carrying out the one act of battle from which a result is expected – the decisive attack.

Marshal of France Ferdinand Foch, *Precepts and Judgments*, 1919

There is always the possibility of accident, of some flaw in materials, present in the general's mind: and the reserve is unconsciously held to meet it.

Colonel T. E. Lawrence (1888–1935)

It is in the use and withholding of their reserves that the great Commanders have generally excelled. After all, when once the last reserve has been thrown in, the Commander's part is played . . . The event must be left to pluck and to the fighting troops.

Sir Winston S. Churchill, *Painting as a Pastime*, 1932

To fight without a reserve is similar to playing cards without capital – sheer gambling. To trust to the cast of dice is not generalship.

Major General J. F. C. Fuller, *Memoirs of an Unconventional Soldier*, 1936

A study of Napoleon's tactics will show us that the first step he took in battle was not to break the enemy's front, and then when his forces were disorganized risk being hit by the enemy's reserves; but instead to draw the enemy's reserves into the fire fight, and directly they were drawn in to break through them or envelop them. Once this was done, security was gained; consequently, a pursuit could be carried out, a pursuit being more often than not initiatied by troops disorganized by victory against troops disorganized by defeat.

Major General J. F. C. Fuller, *Memoirs of an Unconventional Soldier*, 1936

In mechanized warfare the value of a reserve cannot be exaggerated, for increased mobility carries with it the power of effecting innumerable surprises, and the more the unexpected becomes possible, the stronger must be the reserves.

Major General J. F. C. Fuller, *Armoured Warfare*, 1943

RESISTANCE TO REFORM

I have not written this book for military monks, but for civilians, who pay for their alchemy and mysteries. In war there is nothing mysterious, for it is the most common-sense of all the sciences . . . If it possess a mystery, then that mystery is unprogressiveness, or it is a mystery that, in a profession which may, at any moment, demand the risk of danger and death, men are found willing to base their work on the campaigns of Waterloo and Sedan when the only possible war which confronts them is the next war.

Major General J. F. C. Fuller, *The Reformation of War*, 1923

It [the Army] has been in the family for three hundred years, and he is naturally very loath to part with it and inhabit some horrible ferro-concrete house. He cannot afford to modernize it, and, to make both ends meet, he shuts up room after room, and so 'economizes' his reduced income and hopes for better times. He cannot tear himself away from its memories and traditions and family ghosts, and so dry-rot creeps through its foundations and the rain percolates through its roof.

Major General J. F. C. Fuller, *The Reformation of War*, 1923

There are two kinds of treason. The first is constitutional and means betraying one's own country. The second means action whereby a party 'betrays their trust'. The Army and Navy are treasonable under that head for not giving proper improvements to the air service. Of course, I refer to the system and not to any individual.

Brigadier General William 'Billy' Mitchell, 1925, at his court-martial

---

The problem which faces the reformer of armies in peace might be likened to that of an architect called on to alter and modernize an old-fashioned house without increasing its size, with the whole family still living in it (often grumbling at the architect's improvements, since an extra bathroom can only be added at the expense of someone's dressing-room) and under the strictest financial limitations.

Field Marshal Earl Wavell, 'The Army and the Prophets', *Royal United Services Institute Journal (RUSI)*, May 1930

The regular officer has the traditions of forty generations of serving soldiers behind him, and to him the old weapons are the most honoured.

Colonel T. E. Lawrence (1888–1935)

An army is an institution not merely conservative but retrogressive by nature. It has such natural resistance to progress that it is always insured against the danger of being pushed too fast. Far worse and more certain, as history abundantly testifies, is the danger of it slipping backward. Like a man pushing a barrow up a hill, if the soldier ceases to push, the military machine will run back and crush him. To be deemed a revolutionary in the army is merely an indication of vitality, the pulse-beat which shows that the mind is still alive. When a soldier ceases to be a revolutionary it is a sure sign he has become a mummy. (January 1931)

Captain Sir Basil Liddell Hart, *Thoughts on War*, 1944

I found that the technique and practices developed at Benning and Leavenworth would practically halt the development of an open warfare situation, apparently requiring an armistice or some understanding with a complacent enemy.

General of the Army George C. Marshall, 4 December 1933, quoted in Pogue, *George C. Marshall: Education of a General*, 1963

Conventionalism takes a firm stand against innovations. But this inflexibility towards any immediate change is often followed in the long run by the most astounding reversals. When it ultimately changes, it often turns upside down and stands on its head. (April 1936)

Captain Sir Basil Liddell Hart, *Thoughts on War*, 1944

RESOLUTION

I have not the particular shining bauble or feather in my cap for crowds to gaze at or kneel to, but I have power and resolution for foes to tremble at.

Oliver Cromwell (1599–1658)

The true prudence of a general consists of energetic resolve.

Napoleon (1769–1821)

It is not the manner in which the leaders carry out the task of command, of impressing their resolution in the hearts of others, that makes them warriors, far more than all other aptitudes or faculties which theory may expect of them?

General Gerhardt von Scharnhorst, at the time of Bluecher's appointment to command the army of Silesia in 1813, quoted in Foch, *Principles of War*, 1913

I will smash them, so help me god!

Major General Andrew Jackson, 1815, at the Battle of New Orleans

---

Hardship, blood, and death create enthusiasts and martyrs and give birth to bold and desperate resolutions.

Napoleon, at Saint Helena, ed. Herold, *The Mind of Napoleon*, 1955

Great extremities require extraordinary resolution. The more obstinate the resistance of an army, the greater the chances of success. How many seeming impossibilities have been accomplished by men whose only resolve was death!

Napoleon, *Maxims of War*, 1831

Late one night during the war, after a certain serious decision had been taken, my chief coadjutor came to me full of doubts as to whether our decision had been right. 'Let it be,' I answered. 'Only military academies fifty years hence will know for certain whether we did right or wrong.'

Colonel General Hans von Seekt, *Thought of a Soldier*, 1930

RESPONSIBILITY

Responsibility is the test of a man's courage.

Admiral Lord St. Vincent (1735–1833)

This was all my fault, General Pickett. This has been my fight and blame is mine. Your men did all men could do. The fault is entirely my own.

General Robert E. Lee, 3 July 1863, at the Battle of Gettysburg after Pickett's Charge when Lee rode out to meet the remnants of the attacking force, quoted in Stackpole, *They Met at Gettysburg*, 1956

. . . responsibility, the best of educators.

Rear Admiral Alfred Thayer Mahan, *The Life of Nelson*, 1897

I don't know who won the Battle of the Marne, but if it had been lost, I know who would have lost it.

Marshal of France J. C. Joffre, 1919, to the Briey Parliamentary Commission

---

In forty hours I shall be in battle, with little information and on the spur of the moment will have to make most momentous decisions, but I believe that one's spirit enlarges with responsibility and that, with God's help, I shall make them and make them right. It seems that my whole life has been pointed to this moment. When this job is done, I presume I will be pointed to the next step in the ladder of destiny. If I do my full duty, the rest will take care of itself.

General George S. Patton, Jr., 6 November 1942, *War As I Knew It*, 1947

A General Officer who will invariably assume the responsibility for failure, whether he deserves it or not, and invariably give the credit for success to others, whether they deserve it or not, will achieve outstanding success.

General George S. Patton, Jr., *War As I Knew It*, 1947

To men who have been long in battle and have thought about it deeply, there comes at last the awareness of this ultimate responsibility – that one man must go ahead so that a nation may live. No feeling of futility accompanies that thought. At the time one accepts it simply as the rule of life and of death, of struggle and of national survival. But in the long afterglow comes also the realization that a nation may perish because to few of its people have found this truth on the only field where it may be found and because too many of the others who have not found it, unconsciously resist it and rule it out because it has not been part of their experience.

Brigadier General S. L. A. Marshall, *Men Against Fire*, 1947

Hindenburg once said that he was occasionally held responsible for victories, but always for defeats.

Field Marshal Albert Kesselring, *Memoirs of Field Marshal Kesselring*, 1953

Soldiers of an army invariable reflect the attitude of their General. The leader is the essence. Isolated cases of rapine may well be exceptional, but widespread and continuing abuse can only be a fixed responsibility of highest field authority. Resultant liability is commensurate with resulting crime. To hold otherwise would be to prevaricate the fundamental nature of the command function. This imposes no new hazard on a commander, no new limitation on his power. He has always, and properly, been subject to due process of law. Powerful as he may become in time of war, he still is not autocratic or absolute, he still remains responsible before the bar of universal justice. From time immemorial the record of high commanders, of whatever side, has been generally temperate and just.

General Douglas MacArthur, 21 March 1946, in justification of the death penalty for General Homma, *A Soldier Speaks*, 1965

. . . every good soldier in authority should be just as concerned with his responsibility to help those under him make the right turns.

Major General Aubrey 'Red' Newman, *Follow Me*, 1981

When you do a deed, you bear responsibility for it; you decide and that is not a moment to doubt or falter. You are totally engrossed in the deed and with the idea of giving yourself over to that deed and doing all you are capable of. But later, when the deed is over, when you reflect on what you did, you think not merely of the past but of the future, and a sense sharpens of you lacking something, of being short of something, of you not knowing a number of things you ought to know, and this recurrent feeling forces you to think everything over anew and decide for yourself: 'could you not have done better than you did if you had possessed all that you lack?'

Marshal of the Soviet Union Georgi K. Zhukov, quoted in *Soviet Military Review*, 6/1988

RETREAT

No shame in running,
fleeing disaster, even in pitch darkness.
Better to flee from death than feel its grip.

Agamemnon, in Homer, *The Iliad*, c. 800 BC, ix, tr. Fagles

There is no retreat, or to put it more mildly, retirement, which does not bring an infinity of woes to those who make it, to wit: shame, hunger, loss of friends, goods and arms, and death, which is the worst of them, but not the last for infamy endures for ever.

Hernan Cortes, 1520, address to his army, quoted in de Gomara, *Istoria de la Conquista de Mexico*, 1552, tr. Simpson

---

If you have been obliged to withdraw repeatedly before the enemy, the troops get frightened. The same thing happens if you have suffered some check; even when a good opportunity presents itself to you, it must be allowed to escape as soon as you see any wavering among the troops.

Frederick the Great, *Instructions to His Generals*, 1747

Never sound the retreat. Never. Warn the men that if they hear it, it is only a ruse on the part of the enemy.

Field Marshal Prince Aleksandr V. Suvorov (1729–1800), quoted in Blanch, *The Sabres of Paradise*, 1960

There are few generals that have run oftener, or more lustily than I have done. But I have taken care not to run too far, and commonly have run as fast forward as backward, to convince the Enemy that we were like a Crab, that could run either way.

Major General Nathanael Greene, 18 July 1781, to Henry Knox

If we are defeated, we can think about retreating then, and in any case, I shall be dead, so why should I worry?

Marshal of France Pierre Augereau, Duc de Castiglione, 1796, advice to Napoleon who considered retreating before the Battle of Castiglione, quoted in Delderfield, *Napoleon's Marshals*, 1962

Count Peter Ivanovitch. The troops of Count Bellegarde from the Tyrol will come up to Alessandria uninstructed, strangers to the handling of the bayonet and sword . . . visit me, and set off at once to Alessandria, and there reveal to the Bellegarde troops the secret of beating the enemy with cold steel, and carefully adapt them to this conquering attack; two or three lessons will be enough for the instruction of all the detachments, but if there is time, they can study more by themselves; but do you unlearn them how to retreat.

Field Marshal Prince Aleksandr V. Suvorov, 1799, to Prince Bagration during the Italian Campaign on how to deal with allied Austrian troops, quoted in Blease, *Suvorof*, 1920

People who think of retreating before a battle has been fought ought to have stayed home.

Marshal of France Michel Ney, Duc d'Elchingen, Prince de La Moskova, 1805, to Baron Jomini, quoted in Horricks, *Marshal Ney, The Romance and the Real*

Honourable retreats are no ways inferior to brave charges, as having less fortune, more of discipline, and as much valour.

Major General Sir William Napier, *Peninsular War,* 1810

MAXIM 6. A retreat, however skillful the manoeuvres may be, will always produce an injurious moral effect on the army, since by losing the chances of success yourself you throw them into the hands of the enemy. Besides retreats cost far more, both in men and material, than the most bloody engagements; with this difference, that in a battle the enemy loses nearly as much as you, while in a retreat the loss is all on your side.

Napoleon, *The Military Maxims of Napoleon,* 1827, ed. Burnod

---

In a retreat, besides the honour of the army, the loss of life is often greater than in two battles.

Napoleon, *Maxims of War,* 1831

To know when to retreat and to dare to do it.

The Duke of Wellington, when asked what was the best test of greatness in a general, quoted in Fraser, *Words on Wellington,* n.d.

When a battle is lost, the strength of the army is broken – its morale even more than its physical strength. A second battle without the help of new and favourable factors would mean outright defeat, perhaps even absolute destruction. That is a military axiom. It is in the nature of things that a retreat should be continued until the balance of power is re-established – whether by means of reinforcements or the cover of strong fortresses or major natural obstacles or the over-extension of the enemy.

Major General Carl von Clausewitz, *On War,* iii, 1832, tr. Howard and Paret

---

The officers should feel . . . that firmness amid reverses is more honourable than enthusiasm in success, – since courage alone is necessary to storm a position, while it requires heroism to make a difficult retreat before a victorious and enterprising enemy always opposing to him a firm and unbroken front. A fine retreat should meet with a reward equal to that given for a great victory.

Lieutenant General Antoine-Henri Baron de Jomini, *Summary of the Art of War,* 1838

My wounded are behind me and I will never pass them alive.

Major General Zachary Tayor, 1847, upon the suggestion of retreat at Buena Vista

---

Retreat is a concept which fully enters into that of the attack. I retreat over 100 to 200 kilometres so as to go over to the attack on a certain line at a certain moment decided by myself . . . Retreat is one of the movements in the general course of offensive operations . . . Retreat is not flight.

Mikhail Frunze, 1922, quoted in Leites, *The Soviet Style in War,* 1982

It will be readily understood that troops which have once begun to retreat lose heart, discipline slackens, and it is hard to say when or how they will stop and what their conditions will be when they do.

General Aleksei A. Brusilov, *A Soldier's Notebook*, 1931

There had been last-minute reinforcements, a battalion of U.S. Army troops which fought its way through the enemy with heavy losses. Its colonel reported to Puller [USMC] for orders.

'Take your position along those hills and have your men dig in.'

'Yes, sir. Now, where is my line of retreat?'

Puller's voice became low and hard: 'I'm glad you asked me that. Now I know where you stand. Wait one minute.' He took a field telephone and called his tank commander. The Army officer listened to the Marine order:

'I've got a new outfit,' Puller said. He gave its position in detail. 'If they start to pull back from that line, even one foot, I want you to open fire on them.' He hung up the telephone and turned to the Army officer:

'Does that answer your question?'

Lieutenant General Lewis 'Chesty' Puller, 1950, at the Chosin Reservoir, quoted in Davis, *Marine*, 1962

The soldier who has been forced to retreat through no fault of his own loses confidence in the higher command; because he has withdrawn already from several positions in succession he tends to look upon retreat as an undesirable but natural outcome of a battle.

Field Marshal Earl Alexander, *The Alexander Memoirs 1940–1945*, 1961

See also: GOLDEN BRIDGE

RETIREMENT

He says, 'I followed the Kings of Upper and Lower Egypt, the gods; I was with their majesties when they went to the South and North country, in every place where they went; from King Ahmose I, triumphant, King Thutmose I, triumphant, King Thumose II, triumphant, until this Good God, King Thutmose III who is given life forever.

I attained a good old age, having had a life of royal favour, having had honour under their majesties and the love of me having been in court.'

Ahmose-Pen-Nekhbet, Egyptian general of the 18th Dynasty *c.* 1580–1500 BC, quoted in Breasted, *Ancient Records of Egypt*, 1906

The time factor . . . rules the profession of arms. There is perhaps none where the dicta of the man in office are accepted with such an uncritical deference, or where the termination of an active career brings a quicker descent into careless disregard. Little wonder that many are so affected by the sudden transition as to cling pathetically to the trimmings of the past.

Captain Sir Basil Liddell Hart, *Thoughts on War*, 1944

See also: MEMOIRS, REMINISCENCES

REVOLUTION

He who serves a revolution ploughs the sea.

General Simon Bolivar, December 1830, quoted in Madariaga, *Bolivar*, 1852

In time of revolution, with perseverance and courage, a soldier should think nothing impossible.

Napoleon, *Political Aphorisms*, 1848

The right of revolution is an inherent one. When people are oppressed by their government, it is a natural right they enjoy to relieve themselves of the oppression, if they are strong enough, either by withdrawal from it, or by overthrowing it and substituting a government more acceptable.

General of the Army Ulysses S. Grant, *Personal Memoirs of U.S. Grant*, 1885

Every great revolution brings ruin to the old army.

Leon Trotsky, 1921, to Professor Milyukov

We are working for a revolution. If we do not start it by improving the life of the soldiers, all slogans of reforming and improving society are but empty words.

Generalissimo Chiang Kai-shek, 1925, letter to Chou en Lai

People do not make revolution eagerly any more than they do war. There is a difference, however, that in war compulsion plays the decisive role, in revolution there is no compulsion except that of circumstances. A revolution takes place only when there is no other way out. And the insurrection, which rises above a revolution like a peak in the mountain chain of its events, can no more be evoked at will than the revolution as a whole. The masses advance and retreat several times before they make up their minds to the final assault.

Leon Trotsky, *The History of the Russian Revolution*, Vol III, 1932

Revolution is not a dinner party, not an essay, or a painting, nor a piece of embroidery; it cannot be advanced softly, gradually, carefully, considerately, respectfully, politely, plainly and modestly.

Mao Tse-tung, quoted in *Time*, 18 December 1950

The central task and the highest form of a revolution is to seize political power by armed force, to settle problems by force.

Mao Tse-tung, *Problems of War and Strategy*, 1954

One does not necessarily have to wait for a revolutionary situation: it can be created.

Ernesto 'Che' Guevara, *Guerrilla Warfare*, 1961

The fundamental principle of revolutionary war: strike to win, strike only when success is certain; if not, then, don't strike.

General Vo Nguyen Giap, *People's War – People's Army*, 1961

A revolutionary war is never confined within the bounds of military action. Because its purpose is to destroy an existing society and its institutions and to replace them with a completely new state structure, any revolutionary war is a unity of which the constituent parts, invarying importance, are military, political, economic, social, and psychological.

Brigadier General Samuel B. Griffith, II, introduction to Mao Tse-tung, *On Guerrilla Warfare*, 1961

REWARDS AND PUNISHMENTS

One and the same lot for the man who hangs back
and the man who battles hard. The same honour waits
for the coward and the brave. They both go down to Death
the fighter who shirks, the one who works to exhaustion.

Achilles, in Homer, *The Iliad*, ix, *c.* 800 BC, tr. Fagles

Bestow rewards without respect to customary practice; publish orders without respect to precedent. Thus you may employ the entire army as you would one man.

Sun Tzu, *The Art of War*, *c.* 500 BC, tr. Griffith

. . . rewards and punishments must be reliable, otherwise troops will not halt at the sound of the gong or advance at the sound of the drum, so that even if there are million of them, what use are they?

Wu Ch'i, *d.* 381 BC, quoted in Bloodworth, *The Chinese Machiavelli*, 1976

And so by means of such incentives even those who stay at home feel the impulse to emulate such achievements in the field no less than those who are present and see and hear what takes place . . . When we consider this people's [the Romans] almost obsessive concern with military rewards and punishments, and the immense importance which they attach to both, it is not surprising that they emerge with brilliant success from every war in which they engage.

Polybius, *The Rise of the Roman Empire*, *c.* 125 BC

In one day, then, six battles were fought, three at Durazzo and three at the earthwork.

When an accounting was made we found that about 2,000 of the Pompeians had fallen, including many veterans and centurions . . . and that six of their military standards had been brought in. We lost not more than twenty men in all the battles; but in the redoubt every single man was wounded and four centurions of one cohort had lost their eyes. Wishing to produce evidence of their exertion and danger they counted out to Caesar some 30,000 arrows which had been shot into the redoubt, and when the shield of centurion Scaeva was brought forward 120 holes were found in it. In recognition of their services to himself and the state [it was clear that Scaeva was largely responsible for saving the redoubt] Caesar presented him with 200,000 sesterces, read a citation for him, and promoted him from the eighth rank to senior centurion. Later Caesar rewarded the cohort generously with double pay, grain, clothing, and military decorations.

Julius Caesar, *The Civil War*, c. 45 BC, tr. Hadas

On returning from battle, the general should first offer to the gods such sacrifice and festal celebrations as the circumstances permit, promising to offer the customary thank-offerings after complete victory; then he should honour those soldiers who have faced danger most bravely with the gifts and marks of distinction which are usually given, and he should punish those who have shown themselves cowards . . . since these rewards strengthen the self-esteem of those who have deserved well, and encourage others who desire similiar rewards . . . Whenever honour is paid to the brave and punishment of the cowardly is not neglected, then an army must have fair expectation; the latter are afraid to be found wanting, the former are ambitious to show prowess. It is not only necessary in victory to distribute rewards to the individual men but also make a recompense to the army as a whole.

Onasander, *The General*, AD 58

A policy of rewards and penalties means rewarding the good and penalizing wrongdoers. Rewarding the good is to promote achievement; penalizing wrongdoers is to prevent treachery.

It is imperative that rewards and punishments be fair and impartial. When they know rewards are to be given, courageous warriors know what they are dying for; when they know penalties are to be applied, villains know what to fear.

Therefore, rewards should not be given without a reason, and penalties should not be applied arbitrarily. If rewards are given for no reason, those who have worked hard in public service will be resentful; if penalties are applied arbitrarily, upright people will be bitter.

Zhuge Liang (AD 180–234), *The Way of the General*

Paradise is under the shadow of our swords. Forward!

Omar I Ibn Al Khattab, Caliph, AD 637, at the Battle of Qadisiya

---

For soldiers to strive to scale high walls in spite of deep moats and showers of arrows and rocks, or for soldiers to plunge eagerly into the fray of battle, they must be induced by serious rewards; then they will prevail over any enemy.

The rule is 'Where there are serious rewards, there will be valiant men.'

Liu Ji (1310–1375), *Lessons of War*

If a man who serves indolently and a man who serves well are treated in the same way, the man who serves well may begin to wonder why he does so.

Asakura Toshikage (1428–1481), *The 17 Articles of Asakura Toshikage* in Wilson, *Ideals of the Samurai*, 1982

The reward of the general is not a bigger tent, but command.

Justice Oliver Wendell Holmes, Jr., *Law and the Court*, 1913

RISK

Risks should not be taken without necessity or real hope of gain. To do so is the same as fishing with gold as bait.

The Emperor Maurice, *The Strategikon*, c. AD 600

We must risk something for God!

Hernan Cortes, 1519, when he smashed the idol of the God of the Smoking Mirror in the Great Teocalli (Templo Mayor) in Tenochtitlan

---

You say that I run too much risk. I don't do it because I want to but because I am obliged to. If I don't go into danger, nobody else will. They are all volunteers; I can't force them.

Henry IV (Henry of Navarre), King of France (1553–1610)

We should on all occasions avoid a general Action, or put anything to Risque, unless compelled by a necessity, into which we ought never to be drawn.

General George Washington, September 1776, letter to the President of Congress

---

It is true that I must run great risk; no gallant action was ever accomplished without danger.

Admiral John Paul Jones, 1778, letter to the American Commissioners in Paris

---

The rules of conduct, the maxims of action, and the tactical instincts that serve to gain small victories may always be expanded into the winning of great ones with suitable opportunity; because in human affairs the sources of success are ever to be found in the fountains of quick resolve and swift stroke; and it seems to be a law inflexible and inexorable that he who will not risk cannot win.

Admiral John Paul Jones, 1791, letter to Vice-Admiral Kersaint

---

Whatever naval expeditions we have undertaken since I became head of the government have invariably failed because the admirals see double and have picked up the notion, I know not where, that one can make war without taking any risks.

Napoleon, 1804, ed. Herold, *The Mind of Napoleon*, 1955

If the art of war consisted merely in not taking risks glory would be at the mercy of very mediocre talent.

Napoleon (1769–1821)

I was aware that I was risking infinitely too much, but something must be risked for the honour of the Service.

General Sir John Moore, December 1808, letter from Spain

There is always hazard in military movements, but we must decide between possible loss from inaction and the risk of action.

General Robert E. Lee (1807–1870)

First reckon, then risk.

Field Marshal Helmuth Graf von Moltke (1800–1891)

Calculated risk guided by skill is the right way to interpret the motto, 'l'audace, toujours l'audace'. (September 1925)

Captain Sir Basil Liddell Hart, *Thoughts on War*,, 1944

When I consider the state of our armament, and thought of this test – and for soldiers the test is always battle – I repeated to myself: 'Careful, careful, and once more careful with fire. Don't be a child, and don't give rein to your fancy.' But all the character, will, pride and ambition within me rebelled against these cunctatorial 'prudent' reflections. Besides, in my opinion there was no escape. I do not deny that the test of battle was an extraordinarily dangerous undertaking, but we could only win what we needed above all to win, self-confidence and the military respect of those around us, by taking risks, and big risks.

Marshal Joseph Pilsudski, *Memoirs of a Polish Revolutionary and Soldier*, 1931

Dearest Lu,

We've been attacking since the 31st with dazzling success. There'll be consternation amongst our masters in Tripoli and Rome, perhaps in Berlin too. I took the risk against all orders and instructions because the opportunity seemed favourable. No doubt it will all be pronounced good later and they'll all say they'd have done exactly the same in my place. We've already reached our first objective, which we weren't supposed to get to until the end of May. The British are falling over each other to get away. Our casualties small. Booty can't yet be estimated. You will understand that I can't sleep for happiness.

Field Marshal Erwin Rommel, 3 April 1941, letter to his wife, *The Rommel Papers*, 1953

The habit of gambling contrary to reasonable calculations is a military vice which, as the pages of history reveal, has ruined more armies than any other causes.

Captain Sir Basil Liddell Hart, *Thoughts on War*, 1944

Take calculated risks. That is quite different from being rash.

General George S. Patton, Jr., 6 June 1944, letter to his son, *The Patton Papers*, Vol. II, 1974

In making war there must be three parts risk and seven parts security, in addition to one's own subjective effort.

General Lin Piao, 1946, quoted in Bloodworth, *The Chinese Machiavelli*, 1976

War is risk. Either its ends permit of honest differences of opinion about what should best be done, or operations long since would have become an exact science and general staff work would be as routine as logarithms.

Brigadier General S. L. A. Marshall, *Men Against Fire*, 1947

In enterprises in which one risks everything, there usually comes a moment when the person responsible feels that fate is being determined. By a strange convergence the thousand trials in the midst of which he is struggling seem suddenly to blossom into a decisive episode. If it turns out well, fortune will be in his hands.

Charles de Gaulle, *War Memoirs: The Call to Honour 1940–1942*, 1955

RIVER DEFENCE

Never base your defence on rivers unless they cut between rocks and have steep banks. You can defend a river that lies behind the army, but it has yet to be shown how a river in front of the armies can successfully be held. As many times as you take up a position behind a river to keep the enemy from crossing it, that often you will be duped, because sooner or later the enemy, forced to display cunning, finds a suitable moment for stealing his crossing.

Therefore this is what I propose: the only way to defend a river is to keep it behind you.

Frederick the Great, 'Réflexions sur les projets de campagn', and 'Eléments de castramertire et de tactique', *Oeuvres de Frédéric le Grand*, 1846–56, ed., Luvaas, *Frederick the Great on the Art of War*,, 1966

A river must always be regarded as an obstacle which is unable to hold up the enemy more than a few days. One can prevent him from crossing only by holding in readiness strong forces in bridgeheads on the hostile shore, with a view to attacking as soon as the enemy begins to cross . . . Nothing is more dangerous than the attempt at defending seriously a river line by keeping his side of the river occupied; for if the enemy were to cross suddenly with surprise effect – and that he will always be able to do somehow or other – he would find the defender in extensive positions from which the latter will be unable to assemble in time.

Napoleon, 15 March 1813, letter to Jerome, *Correspondance*, No. 19, Vol. XXV, 1858–1870

A river will in any case enable the defender to gain time, require the attacker to make special technical preparations, and warn the latter to be cautious. As a prerequisite for success in effecting a river crossing, the deceit of the defender as to the real intentions has always been considered essential.

General der Infantrie Hugo Baron von Freytag-Loringhoven, *Generalship in the Great War*, Vol. I, 1920

See also: MOUNTAIN FIGHTING, OBSTALCES, TERRAIN

ROUT

When soldiers break and run, good-bye glory, good-bye all defences.

Ajax, in Homer, *The Iliad*, xv, *c.* 800 BC, tr. Fagles

Surely if there is one military maxim of universal value, it is to press hard on a rout.

Colonel T. E. Lawrence, quoted in Liddell Hart, *Colonel Lawrence: The Man Behind the Legend*, 1935

Marshal Timoshenko told me that the army had been so utterly routed by the enemy that the only way to rally the troops was to set up mobile field kitchens and hope that the soldiers would return when they got hungry.

Nikita Khrushchev, of the Soviet disaster at Kiev in 1942, quoted in Messenger, *The Blitzkrieg Story*

See also: PURSUIT, RETREAT

RUSE

Utterance of his majesty to the king's-children, the princes, the king's butlers, and the charioteers: 'Behold ye, the great might of my father, Amon Re. The countries which came from their isles in the midst of the sea, they advanced on Egypt, their hearts relying upon their arms. The net was made ready for them, to ensnare them. Entering stealthily into the harbour-mouth, they fell into it. Caught in their place, they were dispatched, and their bodies stripped. I showed you my might which was in that which my majesty wrought while I was alone. My arrow struck, and none escaped my arms nor my hand. I flourished like a hawk among fowl; my talons descended upon their heads. Amon-Re was upon my right and upon my left, his might and power were in my limbs, a tumult for you; commanding for me that my counsels and my designs should come to pass . . .'

Ramses III, Pharaoh of Egypt, *c.* 1190 BC, the defeat of the Sea Peoples in their descent upon Egypt, quoted in Breasted, *Ancient Records of Egypt*, 1906

Ruses are of great usefulness. They are detours which often lead more surely to the objective than the wide road which goes straight ahead. Animals have only one method of acting, but intelligent men have inexhaustible resources.

You outwit the enemy to force him to fight, or to prevent him from it.

Frederick the Great, *Instructions to His Generals*, 1747

You fear a retreat through Moscow, but I regard it as far-sighted. It will save the army. Napoleon is like a stormy torrent which we are as yet unable to stop. Moscow will be the sponge that sucks him in.

Field Marshal Prince Mikhail I. Kutuzov, 1812, quoted in Parkinson, *Fox of the North*, 1976

We are bred up to feel it a disgrace even to succeed by falsehood; the word spy conveys something as repulsive as slave; we will keep hammering along with the conviction that honesty is the best policy, and that truth always wins in the long run. These pretty little sentiments do well for child's copy book, but a man who acts on them had better sheathe his sword forever.

Field Marshal Viscount Wolseley, *A Soldier's Pocket-Book*, 1869

The conduct of operations is too much influenced by the experience of peacetime exercises. In the conditions of real war, the feeling of uncertainty is magnified, and this makes the opponent much more sensitive to crafty deceptions – so that even the most threadbare ruse has succeeded time after time. (August 1928)

Captain Sir Basil Liddell Hart, *Thoughts on War*, 1944

See also: DECEPTION

SACRIFICE

So fight by the ships, all together. And that comrade
who meets his death and destiny, speared or stabbed,
Let him die! He dies fighting for fatherland –
no dishonour there!
He'll leave behind him wife and sons unscathed,
His house and estates unharmed – once these Argives sail for home,
the fatherland they love.

Hector, rallying the Trojans to storm the Greek camp, in Homer, *The Iliad*, xv, c. 800 BC, tr. Fagles

---

If to die honourably is the greatest
Part of divine virtue, for us fate's done her best
Because we fought to crown Greece with freedom
We lie here enjoying timeless fame.

Simonides (c. 556 BC–468 BC), lyric poet, memorial for the Athenian dead at the Battle of Plataea, 480 BC

God forbid that I should live an Emperor without an Empire! As my city falls, I will fall with it! Whosoever wishes to escape, let him save himself if he can; and whoever is ready to face death, let him follow me!

Emperor Constantine, XI, May 1453, at the fall of Constantinople, quoted in Mijatovich, *Constantine, The Last Emperor of the Greeks*, 1892

Perish yourself but rescue your comrade!

Field Marshal Prince Aleksandr V. Suvorov (1729–1800)

In decisive cases there are moments when victory demands sacrifices and when it becomes necessary to burn your own warships. If military art consisted of always taking a safe position, then glory would become the property of mediocre people.

Napoleon (1769–1821)

Let us stop being selfish . . . To the idea of the common good and of our existence, everything must be sacrificed . . . The lack of funds does not even allow us to take care of the most elementary thing . . . As from today our salaries are cut in half . . . I shall measure patriotism by generosity . . . From this moment on, luxury and comforts must make us ashamed.

General José de San Martin, 1815, quoted in Rojas, *San Martin*, 1957

If the enemy is to be coerced you must put him in a situation that is even more unpleasant than the sacrifice you call on him to make. The hardships of the situation must not of course be merely transient – at least not in appearance. Otherwise the enemy would not give in but would wait for things to improve . . .

Major General Carl von Clausewitz, *On War*, i, 1832, tr. Howard and Paret

---

I offer neither pay, nor quarters, nor provisions; I offer hunger, thirst, forced marches, battles, and death. Let him who loves his country in his heart and not with his lips only, follow me.

Guiseppe Garibaldi (1807–1882)

. . . the loss of property weighs more heavily with the most of mankind; heavier often, than the sacrifices made on the field of battle. Death is popularly considered the maximum punishment of war, but it is not; reduction to poverty brings prayers of peace more surely and more quickly than does the destruction of human life, as the selfishness of man has demonstrated in more than one conflict.

General of the Army Philip H. Sheridan, *Personal Memoirs of P. H. Sheridan,* 1888

I, Nogi Maresuke, commander-in-chief of the Third Imperial Army before Port Arthur, celebrate with sake and many offerings a fete in honour of you . . . I wish to tell you that your noble sacrifice has not been in vain, for the enemy's fleet has been destroyed, and Port Arthur has at last surrendered. I, Nogi Maresuke, took oath with you to conquer or seek oblivion in death. I have survived to receive the Imperial thanks, but I will not monopolize the glory. With you, Spirit of the Dead, who achieved this great result, I desire to share my triumph . . .

General Nogi Maresuke, 1905, quoted in Warner, *The Tide at Sunrise,* 1974

But war is a ruthless taskmaster, demanding success regardless of confusion, shortness of time, and paucity of tools. Exact justice for the individual and a careful consideration of his rights is quite impossible. One man sacrifices his life on the battlefield and another sacrifices his reputation elsewhere, both in the same cause. The hurly-burly of the conflict does not permit commanders to draw fine distinctions; to succeed, they must demand results, close their ears to excuses, and drive subordinates beyond what would ordinarily be considered the limit of human capacity. Wars are won by the side that accomplishes the impossible.

General of the Army George C. Marshall, *Memoirs of My Services in the World War* (written between 1923–26), 1976

The solemn characteristic feature of the soldier's profession is the readiness to die in discharge of duty. Other professions, too, may require the risk of life in discharge of duty, every man may be faced, outside his profession, with the necessity for the last great sacrifice, as an ethical duty; but in no other profession do killing and its corollary, readiness to die, form the essense of professional duty. If the true art of war lies in destroying the enemy, then its exponent must also be prepared to be destroyed himself. This conception of the soldier's function justifies us in speaking of soldiering as something unique. It is the responsibility for life and death which gives the soldier his special character, his gravity, and self-consciousness – not only the responsibility for his own life, which may be sacrificed, not light-heartedly but from a feeling of duty, but the simultaneous responsibility for the lives of comrades and, in the end, for the life of the enemy, whose death is not an act of independent free will on the part of the killer, but an acknowledgement of professional duty. The feeling of responsibility for oneself and others is one of the most vital characteristics of the soldier's life. Responsibility towards oneself demands the most exacting inward and outward training for the military profession, so that the final sacrifice may not be made in vain.

Colonel General Hans von Seekt, *Thoughts of a Soldier,* 1930

The soldier, above all other men, is required to perform the highest act of religious teaching – sacrifice. In battle and in the face of danger and death he discloses those divine attributes which his Maker gave when He created man in his own image. No physical courage and no brute instincts can take the place of the divine annunciation and spiritual uplift which will alone sustain him. However horrible the incidents of war may be, the soldier who is called upon to offer and to give his life for his country is the noblest development of mankind.

General Douglas MacArthur, 14 July 1935, address to the veterans of the Rainbow (42nd) Division, *A Soldier Speaks*, 1965

I would say to the House, as I said to those who have joined this Government: I have nothing to offer but blood, toil, tears and sweat.

We have before us an ordeal of the most grievous kind. We have before us many, many long months of struggle and of suffering. You ask, What is our policy? I will say: It is to wage war, by sea, land, and air, with all our might and with all the strength that God can give us: to wage war against a monstrous tryanny, never surpassed in the dark, lamentable catalogue of human crime. That is our policy. You ask, What is our aim? I can answer in one word: Victory – victory at all costs, victory in spite of all terror, victory however long and hard the road may be; for without victory there is no survival. Let that be realized; no survival for the British Empire; no survival for the urge and impulse of the ages, that mankind will move forward towards its goal. But I take up my task with buoyancy and hope. I feel sure that our cause will not be suffered to fail among men. At this time I feel entitled to claim the aid of all, and I say, Come, then, let us go forward together with our united strength.

Sir Winston S. Churchill, 13 May 1940, first speech to the House of Commons as Prime Minister, *Blood, Sweat and Tears*, 1941

I consider it no sacrifice to die for my country. In my mind we came here to thank God that men like these have lived rather than to regret that they died.

General George S. Patton, Jr., 11 November 1943, speech at an Allied cemetery in Italy, quoted in Semmes, *Portrait of Patton*, 1955

'Not in vain' may be the pride of those who survived and the epitaph of those who fell.

Sir Winston S. Churchill, 1944, speech in the House of Commons

---

. . . the glory of taking the fort is not worth the sacrifices in men which it would demand.

General George S. Patton, Jr., October 1944, remark to General Bradley about Fort Driant at Verdun, *the Patton Papers*, Vol. II, 1974

I want you to remember that no bastard ever won a war by dying for his country. He won it by making the other poor dumb bastard die for his country.

General George S. Patton, Jr., 1944, speech to his troops

If a man is not inclined to risk his life for his country, he should look elsewhere 'till he finds a country he will risk his life for.

Admiral Raymond Spruance (1886–1969)

Archibald Rutledge once wrote that there can be no real love without a willingness to sacrifice. Tuck this away in your inner minds. It may pay off in some crisis coming to you in the years now hidden beyond the horizon. Do you love your country and its flag? Do you love the branch in which you are serving, the men with whom you will be privileged to share service and to command? If you do, then you will be prepared to sacrifice for them, if your responsibilities or the situation demands . . .

General Mathew B. Ridgway, 'Leadership', *Military Review*, 10/1966

SAFETY

For they had learned that true safety was to be found in long previous training, and not in eloquent exhortations uttered when they were going into action.

Thucydides, *The History of the Peloponnesian War*, c. 404 BC

The desire for safety stands against every great and noble enterprise.

Cornelius Tactitus (c. AD 56–c. AD 120), *Annals*

Every attempt to make war easy and safe will result in humiliation and disaster.

General of the Army William T. Sherman, *The Memoirs of General W. T. Sherman*, 1875

'Safety first' is the road to ruin in war.

Sir Winston S. Churchill, 3 November 1940, telegram to Anthony Eden

---

Self-preservation is the keystone in the arch of war, because it is the keystone in that greater arch called life. No normal man wishes to be killed in battle, though he may long to die in battle rather than to die in his bed. He does not wish to do so, because there is no virtue in merely dying, for virtue is to be sought in living and living rightly.

Major General J. F. C. Fuller, *Generalship: Its Diseases and Their Cure*, 1936

Discipline apart, the soldier's chief cares are, first, his personal comfort, i.e., regular rations, proper clothing, good billets, and proper hospital arrangements (square meals and a square deal in fact); and secondly, his personal safety, i.e., that he shall be put into a fight with as good a chance for victory and survival.

Field Marshal Sir Archibald Wavell (1883–1950)

SALUTE

The salute is the mutual greeting of respect and loyalty between members of a fighting organization . . .

General George S. Patton, Jr., May 1941, address to the officers of the 2nd Armored Division, *The Patton Papers*, Vol II, 1974

One morning long ago, as a brand new second lieutenant, I was walking on to parade. A private soldier passed me and saluted; I acknowledged his salute with an airy wave of the hand. Suddenly, behind me, a voice rasped out my name. I spun round. There was my Colonel, for whom I had a most wholesome respect, and with him the Regimental Sergeant-Major, of whom, if truth must be told, I stood in some awe. 'I see', said the Colonel, 'you don't know how to return a salute. Sergeant Major, plant your staff in the ground and let Mr Slim practise saluting it until he does know how to return a salute!' So to and fro I marched in sight of the whole battalion, saluting the Sergeant Major's cane. I could cheerfully have murdered the Colonel, the Sergeant-Major and, more than cheerfully, my fellow subalterns grinning at me. At the end of ten minutes the Colonel called me up to him. All he said was, 'Now remember, discipline begins with the officers!'

Field Marshal Viscount Slim, *Courage and Other Broadcasts*, 1957

A salute from an unwilling man is as meaningless as the moving of a leaf on a tree; it is a sign only that the subject has been caught by a gust of wind. But a salute from the man who takes pride in the gesture because he feels privileged to wear the uniform of the United States, having found military service good, is the epitome of military virtue.

Brigadier General S. L. A. Marshall, *The Officer As Leader*, 1966

THE SCHOOL SOLUTION

In the art of war there are no fixed rules. These can only be worked out according to the circumstances.

Li Ch'uan, Tang Dynasty writer on military affairs and commentator on Sun Tzu, *The Art of War*, c. 500 BC, tr. Griffith

---

In conclusion, then, we have done our part by writing down these things just as our predecessors handed them to us, as well as from our own experience, which goes back a long time. It is up to you, now to apply it to the circumstances that are likely to arise. The outcome of war is not brought about according to the will of men but, just as the affairs of each are weighed, by the providence of God on high.

The Emperor Nikephorus II Phocas, *Skirmishing*, c. AD 969, in Dennis, *Three Byzantine Military Treatises*, 1975

In this art as in poetry and eloquence, there are many who can trace the rules by which a poem or an oration should be composed, and even compose, according to the exactest rule; but for want of that enthusiastic and divine fire, their productions are languid and insipid; so in our profession, many are to be found who know every precept of it by heart; but alas! when called upon to apply them, are immediately at a stand. They then recall their rules and want to make everything, the rivers, the woods, ravines, mountains, etc. subservient to them; whereas their precepts should, on the contrary, be subject to these, who are the only rules, the only guide we ought to follow; whatever manoeuvre is not formed on them is absurd and ridiculous.

These form the great book of war; and he who cannot read it, must be forever content with the title of a brave soldier and never aspire to that of great general.

Major General Henry Lloyd, *History of the Late War in Germany*, 1766–1782

There are no precise or determined rules; everything depends on the character that nature has given to the general, on his qualities, his shortcomings, on the nature of the troops, on the range of the firearms, on the season and on a thousand other circumstances which are never the same.

Napoleon, 'Notes sur l'art de la guerre', *Correspondance*, Vol. XXXI, 1858–1870

Pity the warrior who is contented to crawl about in the beggardom of rules! What genius does must be the best of rules, and theory cannot do better than show how and why it is so.

Major General Carl von Clausewitz, quoted in Montross, *War Through the Ages*, 1960

I have heard in the old army an anecdote very characteristic of [Braxton] Bragg. On one occasion, when stationed at a post of several companies commanded by a field officer, he was himself commanding one of the companies and at the same time acting as post quartermaster and commissary . . . As commander of the company he made a requisition upon the quartermaster – himself – for something he wanted. As quartermaster he declined to fill the requisition, and endorsed on the back of it his reasons for so doing. As company commander he responded to this, urging that his requisition called for nothing but what he was entitled to, and that it was the duty of the quartermaster to fill it. As quartermaster he still persisted that he was right. In this condition of affairs Bragg referred the whole matter to the commanding officer. The latter, when he saw the nature of the matter referred, exclaimed: 'My God, Mr. Bragg, you have quarrelled with every officer in the army, and now you are quarrelling with yourself!'

General of the Army Ulysses S. Grant, *Personal Memoirs of U.S. Grant*, 1885

If man make war in slavish obedience to rules, they will fail.

General of the Army Ulysses S. Grant, *Personal Memoirs of U.S. Grant*, 1885

What is needed is, in the face of specific cases, to appreciate the situation as it is, with its unknown factors, to judge wisely of what is visible, to guess at the unknown, to come quickly to a decision, and finally to act with energy.

Field Marshal Helmuth Graf von Moltke, quoted in Foch, *Principles of War*, 1913

. . . no servitude is more hopeless than that of unintelligent submission to an idea formally correct, yet incomplete. It has all the vicious misleading of a half-truth unqualified by appreciation of modifying conditions; and so seamen who disdained theories, and hugged the belief in themselves as 'practical' became *doctrinaires* in the worst sense.

Rear Admiral Alfred Thayer Mahan, *Types of Naval Officers*, 1901

We must first understand truths, and therefore have an open mind, without prejudice, ready-made ideas, or theories blindly accepted merely because they rest on tradition. One standard alone, that of reason. Then we must apply these truths to specific cases, on the map at first, on the ground later, the battlefield ultimately. Let us not look for similarities, let us not appeal to our memory, it would desert us at the first cannon shot, and let us avoid all charts or formulae. We wish to reach the field with a trained power of judgment; it only need to have us train it, to have us begin training it today. Let us for that purpose seek the reason of things; that will show us how to use them.

Marshal of France Ferdinand Foch, *Principles of War*, 1913

. . . when confronted with a situation, do not try to recall examples given in any particular book on the subject; do not try to remember what your instructor said . . . do not try to carry in your minds patterns of particular exercises or battles, thinking they will fit new cases, because no two sets of circumstances are alike . . .

General of the Armies John J. Pershing, 1918, address to the 1st Infantry Division, quoted in Vandiver, *Black Jack*, 1977

Opposed to genius stand the men of logarithms, who, by their 'if-he-does-this-I-shall-do-that', attempt to exhaust the *cache* of chances which God keeps stocked far away in starland. Genius is God's secret, that is all. Foch thinks he won the war because he 'calculated', as the Yankees say. He forgets that the Germans equally pride themselves on calculations. He either forgets, or he is too modest to tell us, that he has in him like a burning fire a passion for sheer fighting, a fire of passion which burnt up all his sums and figures when the moment came. But if the great Marshal had only told us this, we should have known what weight to attach to his 'if he does this I shall do that'.

General Sir Ian Hamilton, *Soul and Body of an Army*, 1921

The principles which regulate the use of all available means in war – the economic employment by a commander of the forces at his disposal, the building up of a concentration (and, consequently, advances by phases or forward bounds), surprise for the enemy, security for oneself – are of value only (how often has this been said before!) insofar as they are adapted to the circumstances of the given situation. There is nothing specifically military about this generalization; it is as true for politics and industry as it is for armies in the field.

Charles de Gaulle, *The Edge of the Sword*, 1932

Official manuals, by the nature of their compilation, are merely registers of prevailing practice, not the log-books of a scientific study of war.

Captain Sir Basil Liddell Hart, *Thoughts on War*, 1944

There is no approved solution to any tactical situation.

General George S. Patton, Jr., *War As I Knew It*, 1947

Normally, there is no ideal solution to military problems; every course has its advantages and disadvantages. One must select that which seems best from the most varied aspects and then pursue it resolutely and accept the consequences. Any compromise is bad.

Field Marshal Erwin Rommel, *The Rommel Papers*, 1953

SCORCHED EARTH

They make a desert and call it peace.

Cornelius Tacitus (c. AD 56–c. AD 120), *Agricola*

In pushing up the Shenandoah Valley . . . it is desirable that nothing should be left to invite the enemy to return. Take all provisions, forage, and stock wanted for the use of your command. Such as cannot be consumed, destroy. It is not desirable that buildings should be destroyed – they should, rather, be protected; but the people should be informed that so long as an army can subsist among them, recurrences of these raids must be expected, and we are determined to stop them at all hazards.

General of the Army Ulysses S. Grant, 5 August 1864, letter of instruction to General Sheridan

---

Until we can repopulate Georgia, it is useless for us to occupy it; but the utter destruction of its roads, houses and people will crippled their military resources. I can make this march, and make Georgia howl.

General of the Army William T. Sherman, 9 September 1864, telegram to General Grant from Atlanta

---

In case of a forced retreat of Red Army units, all rolling stock must be evacuated; to the enemy must be left not a single engine, a single railway car, not a single pound of grain or gallon of fuel . . . In occupied regions conditions must be made unbearable for the enemy and all his accomplices. They must be hounded and annihilated at every step and all their measures frustrated.

Joseph V. Stalin, 3 July 1941, address to the Soviet people

SEA POWER

He who commands the sea has command of everything.

Themistocles (*c.* 528 BC–462 BC) quoted in Cicero, *Epistolae ad Atticum*

For our naval skill is of more use to us for service on land, than their military skill for service at sea. Familiarity with the sea they will not find an easy acquisition. If you who have been practising at it ever since the Median invasion have not yet brought it to perfection, is there any chance of anything considerable being effected by an agricultural, unseafaring population, who will besides be prevented from practising by the constant presence of strong squadrons of observation from Athens? With a small squadron they might hazard an engagement, encouraging their ignorance by numbers; but the restraint of a strong force will prevent their moving, and through want of practice they will grow more clumsy, and consequently more timid. It must be kept in mind that seamanship, just like everything else, is a matter of art, and will not admit of being taken up occasionally as an occupation for times of leisure; on the contrary, it is so exacting as to leave leisure for nothing else.

Pericles, 431 BC, address to the Athenians on the virtues of sea power, quoted in Thucydides, *The History of the Peloponnesian War*, *c.* 404 BC, tr. Crawley

---

He who rules on the sea will very shortly rule on the land also.

Khayr-Ed-Din (Barbarossa) (*d.* 1546)

. . . this much is certain; that he that commands the sea is at great liberty, and may take as much and as little of the war as he will. Whereas these, that be strongest by land, are many times nevertheless in great straits.

Francis Bacon, *On the True Greatness of Kingdoms and Estates*, 1597

Whosoever commands the sea commands the trade; whosoever commands the trade of the world commands the riches of the world, and consequently the world itself.

Sir Walter Raleigh, *Historie of the Worlde*, 1616

A man of war is the best ambassador.

Oliver Cromwell, (1599–1658)

Without a respectable Navy – alas, America!

Admiral John Paul Jones, 17 October 1776, letter to Robert Morris

Under all circumstances, a decisive naval superiority is to be considered a fundamental principle, and the basis upon which all hope of success must ultimately depend.

General George Washington, 1780, letter

Without a decisive naval force we can do nothing definitive, and with it everything honorable and glorious.

General George Washington, 20 December 1780, letter to Benjamin Franklin

---

. . . it is in our own experience, that the most sincere Neutrality is not a sufficient guard against the depredations of Nations at War. To secure respect to a neutral Flag, requires a naval force, organized, and ready to vindicate it, from insult or aggression. This may even prevent the necessity of going to War, by discouraging the belligerent Powers from committing such violations of the rights of the neutral party, as may first or last, leave no other option.

General George Washington, 7 December 1796, Eight Annual Address to congress, *The Writings of George Washington,* Vol. 35, 1931–44

I do not say the Frenchman will not come; I only say he will not come by sea.

Admiral Lord St. Vincent, 1803, as First Lord of the Admiralty

---

Wherever wood can swim, there I am sure to find this flag of England.

Napoleon, July 1815, at Rochefort

Had I been master of the sea, I should have been lord of the Orient.

Napoleon (1769–1821)

It is not the taking of individual ships or convoys, be they few or many, that strikes down the money power of a nation; it is the possession of that overbearing power on the sea which drives the enemy's flag from it, or allows it to appear only as the fugitive; and by controlling the great common, closes the highway by which commerce moves to and from the enemy's shores. This overbearing power can only be exercised by great navies.

Rear Admiral Alfred Thayer Mahan, *The Influence of Sea Power Upon History,* 1890

Sea power in the broad sense . . . includes not only the military strength afloat that rules the sea or any part of it by force of arms, but also the peaceful commerce and shipping from which alone a military fleet naturally and healthfully springs, and on which it securely rests.

Rear Admiral Alfred Thayer Mahan, *The Influence of Sea Power Upon History,* 1890

The world has never seen a more impressive demonstration of the influence of sea power upon history. Those far distant, storm-beaten ships, upon which the Grand Army never looked, stood between it and the dominion of the world.

Rear Admiral Alfred Thayer Mahan, *The Influence of Sea Power Upon the French Revolution and Empire*, Vol. II, 1892

The Navy is the 1st, 2nd, 3rd, 4th . . . ad infinitum Line of Defence! If the Navy is not supreme, no Army however large is of the slightest use.

Admiral Sir John Fisher, speaking before World War I, quoted in Marder, *The Anatomy of British Sea Power,* 1940

When we speak of command of the seas, it does not mean command of every part of the sea at the same moment, or at every moment. It only means that we can make our will prevail ultimately in any part of the seas which may be selected for operations, and thus indirectly make our will prevail in every part of the seas.

Sir Winston S. Churchill, 11 October 1940, to the House of Commons

. . . history shows that those states which do not have naval forces at their disposal have not been able to hold the status of a great power for very long.

Admiral of the Fleet of the Soviet Union Sergei G. Gorshkov, *Red Sea Rising*, 1974

SECRECY

O divine art of subtlety and secrecy! Through you we learn to be invisible, through you inaudible; and hence hold the enemy's fate in our hands.

Sun Tzu, *The Art of War*, c. 500 BC

Consult with many on proper measures to be taken, but communicate the plans you intend to put in execution to few, and those only of the most assured fidelity. Or better, trust no one but yourself.

Flavius Vegetius Renatus, *Military Institutions of the Romans*, c. AD 378

For what should be done seek the advice of many; for what you will actually do take council with only a few trustworthy people; then off by yourself alone decide on the best and most helpful plan to follow, and stick to it.

The Emperor Maurice, *The Strategikon*, c. AD 600

The general who wants to keep his plans concealed from the enemy should never take the rank and file of his own troops into his confidence.

The Emperor Maurice, *The Strategikon*, c. AD 600

Only those battle plans are successful which the enemy does not suspect before we put them into action.

The Emperor Maurice, *The Strategikon*, c. AD 600

To keep your actions and your plans secret always has been a very good thing. For that reason Metellus, when he was with armies in Spain, replied to one who asked him what he was going to do the next day, that if his shirt knew he would burn it. Marcus Crassus said to one who asked him when he was going to move the army: 'Do you believe that you will be the only one not to hear the trumpet?'

Niccolo Machiavelli, *The Art of War*, 1521

Kalckstein: Your Majesty, am I right in thinking there is going to be a war?
Frederick: Who can tell!
Kalckstein: The movement seems to be directed on Silesia.
Frederick: Can you keep a secret? (Taking him by the hand).
Kalckstein: Oh yes, Your Majesty.
Frederick: Well, so can I!

Frederick the Great, 1740, during his preparation for the seizure of Silesia, quoted in Duffy, *The Military Life of Frederick the Great*, 1985

Secrecy is so necessary to a general that the ancients have even said that there was not a human being able to hold his tongue. But there is a reason for that: If you form the finest plans in the world but divulge them, your enemy will learn about them, and then it will be easy for him to parry them.

Frederick the Great, *Instructions to His Generals*, 1747

Frederick the Great was right when he said that if his night-cap knew what was in his head he would throw it into the fire.

Lieutenant General Antoine-Henri Baron de Jomini, *Summary of the Art of War*, , 1838

No serving soldier can tell his fellow-countryman anything about an Army which is not (1) quite commonplace; (2) an expression of the views of the Authorities at the moment.

General Sir Ian Hamilton, *The Soul and Body of an Army*, 1921

War and truth have a fundamental incompatibility. The devotion to secrecy in the interests of the military machine largely explains why, throughout history, its operations commonly appear in retrospect the most uncertain and least efficient of human activities.

Captain Sir Basil Liddell Hart, *Thoughts on War*, 1944

See also: DECEPTION, SECURITY

SECURITY
(Principle of War)

The ultimate in disposing one's troops is to be without ascertainable shape. Then the most penetrating spies cannot pry in nor can the wise lay plans against you.

It is according to the shapes that I lay the plans for victory, but the multitude does not comprehend this. Although everyone can see the outward aspects, none understands the way in which I have created victory.

Sun Tzu, *The Art of War*, c. 500 BC, tr. Griffith

The course of war cannot be foreseen, and its attacks are generally dictated by the impulse of the moment; and where overweening self-confidence has despised preparations, a wise apprehension has often been able to make head against superior numbers. Not that confidence is out of place in an army of invasion, but in an enemy's country it should also be accompanied by the precautions of apprehension: troops will by this combination be best inspired for dealing a blow, and best secured against receiving one.

Archidamus II of Sparta, 431 BC, speech to the Peloponnesian allied army upon its invasion of Attica, quoted in Thucydides, *The History of the Peloponnesian War*, c. 404, tr. Crawley

---

A general who has been defeated in a pitched battle, although skill and conduct have the greatest share in beating him, may in his defence throw the blame to fortune. But if he has suffered himself to be surprised or drawn into the snares of his enemy, he has no excuse for his fault, because he might have avoided such a misfortune by taking proper precautions and employing spies on whose intelligence he could depend.

Flavius Vegetius Renatus, *Military Institutions of the Romans*, c. AD 378

The person who wants to wage war securely against an enemy must first make sure that his own lands are secure. By secure I mean not only the security of the army but of the cities and the entire country, so that the people who live there may suffer no harm at all from the enemy.

Anonymous Byzantine General, *On Strategy*, c. AD 527–AD 565, quoted in Dennis, *Three Byzantine Military Treatises*, 1985

Even in friendly territory a fortified camp should be set up; a general should never have to say: 'I did not expect it.'

The Emperor Maurice, *The Strategikon*, c. AD 600

After a victory we must not become careless, but be on our guard all the more against surprise attacks by the vanquished.

The Emperor Maurice, *The Strategikon*, c. AD 600

A general who takes nothing for granted is secure in war.

The Emperor Maurice, *The Strategikon*, c. AD 600

No enterprise is more likely to succeed than one concealed from the enemy until it is ripe for execution.

Niccolo Machiavelli, *The Art of War*, 1521

Scepticism is the mother of security. Even though fools trust their enemies, prudent persons never do. The general is the principal sentinel of the army. He should always be careful of its preservation and see that it is never exposed to misfortune. One falls into a feeling of security after battles, when one is drunk with success and when one believes the enemy completely disheartened. Also when a skillful enemy amuses you with pretended peace proposals. One does this through mental laziness and lack of calculation concerning the intentions of the enemy.

Frederick the Great, *Instructions to His Generals*, 1747

Do not neglect the principles of foresight and know that often, puffed up with success, armies have lost the fruit of their heroism through a feeling of false security.

Frederick the Great, *Instructions to His Generals*, 1747

The Principle of Security. The objective in battle being to destroy or paralyse the enemy's fighting strength, consequently the side which can best secure itself against the action of its antagonist will stand the best chance of winning, for by saving its men and weapons, its organization and morale, it will augment its offensive power. Security is, therefore, a shield and not a lethal weapon, consequently the defensive is not the strongest form of war, but merely a prelude to the accomplishment of the objective – the defeat of the enemy by means of the offensive invigorated by defensive measures. The offensive being essential to success, it stands to reason that security without reference to the offensive is no security at all, but merely delayed suicide.

Major General J. F. C. Fuller, *The Reformation of War*, 1923

As danger and the fear of danger are the chief morale obstacles of the battlefield, it follows that the imbuing of troops with sense of security is one of the chief duties of a commander; for, if weapons be of equal power, battles are won by a superiority of nerve rather than by a superiority of numbers. This sense of security, though it may be supplemented by earth works or mechanical contrivances, is chiefly based on the feeling of moral ascendance due to fighting efficiency and confidence in command. Thus, a man who is a skilled marksman will experience a greater sense of security when lying in the open than an indifferent rifleman in a trench.

Given the skilled soldier, the moral ascendancy resulting from his efficiency will rapidly evaporate unless it be skilfully directed and employed. As in all undertakings – civil or military, ultimately we come back to the impulse of the moment, to the brains which control impulse and to each individual nerve which runs through the military body. To give skilled troops to an unskilled leader is tantamount to throwing snow on hot bricks. Skill in command is, therefore, the foundation of security, for a clumsy craftsman will soon take the edge off his tools.

Major General J. F. C. Fuller, *The Foundations of the Science of War*, 1926

See also: DECEPTION, SECRECY

SELF-CONFIDENCE

Never take counsel of your fears.

Lieutenant General Thomas 'Stonewall' Jackson, 18 June 1862, to Major Hotchkiss

---

Wilson, I am a damned sight smarter than Grant. I know a great deal more about war, military history, strategy, and grand tactics than he does; I know more about organization, supply, and administration, and about everything else than he does. But I tell you where he beats me, and where he beats the world. He don't care a damn for what the enemy does out of his sight, but it scares me like hell . . . I am more nervous than he is. I am more likely to change my orders, or to countermarch my command than he is. He uses such information as he has, according to his best judgment. He issues his orders and does his level best to carry them out without much reference to what is going on about him.

General of the Army William T. Sherman, quoted in Williams, *McClellan, Sherman and Grant*, 1962

When I think of the greatness of my job and realize that I am what I am, I am amazed, but on reflection, who is as good as I am? I know of no one.

General George S. Patton, Jr., *The Patton Papers*, Vol. II, 1974

I have never been a self-confident person. My lack of self-confidence has not prevented me from being decisive in my job. When one does a job, when one is responsible for it and makes decisions – here there is no place for doubts of oneself or a lack of confidence. You are completely engrossed in your job and in giving your all to this job and doing everything that you are capable of. But later, when the job is complete, when you reflect on what you have done, when you think not only of the past but also of the future, there is a heightened awareness that you lack something, something is missing, that you had to know a number of things which you did not know and this returning feeling causes you to rethink everything and decide: 'Could you have not done better than what you did, if you had possessed all that you have lacked?'

Marshal of the Soviet Union Georgi K. Zhukov (1896–1974), quoted in *Voyenno istoricheskiy zhurnal*, 12/1987

SELF-CONTROL

Self-control is the chief element in self-respect, and self-respect is the chief element in courage.

Thucydides, *The History of the Peloponnesian War*, c. 404 BC

Let the army see that you are not unduly elated over successes nor utterly cast down by failures.

The Emperor Maurice, *The Strategikon*, c. AD 600

It is well for the general to exercise self-control at all times, but especially in war.

The Emperor Maurice, *The Strategikon*, c. AD 600

The general should be calm in emergencies, prudent in counsel, courteous to his associates. He will be most successful in battle if he charges against the enemy, not like a wild beast, but in a caclulated manner.

The Emperor Maurice, *The Strategikon*, c. AD 600

There is nothing more base than for a man to lose his temper too often. No matter how angry one becomes, his first thought should be to pacify his mind and come to a clear understanding of the situation at hand. Then, if he is in the right, to become angry is correct.

Shiba Yoshimasa (1350–1410), *The Chikubasho*, 1380, in Wilson, *The Ideals of the Samurai*, 1982

But it might be closer to the truth to assume that the faculty known as *self-control* – the gift of keeping calm even under the greatest stress – is rooted in temperament. It is itself an emotion which serves to balance the passionate feelings in strong characters without destroying them, and it this balance alone that assures the dominance of the intellect . . .

Major General Carl von Clausewitz, *On War*, i, 1832, tr. Howard and Paret

---

General Meade was an officer of great merit with drawbacks to his usefulness which were beyond his control . . . He made it unpleasant at times, even in battle, for those around him to approach him with information.

General of the Army Ulysses S. Grant, *Personal Memoirs of U.S. Grant*, 1885

. . . if there are persons with steel nerves on a patrol boat, then their night attacks will be completely successful. Persons with great self-control can achieve miracles, while weak will of the executors and a lack of tenacity to a significant degree diminish the result.

Admiral Stepan O. Makarov (1849–1904), quoted in Albamonov, *Admiral*

But the priceless advantage which may be instilled in the military crowd by a proper training is that it also possesses the means of recovery. That possibility – the resolution of order out of chaos – reposes within every file who has gained within the service a confidence that he has some measure of influence among his fellows. The welfare of the unit machinery depends upon having the greatest possible number of human shock-absorbers – men who in the worst hour are capable of stepping forward and saying: 'This calls for something extra and that means me.' The restoration of control upon the battlefield, and the process of checking fright and paralysis and turning men back to essential tactical duties, does not come simply of constituted authority again finding its voice and articulating its strength to the extremities of the unit boundary. Control is a man-to-man force under fire. No matter how lowly his rank, any man who controls himself contributes to the control of others. A private can steady a general as surely as a cat can look at a king. There is no better ramrod for the back of a senior, who is beginning to buckle, than the sight of a junior who has kept his nerve. Land battles, as to the fighting part, are won by the intrepidity of men in grade from private to captains mainly. Fear is contagious but courage is not less so. The courage of any one man reflects in some degree the courage of all those who are within vision. To the man who is in terror and bordering on panic, no influence can be more steadying than that of seeing some other man near him who is retaining self-control and doing his duty.

Brigadier General S. L. A. Marshall, *The Armed Forces Officer*, 1950

An officer should never speak ironically or sarcastically to an enlisted man, since the latter doesn't have a fair chance to answer back. The use of profanity and epithets comes under the same heading. The best argument for a man keeping his temper is that nobody else wants it; and when he voluntarily throws it away, he loses a main prop to his own position.

Brigadier General S. L. A. Marshall, *The Armed Forces Officer*, 1950

Explosions of temper do not necessarily ruin a general's reputation or influence with his troops; it is almost expected of them ('the privileged irascibility of senior officers', someone has written), and it is not always resented, sometimes even admired, except by those immediately concerned. But sarcasm is always resented and seldom forgiven. In the Peninsula the bitter sarcastic tongue of Craufurd, the brilliant but erratic leader of the Light Division, was much more wounding and feared than the more violent outbursts of Picton, a rough, hot-tempered man.

Field Marshal Earl Wavell, *Soldiers and Soldiering*, 1953

For myself, I do not smoke and I drink no alcohol of any sort; this is purely because I dislike both tobacco and alcohol, and therein I am lucky because I believe one is in far better health without them. In general, I consider that excessive smoking and drinking tend to cloud the brain; when men's lives are at stake this must never be allowed to happen, and it does happen too often. You cannot win battles unless you are feeling well and full of energy.

Field Marshal Viscount Montgomery of Alamein, *The Memoirs of Field Marshal Montgomery*, 1958

Only those who have disciplined themselves can exact disciplined performance from others. When the chips are down, when privation mounts and the casualty rates rise, when the crisis is at hand, which commander, I ask, receives the better response? Is it the one who has failed to share the rough going with his troops, who is rarely seen in the zone of aimed fire, and who expects much and gives little? Or is it the one whose every thought is for the welfare of his men, consistent with the accomplishment of the mission; who does not ask them to do what he has not already done and stands ready to do again when necessary; who with his men has shared short rations, the physical discomforts and rigors of campaign, and will be found at the crisis of action where the issues are to be decided?

General Mathew B. Ridgway, 'Leadership', *Military Review*, 10/1966

SIMPLICITY
(Principle of War)

The art of war is a simple art; everything is in the performance. There is nothing vague in it; everything in it is common sense; ideology does not enter it.

Napoleon, dictation on St. Helena, ed. Herold, *The Mind of Napoleon*, 1955

The art of war is like everything else that is beautiful and simple. The simplest moves are the best. If MacDonald, instead of doing whatever he did, had asked a peasant for the way to Genoa, the peasant would have answered, 'Through Bobbio' – and that would have been a superb move.

Napoleon, 1818, conversation, ed. Herold, *The Mind of Napoleon*, 1955

The whole of military activity must . . . relate directly or indirectly to combat operations. The end for which a soldier is recruited, clothed, armed, and trained, the whole object of his sleeping, eating, drinking, and marching *is simply that he should fight at the right place and the right time.*

Major General Carl von Clausewitz, *On War*, i, 1832, tr. Howard and Paret

---

If one has never personally experienced war, one cannot understand in what the difficulties constantly mentioned really consist, nor why a commander should need any brilliance and exceptional ability. Everything looks simple; the knowledge required does not look remarkable, the strategic options are so obvious that by comparison the simplest problem of higher mathematics has an impressive scientific dignity. Once war has actually been seen the difficulties become clear; but it is still extremely hard to describe the unseen, all-pervading element that brings about this change of perspective.

Major General Carl von Clausewitz, *On War*, i, 1832, tr. Howard and Paret

---

In war only what is simple can succeed.

Field Marshal Paul von Hindenburg (1847–1934)

Spartan simplicity must be observed. Nothing will be done merely because it contributes to beauty, convenience, comfort, or prestige.

Circular from the Officer of the Chief Signal Officer, U.S. Army, 29 May 1945

Simplicity contributes to successful operations. Direct, simple plans and clear, concise orders minimalize misunderstanding and confusion. If other factors are equal, the simplest plan is preferred.

U.S. Army Field Manual 100–5, 19 February 1962

The KISS principle: KEEP IT SIMPLE, STUPID!

US Army Maxim

SKILL

The same man cannot well be skilled in everything; each has his own special excellence.

Euripides (c. 480 BC–406 BC), *Rhesus*, tr. Lattimore

Certes the Frenchmen and Rutters, deriding our new archery . . . will not let in open skirmish, if any leisure serve, to turn up their tails and cry, 'Shoot, English!' and all because our strong shooting is decayed and laid in bed. But if some of our Englishmen now lived that served Edward the Third in his wars with France, the breech of such a varlet should have been nailed to his bum with an arrow and another feathered in his bowels before he should have turned about to see who shot the first.

William Harrison, quoted in Holinshed, *Chronicles*, 1807

Mr. Amery, author of *The Times* History of the War, probed a weak spot in the prevailing

Europeans' theory by arguing that superior skill now counted more than superior numbers, and that its proportionate value would increase with material progress. The same note was struck by General Baden-Powell, who urged that the way to develop it was to give officers responsibility when young – he was left to find his channel for proving this in the Boy Scout movement, and not in the Army.

Captain Sir Basil Liddell Hart, *A History of the World War, 1914–1918*, 1919

Very few people will use skill if brute force will do the trick. The worst thing for a good general is to have superior numbers.

Colonel T. E. Lawrence, 31 March 1935, letter to Liddell Hart, *T. E. Lawrence to His Biographies*

See also: MARKSMANSHIP

SOLDIERS

Just think of how the soldier is treated. While still a child he is shut up in the barracks. During his training he is always being knocked about. If he makes the least mistake he is beaten, a burning blow on his body, another on his eye, perhaps his head is laid open with a wound. He is battered and bruised with flogging. On the march he has to carry bread and water like the load of an ass; the joints of his back are bowed; they hang heavy loads round his neck like that of an ass . . .

An anonymous Egyptian scribe expressing the usual distaste of the pen for the sword

---

. . . remember that zeal, obedience, and a sense of honour mark the good soldier.

Brasidas, 422 BC, speech to the Spartans and their allies before the Battle of Amphipolis

---

For who ought to be more faithful than a man that is entrusted with the safety of his country, and has sworn to defend it to the last drop of his blood? Who ought to be fonder of peace than those that suffer by nothing but war? Who are under greater obligations to worship God than Soldiers, who are daily exposed to innumerable dangers, and have most occasion for his protection?

Niccolò Machiavelli, *The Art of War*, 1521

Love the soldier, and he will love you. That is the secret.

Field Marshal Prince Aleksandr V. Suvorov (1729–1830)

Soldiers generally win battles; generals generally get the credit for them.

Napoleon, 1815, to General Gaspard at St. Helena

Troops are made to let themselves be killed.

Napoleon, 1817, conversation at St. Helena, ed. Herold, *The Mind of Napoleon*, 1955

MAXIM 58. The first qualification of a soldier is fortitude under fatigue and privation. Courage is only the second; hardship, poverty, and want are the best school of the soldier.

Napoleon, *The Military Maxims of Napoleon*, 1831, tr. D'Aguilar

---

MAXIM 59. There are five things the soldier should never be without: his firelock, his ammunition, his knapsack, his provisions (for at least four days), and his entrenching tool. The knapsack may be reduced to the smallest size possible, but the soldier should always have it with him.

Napoleon, *The Military Maxims of Napoleon*, 1831, tr. D'Aguilar

---

The seaman is the main engine on a military ship, and we are only the springs which act on it . . . The seaman controls the sails, it is he too who aims the weapon at the enemy; the seaman does everything . . .

Admiral P. S. Nakhimov (1802–1855)

The only change that breech-loading arms will probably make in the art and practice of war will be to increase the amount of ammunition to be expended, and necessarily to be carried along; to still further 'thin out' the lines of attack, and to reduce battles to short, quick decisive conflicts. It does not in the least affect the grand strategy, or the necessity for perfect organization, drill, and discipline. The companies and battalions will be more dispersed, and the men will be less under the immediate eye of their officers, and therefore a higher order of intelligence and courage on the part of the individual soldier will be an element of strength.

General of the Army William T. Sherman, *The Memoirs of General W. T. Sherman*, 1875

Boys are soldiers in their hearts already – and where the harm? Soldiers are not pugnacious. Paul bade Timothy to be a good soldier. Christ commended the centurion. Milton urged teachers to fit their pupils for all the offices of war. The very thought of danger and self-sacrifice are inspirations.

General Sir Ian Hamilton, *The Soul and Body of an Army*, 1921

As in the shades of a November evening I, for the first time, led a platoon of Grenadiers across the sopping fields . . . the conviction came into my mind with absolute assurance that the simple soldiers and their regimental officers, armed with their cause, would by their virtues in the end retrieve the mistakes and ignorance of Staffs and Cabinets, or Admirals, Generals, and politicians – including, no doubt, many of my own. But, alas, at what a needless cost!

Sir Winston Churchill, *The World Crisis*, 1923

What are the qualities of a good soldier, by the development of which we make the man war-worthy – fit for any war? . . . The following four – in whatever order you place them – pretty well cover the field: discipline, physical fitness, technical skill in the use of weapons, battle-craft.

Field Marshal Earl Wavell, 15 February 1933, lecture at the Royal United Service Institution (RUSI)

The soldier is the Army. No army is better than its soldiers. The soldier is also a citizen. In fact, the highest obligation and privilege of citizenship is that of bearing arms for one's country. Hence it is a proud privilege to be a soldier – a good soldier. Anyone, in any walk of life, who is content with mediocrity is untrue to himself and to American tradition. To be a good soldier a man must have discipline, self-respect, pride in his unit and in his country, a high sense of duty and obligation to his comrades and to his superiors, and self-confidence, borne of demonstrated ability.

General George S. Patton, Jr., *War As I Knew It*, 1947

The soldier is a man; he expects to be treated as an adult, not a schoolboy. He has rights; they must be made known to him and thereafter respected. He has ambition; it must be stirred. He has a belief in fair play; it must be honoured. He has a need of comradeship; it must be supplied. He has imagination; it must be stimulated. He has a sense of personal dignity; it must be sustained. He has pride; it can be satisfied and made the bedrock and respected role. To give a man this is the acme of inspired leadership. He has become loyal because loyalty has been given to him.

General of the Armies George C. Marshall (1880–1959)

Professional soldiers are sentimental men, for all the harsh realities of their calling. In their wallets and in their memories they carry bits of philosophy, fragments of poetry, quotations from the scriptures, which in times of stress and danger speak to them with great meaning.

General Mathew B. Ridgway, *My Battles in War and Peace*, 1956

I have written much of generals and staff officers . . . the war in Burma was a soldier's war. There comes a moment in every battle against a stubborn enemy when the result hangs in the balance. Then the general, however skilful and farsighted he may have been, must hand over to his soldiers, to the men in the ranks and to their regimental officers and leave them to complete what he has begun.

Field Marshal Viscount Slim, *Defeat Into Victory*, 1962

By people I do not mean personnel . . . I mean living, breathing, serving, human beings. They have needs and interests and desires. They have spirit and will, and strengths and abilities. They have weaknesses and faults; and they have means. They are the heart of our preparedness . . . and this preparedness – as a nation and as an Army – depends upon the spirit of our soldiers. It is the spirit that gives the Army . . . life. Without it we cannot succeed.

General Creighton Abrams, quoted in *Military Review*, 4/1985

SPEECHES AND THE SPOKEN WORD

Alexander . . . called a . . . meeting of his officers. There was no need, he said, for any words from him to encourage them to do their duty; there was inspiration enough in the courage they had themselves shown in previous battles, and in the many deeds of heroism they had already performed. All he asked was that every officer of whatever rank, whether he commanded a company, a squadron, a brigade, or an infantry battalion, should urge to their utmost efforts the men entrusted to his command; for they were about to fight, not, as before, for Syria or Phoenicia, or Egypt, but this time the issue at stake was the sovereignty of the whole Asian continent. What need, then, was there for many words of valour, when that valour was already in their own breasts? Let him remind them each for himself to preserve discipline in the hour of danger – to advance, when called upon to do so, in utter silence; to watch the time for a hearty cheer, and, when the moment came, to roar out their battle-cry and put the fear of God into the enemy's hearts. All must obey orders promptly and pass them on without hesitation to their men; and, finally, every one of them must remember that upon the conduct of each depended the fate of all: if each man attended to his duty, success was assured; if one man neglected it, the whole army would be in peril.

With some brief words of exhortation Alexander addressed his officers, and in reply they begged him to have every confidence in them . . .

Flavius Arrianus Xenophon (Arrian), of Alexander's exhortation of his officers before the Battle of Arbela, 331 BC, *The Campaigns of Alexander*, c. AD 150, tr. de Selincourt

---

The general who possesses some skill in public speaking is able, as in the past, to rouse the weak-hearted to battle and restore courage to a defeated army.

The Emperor Maurice, *The Strategikon*, c. AD 600

MAXIM 61. It is not speeches at the moment of battle that render soldiers brave. The veteran scarcely listens to them, and the recruits forgets them at the first discharge. If discourses and harangues are useful, it is during the campaign; to do away with unfavourable impressions, to correct false reports, to keep alive a proper spirit in the camp, and to furnish materials and amusement for the bivouac. All printed orders should keep in view these objects.

Napoleon, *The Military Maxims of Napoleon*, 1831, tr. D'Aguilar

---

Action employs men's fervour, but words arouse it.

Charles de Gaulle, quoted in Richardson, *Fighting Spirit*, 1978

Old man, when you have something to say to officers or men, make it snappy. The fewer words, the better. They won't believe you if you shoot bull. When you face ranks of men and try that, you can hear 'em sigh in despair when you open your mouth, if they sense you're a phoney. They can usually look at you and tell. Maybe it doesn't sound like it, but that's an important thing in a Marine's career.

Lieutenant General Lewis 'Chesty' Puller, quoted in Burke, *Marine*, 1962

Men who can command words to serve their thoughts and feelings are well on their way to commanding men to serve their purposes.

Brigadier General S. L. A. Marshall, *The Armed Forces Officer*, 1950

The commander must have contact with his men. He must be capable of feeling and thinking with them. The solider must have confidence in him. There is one cardinal principle which must always be remembered: one must never make a show of false emotions to one's men. The ordinary soldier has a surprisingly good nose for what is true and what false.

Field Marshal Erwin Rommel, *The Rommel Papers*, 1953

The troops must be brought to a state of wild enthusiasm before the operation begins. They must have that offensive eagerness and that infectious optimism which comes from physical well-being. They must enter the fight with the light of battle in their eyes and definitely wanting to kill the enemy. In achieving this end, it is the spoken word which counts, from the commander to his troops; plain speech is far more effective than any written word.

Field Marshal Viscount Montgomery of Alamein, *The Memoirs of Field Marshal Montgomery*, 1958

SPEED (Principle of War)

Speed is the essence of war. Take advantage of the enemy's unpreparedness; travel by unexpected routes and strike him where he has taken no precautions.

Sun Tzu, *The Art of War*, c. 500 BC, tr. Griffith

Thus, while we have heard of blundering swiftness in war, we have not yet seen a clever operation that was prolonged.

Sun Tzu, *The Art of War*, c. 500 BC, tr. Griffith

He who advances is irresistible; plunges into his enemy's weak positions; he who in withdrawal cannot be pursued moves so swiftly that he cannot be overtaken.

Sun Tzu, *The Art of War*, c. 500 BC, tr. Griffith

Come like the wind, go like the lightning.

Chang Yu, late Sung Dynasty commentator on Sun Tzu, *The Art of War*, c. 500 BC

It is essential to be cautious and take your time in making plans, and once you come to a decision carry it out right away with without any hesitation or timidity. Timidity after all is not caution but the invention of wickedness.

The Emperor Maurice, *The Strategikon*, c. AD 600

An attack may lack ingenuity, but it must be delivered with supernatural speed.

Tu Yu (AD 735–812), commentator on Sun Tzu, *The Art of War*, c. 500 BC

If you move rapidly, above all if the work of your scouts, your intelligence and your couriers is reliable, you can be certain of defeating a battalion with a detachment, an army with a battalion.

Anonymous Byzantine General, 10th-century treatise on tactics, quoted in Guerdan, *Byzantium, Its Triumphs and Tragedy*

In military practice one must plan quickly and carry on without delay, so as to give the enemy no time to collect himself.

Field Marshal Prince Aleksandr V. Suvorov (1729–1800), quoted in Blease, *Suvorof*, 1920

I work in minutes, not in hours.

Field Marshal Prince Aleksandr V. Suvorov (1729–1800), quoted in Blease, *Suvorof*, 1920

. . . it is the moment which gives victory, master it by the swiftness of Caesar, who knew so well how to surprise his enemies, even in broad daylight, to turn them and attack them where and when he wished, without ever being compelled to cut off their supplies of food and fodder . . .

Field Marshal Prince Aleksandr V. Suvorov (1729–1800), quoted in Blease, *Suvorof*, 1920

To do good one must make haste.

Field Marshal Prince Aleksandr V. Suvorov (1729–1800), quoted in Blease, *Suvorof*, 1920

Swiftness and impact are the soul of genuine warfare.

Field Marshal Prince Aleksandr V. Suvorov (1729–1800), quoted in Savkin, *Basic Principles of Operational Art and Tactics*, 1972

Just as lightning has already struck when the flash is seen, so when the enemy discovers the head of the army, the whole should be there, and leave him no time to counteract its dispositions.

Field Marshal François Comte de Guibert (1744–1790)

It is very advantageous to rush unexpectedly on an enemy who has erred, to attack him suddenly and come down on him with thunder before he sees the lightning.

Napoleon (1769–1821)

In order to smash, it is necessary to act suddenly.

Napoleon (1769–1821)

The third rule is never to waste time. Unless important advantages are to be gained from hesitation, it is necessary to set to work at once. By this speed a hundred enemy measures are nipped in the bud, and public opinion is won most rapidly.

Major General Carl von Clausewitz, *Principles of War*, 1812

Speed shall be the motto of this campaign. We shall not allow Morillo any time to threaten our rear, for when he can do something against us, we shall be coming back on him with twice or thrice as much force as we are taking away.

General Simon Bolivar, May 1819, quoted in Madariaga, *Bolivar*

Speed is one of the chief characteristics of strategical marches, as it is of ordinary movements on the battle field. In this one quality reside all the advantages that a fortunate initiative may have procured; and by it we gain in the pursuit all the results that a victory on the battle field has placed in our hands.

Dennis Hart Mahan, *Out Post*, 1847

No great success can be hoped for in war in which rapid movements do not enter as an element. Even the very elements of Nature seem to array themselves against the slow and over-prudent general.

Dennish Hart Mahan, *Out Post*, 1847

Speed is the essential requisite for a first-class ship of war – but essential only to go into action, not out of it.

Rear Admiral John A. Dahlgren (1809–1870)

The true speed of war is not headlong precipitancy, but the unremitting energy which wastes no time.

Rear Admiral Alfred Thayer Mahan, *Lessons of the War With Spain*, 1899

Speed may be either of movement or mind. Speed of movement, again, is subdivided into forward, or momentum, and lateral, or flexibility. The use of mechanical transport, superior marching powers, good organization and staff work, light equipment, all contribute to speed of movement.

Speed of mind to seize an opening and exploit an unforeseen advantage is the product not only of the commander's training and natural ability, but also of the promptness of his information – through personal touch with the progress of the battle, and the instant rendering of reports. (October 1925)

Captain Sir Basil Liddell Hart, *Thoughts on War,* 1944

. . . nothing is more important to the future efficiency of the Army than to multiply its rate of movement. A law of physics that applies with equal force to warfare is that while striking force increases directly with the mass applied, it increases according to the square of the speed of the application. Through proper organization in all echelons, through the development and perfection of reliable combat machines capable of speedy maneuver, and through the improvement of transportation, maintenance, and communication, and supply arrangements, the objective of greater and still greater speed must be pursued.

General of the Army Douglas MacArthur, *Annual Report of the Chief of Staff* June 30, 1935

Speed of action in conjunction with organization, expert manoeuvre and dextrous application to the terrain, with an account of the enemy's air, is a basic guarantee of success in battle. Troops having quickly executed a disposition, quickly regrouped with a changing situation, quickly arising from rest, quickly perfecting a campaign movement, quickly falling out in combat order and opening fire, having quickly attacked and pursued the opponent, can always count on success.

Soviet *1936 Field Service Regulations,* attributed to Marshal of the Soviet Union Mikhail N. Tukhachevskiy, quoted in Garthoff, *Soviet Military Doctrine*

Speed is acquired by making the necessary reconnaissance, providing the proper artillery and other tactical support . . . bringing up every man, and then launching the attack with a predetermined plan so that the time under fire will be reduced to the minimum.

General George S. Patton, Jr., *War As I Knew It,* 1947

Don't Delay: The best is the enemy of the good. By this I mean that a good plan violently executed *now* is better than a perfect plan next week. War is very simple thing, and the determining characteristics are self-confidence, speed, and audacity. None of these things can be perfect, but they can be good.

General George S. Patton, Jr., *War As I Knew It,* 1947

In small operations, as in large, speed is the essential element of success. If the difference between two possible flanks is so small that it requires thought, the time wasted in thought is not well used.

General George S. Patton, Jr., *War As I Knew It,* 1947

One of the first lessons I had drawn from my experience of motorized warfare was that speed of manoeuvre in operations and quick reaction in command are decisive. Troops must be able to carry out operations at top speed and in complete coordination. To be satisfied with norms is fatal. One must constantly demand and strive for maximum performance, for the side which makes the greater effort is the faster – and the faster wins the battle . . .

Field Marshal Erwin Rommel, *The Rommel Papers,* 1953

Except for Napoleon, probably no other general appreciated as fully as Alexander the value of mobility in war. From the opening of his career until its close, speed dominated all his movements, and the result was that, by increasing the time at his disposal, in any given period he could proportionately accomplish more than his opponent.

Major General J. F. C. Fuller, *The Generalship of Alexander the Great,* 1960

There is only one sound way to conduct war as I read history: Deploy to the war zone as quickly as possible sufficient forces to end it at the earliest moment. Anything less is a gift to the other side.

Brigadier General S. L. A. Marshall, 'Thoughts of Vietnam', in Thompson, *The Lessons of Vietnam*, 1977

Lightning speed, more lightning speed; boldness, more boldness.

Motto of the North Vietnamese General Staff, quoted in Summers, *On Strategy*, 1981

SPIRIT

The spirit of the commander is naturally communicated to the troops, and there is an ancient saying that it is better to have an army of deer commanded by a lion than an army of lions commanded by a deer.

The Emperor Maurice, *The Strategikon,* c. AD 600

. . . Perseverance and spirit have done Wonders in all ages.

General George Washington, 20 August 1775, letter to Philip Schuyler, *The Writings of George Washington*, Vol. 3, 1931–44

Do not compare your physical forces with those of the enemy's, for the spirit should not be compared with matter. You are men, they are beasts; you are free, they are slaves. Fight and you shall conquer. God grants victory to constancy.

General Simon Bolivar, 1814, quoted in Madariaga, *Bolivar*, 1952

A warlike spirit, which alone can create and civilize a state, is absolutely essential to national defense and to national perpetuity . . . The more warlike the spirit of the people, the less need for a large standing army . . . Every male brought into existence should be taught from infancy that the military service of the Republic carries with it honour and distinction, and his very life should be permeated with the idea that even death itself may become a boon when a man dies that a nation may live and fulfill its destiny.

General of the Armies Douglas MacArthur, *Infantry Journal*, 3/1927

As long as you are generous and true, and also fierce, you cannot hurt the world or even seriously distress her. She was made to be wooed and won by youth. She has lived and thrived only by repeated subjugations.

Sir Winston S. Churchill, *My Early Life,* 1930

It is the cold glitter in the attacker's eye not the point of the questing bayonet that breaks the line. It is the fierce determination of the drive to close with the enemy not the mechanical perfection of the tank that conquers the trench. It is the cataclysmic ecstasy of conflict in the flier not the perfection of his machine-gun that drops the enemy in flaming ruin.

General George S. Patton, Jr. (1885–1947)

I have always held the view that an army is not merely a collection of individuals, with so many tanks, guns, machine-guns, etc., and that the strength of the army is not just the total of all these things added together. The real strength of an army is, and must be, far greater than the sum total of its parts; that extra strength is provided by morale, fighting spirit, mutual confidence between the leaders and the led and especially with the high command, the quality of comradeship, and many other intangible spiritual qualities.

Field Marshal Viscount Montgomery of Alamein, *The Memoirs of Field Marshal Montgomery*, 1958

To be among them made every day good. Theirs was a beaten, battered division, almost drained of flesh and blood, its rifle companies cut to fifteen or twenty men per unit. They understood the enormity of their misfortune, yet the few who survived glorified it by their spirit.

Brigadier General S. L. A. Marshall, of the men of the 2nd Infantry Division in 1951, *Bringing Up the Rear*, 1979

SPIT AND POLISH

The easiest and quickest path into the esteem of traditional military authorities is by the appeal to the eye rather than to the mind. The 'polish and pipeclay' school is not yet extinct, and it is easier for the mediocre intelligence to become an authority on buttons than on tactics. (March 1925.)

Captain Sir Basil Liddell Hart, *Thoughts on War*, 1944

When spit-and-polish are laid on so heavily that they become onerous, and the ranks cannot see any legitimate connection between the requirements and the development of an attitude which will serve as clear fighting purpose, it is to be questioned that the exactions serve any good object whatever.

Brigadier General S. L. A. Marshall, *The Armed Forces Officer*, 1950

STAFF

But above all, a general should take care to have men of proven fidelity, wisdom, and long experience in military affairs near his person as a sort of council. From such men a general may learn not only the state of his own army, but also that of the enemy's . . .

Niccolo Machiavelli, *The Art of War*, 1521

. . . in all ages and times, all emperors, kings, and formed commonwealths . . . have established a council of men of great sufficiency both in war and peace to assist their generals . . . to the intent that their generals in all important matters should consult with them, the conclusion and resolution of such consultations notwithstanding to remain in the wisdom, judgment, and valour of the generals.

John Smythe (*c.* 1580–1631), *Certain Discourses Military*

Large staffs – small victories.

Field Marshal Prince Aleksandr V. Suvorov (1729–1800)

A good staff has the advantage of being more lasting than the genius of a single man.

Lieutenant General Antoine-Henri Baron de Jomini, *Summary of the Art of War*, 1838

I do not want to make an appointment on my staff except of such as are early risers.

Lieutenant General Thomas 'Stonewall' Jackson, 1862, letter to his wife

---

A bulky staff implies a division of responsibility, slowness of action, and indecision, whereas a small staff implies activity and concentration of purpose. The smallness of Grant's staff throughout the civil war forms the best model for future generations.

General of the Army William T. Sherman, *The Memoirs of General W. T. Sherman*, 1875

The more intimately it [the staff] comes into contact with the troops, the more useful and valuable it becomes. The almost entire separation of the staff from the line, as now practiced by us, and hitherto by the French, has proved mischievous, and the great retinues of staff-officers with which some of our earlier generals began the war were simply ridiculous.

General of the Army William T. Sherman, *The Memoirs of General W. T. Sherman*, 1875

Great captains have no need for counsel. They study the questions which arise, and decide them, and their *entourage* has only to execute their decisions. But such generals are stars of the first magnitude who scarcely appear once in a century. In the great majority of cases, the leader of the army cannot do without advice. This advice may be the outcome of the deliberations of a small number of qualified men. But within this small number, one and only one opinion must prevail. The organization of the military hierarchy must ensure subordination even in thought and give the right and duty of presenting a single opinion for the examination of the general-in-chief to one man and only one.

Field Marshal Helmuth Graf von Moltke, 1862, letter

That is just one illustration of how time may be lost in handling troops, and of the need of an abundance of competent staff officers by the generals in command. Scarcely any of our generals had half of what they needed to keep a *constant & close supervision on the execution of important orders.* And that is always to be done. An army is like a great machine, and in putting it into battle it is not enough for a commander to merely issue the necessary orders. He should have a staff ample to supervise the execution of each step, & to promptly report any difficulty or misunderstanding.

Brigadier General Edward Porter Alexander, of the delays in positioning troops at Gettysburg on 2 July 1863, *Fighting for the Confederacy* (written between 1900–1907 but only published in 1989)

The typical staff officer is the man past middle life, spare, unwrinkled, intelligent, cold, passive, noncommittal; with eyes like a codfish, polite in contact but at the same time unresponsive, cool, calm and as damnably composed as a concrete post or a plaster-of-Paris cast; a human petrification with a heart of feldspar and without charm or the friendly germ; minus bowels, passion or a sense of humour. Happily they never reproduce and all of them finally go to hell.

Anonymous

No military or naval force, in war, can accomplish anything worthwhile unless there is back of it the work of an efficient, loyal, and devoted staff.

Lieutenant General Hunter Liggett (1857–1935)

Without a staff, an army could not peel a potato.

Lieutenant General Hunter Liggett (1857–1935)

There is a type of staff officer who seems to think that it is more important to draft immaculate orders than to get out a reasonably well-worded order in time for action to be taken before the situation changes or the opportunity passes. (June 1933)

Captain Sir Basil Liddell Hart, *Thoughts on War,* 1944

The staff becomes an all-controlling bureaucracy, a paper octopus squirting ink and wriggling its tentacles into every corner. Unless pruned with an axe it will grow like a fakir's mango tree, and the more it grows the more it overshadows the general. It creates work, it creates officers, and, above all, it creates rear-spirit. No sooner is a war declared than the general-in-chief (and many a subordinate general also) finds himself a Gulliver in Lilliput, tied down to his office stool by innumerable threads woven out of the brains of his staff and superior staffs.

Major General J. F. C. Fuller, *Generalship: Its Diseases and Their Cure,* 1933

A yes man on a staff is a menace to a commander. One with the courage of his convictions is an asset.

Major General Orlando Ward, *c.* 1934

The military staff must be adequately composed: it must contain the best brains in the fields of land, air, and sea warfare, propaganda war, technology, economics, politics and also those who know the people's life.

General Erich Ludendorff, *Total War,* 1935

A perfect commander is a man who does things by himself and of his own will [in furtherance of the objectives set down by his government's policy makers]. He will have about him aides to advise, and to put into execution what he demands and without whom the greatest genius cannot prevail. [These assistants comprise the commander's staff]. They are indeed his instruments, who as men understand the machinery of transmitting orders, as the soldier understands his weapons. But they must retire behind the personality of their chief.

Colonel Hermann Foertsch, *The Art of Modern War*, 1940

One of the inevitable problems with the Marine Corps or any other military service is that staff officers take over the minute a war is ended. The combat people run things when the chips are down and the country's life is at stake but when the guns stop, nobody's got a use for a combat man.

The staff officers are like rats; they stream out of hiding and take over. It's true. Just watch what happens to paperwork. God, in peacetime they put out enough to sink a small-size nation into the sea – and when war breaks, most of it just naturally stops. That's the way they do everything. There must always be staff people, of course, or we'd never get anything done – but if we don't stop this empire building of the staff, somebody's going to come along and lick us one of these days. We'll be so knotted in red tape that we can't move.

Lieutenant General Lewis 'Chesty' Puller, June 1946, letter to his wife, quoted in Davis, *Marine*, 1962

There is certainly no black magic about becoming a successful Army staff officer in the field. Youth and brains are the important things. With a comparatively small amount of training and experience most staff appointments can be filled by non-regulars. In 21st Army Group a number of important posts were held by young ex-civilians. One could not have wished for better men.

Montgomery backed youth and the 'clever chap', and this policy paid him enormously.

Major General Sir Francis de Guingand, *Operation Victory*, 1947

Staff work . . . has its own particularly rewards. Chief among them are the broadening of perspective, a more intimate conctact with the views, working methods, and personality characteristics of higher commanders and chance to become acquainted with administrative responsibility from the viewpoint of policy. Although it sounds mysterious and even forbidding until one has done it, the procedures are not more complex or less instructive than in any other type of assignment.

Brigadier General S. L. A. Marshall, *The Armed Forces Officer*, 1950

When you carry a paper in here, I want you to give me every reason you can think of why I should not approve it. If, in spite of your objections, my decision is still to go ahead, then I'll know I'm right.

General of the Army George C. Marshall, instructions to his staff, quoted in Bradley, *A Soldier's Story*, 1951

I now want you to consider the general in relation to his troops. I will begin with a few words about his staff, who are the means by which he controls and directs his army. I will give you two simple rules which every general should observe: first, never to try to do his own staff work; and, secondly, never to let his staff get between him and his troops. What a staff appreciates is that it should receive clear and definite instructions, and then be left to work out the details without interference. What troops and subordinate commanders appreciate is that a general should constantly be in personal contact with them, and should not see everything simply through the eyes of his staff. The less time a general spends in his office and the more with his troops the better.

Field Marshal Earl Wavell, *Soldiers and Soldiering*, 1953

I was evacuated to hospital in England and for some months I took no further part in the war.

I had time for reflection in hospital and came to the conclusion that the old adage was probably correct: the pen was mightier than the sword. I joined the staff.

Field Marshal Viscount Montgomery of Alamein, after being wounded in 1914, *The Memoirs of Field Marshal Montgomery*, 1958

My war experience led me to believe that the staff must be the servants of the troops, and that a good staff officer must serve his commander and the troops but himself be anonymous.

Field Marshal Viscount Montgomery of Alamein, *The Memoirs of Field Marshal Montgomery*, 1958

The qualities of a leader are not limited to commanders. The requirements for leadership are just as essential in the staff officer, and in some respects more exacting, since he does not have that ultimate authority which can be used when necessary and must rely even more than his commander on his own strength of character, his tact and persuasion in carrying out his duties.

General Mathew B. Ridgway, 'Leadership', *Military Review*, 10/1966

It's like dropping a pound of mercury on the floor. It scatters all over but pretty soon you find it all together over in the corner of the room.

General Creighton Abrams, speaking of the difficulty in reducing the Army Staff, quoted in *Military Review*, 4/1985

STRATEGY

The best way to defeat an enemy is to defeat his strategy. The best way to defeat his strategy is to adopt it.

Sun Tzu, *The Art of War*, c. 500 BC

Supreme excellence consists of breaking the enemy's resistance without fighting.

Sun Tzu, *The Art of War*, c. 500 BC

I know well that war is a great evil and the worst of all evils. But since our enemies clearly look upon the shedding of blood as one of their basic duties and the height of virtue, and since each one must stand up for his country and his own people with word, pen, and deed, we have decided to write about strategy. By putting it into practice we shall be able not only to resist our enemies but even to conquer them . . . Strategy is the means by which a commander may defend his own lands and defeat his enemies. The general is the one who practises strategy.

Anonymous Byzantine General, *On Strategy*, c. AD 527–AD 565, in Dennis, *Three Byzantine Military Treatises*, 1985

The most complete and happy victory is this: to compel one's enemy to give up his purpose, while suffering no harm oneself.

Count Belisarius (c. AD 505–AD 565)

Strategy is based on the forces you have, on the strength of the enemy, on the situation of the country where you want to carry the war, and on the actual political condition of Europe . . .

Frederick the Great, 'Pensées et règles générals pour la guerre',*Oeuvres de Frédéric le Grand*, Vol. XXVIII, 1846–56

Strategy means the combination of individual engagements to attain the goal of the campaign.

Major General Carl von Clausewitz, *Principles of War*, 1812

Strategy is the science of making use of space and time. I am more jealous of the latter than of the former. We can always recover lost ground, but never lost time.

Field Marshal August Graf von Gneisenau (1761–1831), quoted in Foertsch, *The Art of Modern War*, 1940

. . . the art of employment of battles as a means to gain the object of the war.

Major General Carl von Clausewitz, *On War*, 1832

Strategy is a system of makeshifts. It is more than a science, it is the application of science to practical affairs; it is carrying through an originally conceived plan under a constantly shifting set of circumstances. It is the art of acting under the pressure of the most difficult kind of conditions. Strategy is the application of common sense in the work of leading an army; its teachings hardly go beyond the first requirement of common sense; its value lies entirely in its concrete application. It is a matter of understanding correctly at every moment a constantly changing situation, and then doing the simplest and most natural thing with energy and determination. This is what makes war an art, an art that is served by many sciences. Like every art, war cannot be learned rationally, but only by experience. In war, as in art, there can be no set standards nor can code of rules take the place of brains.

Field Marshal Helmuth Graf von Moltke, *On Strategy*, 1871

A clever military leader will succeed in many cases in choosing defensive positions of such an offensive nature from the strategic point of view that the enemy is compelled to attack us in them.

Field Marshal Helmuth Graf von Moltke (1800–1891)

The main thing in true strategy is simply this: first deal as hard blows at the enemy's soldiers as possible, and then cause so much suffering to the inhabitants of a country that they will long for peace and press their Government to make it. Nothing should be left to the people but eyes to lament the war.

General of the Army Philip H. Sheridan, quoted in Forbes, *Memoirs and Studies in War and Peace*, 1895

But even if history had not shown that principles of strategy have held good under circumstances so many and various that they may be justly assured of universal value, to sea as well as to land, there would still remain the same mental training afforded by the examination of successive modifications that we have been introduced into the art of war by the great generals.

Rear Admiral Alfred Thayer Mahan, 6 August 1888, address at the U.S. Naval War College

---

As in a building, which, however fair and beautiful the superstructure, is radically marred and imperfect if the foundation be insecure – so, if the strategy be wrong, the skill of the general on the battlefield, the valor of the soldier, the brilliancy of the victory, however otherwise decisive, fail of their effect.

Rear Admiral Alfred Thayer Mahan, *Naval Administration and Warfare*, 1908

Principles of strategy never transcend common sense.

Motto of the highest school of the German Army before 1914, quoted in Trotsky, *Military Writings: Leon Trotsky*, 1967

Where the strategist is empowered to seek a military decision, his responsibility is to seek it under the most advantageous circumstances in order to produce the most profitable results. Hence *his true aim is not so much to seek battle, as to seek a strategic situation so advantageous that if it does not of itself produce the decision, its continuation by a battle is sure to achieve this.* In other words, dislocation is the aim of strategy; its sequel may be either the enemy's dissolution or his disruption in battle. Dissolution may involve some partial measure of fighting, but this has not the character of a battle. (October 1928)

Captain Sir Basil Liddell Hart, *Thoughts on War*, 1944

It is the art of strategy to bring about a battle under the most favourable conditions possible, both as to the ground and the timing, and to force it on the opponent under conditions that are most unpropitious to him.

Colonel Hermann Foertsch, *The Art of Modern War*, 1940

The aim of strategy is to clinch a political argument by means of force instead of words. Normally this is accomplished by battle, the true object of which is not physical destruction, but mental submission on the part of the enemy.

Major General J. F. C. Fuller, *Armoured Warfare*, 1943

Strategy is the determination of the direction of the main blow – The plan of strategy is the plan of the organization of the decisive blow in the direction in which the blow can most quickly give the maximum results.

In other words, *to define the direction of the main blow means to predetermine the nature of operations in the whole period of war, to determine nine-tenths of the fate of the entire war.* In this is the main task of strategy.

Joseph Stalin, quoted in Garthoff, *Soviet Military Doctrine*, 1953

The history of strategy is, fundamentally, a record of the application and evolution of the indirect approach.

Captain Sir Basil Liddell Hart, *Strategy*, 1954

. . . the art of distributing and applying the military means to fulfill the ends of policy.

Captain Sir Basil Liddell Hart, *Strategy*, 1954

. . . the essence of strategy is the abstract interplay which, to use Foch's phrase, springs from the clash between two opposing wills. It is the art which enables a man, no matter what the techniques employed, to master the problems set by any clash of wills and as a result to employ the techniques available with maximum efficiency. It therefore the art of the dialectic of force or, more precisely, *the art of the dialectic of two opposing wills using force to resolve their dispute.*

General Andre Beaufre, *An Introduction to Strategy*, 1965

The laws of strategy are objective and apply impartially to both hostile sides.

Marshal of the Soviet Union V. D. Sokolovskiy, *Soviet Military Strategy*, 3rd ed., 1968

Lesson No. 1 from Fuller: 'To anticipate strategy, imagine.'
Lesson No. 2 from Martel: 'Men, not weapons, will shape the future, so stick with fundamentals.'

Brigadier General S. L. A. Marshall, *Bringing Up the Rear*, 1979

Our strategy to go after this army is very, very simple. First we're going to cut it off, and then we're going to kill it.

General Colin Powell, 23 January 1991, speaking of the Iraqi Army

---

See also: TACTICS, TACTICS VS STRATEGY

STRESS

In this situation the real leaders of the Army stood forth in bold contrast to those of ordinary clay. Men who had sustained a reputation for soliderly qualities, under less trying conditions, proved too weak for the ordeal and became pessimistic calamity howlers. Their organizations were quickly infected with the spirit and grew ineffective unless a more suitable commander was given charge. It was apparent that the combination of tired muscles, physical discomforts, and heavy casualties weakened the backbone of many. Officers who were not in perfect physical condition usually lost the will to conquer and took an exceedingly gloomy view of the situation.

General of the Army George C. Marshall, *Memoirs of My Services in the World War*, written between 1923 and 1926 but not published until 1976

The general is dealing with men's lives, and must have a certain mental robustness to stand the strain of this responsibility.

Field Marshal Earl Wavell, *Generals and Generalship*, 1941

. . . realized how inexorably and inescapably strain and tension wear away at the leader's endurance, his judgment and his confidence. The pressure becomes more acute because of the duty of a staff constantly to present the commander the worst side of an eventuality. In this situation, the commander has to preserve optimism in himself and in his command. Without confidence, enthusiasm and optimism in the command, victory is impossible.

General of the Army Dwight D. Eisenhower, quoted in *Military Review*, 10/1990

The military is a long, hard road, and it makes extraordinary requirements of every individual. In war, particularly, it puts stresses upon men such as they have not known elsewhere, and the temptation to 'get out from under' would be irresistible if their spirits had not been tempered to the ordeal.

Brigadier General S. L. A. Marshall, *The Armed Forces Officer*, 1950

If armies are to hold together under stress, then it must be clear to the troops who comprise the fighting units that their officers are willing to share the ultimate risk of death with them. There is no clearer indication of this willingness than to have officers die in full view of their men.

Paul Savage and Lieutenant Colonel Richard Gabriel, *Crisis in Command*, 1978

One of the common ways that seniors induce self-defeating stress in their juniors is to over-supervise them. Another is to summon them repeatedly to report in person about inconsequential administrative details.

Major General Aubrey 'Red' Newman, *Follow Me*, 1981

Stress is essential to leadership. Living with stress, knowing how to handle pressure, is necessary for survival. It is related to a man's ability to wrest control over his own destiny from the circumstances that surround him. Or, if you like, to prevail over technology. Tied up with this ability is something I can express in one word, 'improvisation'. I mean man's ability to prepare a response to a situation while under pressure.

Admiral James Stockdale, 'Education for Leadership and Survival', *Military Ethics*, 1987

STRIKE WEAKNESS

You may advance and be absolutely irresistible, if you make for the enemy's weak points; you may retire and be safe from pursuit if your movements are more rapid than those of the enemy.

Sun Tzu, *The Art of War*, c. 500 BC

No, an army may be likened to water, for just as flowing water avoids the heights and hastens to the lowlands, so an army avoids strength and strikes weakness.

Sun Tzu, *The Art of War*, c. 500 BC

Go into emptiness, strike voids, bypass what he defends, hit him where he does not expect you.

The 'Martial' Emperor Ts'ao Ts'ao (AD 155–220)

President of the Directoire: 'You often defeated a stronger foe with fewer forces.'

Bonaparte: 'But in this case, too, the lesser forces still suffered defeat at the hands of the larger forces. With an enemy army superior in numbers against me, I dashed like lightning to its flank, smashed it, took advantage of the enemy's disarray, and again rushed with all my forces to other points. Thus, I inflicted defeat piecemeal, and the victory which I won was, as you see, nothing more than the victory of the stronger over the weaker.'

Napoleon, quoted in Colonel V. Ye. Savkin, *Operational Art and Tactics*, 1972

We won the victory where it was easy to win and not over something that was hard to conquer . . . each division of the combined squadrons did its work well and not more. There was nothing remarkable in our bravery. We regard ourselves as a fit example of good fight . . .

Admiral Marquis Togo Heihachiro, remarking on the Battle of Tsushima, quoted in Warner, *The Tide at Sunrise*, 1974

Unless very urgent reasons to the contrary exist, strike at one end rather than at the middle, because both ends can come up to help the middle against you quicker than one can get to help the other; and, as between the two ends, strike at the one upon which the enemy most depends for reinforcements and supplies to maintain his strength.

Rear Admiral Alfred Thayer Mahan, *Sea Power in its Relations with the war of 1812*, 1905

Have been giving everyone a simplified directive of war. Use steamroller strategy; that is, make up your mind on course and direction of action, and stick to it. But in tactics, do not streamroller. Attack weakness. Hold them by the nose and kick them in the pants.

General George S. Patton, Jr., 6 November 1942, *War As I Knew It*, 1947

See also: THE INDIRECT APPROACH

STYLE

. . . in the profession of war, like that of letters, each man has his style.

Napoleon, 6 June 1806, letter to Joseph, *Correspondance*, No. 10325, Vol. XII, 1858–1870

At the head of an army, nothing is more becoming than simplicity.

Napoleon, *Political Aphorisms*, 1848

Style . . . is first and above all the expression of a man's personality, as characteristic as any other trait; or, as some one has said – was it Buffon? – style is the man himself. From this point of view it is susceptible of training, of development, or of pruning; but to attempt to pattern it on that of another person is a mistake. For one chance of success there are a dozen of failure; for you are trying to raise a special product from a soil probably uncongenial, or a fruit from an alien stem – figs from vines.

Rear Admiral Alfred Thayer Mahan, *From Sail to Steam*, 1907

There are apparently two types of successful soldiers. Those who get on by being unobtrusive and those who get on by being obtrusive. I am of the latter type and seem to be rare and unpopular; but it is my method. One has to choose a system and stick to it. People who are not themselves are nobody.

General George S. Patton, Jr., 6 June 1944, *The Patton Papers*, Vol. II, 1974

. . . one of my cardinal rules of battle leadership – or leadership in any profession – is to be yourself, to strive to apply the basic principles of the art of war, and to seek to accomplish your assigned missions by your own methods in your own way.

General Mathew B. Ridgway, *The Korean War*, 1967

SUBORDINATES

Encourage and listen well to the words of your subordinates. It is well known that gold lies hidden underground.

Nabeshima Naoshige (1538–1618), in Wilson, *Ideals of the Samurai*, 1982

Soldiers must never be witnesses to the discussions of their commanders.

Napoleon, 31 March 1805, letter to Marshal De Moncey, *Correspondance*, No. 8507, Vol. X, 1858–1870

A major gulf exists between a commander-in-chief – a general who leads the army as a whole or commands in a theatre of operations

– and the senior generals immediately subordinate to him. The reason is simple: the second level is subjected to much closer control and supervision, and thus gives far less scope for independent thought. People therefore often think outstanding intellectual ability is called for only at the top, and that for all other duties common intelligence will suffice. A general of lesser responsibility, an officer grown grey in the service, his mind well-blinkered by long years of routine, may often be considered to have developed a certain stodginess; his gallantry is respected, but his simplemindedness makes us smile. We do not intend to champion and promote these good men; it would contribute nothing to their efficiency, and little to their happiness.

Major General Carl von Clausewitz, *On War*, i, 1832, tr. Howard and Paret

---

Today I decided to stay in . . . There is nothing that I can do at the front except bother people, as they are all doing a swell job, and I believe advancing very rapidly. One of the hardest things that I have to do to – and I presume any General has to do – is not to interfere with the next echelon of command when the show is going all right.

General George S. Patton, Jr., 18 November 1944, letter to his sister Nita, *The Patton Papers*, Vol. II, 1974

Commanders and staffs should visit units two echelons below their own, and their maps should be so kept. In other words, Corps Commanders or their staffs should visit Division and Regimental Command Posts; the Division Commander should visit Regimental and Battalion Command Posts; the visits above referred to are for command purposes. What might be called inspirational visits should go farther up. The more senior the officer who appears with a very small unit at the front, the better the effect on the troops. If some danger is involved in the visit, its value is enhanced.

General George S. Patton, Jr., *War As I Knew It*, 1947

The more senior the officer, the more time he has. Therefore, the senior should go forward to visit the junior rather than call the junior back to see him. The exception to this is when it is necessary to collect several commanders for the formulation of coordinated plan . . .

General George S. Patton, Jr., *War As I Knew It*,, 1947

There have been great and distinguished leaders in our military Services at all levels who had no particular gifts for administration and little for organizing the detail of decisive action either within battle or without. They excelled because of a superior ability to make use of the brains and command the loyalty of well-chosen subordinates. Their particular function was to judge the goal according to their resources and audacity, and then to hold the team steady until the goal was gained. So doing, they complemented the power of the faithful lieutenants who might have put them in the shade in any IQ test.

Brigadier General S. L. A. Marshall, *The Armed Forces Officer*, 1950

Of course a commander must know in what way to give verbal orders to his subordinates. No two will be the same; each will require different treatment. Some will react differently from the others; some will be happy with a general directive whilst others will like more detail. Eventually a mutual confidence on the subject will grow up between the commander and his subordinates; once this has been achieved there will never be any more difficulties or misunderstandings.

Field Marshal Viscount Montgomery of Alamein, *The Memoirs of Field Marshal Montgomery*, 1958

Closely akin to the relationship with staff officers is keeping in close personal touch with your principal subordinate commanders – in the division, with your brigade and separate battalion commanders; in the corps with your division commanders, their chief of staff, and as many of the commanders of attached corps units as you can; and in the army, with corps and division commanders and their chiefs of staff. There is always time for these visits; administrative work can be done at night. By day you belong with your troops.

General Mathew B. Ridgway, 'Leadership', *Military Review*, 10/1966

I had to walk that tightrope familiar to all top commanders in any profession, civilian as well as military: To maintain the proper balance between according sufficient latitude to the subordinate commander in carrying out broadly stated directives, and in exercising adequate supervision, as befitted the man who would bear the ultimate responsibility for the success or failure of the entire effort.

General Mathew B. Ridgway, *The Korean War*, 1967

See also: DELEGATION OF AUTHORITY

SUCCESS

Success is a matter of planning and it is only careless people who find that Heaven will not help their mortal designs.

Themistocles, 480 BC, the year of the Battle of Salamis, quoted in Herodotus (*c.* 484–420 BC), *The Struggle for Greece*, tr. Cavandar

Whosoever desires constant success must change his conduct with the times.

Niccolo Machiavelli, *Discourses*, 1517

. . . it will be a consolation to you to reflect that the thinking part of Mankind do not form their judgment from events; and that their equity will ever attach equal glory to those actions which deserve success, as to those which have been crowned with it.

General George Washington, 11 September 1778, to General Comte d'Estaing, *The Writings of George Washington*, Vol. 12, 1931–44

Although I cannot insure success, I will endeavour to deserve it.

Admiral John Paul Jones, 1780, letter

I am sure you will deserve success. To mortals is not given the power of commanding it.

Admiral Lord St. Vincent, 14 July 1797, letter to Nelson before sailing against Tenerife

Success is only to be obtained by simultaneous efforts, directed upon a given point, sustained with constancy, and executed with decision.

Archduke Charles of Austria (1771–1841)

. . . in war many roads lead to success, and that they do not all involve the opponent's defeat. They range from *the destruction of the enemy's forces, the conquest of his territory, to a temporary occupation or invasion, to projects with an immediate purpose, and finally passively awaiting the enemy's attacks.* Any one of these may be used to overcome the enemy's will: the choice depends on the circumstances . . .

Major General Carl von Clausewitz, *On War*, i, 1832, tr. Howard and Paret

Sucess in war, like charity in religion, covers a multitude of sins.

General Sir William Napier (1785–1860)

To move swiftly, strike vigorously, and secure all the fruits of the victory is the secret of successful war.

Lieutenant General Thomas 'Stonewall' Jackson (1820–1863) letter, 1861

To have fought my own command daily, on equal terms and in open combats against the thousands that could have been brought against it by the North, would soon have resulted in its entire annihilation. I endeavored to compensate for my limited resources by strategms, surprises, and night attacks, in which the advantage was generally on my side, notwithstanding the superior numbers we assailed. For this reason, the complaint has often been made against me that I would not fight fair . . . in one sense the charge that I did not fight fair is true. I fought for success and not for display. There was no man in the Confederate army who had less of the spirit of knight-errantry in him, or took a more practical view of war than I did.

Colonel John S. Mosby, *Mosby's War Reminiscences,* 1887

Too much success is not wholly desirable; an occasional beating is good for men – and nations.

Rear Admiral Alfred Thayer Mahan, *The Life of Nelson*, 1897

A man who says that his success is due to himself is a fool. Success is one-third ability and two-thirds good fortune – or perhaps more exactly, one-third ability, one third luck (or providence), and one-third the power to 'stick it', to keep on pegging away until fortune turns, as it usually does sooner or later. (October 1932)

Captain Sir Basil Liddell Hart, *Thoughts on War,* 1944

Success is disarming. Tension is the normal state of mind and body in combat. When the tension suddenly relaxes through the winning of a first objective, troops are apt to be pervaded by a sense of extreme well-being and there is apt to ensue laxness in all of its forms and with all of its dangers.

Brigadier General S. L. A. Marshall, *Men Against Fire,* 1947

Possibly the big difference between the basic discipline of the front and rear areas could be best summed this way: when the chips are down, the main question is not how you go about your mission but whether it succeeds.

Brigadier General S. L. A. Marshall, *Men Against Fire*, 1947

No one can guarantee success in war, but only deserve it.

Sir Winston S. Churchill, *Their Finest Hour,* 1949

There are no secrets to success; don't waste time looking for them . . . Success is the result of perfection, hard work, learning from failure, loyalty to those for whom you work and persistence. You must be ready for opportunity when it comes.

General Colin Powell, quoted in *The Washington Post*, 15 January 1989

SUPERIORITY

One who has few must prepare against the enemy; one who has many makes the enemy prepare against him.

Sun Tzu, *The Art of War,* c. 500 BC, tr. Griffith

We must get the upper hand, and if we once have that, we shall keep it with ease, and shall certainly succeed.

The Duke of Wellington, 17 August 1803, despatch

Relative superiority, that is, the skillful concentration of superior strength at the decisive point, is much more frequently based on the correct appraisal of this decisive point, on suitable planning from the start; which leads to appropriate disposition of forces, and on the resolution needed to sacrifice nonessentials for the sake of essentials – that is, the courage to retain the major part of one's forces united. This is particularly characteristic of Frederick the Great and Bonaparte.

Major General Carl von Clausewitz, *On War,* ii, 1832, tr. Howard and Paret

---

The fundamental object in all military combinations is to gain local superiority by concentration.

Rear Admiral Alfred Thayer Mahan, *Naval Strategy*, 1911

One is always liable to be smashed by superior force, but one should never be caught unprepared to do one's best.

Field Marshal Viscount Allenby of Megiddo, 1900, letter to his wife during the Boer War

Armies and nations are mainly composed of normal men, not of abnormal heroes, and once these realize the *permanent* superiority of the enemy they will surrender to *force majeure*.

Captain Sir Basil Liddell Hart, *Thoughts on War*, 1944

SUPPORT

The hardest task in war is to lie in support of some position or battery, under fire without the privilege of returning it; or to guard some train left in the rear, within hearing but out of danger; or to provide for the wounded and dead of some corps which is too busy ahead to care for its own.

General of the Army William T. Sherman, *The Memoirs of General W. T. Sherman*, 1875

Closely allied to the confidence which is due to good leadership is that which is caused by the feeling of being supported. Troops, led well, will forget self, will risk the sacrifice of their own lives by advancing to close with the enemy, but only so long as they feel that other troops are coming on to back them up. Men who are assigned to fill the van of an attack will only sacrifice themselves readily if they feel that their efforts will not be in vain. It is, again, the idea of isolation which undermines morale.

Captain Sir Basil Liddell Hart, *Thoughts on War*, 1944

. . . Nothing will throw an infantry attack off stride as quickly as to promise it support which is not precisely delivered both in time and volume.

On the other hand, an artillery fire which is promptly delivered or an armor which advances steadily and confidently is like a shot in the arm. It moves the men mentally and sometimes bodily, thereby breaking the concentration of fear. But it is wiser to promise nothing than to default on anything. The memory of a default lingers and men consider it proof that the higher command is letting them down.

Brigadier General S. L. A. Marshall, *Men Against Fire*, 1947

Good senior commanders once must be trusted and 'backed' to the limit. Any commander is entitled to help and support from his immediate superior; sometimes he does not get it, a factor to be taken into account if the man fails. If, having received the help he might normally expect, a man fails – then he must go.

Field Marshal Viscount Montgomery of Alamein, *The Memoirs of Field Marshal Montgomery*, 1958

SURPRISE
(Principle of War)

Behold, the wretched, vanquished chief of Kheta [Hittite King] together with numerous allied countries, were stationed in battle array, concealed on the north-west of the city of Kadesh, while his majesty [Ramses II] with his bodyguard, and the division of Amon was marching behind him. The division of Re crossed over the river-bed on the south side of the town of Shabtuna, at the distance of [app. 1.4 miles] from the division of Amon; the division of Ptah was on the south of the city of Aranami; and the division of Sutekh was marching on upon the road. His majesty had formed the first rank of all the leaders of his army . . .

They [Hittites] came forth from the southern side of Kadesh, and they cut through the division of Re in its middle, while they were marching without knowing and without being drawn up for battle. The infantry and the chariotry of his majesty retreated before them. Now, his majesty had halted on the north of the city of Kadesh, on the western side of the Orontes. Then came one to tell his majesty.

His majesty shone like his father Montu, when he took the adornments of war; as he seized his coat of mail, he was like Baal in his hour. The great span [chariot] which bore his majesty called: 'Victory in Thebes', from the great stables of Ramses II, was in the midst of the leaders. His majesty halted the rout; by himself and none other with him. When his majesty went to look behind him, he found 2,500 chariotry surrounding him . . .

The Poem of Pentaur, Ramses II suffering the first great reported surprise attack in history at the Battle of Kadesh (Ramses would have also suffered the first recorded great defeat had he not by personal example fought his way out and rallied his army to pull off at least a draw – PGT), quoted in Breasted, *Ancient Records of Egypt*, 1906

The enemy must not know where I intend to give battle. For if he does not know where I intend to give battle, he must prepare in a great many places. And when he prepares in a great many places, those I have to fight in any one place will be few.

If he prepares to the front his rear will be weak, and if to the rear, his front will be fragile. If he prepares to the left, his right will be vulnerable and if to the right, there will be few on his left. And when he prepares everywhere he will be weak everywhere.

Sun Tzu, *The Art of War*, c. 500 BC, tr. Griffith

The execution of a military surprise is always dangerous, and the general who is never taken off his guard himself, and never loses an opportunity of striking at an unguarded foe, will be most likely to succeed in war.

Thucydides, *History of the Peloponnesian War*, c. 404 BC

Whatever a thing may be, be it pleasant or terrible, the less it has been foreseen the more it pleases or frightens. This is seen nowhere better than in war, where *surprise* strikes with terror even those who are much the stronger party.

Xenophon (c. 431 BC–c. 352 BC)

Among the enemy, when they saw him advancing so unexpectedly, there was a total lack of steadiness. Some were running to take up their positions, others forming into line, others bridling their horses, others putting on their breastplates. The general impression was one of people expecting to suffer rather than cause damage.

Xenophon, of the onset of the Theban army led by Epaminondas at the Battle of Mantineas in 362 BC, *A History of My Times* (*Hellinca*)

Novelty and surprise throw an enemy into consternation but common incidents have no effect.

Flavius Vegetius Renatus, *Military Institutions of the Romans*, c. AD 378

The general must make it one of his highest priorities and concerns to launch secret and unexpected attacks upon the enemy whenever possible. If he is successful in this sort of operation, with only a small group of men, he will put large numbers of the enemy to flight.

Emperor Nikephorus II Phocas, *Skirmishing*, c. AD 969, in Dennis, *Three Byzantine Military Treatises*, 1985

Everything which the enemy least expects will succeed the best. If he relies for security on a chain of mountains that he believes impracticable and you pass these by roads unknown to him, he is confused to start with, and if you press him he will not have time to recover from his consternation. In the same way, if he places himself behind a river to defend the crossing and you find some ford above or below by which to cross, this surprise will derange and confuse him.

Frederick the Great, *Instructions to His Generals*, 1747

Any officer or non-commissioned officer who shall suffer himself to be surprised . . . must not expect to be forgiven.

Major General Sir James Wolfe, 1759, General Orders during the Quebec Expedition

---

Just as lightning has already struck when the flash is seen, so where the enemy discovers the head of the army, the whole should be there, and leave them no time to counteract its dispositions.

Field Marshal François Comte de Guibert, *Essai général de tactique*, 1770, quoted in Liddell Hart, *The Ghost of Napoleon*, 1933

Who can surprise well must conquer.

Admiral John Paul Jones, 10 February 1778, letter to Benjamin Franklin, quoted in Lorenz, *John Paul Jones*, 1943

To astonish is to vanquish.

Field Marshal Prince Aleksandr V. Suvorov (1729–1800), quoted by Chuikov, in Leites, *The Soviet Style in War*, 1982

The enemy reckons we're sixty miles away . . . Suddenly we're on him like a cloudburst. His head whirls. Attack! Cut down, stab, chase, don't let him get away.

Field Marshal Aleksandr V. Suvorov (1729–1800), 1799, quoted in Longworth, *The Art of Victory*, 1966

. . . nothing so pregnant with dangerous consequences, or so disgraceful to an Officer in arms, as a surprise.

Admiral Lord St. Vincent, 1798, Memorandum to the Fleet off Cadiz

---

It is a principle of warfare, that when it is possible to make use of thunderbolts, they should be preferred to cannon.

Napoleon, dictation at St. Helena, ed., Herold, *The Mind of Napoleon*, 1955

To be defeated is pardonable; to be surprised – never!

Napoleon, *Maxims of War*, 1831

MAXIM 95. War is composed of nothing but surprises. While a general should adhere to general principles, he should never lose the opportunity to profit by these surprises. It is the essence of genius. In war there is only one favourable moment. Genius seizes it.

Napoleon, *Military Maxims of Napoleon*, 1831

. . . the universal desire for relative numerical superiority – leads to another desire, which is consequently no less universal: that *to take the enemy by surprise*. This desire is more or less basic to all operations, for without it superiority at the decisive point is hardly conceivable.

We suggest that surprise lies at the root of all operations without exception, though in widely varying degrees depending on the nature and circumstances of the operation.

Major General Carl von Clausewitz, *On War*, iii, 1832, tr. Howard and Paret

---

The two factors that produce surprise are secrecy and speed. Both presuppose a high degree of energy on the part of the government and the commander; on the part of the army, they require great efficiency. Surprise will never be achieved under lax conditions and conduct.

Major General Carl von Clausewitz, *On War*, iii, 1832, tr. Howard and Paret

---

For the side that can benefit from the psychological effects of surprise, the worse the situation is, the better it may turn out, while the enemy finds himself incapable of making coherent decisions. This holds true not only for senior commanders, but for everyone involved; for one peculiar feature of surprise is that it loosens the bonds of cohesion, and individual action can easily become insignificant.

Major General Carl von Clausewitz, *On War*, iii, 1832, tr. Howard and Paret

---

Always mystify. Mislead and surprise the enemy if possible. And when you strike and overcome him, never let up in pursuit so long as your men have strength to follow, for an army routed, if hotly pursued, becomes panic-stricken, and can then be destroyed by half their number.

Lieutenant General Thomas 'Stonewall' Jackson, 1862, quoted in Davis, *They Called Him Stonewall,* 1954

The officers and men who permit themselves to be surprised deserve to die, and the commanding general will spare no efforts to secure them their deserts.

Lieutenant General D. H. Hill, February 1863, address to his troops upon assumption of command

---

The way to destroy the enemy's morale, to show him that his cause is lost, is therefore surprise in every sense of the word bringing into the struggle something 'unexpected and terrible', which therefore has a great effect. It deprives the enemy of the power to reflect, and consequently to discuss.

Under various forms we always find the same principle of *surprise*, seeking to produce on the enemy the same moral result, *terror*; creating in him, by the sudden appearance of something unexpected and overwhelming, the feeling of impotence, the assurance that he cannot win; that is, that he is beaten.

Marshal of France Ferdinand Foch, *Principles of War,* 1913

Inaction leads to surprise, and surprise to defeat, which is after all only a form of surprise.

Marshal of France Ferdinand Foch, *Precepts and Judgments,* 1919

Surprise consists in the hard fact that the enemy suddenly appears in considerable numbers, without his *presence* having been known to be so *near*, for want of *information*, and without it being possible to *assemble*, for want of *protection*; for want, in one word, of a security service.

Marshal of France Ferdinand Foch, *Precepts and Judgments*, 1919

Surprise – the pith and marrow of war!

Admiral Sir John Fisher, *Memories,* 1919

Lack of security, or a false interpretation of the principle of security, leads directly to being surprised. The principle of surprise, like a double-edged tool, is an exceedingly dangerous one in unskilled hands; for, being mainly controlled by psychological factors, its nature is less stable and the conditions affecting it are more difficult to gauge.

Major General J. F. C. Fuller, *The Reformation of War,* 1923

An enemy may be surprised, which implies that he is thrown off balance. This is the best method of defeating him, for it is so economical, one man taking on to himself the strength of many. Surprise may be considered under two main headings: surprise effected by doing something that the enemy does not expect, and surprise effected by doing something that the enemy cannot counter. The first may be denoted as moral surprise, the second as material.

Major General J. F. C. Fuller, *The Reformation of War,* 1923

The subject of surprise is an immense one, and one which influences all forms and modes of war. It is one which is nearly always lost sight of during peace-time, because danger and fear are more often than not abstract quantitites; but in wartime they manifest, and with them manifests surprise – the demoralizing principle.

Major General J. F. C. Fuller, *The Foundations of the Science of War,* 1926

Of all keys to success in war 'unexpectedness' is the most important. By it a commander, whether of an army or a platoon, can often unlock gates which are impregnable to sheer force. (July 1929)

Captain Sir Basil Liddell Hart, *Thoughts on War,* 1944

Movement generates surprise, and surprise gives impetus to movement.

Captain Sir Basil Liddell Hart, 'Strategy', *Encyclopaedia Britannica*, 1929 ed.

Surprise has a stunning effect. Therefore all troop action must be accomplished with the greatest secrecy and speed. Swiftness of action in combination with organization, skilful manoeuvre, and the ability to adopt to the terrain . . . are the basic guarantees of success in battle.

Red Army *1936 Field Service Regulations*, attributed to Marshal of the Soviet Union, Mikhail N. Tukhachevskiy (1893–37)

Absolute superiority *everywhere* is unattainable, hence it must frequently be replaced by relative superiority *somewhere*. To achieve relative superiority somewhere is the main objective of almost all military movements and the essential purpose of generalship. Since relative superiority will hardly be accomplished if the enemy knows the plan of concentration before the hour of attack, the principle of surprise is of importance equal to that of the principle of concentration.

General Waldemar Erfurth, *Surprise*, 1938

Frequently, surprise reduces the unity of the enemy forces and induces the commanders of the enemy army to issue conflicting orders . . . Modern wars offer many examples of panic which led to the frantic flight of whole armies.

General Waldemar Erfurth, *Surprise*, 1938

A military man can scarcely pride himself on having 'smitten a sleeping enemy'; it is more a matter of shame, simply, for the one smitten.

Admiral Yamamoto Isoroku, 9 January 1942, quoted in Agawa, *The Reluctant Admiral*, 1979

We have inflicted a complete surprise on the enemy. All our columns are inserted in the enemy's guts.

Major General Orde Wingate, 11 March 1944, Order of the Day to the 3rd Indian Division

---

. . . to do something which the enemy cannot prevent, and to do something which he does not suspect. The first action may be compared to surprising a man with his eyes open, in the other, the man has his eyes shut.

Major General J. F. C. Fuller, *The Generalship of Alexander the Great*, 1958

Surprises are a commonplace in war – and reconsidered opinions too.

General Robert L. Eichelberger (1886–1961)

Most opponents are at their best if they are allowed to dictate a battle; they are not so good when they are thrown off balance by manoeuvre and are forced to react to your own movements and thrusts. Surprise is essential. Strategical surprise may often be difficult, if not impossible, to obtain; but tactical surprise is always posssible and must be given an essential place in training.

Field Marshal Viscount Montgomery of Alamein, *A History of Warfare*, 1968

A force within striking distance of an enemy must be suitably disposed with regard to its battle positions, being ready *at all times* to fight quickly if surprised.

Field Marshal Viscount Montgomery of Alamein, *A History of Warfare*, 1968

Surprise makes it possible to take the enemy unawares, to cause panic in his ranks, to paralyze his will to resist, to drastically reduce his fighting efficiency, to contain his actions, to disrupt his troops' control, and to deny him the opportunity to take effective countermeasures quickly. As a result, this makes it possible to successfully rout even superior enemy forces with the least possible losses to friendly forces.

Colonel General V. G. Reznichenko, *Taktika*, 1984

Surprise must form the basis of all troop combat activites. Surprise achieved at the beginning of an engagement may become exhausted after a while. Its effect is limited to the time that the enemy needs to eleminate the unequal conditions caused by the unanticipated actions on the part of the opposing side. This is why it is necessary during an engagement to strive both to make maximum use of surprise already achieved and to achieve a new surprise in the actions of all troop elements.

Colonel General V. G. Reznichenko, *Taktika*, 1984

SURRENDER

Merde!

'Le Mot de Cambronne', the actual reply of General Cambronne at Waterloo when called upon to surrender the Old Guard, not the more poetic, 'The Guard dies but does not surrender!'

MAXIM 67. To authorize generals or other officers to lay down their arms in virtue of a particular capitulation, under any other circumstances than when they are composing the garrison of a fortress, affords a dangerous latitutde. It is destructive of all military character in a nation to open such a door to the cowardly, the weak, or even to the misdirected brave. Great extremities require extraordinary resolution. The more obstinate the resistance of an army, the greater the chances of assistance or of success.

How many seeming impossibilities have been accomplished by men whose only resource was death!

Napoleon, *The Military Maxims of Napoleon*, 1831, tr. D'Aguilar

---

MAXIM 68. There is no security for any sovereign, for any nation, or for any general, if officers are permitted to capitulate in the open field, and to lay down their arms in virtue of conditions, favourable to the contracting party, but contrary to the interests of the army at large. To withdraw from danger, and thereby to involve their comrades in greater perils, is the height of cowardice. Such conduct should be proscribed, declared infamous, and made punishable with death. All generals, officers, and soldiers who capitulate in battle to save their own lives, should be decimated.

He who gives the order, and those who obey are alike traitors, and deserve capital punishment.

Napoleon, *The Military Maxims of Napoleon*, 1831, tr. D'Aguilar

---

MAXIM 69. There is but one honourable mode of becoming prisoner of war. That is, by being taken separately; by which is meant, being cut off entirely, and when we can no longer make use of our arms. In this case there can be no conditions, for honour can impose none. We yield to an irresistible necessity.

Napoleon, *The Military Maxims of Napoleon*, 1831, tr. D'Aguilar

---

Tell him to go to hell.

Major General Zachery Taylor, 22 February 1847, reply to Santa Anna's demand for his surrender at the Battle of Buena Vista

---

Sir: Yours of this date proposing armistice and appointment of Commissioners to settle terms of capitulation, is just received. No terms except an unconditional and immediate surrender can be accepted. I propose to move immediately upon your works.

General of the Army Ulysses S. Grant, the origin of the term 'unconditional surrender', and Grant's nickname – when Confederate General Buckner attempted to surrender on terms at Fort Donaldson, *Personal Memoirs of Ulysses S. Grant*, 1885

I know they will say hard things of us. They will not understand how we were overwhelmed by numbers. But that is not the question, Colonel: The question is, is it right to surrender this army. If it is right, then I will take all the responsibility . . . How easily I could be rid of this and be at rest! I have only to ride along the line and all will be over. But it is our duty to live. What will become of the women and children of the South if we are not there to protect them?

General Robert E. Lee, 9 April 1865, upon the surrender at Appomatax, quoted in A. L. Long, *Memoirs of Robert E. Lee,* 1886

Then there is nothing left for me but to go and see General Grant, and I would rather die a thousand deaths.

General Robert E. Lee, 9 April 1865

Now, he told the men in a few words that he had done his best for them & advised them to go home & become as good citizens as they had been soldiers. As he spoke a wave of emotion seemed to strike the crowd & a great many men were weeping, & many pressed to shake his hand & try & express in some way the feelings which shook every heart. As he passed on toward his camp he stopped & spoke to me for a moment & told me that Gen. Grant had very generously agreed that our soldiers could keep their private horses, which would enable them to plant crops before it was too late. This seemed to be a very special gratification to him. Indeed Gen. Grant's conduct toward us in the whole matter is worthy of the very highest praise & indicates a great & broad & generous mind. *For all time it will be a good thing for the whole United States, that of all the Federal generals it fell to Grant to receive the surrender of Lee.*

Brigadier General Edward Porter Alexander, *Fighting for the Confederacy,* written between 1900 and 1907 but not published until 1989

From a purely military standpoint, our cause is not lost. But it is as a *nation*, and not as an *army*, that we are met here, and it is therefore for the nation principally that we must consult . . .

"No! We do not only represent our burghers on commando, the troops over which we are placed in command; we represent also the thousands who have passed away, after making the last sacrifice for their country; the prisoners scattered all the world over; the women and children dying by the thousands in the prison camps of the enemy; we represent the blood and the tears of the whole African nation. From the prisons, the camps, the graves, the veldt, and from the womb of the future, that nation cries out to us to make a wise decision now, to take no step which might lead to the downfall or even to the extermination of our race, and thus make all their sacrifices of no avail.

Brethren, we have vowed to stand fast to the bitter end; but let us be men, and acknowledge that the end has now come, and that it is more bitter than ever we thought it could be. For death itself would be sweet compared with the step which we must now take. But let us bow before the will of God.

General Jan H. Smuts, 30 May 1902, the last meeting of the Orange Free State and Transvaal leadership before the surrender to the British ending the Boer War, quoted in de Wet, *Three Year War,* 1902

. . . there is almost always a way out, even of an apparently hopeless position, if the leader makes up his mind to face the risks.

Major General Paul von Lettow-Vorbeck, *East African Campaigns,* 1957

I would fight without a break. I would fight in front of Amiens. I would fight in Amiens. I would fight behind Amiens. I would never surrender.

Marshal of France Ferdinand Foch, 26 March 1918, to Sir Douglas Haig during the German *Friedenstürm* offensive

---

Helplessness induces hopelessness, and history attests that loss of hope and not loss of life is what decides the issue of war.

Captain Sir Basil Liddell Hart, *The Real War 1914–1918,* 1930

Let no man surrender so long as he is unwounded and can fight.

Field Marshal Viscount Montgomery of Alamein, 30 October 1942, message to the

Eighth Army on the eve of the Battle of Alamein

To receive the unconditional surrender of half a million enemy soldiers, sailors and airmen must be an event which happens to few people in the world. I am very conscious that this was the greatest day of my life.

Admiral Earl Mountbatten, 12 September 1945, upon the Japanese surrender in Singapore, *Personal Diary of Admiral the Lord Louis Mountbatten*, 1988

As I speak there are 100,000 men ashore. This invasion would have taken place on 9th September whether the Japanese had resisted or not. I wish to make this plain; the surrender today is no negotiated surrender. The Japanese are submitting to superior force, now massed here.

I now call upon General Itagaki to produce his credentials.

Admiral Earl Mountbatten, 12 September 1945; Mountbatten's instructions to the Japanese commander at the ceremony of surrender in Singapore, *Personal Diary of Admiral the Lord Louis Mountbatten*, 1988

I was convinced that an effective way really to impress on the Japanese that they had been beaten in the field was to insist on this ceremonial surrender of swords. No Japanese soldier, who had seen his general march up and hand over his sword, would ever doubt that the Invincible Army was invincible no longer. We did not want a repetition of the German First War legend of an unconquered army. With this in mind, I was dismayed to be told that General MacArthur in his over-all instruction for the surrender had decided that the 'archaic' ceremony of the surrender of swords was not to be enforced. I am afraid I disregarded his wishes. In Southwest Asia all Japanese officers surrendered their swords to British officers of similar or higher rank; the enemy divisional and army commanders handed theirs in before large parades of their already disarmed troops. Field Marshal Terauchi's sword is in Admiral Mountbatten's hands; General Kimura's is now on my mantelpiece, where I always intended that one day it should be.

Field Marshal Viscount Slim, *Defeat Into Victory*, 1962

TACTICS

That the army is certain to sustain the enemy's attack without suffering defeat is due to operations of the extraordinary and the normal forces.

Troops thrown against the enemy as a grindstone against eggs is an example of solid acting upon a void.

Generally, in battle, use the normal force to engage; use the extraordinary to win.

Now the resources of those skilled in the use of extraordinary forces are as infinite as the heavens and earth; as inexhaustible as the flow of great rivers.

For they end and recommence; cyclical, as are the movements of the sun and moon. They die away and are reborn; recurrent, as the passing seasons.

The musical notes are only five in number but their melodies are so numerous that one cannot hear them all.

The primary colours are only five in number but their combinations are so infinite that one cannot visualize them all.

The flavours are only five in number but their blends are so various that one cannot taste them all.

In battle there are only the normal and extraordinary forces, but their combinations are limitless; none can comprehend them all.

For these two forces are mutually reproductive; their interaction as endless as that of interlocked rings. Who can determine where one ends and the other begins?

Sun Tzu, *The Art of War*, c. 500 BC, tr. Griffith
Note: The normal forces could be described as the direct approach and the extraordinary forces the indirect approach – PGT.

Tactics is a science which enables one to organize and manoeuvre a body of armed men in an orderly manner. Tactics may be divided into four parts: proper organization of men for combat; distribution of weapons according to the needs of each man; the occasion; the management of war, of personnel and materials, including an examination of ways and causes as well as of what is advantageous.

Anonymous Byzantine general (Anonymous Vari), *Strategy*, c. AD 527–AD 565

Grand tactics is the art of forming good combinations preliminary to battles as well as during their progress. The guiding principle in tactical combinations, as in those of strategy, is to bring the masses of the force at hand against a part of the opposing army and upon that point the possession of which promises the most important results.

Lieutenant General Antoine-Henri Baron de Jomini, *Summary of the Art of War*, 1838

How different is almost every military problem, except in the bare mechanism of tactics. In almost every case the data on which a solution depends is lacking . . . Too often the general has only conjectures to go on, and these based on false premises . . . What is true now, at the next moment may have no existence, or exist in a contrary sense . . . These considerations explain why history produces so few captains.

Dennis Hart Mahan, *Advance Guard, Outpost and Detachment Service of Troops, with the Essential Principles of Strategy and Tactics*, 1863

Tactics is an art based on the knowledge of how to make men fight with maximum energy against fear, a maximum which organization alone can give.

Colonel Charles Ardnant du Picq, *Battle Studies*, 1880

Nine-tenths of tactics are certain, and taught in books; but the irrational tenth is like the kingfisher flashing across the pool and that is the test of generals. It can only be ensured by instinct, sharpened by thought practising the stroke so often that at the crisis it is as natural as a reflex.

Colonel T. E. Lawrence, 'The Science of Guerrilla Warfare', *Encylopaedia Britannica*, 1929

Tactics are the cutting edge of strategy, the edge which chisels out the plan into action; consequently the sharper this edge is the clearer will be the result.

Major General J. F. C. Fuller, *Grant and Lee, A Study in Personality and Generalship*, 1933

There is only one tactical principle which is not subject to change. It is: to use the means at hand to inflict the maximum amount of wounds, death and destruction on the enemy in the minimum of time.

General George S. Patton, Jr., *War As I Knew It*, 1947

It seems also that he who devises or develops a new system of tactics deserves special advancement on the military honour role of fame. All tactics since the earliest days have been based on evaluating an equation in which x=mobility, y=armour, and z=hitting power. Once a satisfactory solution has been found and a formula evolved, it tends to remain static until some thinking soldier (or possibly civilian) recognizes that the values of x, y, z have been changed by the progress of inventions since the last formula was accepted and that a new formula and new systems of tactics are required.

Field Marshal Earl Wavell, *Soldiers and Soldiering*, 1953

On the opening page of his great work *On War*, Clausewitz makes a very simple yet profound remark. It is, that 'War is nothing but a duel on an extensive scale', and he likens it to a struggle between two wrestlers; between two pugilists would be a more apt comparison. If so, then the primary elements of tactics are to be seen in their simplest form in a fight between two unarmed men. They are: to think, to guard, to move and to hit.

Before a bout opens, each man must consider how best to knock out his adversary, and though as the fight proceeds he may be compelled to modify his means, he must never abandon his aim. At the start he must assume a defensive attitude until he has measured up his opponent. Next, he must move under cover of his defence towards him, and lastly by foot-play, and still under cover of his defence, he must assume the offensive and attempt to knock him out. In military terms, the four primary tactical elements are: the aim or object, security, mobility and offensive power.

If two pugilists are skilled in their art, they will recognize the value of three accentuating elements. They will economize their physical force, so as not to exhaust themselves prematurely; they will concentrate their blows against the decisive point selected, the left or right of their opponent's jaw, or his solar plexus, and throughout will attempt to surprise him – that is, take him off-guard, or do something which he does not expect, or cannot guard against. In military terms these accentuating elements are: economy of force, concentration of force, and surprise.

Major General J. F. C. Fuller, *The Generalship of Alexander the Great,* 1958

TACTICS AND STRATEGY

All men can see these tactics whereby I conquer, but what none can see is the strategy out of which victory evolved.

Sun Tzu, *The Art of War, c.* 500 BC

The conduct of war . . . consists in the planning and conduct of fighting . . . It consists of a greater or lesser number of single *acts each complete in itself*, which . . . are called 'engagements' and which form new entities. This gives rise to the completely different activity of *planning and executing these engagements themselves*, and of *coordinating* each of them with the others in order to further the object of the war. One has been called *tactics*, and the other *strategy*.

According to our classification, then, tactics teaches *the use of armed forces in the engagement*; strategy, *the use of engagements for the object of the war.*

Major General Carl von Clausewitz, *On War,* 1832, tr. Howard and Paret

The original means of strategy is victory – that is, tactical success; its ends, in the final analysis, are those objects which will lead directly to peace. The application of these means for these ends will also be attended by factors that will influence it to a greater or lesser degree.

These factors are the geographical surroundings and the nature of the terrain . . ., the time of day . . . ; and the weather . . .

Strategy, in connecting these factors with the outcome of an engagement, confers a special significance on that outcome and thereby on the engagement: *it assigns a particular aim to it.* Yet, insofar as that aim is not the one that will lead directly to peace, it remains subsidiary and is also to be thought of as a means. Successful engagements or victories in all stages of importance may therefore be considered as strategic means.

Major General Carl von Clausewitz, *On War,* ii, 1832, tr. Howard and Paret

---

Tactical success, those attained *in the course* of the engagement, *usually occur during the phase of disarray and weakness*. On the other hand, the strategic success, the overall effect of the engagement, the completed victory, whether great or insignifant, *already lies beyond that phase*. The strategic outcome takes shape only when the fragmented results have combined into a single independent whole . . .

Major General Carl von Clausewitz, *On War,* iii, 1832, tr. Howard and Paret

---

Strategy is the art of making war upon the map, and comprehends the whole theatre of operations. Grand tactics is the art of posting troops upon the battlefield according to the characteristics of the ground, of bringing them into action, and of fighting upon the ground, in contradistinction to planning upon a map. Logistics comprises the means and arrangements which work out the plans of strategy and tactics. Strategy decides where to act; logistics brings the troops to this point; grand tactics decides the manner of execution and the employment of the troops.

Lieutenant General Antoine-Henri Baron de Jomini, *Summary of the Art of War,* 1838

Strategy furnishes tactics with the opportunity to strike and with the prospect of success, through its conduct of the armies and of their concentration on the field of battle. On the other hand, however, it accepts the results of every single engagement, and builds upon them. Strategy retires when a tactical victory is in the making, in order to exploit the newly created situation.

Field Marshal Helmuth Graf von Moltke, quoted in Foertsch, *The Art of Modern War*, 1940

A tactical success is only really decisive, if it is gained at the strategically correct spot.

Field Marshal Helmuth Graf von Moltke, quoted in Foertsch, *The Art of Modern War*, 1940

. . . *Contact* (a word which perhaps better than any other indicates the dividing line between strategy and tactics) . . .

Rear Admiral Alfred Thayer Mahan, *The Influence of Sea Power Upon History*, 1890

As a rule, strategy concerns itself with those large-scale measures which serve to bring the forces into play at the decisive point under the most favourable conditions possible, while tactics relates to what is done in the engagement itself. Strategy might be called the science of generalship, while tactics is that of handling troops.

Field Marshal Colmar Baron von der Goltz (1843–1916)

It seems to me personally that attempts to define and separate these two fields are mistaken. I don't believe that strategy and tactics can be isolated in any event. Strategy and tactics are attributes of a single phenomenon; the difference is only one of degree.

Mikhail V. Frunze (1885–1925), quoted in Gareyev, *M. V. Frunze, Military Theorist*, 1985

The greater the certainty with which the success of the main force attack can be predicted in advance, the easier the task set that main force will be. One must not count on the heroism of the troops. *The strategy must ensure that the tactical task is a readily feasible one.*

Marshal of the Soviet Union Mikhail N. Tukhachevskiy, 1924, quoted in Simpkin, *Deep Battle*, 1987

As regards the relation of strategy to tactics, while in execution the borderline is often shadowy, and it is difficult to decide exactly where a strategical movement ends and a tactical movements begins, yet in conception the two are distinct. Tactics lies in, and fills the province of, fighting. Strategy not only stops on the frontier, but has for its purpose the reduction of fighting to the slenderest proportions . . . (October 1928)

Captain Sir Basil H. Liddell Hart, *Thoughts on War*, 1944

I've read your book during these holidays, which I've had to spend in camp, on Fire Picquet. It's uncommonly good: so lively and interesting, and wide. You establish your thesis: but I fear that you could equally have established the contrary thesis, had the last war been a manoeuvre war and not a battle war [World War I]. These pendulums swing back and forward. If they rested still that would be absolute truth: but actually when a pendulum stands still it's that the clock has stopped, not that it has achieved absolute time!

Colonel T. E. Lawrence, letter to Liddell Hart, 31 May 1929, quoted in *T. E. Lawrence to His Biographers*, 1963

In peace we concentrate so much on tactics that we are apt to forget that it is merely the handmaiden of strategy. (September 1930)

Captain Sir Basil H. Liddell Hart, *Thoughts on War*, 1944

So far as I can see strategy is eternal, and the same and true: but tactics are the ever-changing languages through which it speaks. A general can learn as much from Belisarius as from Haig – but not a soldier. Soldiers have to know their means.

Colonel T. E. Lawrence, Whitmonday 1933, quoted in *T. E. Lawrence to His Biographers*, 1963

. . . We do not want any of our commanders in the war to detach himself from the objective conditions and become a blundering hothead, but we decidedly want every commander to become a general who is both bold and sagacious. Our commanders should have not only the boldness to overwhelm the enemy but also the ability to remain masters of the situation throughout the changes and vicissitudes of the entire war. Swimming in the ocean of war, they must not flounder but make sure of reaching the opposite shore with measured strokes. Strategy and tactics, as the laws for directing war, constitute the art of swimming in the ocean of war.

Mao Tse-tung, *On Protracted War*, May 1938

Tactics is the art of handling troops on the battlefield, strategy is the art of bringing forces to the battlefield in a favourable position.

Field Marshal Earl Wavell, quoted in Chandler, *The Atlas of Military Strategy*, 1980

I rate the skilful tactician above the skilful strategist, especially he who plays the bad cards well.

Field Marshal Earl Wavell, *Soldiers and Soldiering*, 1953

The general principles of strategy apply to tactics too, only that their application is different. A glance back at history will show that while tactics, like strategy, is based on certain truths and laws of permanent worth, the application of these laws will show that tactical forms have changed and developed along with the times, and in accordance with the means which the times provide.

Colonel Hermann Foertsch, *The Art of Modern War*, 1940

. . . definition of strategy as – 'the art of distributing and applying military means to fulfill the ends of policy'. For strategy is concerned not merely with the movement of forces – as its role is often defined – but with the effect. When the application of the military instrument merges into actual fighting, the dispositions for and control of such direct action are termed 'tactics'. The two categories, although convenient for discussion, can never be truly divided into separate compartments because each not only influences but merges into the other.

Captain Sir Basil Liddell Hart, *Strategy*, 1954

In strategy . . . calculation is simpler and a closer approximation to truth than is possible in tactics. For in war the chief incalculable is the human will, which manifests itself in resistance, which in turn lies in the province of tactics. Stragegy has not to overcome resistence, except from nature. *Its purpose is to diminish the possibility of resistence,* and it seeks to fulfill this purpose by exploiting the elements of *movement* and *surprise*.

Captain Sir Basil H. Liddell Hart, *Strategy*, 1954

TALENT FOR WAR

Deadly going –
then and there the Trojans might have been rolled back,
far away from the ships and tents to wind-torn Troy
if Polydamus had not rushed to headstrong Hector:
'Impossible man! Won't you listen to reason?
Just because some god exhalts you in battle
you think you can beat the rest at tactics too.
How can you hope to garner all the gifts at once?
One man is a splendid fighter – a god has made him so –
one's a dancer, another skilled at lyre and song,
and deep in the next man's chest far seeing Zeus
plants the gift of judgment, good clear sense.
And many reap the benefits of that treasure:

troops of men he saves, as he himself knows best.'

Homer, *The Iliad*, c. 800 BC, tr. Fagles

The best general is not the man of noble family, but the man who can take pride in his own deeds.

The Emperor Maurice, *The Strategikon*, c. AD 600

Unless a man is born with a talent for war, he will never be other than a mediocre general. It is the same with all talents; in painting, or in music, or in poetry, talent must be inherent for excellence. All sublime arts are alike in this respect. That is why we see so few outstanding men in science. Centuries pass without producing one. Application rectifies ideas but does not furnish a soul for that is the work of nature.

Field Marshal Maurice Comte de Saxe, *My Reveries*, 1732

It is bad to lack good fortune, but it is a misfortune to lack talent . . . The fortune of war is on the side of the soldier of talent.

Field Marshal Aleksandr V. Suvorov (1729–1800), quoted in Ossipov, *Suvorov*, 1945

War is a serious sport, in which one can endanger his reputation and his country: a rational man must feel and know whether or not he is cut out for this profession.

Napoleon, to Eugene, 30 April 1809, *Correspondance*, No. 15144, Vol. XVIII, 1858–1870

The capital role played by instinct in this matter of conception is recognized in our common parlance. We say of a statesman, a soldier, or a man of business who conceives rightly, that is to say in accordance with what is, that he has a 'sense of reality', that he has 'a gift', or that he is possessed of 'vision' or 'flair'. Where action is concerned, nothing can take the place of this effort supplied by nature itself. To it alone is due the fact that men whose intelligence does not call attention to itself, who do not excel at speculation, who do not shine in 'paper work' or in theoretical argument, are transformed into masters of the battlefield.

Charles De Gaulle, *The Edge of the Sword*, 1932

THE TANK

I managed to get astride one of the German trenches . . . and opened fire with Hotchkiss machine-guns. There were some Germans in the dug-outs and I shall never forget the look on their faces when they emerged . . .

Captain H. W. Mortimore, 15 September 1916, commander of the first tank in action, quoted in Wilson, *The Myriad Faces of War*, 1986

Employment of tanks in mass is our greatest enemy.

General Erich Ludendorff, quoted in Simpkin, *Race to the Swift*, 1985

That the application of petrol to land warfare will prove as great a step in tactics as that of steam in naval warfare.

That the characteristics of security, offensive power and mobility, which the tank combines in a higher degree than any single other weapons, are those which are fundamental to success in war.

That the tank enables the main advantages of sea warfare, unrestricted movement, to be to a great extent superimposed on that of land warfare . . .

That the application of machinery to land warfare is as great a saver of man-power as its application to manufacture. That is: the tank does not create another man-power problem, but is a solution to existing problems.

Major General J. F. C. Fuller (memorandum of 1918), *Memoirs of an Unconventional Soldier*, 1936

There are two kinds of infantry: men who have gone into action with Tanks, and men who have not; and the former never want to go into action without Tanks again.

Monsieur Loucheur, French Minister of Munitions, January 1919, cited in Williams–Ellis, *The Tanks Corps*, 1919

But it is not, of course, enough that the Tank offers protection to those who fight in it. A trench or a hole in the ground will do the same. But the Tank is essentially a mobile weapon of *offence*. It is the weapon for the nation which does not fight willingly, but when it fights, fights to win, and to win quickly with as little bloodshed as possible. It is the weapon for men who, if they must fight, like to fight like intelligent beings still subjecting the material world to their will, and who are most unwillingly reduced to the roles of mere marching automata, bearers of burdens and diggers of the soil, roles from which the patient German did not seem adverse.

Major Clough Williams–Ellis, *The Tank Corps*, 1919

The tank marks as great a revolution in land warfare as an armoured steamship would have marked had it appeared amongst the toilsome triremes of Actium.

General Sir Ian Hamilton, *The Soul and Body of an Army*, 1921

There never was a moment when it was possible to say that a tank had been 'invented'. There never was a person about whom it could be said 'this man invented the tank'. But there was a moment when the actual manufacture of the first tanks was definitely ordered, and there was a moment when an effective machine was designed as the direct outcome of this authorization.

Winston S. Churchill, *The World Crisis*, 1923

'It was Mr. Churchill who, as First Lord of the Admiralty, gave the first order for eighteen tanks, or 'landships' as they were then called, on 26 March 1915. He did not inform either the War Office or the Treasury – an almost unprecedented and certainly unconstitutional reticence, dictated by fear that conventional minds might stifle a great idea.' – Colin Coote, ed., Winston S. Churchill, *Maxims and Reflections*, 1949.

Once appreciate that tanks are not an extra arm or mere aid to infantry but the modern form of heavy cavalry and their true military use is obvious – to be concentrated and used in as large masses as possible for a decisive blow against the Achilles' heel of the enemy army, the communications and command centres which form its nerve system.

Captain Sir Basil Liddell Hart, writing in the *Westminster Gazette*, 1925

The reader will find in Fuller many interesting ideas about tank actions. He should pay particular attention to the actions of tanks in the enemy's rear, which, together with a simultaneous frontal assault, must undoubtedly result in more intensive manoeuvre and more decisive tactical action in modern battles. To the turning movement and the flank and rear attack (the sphere of classical, conventional manoeuvre) are added the seizure of rear boundaries, complete penetrations and sudden machine-gun attacks in depth (*desanty*) by tanks.

Marshal of the Soviet Union Mikhail N. Tukhachevskiy, Preface to the 1931 Russian-language edition of J. F. C. Fuller's *The Reformation of War*, 1931

The setting up of a deep battle – that is the simultaneous disruption of the enemy's tactical layout over its entire depth – requires two things of tanks. On the one hand they must help the infantry forward and accompany it; on the other they must penetrate into the enemy's rear, both to disorganize him and to isolate his main forces from the reserves at his disposal. This deep penetration by tanks must create in the enemy's rear an obstacle for him, on to which he must be forced back and on which his main forces must be destroyed.

Marshal of the Soviet Union Mikhail N. Tukhachevskiy, 'New Questions of War', written 1931–2, published finally in *Voyenno istorichiskiy zhurnal*, 2/1962, quoted in Richard Simpkin, *Deep Battle*, 1987

. . . the internal-combustion engine which is ready to carry whatever one wants, wherever it is needed, at all speeds and distances; . . . the internal-combusion engine which, if it is armoured, possesses such a fire power and shock power that the rhythm of the battle corresponds to that of its movements.

Charles de Gaulle, *The Army of the Future* (*Vers l'armée de métier*), 1934

On the other side of the coin, a group of comrades looked not a little askance at the introduction into the army of tanks in mass. They were scared of rain, snow, autumn, spring and the like. Some comrades said that, because of our roads and climatic conditions, tanks would only be able to operate 'for something like six weeks in the year'. However the tank's talent for lively action soon stood the misgivings of these timid theorists on their heads. Tanks perform perfectly well in summer, winter, spring and autumn.

Marshal of the Soviet Union Mikhail N. Tukhachevskiy, *Red Army Field Service Regulations, 1936*, cited in Richard Simpkin, *Deep Battle*, 1987

Until our critics can produce some new and better method of making a successful land attack other than self-massacre, we shall continue to maintain our belief that tanks – properly employed, needless to say – are today the best means available for a land attack.

Colonel General Heinz Guderian, *Achtung! Panzer!*, 1937

. . . [The tank] is therefore the weapon of potentially decisive attack. Mobility and firepower will only be exploited to the full if the attack achieves deep penetration and the armoured force, having broken out, can go over to the pursuit . . . The higher the concentration of tanks, the faster, greater and more sweeping will be the success – and the smaller our own losses . . . Tanks must attack with surprise, and as far as possible where the enemy is known or presumed to be weak . . . The tank needs supporting arms which complement it and can go everywhere with it . . . Even in defence, the tank must be employed offensively. Concentration is even more important here, so that the enemy's superiority can be offset at least in one spot.

Colonel General Heinz Guderian, *Panzer Marsch*! n.d., cited in Simpkin, *The Race to the Swift*, 1985

The shorter the battle, the fewer men will be killed and hence the greater their self-confidence and enthusiasm. To produce a short battle, tanks must advance rapidly but not hastily . . . Mobile forces should be used in large groups and [be] vigorously led. They must attempt the impossible and dare the unknown.

General George S. Patton, Jr., 1939, after reading an article by Heinz Guderian on tanks, cited in Blumenson, ed., *The Patton Papers*, Vol II, 1974

I want to bring to the attention of every officer here the professional significance which will attach to the success or failure of the 2nd Armored Division in the Tennessee maneuvers. There are a large number of officers, some of them in high places in our country, who through lack of knowledge as to the capability of an armored division are opposed to them and who would prefer to see us organize a large number of old fashioned divisions about whose ability the officers in question have more information. It is my considered opinion that the creation of too many old type divisions will be distinctly detremental and that the future of our country may well depend on the organization of a considerably larger number of armored divisions than are at present visualized. Therefore it behooves everyone of us to do his uttermost to see that in these forthcoming maneuvers we are not only a success but such an outstanding success that there could be no possible doubt in the minds of anyone as to the effectiveness of the armored divisions. Bear this in mind every moment.

General George S. Patton, Jr., address to the officers of the 2nd Armored Division, May 1941, *The Patton Papers*, Vol. II, 1974

. . . Although the tactical consequences of motorisation and armour had been pre-eminently demonstrated by British military critics, the responsible British leaders had not taken the risk either of using this hitherto untried system as a foundation for peacetime training, or of applying it in war.

Field Marshal Erwin Rommel, *The Rommel Papers,* 1953 *Note by General Bayerlein.* Rommel was here referring to Captain Liddell Hart and General Fuller. In his opinion the British could have avoided most of their defeats if only they had paid more heed to the modern theories expounded by these two writers before the war. During the war, in many conferences and personal talks with Field Marshal Rommel, we discussed Liddell Hart who made the deepest impression on the Field Marshal – and greatly influenced his tactical and strategic thinking. He, like Guderian, could in many respects be termed Liddell Hart's 'pupil'.

The officers of a panzer division must learn to think and act independently within the framework of the general plan and not wait until they receive orders.

Field Marshal Erwin Rommel, *The Rommel Papers*, 1953

. . . It was principally the books and articles of the Englishmen, Fuller, Liddell Hart, and Martel, that excited my interest and gave me food for thought. These far-sighted soldiers were even then trying to make of the tank something more than just an infantry support weapon. They envisaged it in relationship to the growing motorization of our age, and thus they became the pioneers of a new type of warfare on the largest scale.

I learned from them the concentration of armour, as employed in the battle of Cambrai. Further, it was Liddell Hart who emphasized the use of armoured forces for long-range strokes, operations against the opposing army's communications, and also proposed a type of armoured division combining panzer and panzer-infantry units. Deeply impressed by these ideas I tried to develop them in a sense practicable for our own army. So I owe many suggestions of our further development to Captain Liddell Hart.

Colonel General Heinz Guderian, *Panzer Leader*, 1953

I was brought up in the Israeli Army which no doubt, was very much influenced by your unorthodoxic [sic] school of thought, and of course I am strongly in favour of your ideas.

General Ariel Sharon, letter to B. H. Liddell Hart

Armor – The Combat Arm of Decision.

U S Army Motto

TEAM/TEAMWORK

Four brave men who do not know each other will not dare to attack a lion. Four less brave, but knowing each other well, sure of their reliability and consequently of their mutual aid, will attack resolutely.

Colonel Charles Ardnant du Picq, *Battle Studies*, 1880

The work carried through in the munitions factories, and the ingenuity and solid labour that backed the efforts of the soldier in the field, are perhaps not yet fully appreciated by the fighting men. In France one might hear of sporadic unrest, but till one met with it, one realized nothing of the faithful grind at production of objects of whose destination the worker often knew nothing, of the blind patience under duress of shortage, and of crowded accommodation; of hope deferred.

The Tank Corps was fortunate indeed in having established at an early date close relations with its workers, and more fortunate still at a critical time in being able to declare a substantial dividend on the capital of wealth, labour and brains entrusted to it by its section of industrial Britain.

Major General Hugh Elles (Commander of the Royal Tank Corps in World War I) in the preface, Williams–Ellis, *The Tank Corps*, 1919

But after all, no leadership or expert naval direction could be successful unless it were supported by the whole body of officers and men of the Navy. It is upon these faithful, trusty servants in the great ships and cruisers that the burden falls directly day after day. In particular, the flotillas of destroyers, of submarines watching at the throat of the Elbe, of anti-submarine craft, of minesweepers multiplying on all our coasts – all these have undergone, and are undergoing, a toil and strain which only those who are informed in detail, can understand. Many vexatious tasks lie before the Royal Navy and before its comrades in the Merchant Navy, and I, as always warn you, rough and violent times lie ahead, but everything that has happened since the beginning of this war should give the nation confidence that in the end the difficulties will be surmounted, the problems solved, and duty done.

Sir Winston S. Churchill, speech in the House of Commons, 18 December 1939, *Blood, Sweat, and Tears*, 1941

An Army is a team. It eats, sleeps, lives and fights as a team. All this stuff you've been hearing about individuality is a bunch of crap. The billious bastard who wrote that kind of stuff for the *Saturday Evening Post* didn't know any more about real battle than he did about fucking . . .

. . . Every man, every department, every unit, is important to the vast scheme of things. The ordnance men are needed to supply the guns, the QM to bring up the food and the clothes for us, for where we are going, there isn't a hell of a lot for us to steal. Every man in the mess hall, even the one who heats the water to keep us from getting diarrhea, has a job to do. Even the chaplain is important, for if we get killed and he isn't there to bury us, we would all go to hell.

General George S. Patton, Jr., speech to the Third Army, June 1944, 'A General Talks to His Army', 1944

Our military forces are one team – in the game to win regardless of who carries the ball. This is no time for 'fancydans' who won't hit the line with all they have on every play, unless they can call the signals. Each player on the team – whether he shines in the spotlight of the backfield or eats dirt in the line – must be All-American.

General of the Army Omar N. Bradley, 19 October 1949, testimony to Congress

---

See also: COHESION, COMRADESHIP, ESPRIT DE CORPS

TEMPO/MOMENTUM

In battle, momentum means riding on the force of the tide of events. If enemies are on the way to destruction, then you follow up and press them; their armies will surely collapse.

The rule is 'Use the force of momentum to defeat them . . . '

Liu Ji (AD 1310–1375), *Lessons of War*

Never pull up during an attack.

Field Marshal Prince Aleksandr V. Suvorov, *The Science of Victory*, 1976

Defeat the enemy with cold steel, bayonets, sword, and picks . . . Don't slow down during an attack. When the enemy is broken, shattered, then pursue him at once, and don't give him time either to collect or re-form. If he surrenders, spare him; only order him to throw down his arms. During the attack call on the enemy to surrender . . . Spare nothing, don't think of your labours; pursue the enemy night and day, so long as anything is left to be destroyed.

Field Marshal Prince Alexsandr V. Suvorov, *The Science of Victory*, 1796

[The balance of rest and activity, the tricks to keep up the momentum] With this you get speed and the men don't get tired. The enemy reckons we're sixty miles away . . . Suddenly we're on him like a cloudburst. His head whirls. Attack! That's why we came, that's why God sent us. Cavalry! Charge! Cut down, stab, chase, don't let them get away.

Field Marshal Prince Aleksandr V. Suvorov, *The Science of Victory*, 1796

MAXIM 9. The strength of an army, like the power in mechanics, is estimated by multiplying the mass by the rapidity; a rapid march augments the *morale* of an army, and increases all the chances of victory.

Napoleon, *The Military Maxims of Napoleon*, 1831, tr. D'Aguilar

---

Abstaining from follow-on operations until the enemy army is completely destroyed deprives the victor of continued control of the situation. A pause faces him with the need to fight a new battle, in which the chances of success are more or less equal for both sides, just as they are in the initial operation. Above all a pause of this kind leads to indecisiveness of command, likewise to unpreparedness of service support – physical communications, signals and supplies.

Marshal of the Soviet Union Mikhail N. Tukhachevskiy, 1924, quoted in Simpkin's *Deep Battle*, 1987

An acceleration of pace began last century, with the development of the railway and the steamship. It became marked in the last war, when the motor was added, yet the misunderstanding of its meaning and the misapplication of its qualities helped to produce the paradoxical situation that mobility in particular elements had issue in a state of general immobility. During the War aircraft were only in infancy; since then they have multiplied in quantity, range, and speed. There is reason to ask whether we have yet grasped the many-sided effects on war in general when the operations of war are carried out, on different yet conjoint planes, at speeds that vary from three miles an hour to three hundred miles an hour. (June 1936)

Captain Sir Basil Liddell Hart, *Thoughts on War*, 1944

Two fundamental lessons of war experience are – never to check momentum; never to resume mere pushing.

Captain Sir Basil H. Liddell Hart, *Thoughts on War*, 1944

A certain degree of disorganization attends every stage of an advance. Immobilization is ever the child of disorganization. Troops start an attack in a certain formation; movement and enemy fire gradually destroy its integrity; unity of action becomes dissipated at the same rate. The result is that unless the commander is thinking always beyond the immediate objective and planning the means by which he can restore impetus after the object is won, the attack will bog down even though the unit hasn't suffered critical losses in man strength.

Brigadier General S. L. A. Marshall, *Men Against Fire*, 1947

A commander must accustom his staff to a high tempo from the outset, and continuously keep them up to it. If he once allows himself to be satisfied with norms, or anything less than an all-out effort, he gives up the race from the starting post, and will sooner or later be taught a bitter lesson.

Field Marshal Erwin Rommel, *The Rommel Papers*, 1953

TENACITY

We fight, get beaten, and fight again.

Major General Nathanael Greene, 22 June 1781, of his tactics in the Southern Department during the American Revolution

---

If the military leader is filled with high ambition and if he pursues his aims with audacity and strength of will, he will reach them in spite of all obstacles.

Major General Carl von Clausewitz, *Principles of War,* 1812

Tenacity of purpose and untiring energy in execution can repair a first mistake and baffle deeply laid plans.

Rear Admiral Alfred Thayer Mahan (1840–1914)

Victory will come to the side that outlasts the other.

Marshal of France Ferdinand Foch, 7 September 1914, order during the First Battle of the Marne

---

I knew my ground, my material and my allies. If I met fifty checks I could yet see a fifty-first way to my object.

Colonel T. E. Lawrence (1888–1935)

Once an action is begun, it should be carried through regardless of difficulties. If the going is hard, one should think of the enemy's losses and his fear, and so carry on to victory. A commander should not ignore the possibility of help coming to the enemy, nor should he be overly worried about it. An action begun and broken off when tenaciousness might have secured victory only breaks down morale. To persist and win through reveals a commander's strong will to win. On the other hand, when an action has been ordered, or even initiated, and circumstances so change the situation as to make it hopeless, a reversal of the order reveals a commander's versatility.

General Lin Piao, 1946, quoted in Ebon, *Lin Piao,* 1970

Positions are seldom lost because they have been destroyed, but almost invariably because the leader has decided in his own mind that the position cannot be held.

General Alexander A. Vandegrift (1887–1973), quoted in *Warfighting,* Fleet Marine Force Manual (FMFM), 1, 1990

TERRAIN

Confirmation of the ground is of the greatest assistance in battle. Therefore, to estimate the enemy situation and to calculate distances and the degree of difficulty of the terrain so as to control victory are virtues of the superior general. He who fights with full knowledge of these factors is certain to win; he who does not will surely be defeated.

Sun Tzu, *The Art of War,* c. 500 BC, tr. Griffith

The nature of the ground is often of more consequence than courage.

Flavius Vegetius Renatus, *The Military Institutions of the Romans,* c. AD 378

We should choose the terrain not only to suit our armament, but also with a view to the various peoples. Parthians and Gauls handle themselves well on the Plains. The Spanish and Ligurians fight better in the mountains and hills, and the Britons in the woods, while the Germans are more at home in the swamps.

Whatever terrain the general chooses, he should make his troops familiar with it. They will then be able to avoid the rough spots and because of their knowledge of the area will fight the enemy with confidence.

The Emperor Maurice, *The Strategikon,* c. AD 600

A general should possess a perfect knowledge of the localities where he is carrying on a war.

Niccolo Machiavelli, *Discourses,* 1517

Knowledge of the country is to a general what a rifle is to an infantryman and what the rules of arithmetic are to a geometrician. If he does not know the country he will do nothing but make gross mistakes. Without this knowledge his projects, be they otherwise admirable, become ridiculous and often impracticable. Therefore study the country where you are going to act!

Frederick the Great, *Instructions to His Generals*, 1747

. . . terrain for the military man is the same as the chess board for the player who wants to deploy and move his pawns, knights and elephants in the most effective way.

Frederick the Great, *Testament Politique*, 1768

. . . On every occasion where it is a question of attacking the enemy, the way in which you must fight depends upon the terrain and the advantages that the enemy is skilful enough to procure for himself . . . Terrain is the foremost oracle that one must consult, after which he can fathom the enemy dispositions by his own knowledge of the rules of war.

Frederick the Great, 'Pensées et règles générales pour la guerre', 1775, *Oeuvres de Frédéric le Grand*, Vol. XXVII, 1846–56, quoted in Luvaas, ed., *Frederick the Great on the Art of War*, 1966

As many different terrains as there are in existence, just so many different battles will there be.

Frederick the Great (1712–1786), quoted in von Freytag-Loringhoven, *Generalship in the Great War*, 1920

Where a deer can cross, a soldier can cross.

Field Marshal Prince Aleksandr V. Suvorov (1729–1800)

MAXIM 104. An army can march anywhere and at any time of the year, wherever two men can place their feet.

MAXIM 105. Conditions on the groud should not alone decide the organization for combat, which should be determined from consideration of all circumstances.

MAXIM 111. The geographical conditions of a country, life in plains or mountains, education or discipline, have more influence than climate on the character of the troops.

Napoleon, *Military Maxims of Napoleon*, 1827, ed., Burnod

. . . Now we must address ourselves to a special feature of military activity – possibly the most striking even though it is not the most important – which is not related to temperament, and involves merely the intellect. I mean the relationship between warfare and terrain.

This relationship . . . is *a permanent factor* – so much so that one cannot conceive of a regular army operating except in a definite space. Second, its importance is *decisive in the highest degree*, for it affects the operations of all forces, and at times entirely alters them. Third, its influences may be felt in the *very smallest feature of the ground*, but it can also dominate *enormous areas*.

Major General Carl von Clausewitz, *On War*, i, 1832, tr. Howard and Paret

---

Could a dog pass? Then that's enough. Where a dog can go, so can a Russian soldier.

General Veliaminov, 1830s, describing warfare in the Caucasus, quoted in Blanch, *The Sabres of Paradise*, 1960

Every day I feel more and more in need of an atlas, as the knowledge of the geography in its minutest details is the essential to a true military education.

General of the Army William T. Sherman, quoted in Chandler, ed., *Atlas of Military Strategy*, 1980

Natural hazards, however formidable, are inherently less dangerous and uncertain than fighting hazards.

Captain Sir Basil H. Liddell Hart, *Strategy*, 1954

THEORY AND PRACTICE

. . . it is simply not possible to construct a model for the art of war that can serve as scaffolding on which the commander can rely for support at any time. Whenever he has to fall back on his innate talent, he will find himself outside the model and in conflict with it; no matter how versatile the code, the situation will always lead to the consequences we have already alluded to: *talent and genius operate outside the rules, and theory conflicts with practice.*

Major General Carl von Clausewitz, *On War*, ii, 1832, tr. Howard and Paret

---

. . . Theory will have fulfilled its main task when it is used to analyse the constituent elements of war, to distinguish precisely what at first sight seems fused, to explain in full the properties of the means employed and to show their probable effects, to define clearly the nature of the ends in view, and to illuminate all phases of warfare in a thorough critical inquiry. Theory then becomes a guide to anyone who wants to learn about war from books; it will light his way, ease his progress, train his judgment and help him to avoid pitfalls.

Major General Carl von Clausewitz, *On War*, ii, 1832, tr. Howard and Paret

---

Every theory becomes infinitely more difficult from the moment it touches on the province of moral quantities.

Major General Carl von Clausewitz, quoted in Liddell Hart, *The Sword and the Pen*, 1976

One of the surest ways of forming good combinations in war would be to order movements only after obtaining perfect information of the enemy's proceedings. In fact, how can any man say what he should do himself, if he is ignorant what is adversary is about? Even as it is unquestionably of the highest importance to gain this information, so it is a thing of the utmost difficulty, not to say impossibility. This is one of the chief causes of the great difference between the theory and the practice of war.

Lieutenant General Antoine-Henri Baron de Jomini, *Summary of the Art of War*, 1838

For us theory is the basis of confidence in the actions being taken . . .

V. I. Lenin (1870–1924)

To those comrades who wish to build in military affairs by means of the Marxist method, I recommend that they review our field statutes in this light and indicate just what changes – from the standpoint of Marxism – should be introduced into the rules for gathering intelligence, for securing one's lines, for artillery preparation, or for attack. I should very much like to hear at least a single new word in this sphere arrived at through the Marxist method – not just 'an opinion or so' but something new and practical.

Leon Trotsky, 8 May 1922, opening remarks to the Military Scientific Society, *Military Writings*, 1969

Due to the facility with which men, material, and even situations may be created on paper, many writers produce theses on war as admirable as they are impractical. Due also to our school experiences where such intangible factors as morale, training, discipline, fatigue, equipment, and supply cannot be considered, we adopt the simple course of first omitting reference to these factors, and later of assuming them all satisfactory . . .

Duped by the historians who explain defeat and enhance victory by assuring us that both result from the use of PERFECT armies, the fruit of super-thinking, we never stop to consider the inaptness of the word perfect as a definition for armies.

General George S. Patton, Jr., January 1928, *The Patton Papers*, Vol. I, 1972

I have nothing to say against theoretical training, and certainly nothing against practical training. Whoever would become master of his craft must have served as apprentice and journeyman; only a genius can bridge gaps in this sequence of instruction. Every man of action is an artist, and he must know the material with which, in which, and against which he works before he begins his task.

Colonel General Hans von Seekt, *Thoughts of a Soldier*, 1930

The objective value of a broad survey of war is not limited to the research for new and true doctrine. If a broad survey is an essential foundation for any theory of war, it is equally necessary for the ordinary military student who seeks to develop his own outlook and judgment. Otherwise his knowledge of war will be like an inverted pyramid balanced precariously on a slender apex.

Captain Sir Basil Liddell Hart, *Strategy*, 1954

TIME/TIMING

If one knows where and when a battle will be fought his troops can march a thousand *li* and meet on the field. But if one knows neither the battleground nor the day of battle, the left will be unable to aid the right, or the right, the left; the van to support the rear, or the rear, the van. How much more is this so when separated by several tens of *li*, or, even a few.

Sun Tzu, *The Art of War*, *c.* 500 BC, tr. Griffith

It is better to be at the right place with ten men than absent with ten thousand.

Tamerlane (AD 1336–1405), quoted in Lamb, *Tamerlane, The Earth Shaker*, 1928

The advantage of time and place in all martial actions is half a victory, which being lost is irrecoverable.

Sir France Drake, 1558, Letter to Elizabeth I

The moment provides the victory. One moment decides the outcome of a battle and one – the success of a campaign. I do not operate in hours, but rather in minutes.

Field Marshal Prince Aleksandr V. Suvorov (1729–1800), quoted in *Voyenny Vestnik*, 11/1986

In military operations, hours determine success and campaigns.

Napoleon, 20 March 1800, to Admiral Mazarredo, *Correspondance*, No. 4689, Vol. VI, 1858–1870

In military operations, time is everything.

The Duke of Wellington, 30 June 1800, dispatch

Go, sir, gallop, and don't forget that the world was made in six days. You can ask me for anything you like, except time.

Napoleon, 1803, to a staff officer

Time is everything; five minutes make the difference between victory and defeat.

Admiral Lord Nelson (1758–1805)

The loss of time is irretrievable in war; the excuses that are advanced are always bad ones, for operations go wrong only through delays.

Napoleon, 20 March 1806, to Joseph, *Correspondance*, No. 9997, Vol. XII, 1858–1870

It may be that in the future I may lose a battle, but I shall never lose a minute.

Napoleon, quoted in Liddell Hart, *Strategy*, 1954

. . . force and time in this kind of operation amount to almost the same thing, and each can to a very large extent be expressed in terms of each other.

A week lost was about the same as a division. Three divisions in February could have occupied the Gallipoli Peninsula with little fighting. Five could have captured it after March 18. Seven were insignificant at the end of April, but nine just might have done it. Eleven might have sufficed at the beginning of July. Fourteen were to prove insufficient on August 7.

Sir Winston S. Churchill, of the Gallipoli Campaign in 1915

The war of 1914–1918 changed all ideas of time, in a military sense, and especially in the duration of its battles. For several thousand years of warfare a battle, however great its scale, had been a matter of hours. With the

World War the standard became months – because the battles became sieges, without being recognized or scientifically treated as such. The change, it is to be hoped, is a transitory one, for quantity does not imply quality, whereas duration does imply immobility and indecisiveness – which are the negation of generalship. Whether from the standpoint of military science or from that of the drain of human life, long battles are bad battles. (March 1926)

Captain Sir Basil Liddell Hart, *Thoughts on War*, 1944

Mobility of troops economizes time in battle. *Time in war is all.*

Marshal of the Soviet Union Aleksandr I. Yegorov (1883–1939), quoted in Garthoff, *Soviet Military Doctrine*, 1953

Our cards were speed and time, not hitting power, and these gave us strategical rather than tactical strength. Range is more to strategy than force.

Colonel T. E. Lawrence, 'The Science of Guerrilla Warfare', *Encyclopaedia Britannica*, 1929

One of the deeper truths of war has been aptly expressed by Lawrence – 'timing in war depends on the enemy as much as upon yourself'. Timetables too often enable the enemy to turn the tables on you. (August 1933.)

Captain Sir Basil Liddell Hart, *Thoughts on War*, 1944

. . . The history of failure in war can almost be summed up in two words: too late. Too late in comprehending the deadly purpose of a potential enemy; too late in realizing the mortal danger; too late in preparedness; too late in uniting all possible forces for resistence; too late in standing with one's friends. Victory in war results from no mysterious alchemy of wizardry but entirely upon the concentration of superior force at the critical points of combat. To face an adversary in detail has been the prayer of every conqueror in history.

General of the Army Douglas MacArthur, 16 September 1940, statement for the Committee to Defend America by Aiding the Allies, *A Soldier Speaks*, 1965

Death in battle is a function of time. The longer the troops remain under fire, the more men get killed. Therefore, everything must be done to speed up movement.

General George S. Patton, Jr., April 1943, *The Patton Papers*, Vol. II, 1974

A good solution applied with vigor *now* is better than a perfect solution ten minutes later.

General George S. Patton, Jr., *War As I Knew It*, 1947

. . . Bridgeheads in the hands of the Russians are a grave danger indeed. It is quite wrong not to worry about bridgeheads, and to postpone their elimination. Russian bridgeheads, however small and harmless they may appear, are bound to grow into formidable danger-points in a very brief time and soon become insuperable strong-points. A Russian bridgehead, occupied by a company in the evening, is sure to be occupied by at least a regiment the following morning and during the night will become a formidable fortress, well-equipped with heavy weapons and everything necessary to make it almost impregnable . . . There is only one sure remedy which must become a principle: If a bridgehead is forming . . . attack, attack at once, attack strongly. Hesitation will always be fatal. A delay of an hour may mean frustration, a delay of a few hours does mean frustration, a delay of one day may mean a major catastrophe. Even if there is no more than one infantry platoon and one single tank available, attack! Attack when the Russians are still above ground, when they can still be seen and tackled, when they have had no time as yet to organize their defence, when there are no heavy weapons available. A few hours later will be too late. Delay means disaster: resolute energetic and immediate action means success.

Major General F. W. von Mellenthin, *Panzer Battles*, 1955

Time is the essence in war, and while a defeat may be balanced by a battle won, days and hours – even minutes – frittered away, can never be regained.

Brigadier General Samuel B. Griffith II, *The Battle for Guadalcanal*, 1963

A battalion this week can often be more effective than a division next month.

British Army motto

TIMIDITY

A general good at commanding troops is like one sitting in a leaking boat or lying under a burning roof. For there is not time for the wise to offer counsel nor the brave to be angry. All must come to grips with the enemy. And therefore it is said that of all the dangers in employing troops, timidity is the greatest and that the calamities which overtake an army arise from hesitation.

Wu Ch'i (430–381 BC) commentary in Sun Tzu, *The Art of War*, c. 500 BC, tr. Griffith

---

It is essential to be cautious and take your time in making plans, and once you come to a decision to carry it out right away without any hesitation or timidity. Timidity, after all is not caution, but the invention of wickedness.

The Emperor Maurice, *The Strategikon*, c. AD 600

He either fears his fate too much,
  or his deserts are small,
That puts it not unto the touch,
  to win or lose it all.

James Graham, Marquise of Montrose (1612–1650)

'Help, danger!' and other figments of the imagination are all right for old women, who are afraid to get off the stove because they may break their legs, and for lazy, luxurious people, and blockheads – for miserable self-protection, which in the end, whether good or bad, in fact, always passes for bravery with the story tellers.

Field Marshal Prince Aleksandr V. Suvorov (1729–1800), quoted in Blease, *Suvorof*,

---

The torment of precautions often exceeds the dangers to be avoided. It is sometimes better to abandon one's self to destiny.

Napoleon (1769–1821)

Given the same amount of intelligence, timidity will do a thousand times more damage in war than audacity.

Major General Carl von Clausewitz, *On War*, iii, 1821, tr. Howard and Paret

---

. . . When told by Hamilton of [Douglas] Haig's appointment to take charge of a column, he remarked, 'Haig will do nothing! He's quite all right, but he's too ——— cautious: he will be so fixed on not giving the Boers a chance he'll never give himself one. If I were to go to him one evening and offer to land him at daybreak next morning within galloping distance of 1,000 sleeping Boers, I know exactly what he'd do: he'd insist on sending out someone else to make sure the Boers were really there – to make sure no reinforcements were coming up to them, and to make so dead sure in fact, that when he did get there not a single d———d Boer would be within ten miles of him.

Colonel Wools-Sampson, comment during the Boer War (1899–1902) on Douglas Haig who was later to command the British Expeditionary Force (BEF) in World War I, quoted in Liddell Hart, *Through the Fog of War*, 1938

TRADITION

The exposition that I have been giving is based on the teachings of experience, but what you must do is to make your dispositions in accordance with the requirements of the situation with which you are confronted by the urgency of the actual moment. Tradition does not give the answer. The degree of the strength of God's help [i.e., of the commander's capacity for intuitive improvisation – Arnold Toynbee] is the determining factor in deciding the outcome of war.

Anonomous Byzantine General, *Peri Parathromis Polemon*, 10th century

---

To give reputation to the army of any state, it is necessary to revive the discipline of the ancients, cherish and honour it, and give it life, so that in return it may give reputation to the state.

Niccolo Machiavelli, *Discourses*, 1517

. . . Nothing is so disgraceful as slavishness to custom; this is both a result of ignorance and a proof of it.

Field Marshal Maurice Comte de Saxe, *My Reveries*, 1732

The value of tradition to the social body is immense. The veneration for practices, or for authority, consecrated by long acceptance, has a reserve of strength which cannot be obtained by any novel device.

Rear Admiral Alfred Thayer Mahan, 'The Military Rule of Obedience', *National Review*, 3/1902

. . . as descendants of the men who gained such splendid victories in so many battles from 1702 onwards we are simply unable to be cowardly. We've got to win our battles, whatever the cost, so that people will say 'They were worthy descendants of the 32nd' and that's saying a hell of a lot.

Unidentified colonel of the regiment, quoted in Richardson, *Fighting Spirit*, 1978

It takes the Navy three years to build a ship. It would take three hundred to rebuild a tradition.

Admiral Sir Andrew Browne Cunningham, May 1941, when he disapproved his staff's recommendation that the Royal Navy abandon British ground forces on Crete in order to protect the Royal Navy's ships

---

The spirit of discipline, as distinct from its outward and visible guises, is the result of association with martial traditions and their living embodiment.

Captain Sir Basil Liddell Hart, *Thoughts on War*, 1944

However good and well trained a man may be as an individual, he is not a good soldier till he has become absorbed into the cooperate life of his unit and has been entirely imbued with its traditions.

Field Marshal Earl Wavell, *Soldiers and Soldiering*, 1953

However praiseworthy it may be to uphold tradition in the field of soldierly ethics, it is to be resisted in the field of military command.

Field Marshal Erwin Rommel, *The Rommel Papers*, 1953

Every European nation has a strong tendency to be tied by tradition even in matters of science. It was not, therefore, surprising that after World War I, in which the respective commands had exploited the then existing means of war to their utmost limit, many of the European General Staffs became rigidly doctrinaire in their outlook. While unquestioningly accepting the views of great men in matters of principle, they themselves became lost in the detail, tangled it all up into a dreadful complexity, turned warfare into an exchange of memoranda and stuck to their ideas through thick and thin. Soon they were unable to see even the simplest possibilities of a situation, and they never failed to find in their followers and fellow-thinkers a sounding-board for their theories.

Field Marshal Erwin Rommel, *The Rommel Papers*, 1953

There can be no doubt that old traditions are in theory of great value to any army. The characters of the more eminent General Staff officers of the past should have provided a younger generation with good models without, at the same time, hindering or perhaps preventing contemporary development. But in practice tradition is not always regarded as simply supplying ideals of behaviour, but rather as a source of practical example, as though an imitation of what was done before could reproduce identical results despite the fact that meanwhile circumstances and methods have completely altered. Hardly any mature institution can avoid this fallacious aspect of tradition.

Colonel General Heinz Guderian, *Panzer Leader*, 1953

. . . Some people scoff at tradition. They are right if tradition is taken to mean that you must never do something for the first time; but how wrong they are if you regard tradition as a standard of conduct, handed down to you, below which you must never fall. Then tradition, instead of being a pair of handcuffs to fetter you, will be a handrail to steady and guide you in steep places. One of those traditions, perhaps the greatest of them, is leadership, for the be-all and the end-all of an officer is to be a leader.

Field Marshal Viscount Slim, *Courage and Other Broadcasts*, 1957

TRAINING: The Conduct of Training

Now men generally die when they cannot help it and are defeated by a disadvantageous situation. Therefore in employing troops, instructing them and warning them is of the first importance. If one man studies war he can successfully instruct ten; if ten men study they can successfully instruct a hundred; one hundred can successfully instruct one thousand, and one thousand can successfully instruct ten thousand. Ten thousand can instruct the entire army.

Wu Ch'i (430–381 BC), commentary to Sun Tzu, *The Art of War*, c. 500 BC, tr. Griffith

---

We are informed by the writings of the ancients that, among their other exercises was that of the post. They gave their recruits round bucklers woven with willows, twice as heavy as those used on real service, and wooden swords double the weight of the common ones. They excercised them with these at the post both morning and afternoon.

. . . Every soldier, therefore fixed a post firmly in the ground, about the height of six feet. Against this, as against a real enemy, the recruit was exercised with the above mentioned arms, as with the common shield and sword, sometimes aiming at the head or face, sometimes at the sides, at others endeavouring to strike at the thighs or legs. He was instructed in what manner to advance and retire, and in short how to take every advantage of his adversary; but was thus above all particularly cautioned not to lay himself open to his antagonist while aiming his stroke at him.

. . . This was the method of fighting principally used by the Romans, and their reason for exercising recruits with arms of such a weight at first was, that then they came to carry the common ones, which were so much lighter, the difference might enable them to act with greater security and alacrity in time of action.

Flavius Vegetius Renatus, *Military Institutions of the Romans*, c. AD 378

The man who spends more sleepless nights with his army and who works harder in drilling his troops runs the fewest risks in fighting the foe.

The Emperor Maurice, *The Strategikon*, c. AD 600

There is no other possible way, as far as strategy and experience are concerned, for you to prepare for warfare except by first exercising and training the army under your command. You must accustom them to and train them in the handling of weapons and get them to endure bitter and wearisome tasks and labours. They should not be allowed to become slack or lazy or to give themselves completely to drunkenness, luxury, or other kinds of debauchery. They certainly ought to receive their salaries and money for provisions regularly, as well as gifts and bonuses, more than are customary or stipulated.

The Emperor Nikephorus II Phokas, *Skirmishing*, c. AD 969, in Dennis, *Three Byzantine Military Treatises*, 1985

. . . When Emperor Manuel took over the imperial office, he became concerned as to how the Romans might improve their armament for the future. It had previously been customary for them to be armed with round-shields and for the most part to carry quivers and decide battle by bows, but he taught them to hold ones [shields] reaching to their feet, and trained them to wield long lances and skilfully practise horsemanship. Desiring to make respites from preparation for war, he was frequently accustomed to practise riding; making a pretence of battle, he placed formations opposite one another. Thus charging with blunted lances, they practised manoeuvring in arms. So in brief time the Roman [Byzantine] excelled the French and Italians. Nor did the emperor himself hold aloof from these conflicts, but he was arrayed in the front ranks, wielding a lance incomparable in length and size. For in addition to what has been said, a great length of banner was fastened to it, which, as it is divided in eight parts, is customarily called an 'eight-footer'. Indeed they say that Raymond, a man like the legendary Herakleis, when he came to Byzantion, was astonished and thought the matter some trick. Approaching the emperor, he asked for the very spear and shield; taking them, he realized the truth and announced his discovery with surprise.

General John Kinnamos, *Deeds of John and Manuel Comnenus*, AD 1176

Again, all wise and sufficient generals and colonels have always had special regard, when the enemy hath not been near at hand, that their sergeants-major, captains, and other officers should oftentimes in the field reduce their bands and regiments into diverse forms, and to teach their soldiers all orders military, with the use of their weapons in every degree, time, and place; as also how to lodge in their quarters orderly, and therewithal to understand the orders of watches, bodies of watches, sentinels, rounds, and counter-rounds, with many other matters military whereby they might be made prompt and ready to encounter the enemy.

John Smythe (*c.* 1580–1631), *Certain Discourses Military*

. . . the troops should be exercised frequently, cavalry as well as infantry, and the general should often be present to praise some, to criticize others, and to see with his own eyes that the orders . . . are observed exactly.

Frederick the Great, *Instructions for His Generals*, 1747

The soldiers like training provided it is carried out sensibly.

Field Marshal Prince Aleksandr V. Suvorov (1729–1800), *The Suzdal Manual*, 1768, quoted in Ossipov, *Strategy*, 1945

. . . Training is hard work, but it does not have to be dull. Much military training is presented in boring fashion. The troops lose interest and do not absorb the instruction, the training program fails, and the morale of the troops drops. Good training requires a lot of mental effort; the commander must devise ways to make training intellectually and physically challenging to the troops. The unfortunate thing is that so many commanders don't recognize dull training. But their troops do.

Lieutenant General Arthur S. Collins, Jr., *Common Sense Training*, 1978

TRAINING: Train As You Fight

. . . if any one does but attend to the other parts of their [Roman] military discipline, he will be forced to confess that their obtaining so large a dominion, hath been the acquisition of their valour, and not the bare gift of fortune; for they do not begin to use their weapons first in time of war, nor do they then put their hands first into motion, while they avoided so to do in times of peace; but as if their weapons did always cling to them, they have never any truce from warlike exercises; nor do they stay till times of war admonish them to use them; for their military exercises differ not at all from the real use of their arms, but every soldier is every day exercised, and that with great dilligence, as if it were in time of war which is the reason why they bear the fatigue of battles so easily; for neither can any disorder remove them from their usual regularity, nor can fear affright them out of it, nor can labour tire them; which firmness of conduct makes them always to overcome those that have not the same firmness; nor would he be mistaken that should call those their exercises unbloody battles, and their battles bloody exercises.

Flavius Josephus (AD 37–c. AD 100), *The Wars of the Jews*

Make peace a time of training for war, and battle an exhibition of bravery.

The Emperor Maurice, *The Strategikon*, c. AD 600

My troops are good and well-disciplined, and the most important thing of all is that I have thoroughly habituated them to perform everything that they are required to execute. You will do something more easily, to a higher standard, and more bravely when you know that you will do it well.

Frederick the Great, *Principes Généraux*, 1748

. . . the troops should learn in peacetime only what must done in wartime.

General Mikhail I. Dragomirov, quoted in *Voyenno istorichiskiy zhurnal*, 9/1985

There is no studying the battlefield. It is then simply a case of doing what is possible, to make use of what one knows and, in order to make a little possible, one must know much.

Marshal of France Ferdinand Foch, quoted in Marshall, *Infantry in Battle*, 1939

There is a tendency in peace time to conduct training by use of stereotyped situations which are solved by stereotyped solutions. In war, however, we cannot say, 'This situation is so and so according to the rules which I have learned, I must attack or defend'. The situations that confront one in war are generally obscure, highly complicated and never conform to type. They must be met by an alert mind, untrammelled by set forms and fixed ideas.

In our peace-time tactical training we should use difficult, highly imaginative situations and require clear, concise and simple orders. The more difficult the situations, the more simple the order must be. Above all let us kill everything steroetyped; otherwise it will kill us.

Captain Adolf von Schell, *Battle Leadership*, 1933

Troops must be toughened mentally so that adverse conditions will not divert them from their mission. Fatigue, loss of sleep, limited rations, adverse weather conditions and other hardships must not weaken the determination to find and destroy the enemy.

General Lesley J. McNair, quoted in Kahn, *McNair, Educator of an Army*, 19??

The commander must be at constant pains to keep his troops abreast of all the latest tactical experience and developments, and must insist on their practical application. He must see to it that his subordinates are trained in accordance with the latest requirements. The best form of welfare for the troops is first-class training, for this saves unnecessary casualties.

Field Marshal Erwin Rommel, *The Rommel Papers*, 1953

Training is all-encompassing and should be related to everything a unit does or can have happen to it.

Lieutenant General Arthur S. Collins, Jr., *Common Sense Training*, 1978

TRAINING: General

For they had learned that true safety was to be found in long previous training, and not in eloquent exhortations uttered when they were going into action.

Thucydides, of the Spartans at the Battle of Mantinea, 418 BC, *The History of the Peloponnesian War*, v, c. 404 BC

Marching well and keeping ranks are skills which can only be acquired through long drill. The most valiant men in the world would fight very badly if they were not trained, and it is clear that men of the most moderate courage can become warlike by working hard at this profession . . . In truth, you will be able to make brave through practice those who are not so by nature and to show that with care one can make good soldiers out of all sorts of men . . .

Louis XIV, King of France, *Mémoires de Louis XIV*, 1860, in Ralston, *Soldiers and States*, 1966

To bring men to a proper degree of Subordination, is not the work of a day, a month, or even a year.

General George Washington, 24 September 1776, letter to the President of Congress

---

Training is light, and lack of training is darkness. The problem fears the expert. If a peasant doesn't know how to plough, he can't grow bread. A trained man is worth three untrained: that's too little – say six – six is too little – say ten to one. We will beat them all, roll them up, take them prisoner! In the last campaign the enemy lost 75,000 counted, but more like 100,000 in fact. He fought with skill and desperation, but we didn't lose 500. You see, lads! Military training! Gentlemen, what a marvellous thing it is!

Field Marshal Prince Aleksandr V. Suvorov, *The Science of Victory*, 1796

. . . Military sway lies not merely in warships and arms, but also in the immaterial power that wields them. If we consider that a gun whose every shot tells can hold its own against a hundred guns which can hit only one shot in a hundred, we seamen must seek military power spiritually. The cause of the recent victory of our navy, though it was in a great degree due to the Imperial virtue, must also be attributed to our training in peaceful times which produced its fruit in war.

Admiral Marquis Togo Heihachiro, 21 December 1905, his farewell to the Japanese Fleet, quoted in Ogasawara, *Life of Admiral Togo*, 1934

. . . in war the issues are decided not by isolated acts of heroism but by the general training and spirit of an army.

Field Marshal Albert Kesselring, *Memoirs of Field Marshal Kesselring*, 1953

The definition of military training is success in battle. In my opinion that is the only objective of military training. It wouldn't make any sense to have a military organization on the backs of the American taxpayers with any other definition. I've believed that ever since I've been a Marine.

Lieutenant General Lewis 'Chesty' Puller, 2 August 1956, quoted in Burke, *Marine*, 1962

. . . The essential characteristics of a good army are that it be well trained and well disciplined. These two characteristics are apparent in every unit achievement, whether in peace or in war. Discipline derives and flows from training and serves to emphasize a fundamental point essential to a philosophy of training: that training is all-encompassing. Training permeates everything a military organization does.

Lieutenant General Arthur S. Collins, Jr., *Common Sense Training*, 1978

. . . It is astounding what well-trained and dedicated soldiers can accomplish in the face of death, fear, physical privation, and an enemy determined to kill them.

Lieutenant General Arthur S. Collins, Jr., *Common Sense Training*, 1978

Perhaps somewhere in primal reaches of our Army's memory, left over from the days ten thousand years ago when armies first began, there's a simple and fundamental formula: SKILL + WILL = KILL.

Colonel Dandridge M. Malone, *Army Magazine*, 9/1979

TRAINING: Lack of Training

No speech of admonition can be so fine that it will at once make those who hear it good men if they are not good already; it would surely not make archers good if they had not had previous practice in shooting; neither could it make lancers good, nor horsemen; it cannot even make men able to endure bodily labour, unless they have been trained to do it before.

Cyrus the Younger, *d.* 401 BC

. . . for idleness makes the body soft and weak, while relaxation makes the soul cowardly and worthless; since pleasures capturing the passions by enticement of daily habit, corrupt even the most courageous man. For this reason the soldiers must never be without occupation. When after some time spent in idleness they are compelled to go against the enemy, they do not go willingly nor do they stand their ground, but because they have departed from their formal habits, they quickly become dismayed, even before making trial of danger, and even if they do not make trial, they quickly retreat, being incapable either of feeling hope or of sustaining the stress of battle.

Onasander, *The General* AD 58

If officers are unaccustomed to rigorous drilling they will be worried and hesitant in battle. If generals are not thoroughly trained they will inwardly quail when they face the enemy.

Wang Ling, 2nd Century AD, commentator on Sun Tzu, *The Art of War*, *c.* 500 BC, tr. Griffith

---

With a raw army it is possible to carry a formidable position, but not to carry out a plan or design.

Napoleon, quoted in Liddell Hart, *Reputations Ten Years After*, 1928

The greatest difficulty I find is in causing orders and regulations to be obeyed. This arises not from a spirit of disobedience, but from ignorance.

General Robert E. Lee, quoted in Montross, *War Through the Ages*, 1960

Untutored courage is useless in the face of educated bullets.

General George S. Patton, Jr., in the *Cavalry Journal*, 4/1922

In no other profession are the penalties for employing untrained personnel so appalling or so irrevocable as in the military.

General of the Army Douglas MacArthur, *Annual Report of the Chief of Staff, U.S. Army*, 1933

We have verified the inevitable – that inadequately trained officers cannot train troops effectively.

General Lesley J. McNair, 1943, quoted in Kahn, *McNair, Educator of an Army*

Probably our most fundamental and important advantage over the enemy in North Africa was that when my army arrived in Africa in 1941, it was in a better position to benefit from further training on modern lines than were the British. My officers, particularly the younger commanders and General Staff Officers, were up to date in their thinking and not hampered by the conservatism of the British officer.

Field Marshal Erwin Rommel, *The Rommel Papers*, 1953

One of the biggest reasons for failure in the field of battle is not knowing what to do next and, in most cases, this is the result of not having been trained thoroughly in what to expect on the battlefield.

General Orlando Ward, quoted in Gubler, *Combat Actions in Korea*, 1954

The army is an instrument of war: its sole purpose is to fight the country's enemies, and tactical preparedness is an essential element of success in this struggle. An army lacking in tactical training is forced to learn in the course of fighting at the cost of unnecessary losses.

Marshal of the Soviet Union Georgi K. Zhukov, *Reminiscences and Reflections*, 1974

TRAINING: Individual and Unit Training

We must remember that one man is much the same as another, and that he is best who is trained in the severest school.

Thucydides, *The History of the Peloponnesian War*, i, c. 404 BC

Few men are born brave; many become so through training and force of discipline.

Flavius Vegetius Renatus, *Military Institutions of the Romans*, c. AD 378

The ancients have passed on to us the necessity of training and organizing the army, which is obviously useful and quite fundamental. They would train not only the army as a unit, but they would also teach each individual soldier and have him practise how to use his weapons skillfully. In actual combat, then, bravery, assisted experience and skill in handling weapons, should make him invincible. There is, assuredly, a need for exercises and for careful attention to weapons. For many of the Romans and Greeks of old with small armies of trained and experienced men put to flight armies of tens of thousand of troops.

General Nikephorus Ouranos, *Campaigns Organization and Tactics*, c. AD 994, in Dennis, *Three Byzantine Military Treatises*, 1985

Every soldier must be brought to the state where it can be said of him: 'You have learned all there is to be learned, take care not to forget it.'

Field Marshal Aleksandr V. Suvorov (1729–1800), favorite maxim on training, quoted in Ossipov, *Suvorov*, 1945

The training of an infantry company for war, considered by the uninitiated as one of the simplest things in the world, is in reality the most complex; it is one constant struggle against human nature, and incessant variations of the tactical situation and of the ground, to say nothing of the frequent changes in the company as regards junior officers and non-commissioned officers.

Captain R. C. B. Haking, *Company Training*, 1917

Body and spirit I surrendered whole
To harsh instructors – and received a soul.

Rudyard Kipling, 'The Wonder', *Epitaphs*, 1919

The first day I was at camp I was afraid I was going to die. The next two weeks my sole fear was that I wasn't going to die. And after that I knew I'd never die because I'd become so hard nothing could kill me.

Anonymous American soldier, quoted in Cowing, *Dear Folks at Home*, 1919

Battles are fought by platoons and squads. Place emphasis on small unit combat instruction so that it is conducted with the same precision as close-order drill.

General George S. Patton, Jr., *War As I Knew It*, 1947

. . . participation in sport may help turn a mild bookkeeper into a warrior if it has conditioned his mind so that he relishes the contest. The act of teaching one man to participate with other men in any training endeavor is frequently the first step in the development of new traits of receptiveness and outward giving in his character. It is from the acquiring of the habit of working with the group and of feeling responsible to the group that his thoughts are apt to turn ultimately to the welfare of the group when tactical disintegration occurs in battle; the more deeply this is impressed into his consciousness, the quicker will he revert under pressure to thinking and acting on behalf of the group.

Brigadier S. L. A. Marshall, *Men Against Fire*, 1947

. . . Every normal man needs to have some sense of a contest, some feeling of resistance overcome, before he can make the best use of his facilities. Whatever experience serves to give him confidence that he can compete with other men helps to increase his solidarity with other men.

Brigadier General S. L. A. Marshall, *The Armed Forces Officer*, 1950

One great difficulty of training the individual soldier in peace is to instil discipline and yet to preserve the initiative and independence needed in war. The best soldier in peace (officer or man) is not necessarily the best soldier in war – though he is so more often than not – and it is not always easy in peace conditions to recognize the man who will make good in war. The soldier who is a thorough nuisance in barracks is occasionally a treasure in the field, though not nearly as often as Hollywood and the sentimental novelists would have us believe.

Field Marshal Earl Wavell, *Soldiers and Soldiering*, 1953

. . . In the matter of training, all higher headquarters must seek to cut to a proper minimum the required hours for specified subjects, in an endeavor to leave the company or battery commander a reasonable latitude in the training of his unit in subjects which he selects as requiring greater emphasis . . . In training it is also necessary to avoid over-supervision. By this I do not mean that battalion, regimental, or higher commanders should spend any less time in the field and in inspection of training, but the company commander must have the feeling that he is being trusted with the training of his company and that each move he makes is not being supervised by a hovering senior . . . We have made the routine tactical problem a pretty dull affair. Troops spend long periods doing nothing, thinking nothing, learning nothing . . . At all costs we must avoid a false front.

General Anthony McAuliffe, 1954, quoted in Collins, *Common Sense Training*, 1978

The whole training of an officer seeks to accomplish one purpose – to instill in him the ability to take over in battle in a time of crisis.

General Mathew B. Ridgway, *Soldiers*, 1956

During training, the [Israeli] soldier is thrown more on his own than under the United States system. All instruction is pointed toward sharpening the power of decision in the average individual. Physical exercise and lecture courses are aimed to test and increase personal initiative. Israeli trainers believe that teaching the man to think clearly, observe keenly and report accurately is the main object of the school of the soldier.

Brigadier General S. L. A. Marshall, *Sinai Victory*, 1958

Individual training is the foundation on which unit effectiveness is built. It is the source of a soldier's confidence and trust in the Army.

Lieutenant General Arthur S. Collins, Jr,. *Common Sense Training*, 1978

. . . Defeats, collapses, and panics start with small units, and victories are determined by their measure of success. Skillful senior commanders can bring their armies into battle under favorable conditions, but the small-unit leaders at the company level are the ones who win the battle. Given anything like equal terms, the best-trained army will win. Even if the odds are markedly against an army which has well-trained small units and small-unit commanders, that army will often defeat a force superior in numbers and equipment. At a minimum, it will not be routed and will survive to fight on.

Lieutenant General Arthur S. Collins, Jr., *Common Sense Training*, 1978

PHYSICAL TRAINING

. . . the young recruits in particular must be exercised in running, in order to charge the enemy with vigour, to occupy on occasion an advantageous post with greater expedition, and balk the enemy in their designs upon the same; and that they may, when sent to reconnoitre, advance with speed, and return with celerity and more easily overtake the enemy in pursuit.

Flavius Vegetius Renatus, *Military Institutions of the Romans*, c. AD 378

The foundation of training depends on the legs and not the arms. All the mystery of manoeuvres and combats is in the legs, and it is to the legs that we should apply ourselves. Whoever claims otherwise is but a fool and not only in the elements of what is called the profession of arms.

Field Marshal Maurice Comte de Saxe, *My Reveries*, 1732

Is it really true that a seven-mile cross-country run is enforced upon all in this division, from generals to privates? . . . It looks to me rather excessive. A colonel or a general ought not to exhaust himself in trying to compete with young boys running across country seven miles at a time. The duty of officers is no doubt to keep themselves fit, but still more to think of their men, and to take decisions affecting their safety or comfort. Who is the general of this division, and does he run the seven miles himself? If so, he may be more useful for football than for war. Could Napoleon have run seven miles across country at Austerlitz? Perhaps it was the other fellow he made run. In my experience, based on many years' observation, officers with high athletic qualifications are not usually successful in the higher ranks.

Sir Winston S. Churchill, 4 February 1941, note for the Secretary of State for War

---

A pint of sweat will save a gallon of blood.

General George S. Patton, Jr., 8 November 1942, message to his forces before the landing at Casablanca

---

A man who takes a lot of exercise rarely exercises his mind adequately.

Captain Sir Basil Liddell Hart, *Thoughts on War*, 1944

More emphasis will be placed on the hardening of men and officers. All soldiers and officers should be able to run a mile with combat pack in ten minutes and march eight miles in two hours. When soldiers are in actual contact with the enemy, it is almost impossible to maintain physical condition, but if the physical condition is high before they gain contact, it will not fall off sufficiently during contact to be detrimental.

General George S. Patton, Jr., *War As I Knew It*, 1947

Truly then, it is killing men with kindness not to insist upon physical standards during training which will give them a maximum fitness for the extraordinary stresses of campaigning in war.

Brigadier General S. L. A. Marshall, *Men Against Fire*, 1947

When troops lack the coordinated response which comes of long, varied and rigorous exercises, their combat losses will be excessive, and they will lack cohesion in their action against the enemy, and they will uselessly expend much of their initial velocity.

Brigadier General S. L. A. Marshall, *The Armed Forces Officer*, 1950

The highest form of physical training that an officer can undergo is the physical conditioning of his own men.

Brigadier General S. L. A. Marshall, *The Armed Forces Officer*, 1950

All military forces remain relatively undisciplined until physically toughened and mentally conditioned to unusual exertion.

Brigadier General S. L. A. Marshall, *The Armed Forces Officer*, 1950

An intramural company athletic program constitutes training, which is everybody's business. The development of greater endurance, improved agility, and coordination results in a better state of health and a better mental attitude. Commanders' reluctance to take the time to improve both physical conditioning and mental attitude through this program must be overcome. Too many of them wear blinders, particularly at battalion and brigade level.

Lieutenant General Arthur S. Collins, Jr., *Common Sense Training*, 1978

TREATIES/DIPLOMACY

All treaties are broken from considerations of interest; and in this respect republics are much more careful in the observance of treaties than princes . . . I speak of the breaking of treaties from some extraordinary cause; and here I believe, from what has been said, that the people are less frequently guilty of this than princes, and are therefore more to be trusted.

Niccolo Machiavelli, *Discourses*, 1517

Diplomacy without arms is music without instruments.

Frederick the Great (1712–1786)

Treaties are observed as long as they are in harmony with interests.

Napoleon, *The Military Maxims of Napoleon*, 1827, ed. Burnod

---

Does the cessation of diplomatic notes stop the political relations between different nations and governments: Is not war merely another kind of writing and language for their thought: It has, to be sure, its own grammar, but not its own logic.

Major General Carl von Clausewitz, *On War*, 1832

Diplomacy has rarely been able to gain at the conference table what cannot be gained or held on the battlefield.

General Walter Bedell Smith, 1954, after a conference on Indo–China

---

Treaties are like flowers and young girls – they last while they last.

Charles de Gaulle, on a Franco–German treaty, *Time*, 12 July 1963

TRUST

. . . for if men have a spontaneous and natural love for their general, they are quick to obey his commands, they do not distrust him, and they co-operate with him in case of danger.

Onasander, *The General*, AD 58

What makes soldiers in battle prefer to charge ahead rather than retreat even for survival is the benevolence of the military leadership. When soldiers know their leaders care for them as they care for their own children, then the soldiers love their leaders as they do their own fathers. This makes them willing to die in battle, to requite the benevolence of their leaders.

Liu Ji (1310–1375), *Lessons of War*

. . . the commander stakes his reputation, his career, all that he is and may be in his profession, on the issue of a fight in which his instinct must derive so much that is unknown and his science weigh all that is imponderable: above all, he stakes the coin he values most – the trust that his officers and men have in him.

Lieutenant General Sir Francis Tucker, in Horner, *The Commanders*, 1984

Trust is a many-splendored thing which

pervades human relations at all levels in the military service.

Major General Aubrey 'Red' Newman, *Follow Me*, 1981

UNCERTAINTY/THE UNEXPECTED/ THE UNKNOWN

War involves in its progress such a train of unforeseen and unsupposed circumstances that no human wisdom can calculate the end. It has but one thing certain, and that is to raise taxes.

Thomas Paine (1737–1809) *Prospects on the Rubicon*, 1787

Although our intellect always longs for clarity and certainty, our nature often finds uncertainty fascinating. It prefers to day-dream in the realms of chance and luck rather than accompany the intellect on its narrow and tortuous path of philosophical inquiry and logical deduction only to arrive – hardly knowing how – in unfamiliar surroundings where all the usual landmarks seem to have disappeared. Unconfirmed by narrow necessity, it can revel in a wealth of possibilities; which inspire courage to take wing and dive into the element of daring and danger like a fearless swimmer into the current.

Major General Carl von Clausewitz, *On War*, i, 1832, tr. Howard and Paret

---

The commander in war must work in a medium which his eyes cannot see; which his best deductive powers cannot always fathom; and with which because of constant changes he can rarely become familiar.

Major General Carl von Clausewitz, *On War*, i, 1832, tr. Howard and Paret

---

War is the realm of uncertainty; three-quarters of the factors on which action in war is based are wrapped in a fog of greater or lesser uncertainty. A sensitive and discriminating judgment is called for; a skilled intelligence to scent out the truth.

Major General Carl von Clausewitz, *On War*, i, 1832, tr. Howard and Paret

---

He who wars walks in a mist through which the keenest eye cannot always discern the right path.

General Sir William Napier, *History of the War in the Peninsula*, 1840

You will usually find that the enemy has three courses of action open to him, and of these he will adopt the fourth.

Field Marshal Helmuth Graf von Moltke (1800–1891)

The *unknown* is the governing condition of war.

Everybody is familiar with the principle (so you might think), and being familiar with it will distrust the unknown and master it; the unknown will no longer exist.

This is not true in the least. All armies have lived and marched amidst the unknown.

Marshal of France Ferdinand Foch, *Precepts and Judgments*, 1919

No other art is so founded on uncertainties as the art of war. A lifetime may be put into its preparation, where its exercise takes but a brief while. Experience cannot be gained at any time, or from the study of any age, and experience once gained may be out of date tomorrow. Training, as part of the intellectual life of a soldier, therefore, takes precedence over experience.

Colonel Hermann Foertsch, *The Art of Modern War*, 1940

UNIFORM

And now, bursting with rage against the men
of Troy,
he donned Hephaestus' gifts – magnificent
armour
the god of fire forged with all his labour.
First he wrapped his legs with well-made
greaves,

fastened behind his heels with silver ankle-clasps,
next he strapped the breastplate around his chest
then over his shoulder Achilles slung his sword,
the fine bronze blade with its silver-studded hilt,
then hoisted the massive shield flashing far and wide
like a full round moon – gleaming bright as the light
that reaches the sailors out at sea, the flare of a watch fire
burning strong in a lonely sheepfold up some mountain slope
when the gale winds hurl the crew that fights against them
far over the fish-swarming sea, far from loved ones –
so the gleam from Achilles' well-blazoned shield
shot up and hit the skies. Then lifting his rugged helmet
he set it down on his brows, and the horsehair crest
shone like a star and the waving golden plumes shook
that Hephaestus drove in bristling thick along its ridge.
And brilliant Achilles tested himself in all his gear,
Achilles spun on his heels to see if it fit tightly,
see if his shining limbs ran free within it, yes,
and if felt like buoyant wings lifting the great captain.

Homer, *The Iliad*, xix, c. 800 BC, tr. Fagles

On 31 May, 1740, Frederick William I died, and when those around him sang the hymn, 'Naked came I into the world and naked I shall go', he had just sufficient strength to mutter, 'No, not quite naked; I shall have my uniform on.'

Major General J. F. C. Fuller, *A Military History of the Western World*, Vol. II, 1955

I have an insuperable bias in favor of an elegant uniform and a soldierly appearance, so much so that I would rather risk my life and reputation at the head of the same men in an attack, clothed and appointed as I could wish, merely with bayonets and a single charge of ammunition, than to take them as they appear in common with sixty rounds of cartridges.

Major General 'Mad' Anthony Wayne to George Washington, 1776

A well-dressed soldier has more respect for himself. He also appears more redoubtable to the enemy and dominates him; for a good appearance is itself a force.

General Barthelemy-Catherine Joubert (1769–99)

I think it indifferent how a soldier is clothed, provided it is in a uniform manner; and that he is forced to keep himself clean and smart, as a soldier ought to be.

The Duke of Wellington, letter to the War Office, 1811

. . . dress them in red, blue, or green – they'll run away just the same.

King of the Two Sicilies, c. 1800, in response to a suggestion that new uniforms might improve the martial quality of his troops

---

A soldier must learn to love his profession, must look to it to satisfy all his tastes and his sense of humour. That is why handsome uniforms are useful.

Napoleon to General Gaspard Gourgaud, St. Helena, 1815

The secret of uniform was to make a crowd solid, dignified, impersonal: to give it the singleness and tautness of an upstanding man.

General T. E. Lawrence, *Revolt in the Desert*, 1927

. . . Self-respect and *esprit de corps* demand some form of outward expression. Hence the justification of the uniform as the mark of a special class. The uniform indicates the soldier's responsibility; it is the outward sign of inward comradeship; it supports and confirms discipline.

Colonel General Hans von Seekt, *Thoughts of a Soldier*, 1930

It is proverbial that well-dressed soldiers are usually well-behaved soldiers.

Major General John A. Lejeune, *Reminiscences of a Marine*, 1930

UNITY OF COMMAND

For what should be done seek the advice of many; for what you will actually do take counsel with only a few trustworthy people; then off by yourself alone decide on the best and most helpful plan to follow, and stick to it.

The Emperor Maurice, *The Strategikon*, c. AD 600

. . . Now in war there may be one hundred changes in each step. When one sees he can, he advances; when he sees that things are difficult, he retires. To say that a general must await commands of the sovereign in such circumstances is like informing a superior that you wish to put out a fire. Before the order to do so arrives the ashes are cold. And it is said one must consult the Army Supervisor in these matters! This is as if in building a house beside the road one took advice from those who pass by. Of course the work would never be completed!

To put a rein on an able general while at the same time asking him to suppress a cunning enemy is like tying up the Black Hound of Han and then ordering him to catch elusive hares. What is the difference?

Ho Yen-hsi, Chinese military theorist of the Sung Dynasty, c. AD 1000, commentary in Sun Tzu, *The Art of War*, c. 500 BC, tr. Griffith

---

An army should have but one chief; a greater number is detrimental.

Niccolo Machiavelli, *Discourses*, 1517

. . . The first security for success is to confer the command on one individual. When the authority is divided, opinions are divided likewise, and the operations are deprived of that *ensemble* which is the first essential of victory. Besides, when an enterprise is common to many, and not confined to a single person, it is conducted without vigour, and less interest is attached to the result.

Field Marshal Count Raimond Montecuccoli (1609–1680), quoted in ed., Chandler, *The Military Maxims of Napoleon*, 1988

It is generally agreed that of all situations in which the uncontested authority of one man may be necessary to the public good, it is most evidently so in war. Here everyone knows that resolutions must be prompt, discipline exact, orders unquestioned, and obedience absolute, that the least instant that is lost in discussion may cause a golden opportunity to escape forever and that the slightest mistakes that one commits are often paid for in blood . . .

Indeed what may one expect but tumult and confusion in a body where those who must obey cannot be sure who has the right to give them commands, where those who aspire to authority think more of settling their own private disputes than of looking out for the welfare and the security of the troops that have been entrusted to them.

Louis XIV, King of France, *Mémoires de Louis XIV*, 1860, in Ralston, *Soldiers and States*, 1966

To the Executive Directory

HQ, Lodi
25 Floreal, Year IV (14 May, 1796)

That needs only a single general, but also that nothing shall interfere with his movements and operations. I have carried on this campaign without consulting anyone; I should have done no good had I had to come to terms with the viewpoint of another. Lacking everything, I have got the better of far superior forces because, in the belief that I had your confidence, my movements have been as rapid as my thought.

Napoleon (1769–1821)

In war the commander alone understands the importance of certain things. He alone by his will and superior insight can conquer and overcome all difficulties.

Napoleon (1769–1821)

Nothing in war is more important than unity of command. Thus, when war is waged against a single power there must be but one army, acting on one line and led by one chief.

Napoleon (1769–1821)

Better one bad general than two good ones . . .

Napoleon (1769–1821)

The inevitable end of multiple chiefs is that they fade and disappear for lack of unity.

Napoleon (1769–1821)

An army is a collection of armed men obliged to obey one man. Every change in the rules which impairs the principle weakens the army.

General of the Army William T. Sherman (1820–1891)

I never held a council of war in my life.

General of the Army Ulysses S. Grant (1822–1885)

You can't run a military operation with a committee of staff officers in command. It will be nonsense.

Field Marshal Viscount Montgomery of Alamein, quoted in de Guingand, *Operation Victory*, 1947

Now with the United States, it's a very simple thing. The War Department and the government decide that they have to do a certain thing with military forces. They look at the problem and say, 'This is the mission for our forces.' They pick a man. They confer with him about how much he'll need: How many divisions, how much air and navy, and all the rest of it. Then they say, all right, it's *your* job.

General of the Army Dwight D. Eisenhower, *General Eisenhower on The Military Churchill*, 1970

VACILLATION/HESITATION/IRRESOLUTION

The god of war hates those who hesitate.

Euripides, *Heraclidae*, c. 425 BC

Vacillation is nothing less than disaster for an army.

Takeda Nobushige, *Opinions in 99 Articles*, 1558, in Wilson, *Ideals of the Samurai*, 1982

Postponement breeds difficulties.

Field Marshal Prince Aleksandr V. Suvorov (1829–1800), quoted in Longworth, *The Art of Victory*, 1966

Hesitation and half measures lose all in war.

Napoleon, *Maxims of War*, 1831

Irresolution in war is the most dangerous defect in a commander, more especially when the enemy is approaching. He must make his mind up without long deliberation, and above all things prevent French soldiers from giving way to their propensity for criticism. The most distinguished men in the career of arms have never ceased repeating this axion: 'Make your preparation for attack or defence *instantly* on the enemy's approach; should you even be obliged to execute them with disadvantage, do not hesitate.' The enemy, who is a good observer, would take advantage of your indecision. It is often better to come to a bad decision immediately, than to hesitate between several good ones; for the bad one has always some favourable side by which success may be obtained. Moreover a vigilant mind is never embarrassed by the presence of the enemy, which on the contrary will tend to facilitate the boldness of its conceptions.

Marshal of France Michel Ney, Duc d'Elchingen, Prince de la Moskova, *The Memoirs of Marshal Ney*, 1834

Procrastination is like death.

V. I. Lenin (1870–1924)

He who will delay and shilly-shally in his place awaiting a clarified situation will himself inform the enemy and lose the initiative.

Soviet 1936 Cavalry Combat Regulations

VALOUR

Hidden valour is as bad as cowardice.

Latin proverb

In valour there is hope.

Cornelius Tacitus (c. AD 56–c. AD 120), *Annals*

Valour is superior to numbers.

Flavius Vegetius Rennatus, *Military Institutions of the Romans*, c. AD 378

Valour lies halfway between rashness and cowheartedness.

Miguel de Cervantes, *Don Quixote*, 1615

Love of fame, fear of disgrace, schemes for advancement, desire to make life comfortable and pleasant, and the urge to humiliate others are often at the root of the valour men hold in such high esteem.

Perfect valour consists in doing without witnesses that which we would be capable of doing before everyone else.

François Duc de la Rochefoucauld, *Réflexions ou sentences et maximes morales*, 1665

If valour can make amends for the want of numbers, we shall probably succeed.

Major General Sir James Wolfe, 1759, letter to the Elder Pitt from Halifax before the attack on Quebec

---

Among the men who fought on Iwo Jima, uncommon valor was a common virtue.

Admiral Chester W. Nimitz, March 1945

VETERANS

Campaign Against the Hyksos:
Siege of Avaris

One besieged the city of Avaris; I showed valour on foot before his Majesty; then I was appointed (to the ship) 'Shining-in-Memphis'.

Second Battle of Avaris

One fought on the water in the canal . . . Then I fought hand to hand, I brought away a hand. It was reported to the royal herald. One gave me the gold of valour.

Third Battle of Avaris

Then there was again fighting in this place. I again fought hand to hand there; I brought away a hand. One gave me the gold of bravery in the second place.

First Rebellion, Interrupting Siege of Avaris

One fought in this Egypt, south of this city; then I brought away a living captive, a man; I descended into the water; behold, he was brought a seizure upon the road of this city . . . It was announced to the royal herald. Then one presented me with gold in double measure.

Capture of Avaris

One captured Avaris; I took captive there one man and three women, total four heads, his majesty gave them to me for slaves.

Siege of Sharuhen

One besieged Sharuhen for 6 years and his majesty took it. Then I took captive there two women and one hand. One gave me the gold of bravery besides giving me the captives as slaves.

Campaign against Nubia

Now, after this majesty had slain the Asiatics, he ascended the river . . . to destroy the

Nubian Troglodytes; his majesty made a great slaughter among them. Then I took captive there, two living men, and three hands. One presented me with gold in double measures besides giving me two female slaves. His majesty sailed down-stream, his heart joyous with the might of victory for he had seized Southerners and Northerners.

Second Rebellion

There came an enemy of the South; his fate, his destruction approached; the gods of the South seized him, and his majesty found him in Tintto-emu. His majesty carried him off a living prisoner, and all his people carried captive. I carried away two archers as a seizure in the ship of the enemy; one gave to me five heads besides pieces of land . . . It was done to all the sailors likewise.

Third Rebellion

Then came that fallen one, whose name was Teti-en; he had gathered to himself rebels. His majesty slew him and his servants, annihilating them. There was given to me three heads, and fields . . . in my city.

Ahmose son of Ebana, naval officer of Amenhotep I, 1557–1501 BC

As long as there are a few veterans, you can do what you want with the rest.

Field Marshal Maurice Comte de Saxe, *My Reveries*, 1732

As for the cavalry, it should never be touched; old troopers and old horses are good, and recruits of either are absolutely useless. It is a burden, its is an expense, but it is indispensible.

In regard to the infantry, as long as there are a few old heads you can do what you want with the tails; they are the greatest number, and the return of these men in peace is a noticeable benefit to the nation, without a serious diminution of the military forces.

Field Marshal Maurice Comte de Saxe, *My Reveries*, 1732

There are no greater patriots than those good men who have been maimed in the service of their country.

Napoleon, *Political Aphorisms*, 1848

Thus terminated the war, and with it all remembrance of the veteran's services.

General Sir William Napier, *History of the War in the Peninsula*, 1850

. . . I believe that five hundred new men added to an old and experienced regiment were more valuable than a thousand men in the form of a new regiment, for the former by association with good, experienced captains, lieutenants, and non-commissioned officers, soon became veterans, whereas the latter were generally unavailable for a year.

General of the Army William T. Sherman, *Memoirs of General W. T. Sherman*, 1875

The citizen soldiers were associated with so many disciplined men and professionally educated officers, that when they went into engagements it was with a confidence they would not have felt otherwise. They became soldiers themselves almost at once.

General of the Army Ulysses S. Grant, *Personal Memoirs of U.S. Grant*, 1885

A man who is good enough to shed his blood for his country is good enough to be given a square deal afterwards. More than that no man is entitled to, and less than that no man shall have.

President Theodore Roosevelt, 7 December 1903

There is no sight in all the pageant of war like young, trained men going up to battle. The columns look solid and business-like . . . They go on like a river that flows very deep and strong. Uniforms are drab these days, but there are points of light on the helmets and bayonets, and light in the quick steady eyes and the brown young faces, greatly daring. There is no singing – veterans know, and they do not sing much – and there is no excitement at all; they are schooled craftsmen, going up to impose their will, with the tools of their trade, on another lot of fellows; and there is nothing to make a fuss about.

Colonel John W. Thompson, Jr., *Fix Bayonets*, 1926

It takes very little yeast to leaven a lump of dough . . . It takes a very few veterans to leaven a division of doughboys.

General George S. Patton, Jr., *War As I Knew It*, 1947

VICTORY

Great joy has come in Egypt, rejoicing comes forth from the towns of Tomeri. They converse of the victories which Merneptah has achieved among the Tehenu: 'How amiable is he, the victorious rule! How magnified is the king among the gods! How fortunate is he, the commanding lord! Sit happily down and talk, or walk far out upon the way, for there is no fear in the heart of the people. The strongholds are left to themselves, the wells are opened again. The messengers skirt the battlements of the walls, shaded from the sun, until their watchmen wake. The soldiers lie sleeping, and the border scouts are in the field at their own desire. The herds of the field are left as cattle sent forth, without herdsmen, crossing at will the fulness of the stream. There is no uplifting of a shout in the night: 'Stop! Behold, one comes, one comes with the speech of strangers!' One comes and goes with singing, and there is no lamentation of mourning people. The towns are settled again anew; as for the one that ploweth his harvest, he shall eat it. Re has turned himself to Egypt; he was born, destined to be her protector, the King Merneptah.

Merneptah, Pharaoh of Egypt (1225 BC–1215 BC), celebration of victory over the Libyans, quoted in Breasted, *Ancient Records of Egypt*, 1906

Thus we may know that there are five essentials for victory: (1) He will win who knows when to fight and when not to fight. (2) He will win who knows how to handle both superior and inferior forces. (3) He will win whose army is animated by a same spirit throughout all the ranks. (4) He will win who, has prepared himself, waits, to take the enemy unprepared. (5) He will win who has military capacity and is not interfered with by his sovereign. Victory lies in the knowledge of these five points.

Hence the saying: If you know the enemy and know yourself, you need not fear the result of a hundred battles. If you know yourself but not the enemy, for every victory gained you will also suffer a defeat. If you know neither the enemy nor yourself, you will succumb in every battle.

Sun Tzu, *The Art of War*, c. 500 BC, tr. Giles

Dead men have no victories.

Euripides (c. 484 BC–406 BC), *The Phoenician Women*

A good general not only sees the way to victory; he also knows when victory is impossible.

Polybius, *Histories*, c. 125 BC

Victory in war does not depend entirely upon numbers or mere courage; only skill and discipline will ensure it.

Flavius Vegetius Renatus, *The Military Institutions of the Romans*, c. AD 378

The greatest happiness is to vanquish your enemies, to chase them before you, to rob them of their wealth, to see those dear to them bathed in tears, to clasp to your bosom their wives and daughters.

Genghis Khan (AD 1162–1227)

The pursuit of victory without slaughter is likely to lead to slaughter without victory.

The Duke of Marlborough (1650–1722)

Victory is achieved only through the combination of courage and military art.

Field Marshal Prince Aleksandr V. Suvorov (1729–1800), quoted in Gareyev, *Frunze, Military Theorist*, 1985

If in conclusion we consider the total concept of a victory, we find that it consists of three elements:

1. The enemy's greater loss of material strength
2. His loss of morale
3. His open admission of the above by giving up his intentions.

Major General Carl von Clausewitz, *On War*, v, 1832, tr. Howard and Paret

---

. . . Since war contains a host of interactions since the whole series of engagements is, strictly speaking, linked together, since in every victory there is a culminating point beyond which lies the realm of losses and defeats – in view of all these intrinsic characteristics of war, we say there is only one result that counts: *final victory*. Until then, nothing is decided, nothing won, and nothing lost. In this form of war we must always keep in mind that it is the end that crowns the work. Within the concept of absolute war, then, war is indivisible, and its component parts (individual victories) are of value only in their relation to the whole.

Major General Carl von Clausewitz, *On War*, viii, 1832, tr. Howard and Paret

---

For the victor the engagement can never be decided too quickly, for the vanquished it can never last too long. The speedy victory is a higher degree of victory.

Major General Carl von Clausewitz, *On War*, 1832

Man does not enter battle to fight, but for victory. He does everything he can to avoid the first and obtain the second.

Colonel Charles Ardnant du Picq, *Battle Studies*, 1880

A victory on the battlefield is of little account if it has not resulted either in breakthrough or encirclement. Though pushed back, the enemy will appear again on different ground to renew the resistance he momentarily gave up. The campaign will go on.

Field Marshal Alfred Graf von Schlieffen (1833–1913)

The laurels of victory are at the point of the enemy bayonets. They must be plucked *there*; they must be carried by a hand-to-hand fight if one really means to conquer.

Marshal of France Ferdinand Foch, *Precepts and Judgments*, 1919

Victory is an inclined plain. On condition that you do not check your movement the moving mass perpetually increases its speed.

Marshal of France Ferdinand Foch, *Precepts and Judgments*, 1919

Faith in victory determines victory.

Marshal of France Louis-Hubert Lyautey, 8 July 1920

There is nothing certain about war except that one side won't win.

General Sir Ian Hamilton, *Gallipoli Diary*, 1920

Final victory has never been held the acid test of generalship, even where two professional armies have met. Ultimate defeat does not blind us to the skill of Hannibal, Napoleon and Lee. (January 1924)

Captain Sir Basil Liddell Hart, *Thoughts on War*, 1944

The true national object in war, as in peace, is a more perfect peace. The experience of history enable us to deduce that gaining military victory is not in itself equivalent to gaining the object of war. (October 1925)

Captain Sir Basil Liddell Hart, *Thoughts on War*, 1944

Victory, in the true sense, implies that a nation's prospects after the war is better than if it had not made war. Victory, in this sense, is only possible if the result is quickly gained –

for which only an aggressor can hope – or the effort is economically proportioned to the national resources. Favoured by geography, it has been Britain's distinction to excel in this wise economy of force. In a far-sighted fulfillment of the principle, looking beyond war to peace, lay the secret of her unbroken prosperity during three centuries. (January 1931)

Captain Sir Basil Liddell Hart, *Thoughts on War,* 1944

If you concentrate exclusively on victory, with no thought for the after effect, you may be too exhausted to profit by the peace, while it is almost certain that the peace will be a bad one, containing the germs of another war. (April 1939)

Captain Sir Basil Liddell Hart, *Thoughts on War,* 1944

Victory at all costs, victory in spite of terror, victory however long and hard the road may be; for without victory there is no survival.

Sir Winston S. Churchill, 13 May 1940, in the House of Commons

---

A *conversation between this Commander and his political commissar, the latter saying*: 'I was with the troops today, spoke with people. They have only one thought: we shall stay to the death, we shall not let the Fascists go through.'

– To stay is little. The Germans . . . can go through on our corpses. It is necessary to beat them . . .

General I. A. Belov, The Fall of 1941, on the approaches to Moscow, quoted in Leites, *The Soviet Style in War,* 1982

The problems of victory are more agreeable than those of defeat, but they are no less difficult.

Sir Winston S. Churchill, in the House of Commons, 11 November 1942

No one can guarantee success in war, but only deserve it.

Sir Winston S. Churchill, *Their Finest Hour,* 1948

In war there is no second prize for the runner-up.

General of the Army Omar N. Bradley, *Military Review,* 2/1950

The vengeful passions are uppermost in the hour of victory.

Captain Sir Basil Liddell Hart, *Defence of the West,* 1950

In war, there is no substitute for victory.

General of the Army Douglas MacArthur, 19 April 1951, Address to Congress

---

Victory in the true sense implies that the state of peace, and of one's people, is better after the war than before. Victory in this sense is only possible if a quick result can be gained or if a long effort can be economically proportioned to the national resources. The end must be adjusted to the means. Failing a fair prospect of such a victory, wise statesmanship will miss no opportunity for negotiating peace. Peace through stalemate, based on a coincident recognition by each side of the opponent's strength, is at least preferable to peace through common exhaustion – and has often provided a better foundation for lasting peace.

Captain Sir Basil Liddell Hart, *Strategy,* 1954

WAR: The Causes of War

Rage – Goddess, sing the rage of Peleus' son Achilles,
murderous, doomed, that cost the Achaeans countless losses,
hurling down to the house of Death so many sturdy souls,
great fighters' souls, but made their bodies carrion,
feasts for dogs and birds,
and the will of Zeus was moving towards its end.

Begin, Muse, when the two first broke and clashed,
Agamemnon lord of men and brilliant Achilles.

The invocation of the Muse by Homer, *The Iliad*, c. 800 BC, tr. Fagles

---

Wars spring from unseen and generally insignificant causes.

Thucydides, *The Peloponnesian War*, c. 404 BC

And therefore let none hesitate to accept war in exchange for peace. Wise men refuse to move until wronged, but brave men as soon as they are wronged go to war, and when there is a good opportunity make peace again. They are not intoxicated by military success; but neither will they tolerate injustice from a love of peace and ease. For he whom pleasure makes a coward will quickly lose, if he continues inactive, the delights of ease which he is so unwilling to renounce . . .

Speech of the Corinthians urging war with Athens at the Second Peloponnesian Conference, 431 BC, in Thucydides, *The Peloponnesian War*, c. 404 BC

The cause of war must be just.

The Emperor Maurice, *Strategikon*, c. AD 600

A necessary war is a just war.

Niccolo Machiavelli, *The Prince*, 1513

. . . it is most necessary for a general in the first place to approve his cause, and settle an opinion of right in the minds of his officers and soldiers.

General George Monck (1608–1670)

The most just war is one which is founded upon undoubted rights and which, in addition, promises to the state advantages commensurate with the sacrifices required and the hazards incurred.

Lieutenant General Antoine-Henri Baron de Jomini, *Summary of the Art of War*, 1838

States which are weak from a military standpoint, and which are surrounded by strong neighbours invite war, and if they neglect their military organizations from false motives, they court dangers by their own supineness.

Field Marshal Colmar Baron von der Goltz, *Nation in Arms*, 1883

Where evil is mighty and defiant, the obligation to use force – that is war – arises.

Rear Admiral Alfred Thayer Mahan, *Naval Strategy*, 1911

Once in a generation a mysterious wish for war passes through the people. Their insinct tells them that a *there is no other way* of progress and of escape from habits that no longer fit them. Whole generations of statesmen will fumble over reforms for a lifetime which are put into full-blooded execution within a week of a declaration of war. There is *no other way*. Only by intense sufferings can the nations grow, just as a snake once a year must with anguish slough off the once beautiful coat which has now become a strait-jacket.

General Sir Ian Hamilton, *Gallipoli Diary*, 1920

The morale of an Army is compounded of enthusiasm for the national cause and of belief in its own arms. The value to an Army of a righteous cause is no new discovery. William the Conqueror manoeuvred for years to put Harold morally in the wrong before he began to move his Knights; he succeeded and robbed Harold of his Bishops before the game began. William, in fact, made much the same use of Harold's oath on the relics of the saints as Northcliffe did of Kaiser William's solemn engagement to protect Belgium. The Papal Bull secured by the art of William the Conqueror was worth to him five thousand coats of mail, and the loss William the Conqured suffered by the scrapping of his signature was five million fighting men – no less! The materialistic Germans of 1914 deliberately, quite deliberately, put geographical and technical advantage above clean consciences; had they not done so they would have won the war.

General Sir Ian Hamilton, *Soul and Body of an Army*, 1921

National insecurity is one of the fundamental causes of war, especially if the nation concerned is militarily powerful. All nations are impelled by the instinct of national preservation to seek secure frontiers, and, if secure frontiers cannot be gained by peaceful methods, powerful nations will seek to secure them by war. A strong frontier is nothing else than a natural fortress, which, when garrisoned, secures the nation against attack. The object is the security of the nation, consequently . . . the breaking down of the national will is the surest means of forcing the fortress to collapse.

Major General J. F. C. Fuller, *The Foundation of the Science of War*, 1926

The more I study war, the more I come to feel that the cause of war is fundamentally psychological rather than political or economic. When I come in contact with a militarist, his stupidity depresses me and makes me realize the amount of human obtuseness that has to be overcome before we can make progress towards peace. But contact with pacifists too often has the effect of making me almost despair of the elimination of war, because in their very pacifism the element of pugnacity is so perceptible. Moreover, with this active dislike, which in itself hinders understanding, there is mingled an attitude toward – rather like the Arabs' attitude to disease – that it is an affliction of Providence, against which 'charms' may avail, but not scientific study. Until we understand war in the fullest sense, which involves an understanding of men in war, among other elements, it seems to me that we can have no more prospect of preventing war than the savage has of preventing plague. (May 1934)

Captain Sir Basil Liddell Hart, *Thoughts on War*, 1944

In this solemn hour it is a consolation to recall and to dwell upon our repeated efforts for peace. All have been ill-starred, but all have been faithful and sincere. This is of the highest moral value – and not only moral value, but practical value – at the present time, because of the wholehearted concurrence of scores of millions of men and women, whose cooperation is indispensable and whose comradeship and brotherhood are indispensable, is the only foundation upon which the trial and tribulation of modern war can be endured and surmounted. This moral conviction alone affords that ever-fresh resilience which renews the strength and energy of people in long, doubtful and dark days. Outside, the storms of war may blow and the lands may be lashed with the fury of its gales, but in our own hearts this Sunday morning there is peace. Our hands may be active, but our consciences are at rest.

Sir Winston S. Churchill, *Blood, Sweat, and Tears*, 1941

There is no merit in putting off a war for a year, if when it comes, it is a far worse war or one much harder to win.

Sir Winston S. Churchill, *The Gathering Storm*, 1948

War for an aggressor nation is actually a nearly complete collapse of policy. Once war comes, then nearly all prewar policy is utterly invalid because the setting in which it was designed to function no longer corresponds with the facts of reality. When war comes, we at once move into a radically different world.

Admiral J. C. Wylie, USN

From the beginning of man's recorded history physical force, or the threat of it, has been freely and incessantly applied to the resolution of social problems. It persists as an essential element in the social pattern. History suggests that as a society of men grows more orderly the application of forces tends to become better ordered. The requirement for it has shown no sign of disappearing.

General Sir John Hackett, *The Profession of Arms*, 1963

WAR: Civil War

Lost to the clan,
lost to the hearth, lost to the old ways, that one
who lusts for all the horrors of war with his own people.

Speech by King Nestor of Pylos, Homer, *The Iliad*, ix, *c.* 800 BC, tr. Fagles

---

Now they were thoroughly convinced that civil dissentions were much more to be dreaded than a war carried out in a foreign country against a foreign enemy.

Polybius, *The Histories*, *c.* 125 BC

It was easy to begin the war, but no man knew where it would end.

The Earl of Manchester, 1644, after the Battle of Marston Moor during the English Civil War

---

I am one of those who have probably passed a longer period of my life engaged in war than most men, and principally in civil war; and I must say this, that if I could avoid, by any sacrifice whatever, even one month of civil war in the country to which I was attached, I would sacrifice my own life in order to do it.

The Duke of Wellington, March 1829, to the House of Lords

This war was never really contemplated in earnest. I believe if either the North or the South had expected that their differences would result in this obstinate struggle, the cold-blooded Puritan and the cock-hatted Huguenot and Cavalier would have made a compromise.

Major General George Pickett, 27 June 1862, letter to his fiancée during the American Civil War

WAR: Conduct of War

In war events of importance are the results of trivial causes.

Julius Caesar, *The Gallic War*, *c.* AD 51

. . . war is waged not with numbers but with wits.

Field Marshal Prince Aleksandr V. Suvorov (1729–1800), quoted in Gareyev, *Frunze, Military Theorist*, 1985

War justifies everything.

Napoleon, 1808, ed., Herold, *The Mind of Napoleon*, 1955

The conduct of war resembles the workings of an intricate machine with tremendous friction, so that combinations which are easily planned on paper can be executed only with effort.

Major General Carl von Clausewitz, *Principles of War*, 1812

The conduct of war branches out in almost all directions and has no definite limits; while any system, any model, has the finite nature of a synthesis. An irreconcilable conflict exists between this type of theory and actual practice.

Major General Carl von Clausewitz, *On War*, ii, 1832, tr. Howard and Paret

---

And so it is that intelligence, instinct, and the leader's authority combine to make the conduct of war what it is. But what are these faculties if not the expression of a powerful and resourceful personality? Other things being equal, it is safe to say that the value of the fighting man is in exact proportion to the value, in terms of personality, of those under whose command he serves. Training for war is, first and foremost, training in leadership, and it is literally true, for armies as well as for nations, that where the leadership is good, the rest shall be added to them.

Charles de Gaulle, *The Edge of the Sword*, 1932

There is only one way to conduct war as I read history: Deploy to the war zone as quickly as possible sufficient forces to end it at the earliest moment. Anything else is a gift to the enemy.

Brigadier General S. L. A. Marshall, 'Some Thoughts on Vietnam', in ed., Thompson, *The Lessons of Vietnam*,

WAR: The Economy of War

Where the army is, prices are high; when prices rise the wealth of the people is exhausted. When wealth is exhausted the peasantry will be afflicted with urgent exactions.

Sun Tzu, *The Art of War*, *c.* 500 BC, tr. Griffith

As to government expenditures, those due to broken-down chariots, worn-out horses, armour and helmets, arrows and crossbows, lances, hand and body shields, draft animals and supply wagons will amount to sixty per cent of the total.

Hence the wise general sees to it that his troops feed on the enemy, for one bushel of the enemy's provisions is equivalent to twenty of his; one hundredweight of enemy fodder to twenty hundredweight of his.

Sun Tzu, *The Art of War*, ii, *c.* 500 BC, tr. Griffith

War is a matter not so much of arms as of expenditure, through which arms may be made of service.

Thucydides, *History of the Peloponnesian War*, i, *c.* 404 BC

Fight thou with shafts of silver and thou shalt conquer in all things.

Response of the Delphic Oracle to Philip of Macedon's question how he might be victorious in war, quoted from Plutarch, *Apothegms*

War cannot be put on a certain allowance.

Archidamus III, King of Sparta, *d.* 328 BC

Money, more money, always money.

Marshal Gian Giacomo de Trivulce (1441–1518), when François I, King of France, asked him what he needed to make war

---

It is not gold, but good soldiers, that ensure success in war . . . for it is impossible that good soldiers should not be able to procure gold, as it is impossible for gold to procure good soldiers.

Niccolo Machiavelli, *Discourses*, 1517

Men, arms, money and provisions are the sinews of war, but of these four, the first two are the most necessary; for men and arms will always find money and provisions, but money and provisions cannot always raise men and arms.

Niccolo Machiavelli, *The Art of War*, 1521

The army is a school in which the miser becomes generous, and the generous prodigal; miserly soldiers are like monsters, very rarely seen.

Miguel de Cervantes, *Don Quixote*, 1615

The army is a sack with no bottom.

General Simon Bolivar (1783–1830), quoted in *Military History*, 8/1988

Clausewitz had observed (and Vietnam was to prove) the economic approach to military strategy 'stood in about the same relationship to combat as the craft of the swordsmith to the art of fencing'.

Colonel Harry Summers, *Parameters*, Summer 1987

WAR: Guerrilla/Irregular Warfare

Advance like foxes, fight like lions, and fly like birds.

North-eastern Indian tactical maxim

. . . It is not to be expected that the rebel Americans will risk a general combat or a pitched battle, or even stand at all, except behind intrenchments as at Boston. Accustomed to felling of timber and to grubbing up trees, they are very ready at earthworks and palisading, and will cover and intrench themselves wherever they are for a short time left unmolested with surprising alacrity . . . Composed as the American Army is, together with the strength of the country, full of woods, swamps, stone walls, and other inclosures and hiding places, it may be said of it that every private man will in action be his own general, who will turn every tree and bush into a kind of temporary fortress, and from whence, when he hath fired his shot with all deliberation, coolness, and uncertainty which hidden safety inspires, he will skip as it were to the next, and so on for a long time till dislodged either by cannon or by a resolute attack of light infantry.

Lieutenant General John 'Gentleman Johnny' Burgoyne (1722–1792), quoted in Lloyd, *A Review of the History of Infantry*, 1908

According to our idea of a People's War, it should, like a kind of nebulous vapoury essence, never condense into a solid body . . . Still, however, it is necessary that this mist should collect at some points into denser masses, and form threatening clouds from which now and again a formidable flash of lightning may burst forth.

Major General Carl von Clausewitz, *On War*, 1832, tr. Graham

---

The more the enemy extends himself, the greater is the effect of an armed populace. Like a slow, gradual fire it destroys the bases of the enemy force.

Major General Carl von Clausewitz, *On War*, 1832, tr. Graham

---

The military value of a partisan's work is not measured by the amount of property destroyed or the number of men killed or captured, but by the number he keeps watching.

Colonel John S. Mosby, *Mosby's War Reminiscences*, 1887

To oppose successfully such bodies of men as our burghers had to meet during this war demanded *rapidity of action* more than anything else. We had to be quick at fighting, quick at reconnoitring, quick (if it became necessary) at flying!

Christiaan R. De Wet, *Three Years War*, 1902

Guerrilla war is far more intelligent than a bayonet charge.

Colonel T. E. Lawrence, 'The Science of Guerrilla Warfare', *Encyclopaedia Britannica*, 1929

The ability to run away is precisely one of the characterics of guerrillas. Running away is the chief means of getting out of passivity and regaining the initiative.

Mao Tse-tung (1893–1976), quoted in Simpkin, *The Race to the Swift*, 1985

[Guerrilla War] must have a friendly population, not actively friendly, but sympathetic to the point of not betraying rebel movements to the enemy. Rebellions can be made by two per cent active in a striking force, and 98 passively sympathetic.

Colonel T. E. Lawrence, 'The Science of Guerrilla Warfare', *Encylopaedia Britannica*, 1929

The advantages are nearly all on the side of the guerrilla in that he is bound by no rules, tied by no transport, hampered by no drill-books, while the soldier is bound by many things, not the least by his expectation of a full meal every so many hours. The soldier usually wins in the long run, but very expensively.

Field Marshal Earl Wavell, 30 August 1932, critique on a counter-guerrilla exercise, Blackdown

---

. . . Governments saw men only in mass; but our men, being irregulars, were not formations, but individuals. An individual death, like a pebble dropped in water, might make but a brief hole; yet rings of sorrow widened out therefrom. We could not afford casualties.

Materials were easier to replace. It was our obvious policy to be superior in some one tangible branch; gun-cotton or machine-guns or whatever could be made decisive. Orthodoxy had laid down the maxim, applied to men, of being superior at the critical point and moment of attack. We might be superior in equipment in one dominant moment or respect; and for both things and men we might give the doctrine a twisted negative side, for cheapness' sake, and be weaker than the enemy everywhere except in that one point or matter. The decision of what was critical would always be ours. Most wars were wars of contact, both forces striving into touch to avoid tactical surprise. Ours should be a war of detachment. We were to contain the enemy by the silent threat of a vast unknown desert, not disclosing ourselves till we attacked. The attack might be nominal, directed not against him, but against his stuff . . .

Colonel T. E. Lawrence, *The Seven Pillars of Wisdom*, 1926

Flexible employment of forces is more indispensable in guerrilla warfare than in regular warfare . . . and the chief ways of employing forces consist in dispersing, concentrating and shifting them . . . In general the shifting of forces should be done secretly and swiftly.

Mao Tse-tung (1893–1976), quoted in Simpkin, *The Race to the Swift*, 1985

When the enemy advances, we retreat.
When he escapes we harass.
When he retreats we pursue.
When he is tired we attack.
When he burns we put out the fire.
When he loots we attack.
When he pursues we hide.
When he retreats we return.

Mao Tse-tung (1893–1976)

Guerrilla war, too, inverts one of the main principles of orthodox war, the principle of 'concentration' – and on both sides. Dispersion is an essential condition of survival and success on the guerrilla side, which must never present a target and thus can operate in minute particles, though these may momentarily coagulate like globules of quicksilver to overwhelm some weakly guarded objective. For guerrillas as the principle of 'concentration' has to be replaced by that of 'fluidity of force' . . . dispersion is also a necessity on the side opposed to the guerrillas, since there is no value in a narrow concentration of force against such elusive forces, nimble as mosquitoes. The chance of curbing them lies largely in being able to extend a fine but closely woven net over the widest possible area. The more extensive the controlling net, the more likely that anti-guerrilla drives will be effective.

Captain Sir Basil Liddell Hart, *Strategy*, 1954

In many ways this war [against the Sinn Fein] was far worse than the Great War which had ended in 1918. It developed into a murder campaign in which, in the end, the soldiers became very skilfull and more than held their own. But such a war is thoroughly bad for officers and men; it tends to lower their standards of decency and chivalry, and I was glad it was over.

Field Marshal Viscount Montgomery of Alamein, *The Memoirs of Field Marshal Montgomery*, 1958

. . . If we wanted to win swiftly, success could not be ensured. For that reason, in the process of making preparations, we continued to follow the enemy's situation and checked and rechecked our potentialities again. And we came to the conclusion that we could not secure success if we struck swiftly. In consequence, *we resolutely chose the other tactic: to strike surely and advance surely.* In taking this correct decision, we strictly followed this fundamental principle of the conduct of revolutionary war: strike to win, strike only when success is certain; if it is not, then don't strike.

General Vo Nguyen Giap, *People's War, People's Army*, 1961

Concentration of troops to realize an overwhelming superiority over the enemy where he is sufficiently exposed in order to destroy his manpower; initiative, suppleness, rapidity, surprise, suddenness in attack and retreat. As long as the strategic balance of forces remains disadvantageous, resolutely to muster troops to obtain absolute superiority in combat in a given place, and at a given time. To exhaust little by little by small victories the enemy forces and at the same time to maintain and increase ours. In these concrete conditions it proves absolutely necessary not to lose sight of the main objective of the fighting that is the destruction of the enemy manpower. Therefore losses must be avoided even at the cost of losing ground . . .

General Vo Nguyen Giap, *People's War, People's Army*, 1961

There is a malevolent definition that says: 'The guerrilla fighter is the Jesuit of warfare.' By this is indicated a quality of secretiveness, of treachery, of surprise that is obviously an essential element of guerrilla warfare. It is a special kind of Jesuitism, naturally prompted by circumstances, which necessitates acting at certain moments in ways different from the romantic and sporting conceptions with which we are taught to believe war is fought.

Che Guevara, *Guerrilla Warfare*, 1961

WAR: The Love of War

Men grow tired of sleep, love, singing and dancing, sooner than of war.

Homer, *The Iliad*, c. 800 BC

War – I know it well, and the butchery of men.
Well I know, shift to the left, shift to the right
my tough tanned shield. That's what the real drill
defensive fighting means to me. I know it all,
how to charge in the rush of plunging horses –
I know how to stand and fight to the finish,
twist and lunge in the War-god's deadly dance.

Hector to Ajax, in Homer, *The Iliad*, vii, 800 BC, tr. Fagles

---

. . . a man whose passion was war. When he could have kept at peace without shame or damage, he chose war; when he could have been idle, he wished for hard work that he might have war; when he could have kept wealth without danger, he chose to make it less by making war; there was a man who spent upon war as if it were a darling lover or some other pleasure.

Xenophon, describing the Spartan mercenary general Clearchus (*c.* 431 BC–*c.* 352 BC), *The March Up Country*, tr. Rouse

---

That shall be my music in the future!

Charles XII, King of Sweden, 1700, at Copenhagen the first time under fire as he listened to the bullets' whistle

---

A shot through the mainmast knocked a few splinters about us. He observed to me, with a smile, 'It is warm work, and this day may be the last to any of us at a moment'; and then, stopping short at the gangway, he used an expression never to be erased from my memory, and said with emotion, 'But mark you, I would not be elsewhere for thousands.'

Admiral Lord Nelson, 2 April 1801, during the Battle of Copenhagen, as related by Colonel William Parker, quoted in Mahan, *The Life of Nelson*, Vol. II, 1897

It is well that war is so terrible – we would grow too fond of it.

General Robert E. Lee to Longstreet, Battle of Fredericksburg, 13 December 1862.

To ride boldly at what is in front of you, be it fence or enemy; to pray, not for comfort, but for combat; to remember that duty is not to be proved in the evil day, but then to be obeyed unquestioning; to love glory more than the temptations of wallowing at ease.

Justice Oliver Wendell Holmes, Jr., (1841–1935)

I'm afraid the war will end very soon now, but I suppose all good things come to an end sooner or later, so we mustn't grumble.

Field Marshal Earl Alexander (as a junior officer), 1917, letter to his mother

---

It is not the horrors of war that will deter any virile young man from welcoming it, but the plain truth that, instead of a gallant adventure, he is setting out on a farcial futility. The best antidote to war is a widening sense of humour and a keener sense of the ridiculous. (November 1934)

Captain Sir Basil Liddell Hart, *Thoughts on War,* 1944

If the Germans are a warrior race, they are certainly militarist also. I think they love the military pageant and the panoply of war; and the feeling of strength and power that a well-organized and disciplined unit gives to each and every individual member of that unit. I am quite willing to admit that I myself share this curious attraction for the strength and elegance of beautifully trained and equipped formations, with all the art and subtlety of their movements in action against an enemy. I can well understand the enthusiasm which the soldiers – from marshals to the private soldier – showed for Napoleon; and why they followed their leader without doubt or question in his victorious campaigns. Feeling thus, they shared the glory of his conquests.

Field Marshal Earl Alexander, *The Alexander Memoirs*, 1961

WAR: Means of War

. . . a king's raw materials and instruments of rule are a well-peopled land, and he must have men of prayer, men of war and men of work.

Alfred the Great, King of England (AD 849–901)

Every one may begin a war at his pleasure, but cannot so finish it . . . And money alone, so far from being a means of defence, will only render a prince the more liable to being plundered. There cannot, therefore, be a more erroneous opinion than that money is the sinews of war.

I maintain, then, contrary to the general opinion, that the sinews of war are not gold, but good soldiers; for gold alone will not procure good soldiers, but good soldiers will always procure gold. By making their wars with iron, they never suffered for the want of gold.

Niccolo Machiavelli, *Discourses*, 1517

The means of waging war are the means of waging peace. Practically everything which in peace time possesses energy, can in war time be turned into war energy – newspapers, sewing-machines, motor cars, dye works, watch factories, royal academicians; in fact all industries, trades, and professions; all the moral physical, economic, intellectual, and even spiritual, forces of the nation. At the apex of these we have the navy, army and air force which can attack the enemy physically, economically and morally, that is by fighting, starvation, and terror.

Major General J. F. C. Fuller, *Lectures of F. S. R. II,* 1931

For weeks and weeks now I have been trying to make bricks without straw, which in itself is bad enough, but which is made much worse when others believe you have the straw.

Major General George Alan Vasey, quoted in Horner, *The Commanders, 1984*

WAR: Nature of War

War is a matter of vital importance to the state; the province of life or death; the road to survival or ruin. It is mandatory that it be thoroughly studied.

Sun Tzu, *The Art of War,* c. 500 BC

War is the last of all things to go according to programme.

Thucydides, *The Peloponnesian War,* c. 404 BC

It is not the object of war to annihilate those who have given provocation for it, but to cause them to mend their ways; not to ruin the innocent and guilty alike, but to save both.

Polybius, *The Histories*, v, c. 125 BC

The circumstances of war are sensed rather than explained.

Field Marshal Maurice Comte de Saxe (1696–1750)

War is a trade for the ignorant and a science for the expert.

Chevalier Jean Charles Follard (1669–1752)

I have said that the art of war is divided between force and strategem. What cannot be done by force, must be done by strategem.

Frederick the Great, *Frederick's Instructions for His Generals*, 1747

War is an option of difficulties.

Major General Sir James Wolfe (1727–1759)

War admittedly has its own grammar, but not its own logic.

Major General Carl von Clausewitz, *On War*, 1832, tr. Howard and Paret

---

War is an act of violence whose object is to constrain the enemy to accomplish our will.

Major General Carl von Clausewitz, *On War*, 1832, tr. Howard and Paret

---

. . . a trinity of violence, chance, and reason.

Major General Carl von Clausewitz, *On War*, 1832, tr. Howard and Paret

---

War is an act of violence pushed to its umost limits.

Major General Carl von Clausewitz, *On War*, 1832, tr. Howard and Paret

War is not only chameleon-like in character, because it changes colours in some degree in each particular case, but it is also, as a whole, in relation to the predominant tendencies that are in it, a wonderful trinity, composed of the original violence of its elements, hatred and animosity, which may be looked upon as blind instinct; of the play of probabilities and chance, which make a free activity of the soul; and of the subordinate nature of a political instrument, by which it belongs purely to reason.

Major General Carl von Clausewitz, *On War*, 1832, tr. Howard and Paret

---

. . . War is nothing but a duel on a larger scale. Countless duels go to make up war, but a picture of it as a whole can be formed by imagining a pair of wrestlers. Each tries through physical force to compel the other to do his will; his *immediate* aim is to *throw* his opponent in order to make him incapable of further resistance.

War is thus an act of force to compel our enemy to do our will.

Major General Carl von Clausewitz, *On War*, 1832, tr. Howard and Paret

---

. . . war is not an exercise of the will directed at inanimate matter, as is the case with the mechanical arts, or at matter which is animate but passive and yielding, as is the case with the human mind and emotions in the fine arts. In war, the will is directed at an animate objects that *reacts*.

Major General Carl von Clausewitz, *On War*, 1832, tr. Howard and Paret

---

If we now consider briefly, the *subjective nature* of war – the means by which war has to be fought – it will look more than ever like a gamble. The element in which war exists is danger. The highest of all moral qualities in time of danger is certainly *courage*. Now courage is perfectly compatible with prudent calculation but the two differ none the less, and pertain to different psychological forces. Daring, on the other hand, boldness, rashness, trusting in luck are only variants of courage, and all these traits of character seek their proper element – chance.

Major General Carl von Clausewitz, *On War*, 1832, tr. Howard and Paret

---

. . . War is not pastime; it is no mere joy in daring and winning, no place for irresponsible enthusiasts. It is a serious means to a serious end, and all its colourful resemblance to a game of chance, all the vicissitudes of passion, courage, imagination and enthusiasm it includes are merely its special characteristics.

Major General Carl von Clausewitz, *On War*, 1832, tr. Howard and Paret

---

. . . War is a pulsation of violence, variable in strength and therefore variable in the speed with which it explodes and discharges its energy. War moves on its goal with varying speeds; but it always lasts long enough for influence to be exerted on the goal and for its own course to be changed in one way or another – long enough, in other words, to remain subject to the action of a superior intelligence.

Major General Carl von Clausewitz, *On War*, 1832, tr. Howard and Paret

---

All war presupposes human weakness, and seeks to exploit it.

Major General Carl von Clausewitz, *On War*, 1832, tr. Howard and Paret

---

You might as well appeal against the thunderstorm as against the terrible hardships of war.

General of the Army William T. Sherman, 12 September 1864, letter to the mayor of Atlanta

---

You cannot qualify war in harsher terms than I will. War is cruelty, and you cannot refine it.

General of the Army William T. Sherman, 12 September 1864, letter to the mayor of Atlanta

---

War is at best barbarism . . . Its glory is all moonshine. It is only those who have neither fired a shot nor heard the shrieks and groans of the wounded, who cry aloud for blood, more vengeance, more desolation. War is hell.

General of the Army William T. Sherman (1820–1891)

War is an integral part of God's ordering of the universe. In war, man's noblest virtues come into play. Courage and renunciation, fidelity to duty and a readiness for sacrifice that does not stop short of offering up life itself. Without war the world would become swamped in materialism.

Field Marshal Helmuth Graf von Moltke (1800–1891), quoted in Toynbee, *War and Civilization*, 1950

War is a perpetual struggle with embarrassments.

Field Marshal Colmar Baron von der Goltz, *The Nation in Arms*, 1883

War is a game to be played with a smiling face.

O, horrible war! Amazing medley of the glorious and the squalid, the pitiful and the sublime! If modern men of light and leading saw your face closer, simple folk would see it hardly ever.

Sir Winston S. Churchill, *London to Ladysmith*, 1900

From Jomini also I imbibed a fixed disbelief in the thoughtlessly accepted maxim that the statesman and general occupy unrelated fields. For this misconception I substituted a tenet of my own, that war is simply a violent political movement; and from an expression of his, 'The sterile glory of fighting battles merely to win them', I deduced, what military men are prone to overlook, that 'War is not fighting, but business'.

Rear Admiral Alfred Thayer Mahan, *From Sail to Steam*, 1907

Victory smiles upon those who anticipate the changes in the character of war, not those who wait to adapt themselves after they occur.

General Giulio Douhet, *Command of the Air*, 1921

War will not suffer any set form. Measures which on one section of the front and on a certain occasion – be it in attack or defence – resulted in a complete success, proved in a different place and under different circumstances failures. The reason for this is found in the fact that war has been carried on, from time immemorial, by men against men, and this holds true even today, in spite of the progress made in technical means. *Therefore, in the last analysis, it is the moral qualifications which, as in the past, will also decide the issue today.*

General der Infantrie Hugo Baron von Freytag-Loringhoven, *Generalship in the Great War*, 1921

War is a specific form of relations between men. In consequence, war methods and war usages depend upon the anatomical and physical qualities of individuals, upon the form of organization of the collective man, upon his technology, his physical and cultural-historical environment, and so on. The usages and methods of warfare are thus determined by changing circumstances and, therefore, they themselves can in nowise be eternal.

Leon Trotsky, 1922, *Military Writings*, 1967

War, which used to be cruel and magnificent, has now become cruel and squalid. It is all the fault of democracy and science. From the moment that either of these meddlers and muddlers was allowed to take part in actual fighting, the doom of War was sealed. Instead of a small number of well-trained professionals championing their country's cause with ancient weapons and a beautiful intricacy of archaic movement, we now have entire populations, including even women and children, pitted against each other in brutish mutual extermination, and only a set of blear-eyed clerks left to add up the butcher's bill. From the moment when Democracy was admitted to, or rather forced itself upon, the battlefield, War ceased to be a gentleman's pursuit.

Sir Winston S. Churchill, *My Early Life*, 1930

The story of the human race is war.

Sir Winston S. Churchill, *Marlborough*, 1933

War is a series of local emergency measures.

Major General J. F. C. Fuller, *Memoirs of an Unconventional Soldier*, 1936

War is a contest of strength, but the original pattern of strength changes in the course of the war. Here the decisive factor is subjective effort – winning more victories and committing fewer errors. The objective factors provide the possibility for such change, but in order to turn this possibility into actuality both correct policy and subjective effort are essential. It is then that the subjective plays the decisive role.

Mao Tse-tung, *On Protracted War*, 1938

. . . War and truth have a fundamental incompatibility. The devotion to secrecy in the interests of the military machine largely explains why, throughout history, its operations commonly appear in retrospect the most uncertain and least efficient of human activities.

Captain Sir Basil Liddell Hart, *Through the Fog of War*, 1938

There is no working middle course in wartime.

Sir Winston S. Churchill, 2 July 1942, speech in the House of Commons

---

War creates such a strain that all the pettiness, jealousy, ambition and greed, and selfishness begin to leak out of the seams of the average character. On top of this are the problems created by the enemy.

General of the Army Dwight D. Eienhower, 16 December 1942, *Letters to Mamie*, 1978

War is very simple, direct, and ruthless. It takes a simple, direct, and ruthless man to wage war.

General George S. Patton, Jr., 15 April 1943, diary entry

War, like politics, is a series of compromises.

Captain Sir Basil Liddell Hart, *Thoughts on War*, 1944

The main ethical objection to war for intelligent people is that it is so deplorably dull and usually so inefficiently run . . . Most people seeing the muddle of war forget the muddles of peace and the general inefficiency of the human race in ordering its affairs.

Field Marshal Earl Wavell, unpublished *Recollections*, 1947

. . . war in its technical aspect is always a movement of societies. The passion that animate it and the pretexts it invokes unfailingly cloak a dispute over the material or spiritual destiny of men. Alexander's victories were those of a civilization. It was the barbarian's passionate hunger which caused the fall of the Roman Empire. There would have been no Arab invasion without the Koran. No crusades without the Gospels. The *ancien régime* in Europe rose against France when the Assembly proclaimed, 'Men are born free and equal by law'.

Charles de Gaulle, *Salvation 1944–1946*, 1960

War is a very simple thing, and the determining characteristics are self-confidence, speed, and audacity. None of these things can ever be perfect, but they can be good.

General George S. Patton, Jr., *War As I Knew It*, 1947

War is always a matter of doing evil in the hope that some good may come of it.

Captain Sir Basil Liddell Hart, *Defence of the West*, 1950

War has changed little *in principle* from the beginning of recorded history. The mechanized warfare of today is only an evolution from the time when men fought with clubs and stones, and its machines are as nothing without the men who invent them, man them and give them life. War is force – force to the utmost – force to make the enemy yield to our own will – to yield because they see their comrades killed and wounded – to yield because their own will to fight is broken. War is men against men. Mechanized war is still against men, for machines are masses of inert metal without the men who control – or destroy them.

Admiral of the Fleet Ernest J. King, *Fleet Admiral King: A Naval Record*, 1952

As every combat veteran knows, war is primarily sheer boredom punctuated by moments of stark terror.

Colonel Harry G. Summers, Jr., *On Strategy*, 1981

WAR: The Next War

But since the discovery of gunpowder has changed the art of war, the whole system has, in consequence, been changed. Strength of body, the first quality among the heroes of antiquity, is at present of no significance. Strategem vanquishes strength, and art overcomes courage. The understanding of the general has more influence on the fortunate or unfortunate consequences of the campaign than the prowess of the combatants. Prudence prepares and traces the route that valour must pursue; boldness directs the execution, and ability, not good fortune, wins the applause of the well informed.

Frederick the Great (1712–1786), 'Reflections on Charles XII', *Posthumous Works of Frederick the Great*, tr. Holcroft, 1789, in ed. Luvaas, *Frederick the Great on the Art of War*, 1966

Every age has its own kind of war, its own limiting conditions and its own peculiar preconceptions.

Major General Carl von Clausewitz, *On War*, 1832, tr. Howard and Paret

The laws of successful war in one generation would ensure defeat in another.

General of the Army Ulysses S. Grant (1822–1885)

The old wars were decided by their episodes rather than by their tendencies. In this (modern) war the tendencies are far more important that the episodes. Without winning any sensational victories, we may win . . . Germany may be defeated more fatally in the second or third year of the war than if the Allied armies had entered Berlin in the first.

Sir Winston S. Churchill, November 1915, speech in the House of Commons

---

The idea that every war has been different from the last is the delusion of those who know no history. The next war has normally begun where the last left off, with, perhaps, a slight modification, governed not by the actual development of weapons in the interval but by such fractions of that development as have been recognized and incorporated by the peacetime armies. The generals, however, have usually begun more nearly where the last war began. In consequence, they are discomfited. And public opinion complains that they have stood still while warfare has changed. It fails to realize that the generals have moved – backwards. (April 1930)

Captain Sir Basil Liddell Hart, *Thoughts on War*, 1944

Organizations created to fight the last war better are not going to win the next.

General James M. Gavin (1907–1987)

. . . History knows many examples when the armed forces of individual countries prepared for a future war, taking into account only the clearly manifested relationships of the past war, without examining the dialectical relationships and their dependence on changes which are taking place in the world. Such an approach usually leads to major errors.

Marshal of the Soviet Union Nikolai V. Ogarkov, *History Teaches Vigilance*, 1985

WAR: Politics/Policy and War

A great country can have no such thing as a little war.

The Duke of Wellington, 1815

It is clear . . . that war is not a mere act of policy but a true political instrument, a continuation of political activity by other means. What remains peculiar to war is simply the peculiar nature of its means. War in general, and the commander in any specific instance, is entitled to require that the trend and designs of policy shall not be inconsistent with these means. That, of course, is no small demand; but however much it may affect political aims in a given case, it will never do more than modify them. The political object is the goal, war is the means of reaching it, and means can never be considered in isolation from their purpose.

Major General Carl von Clausewitz, *On War*, i, 1832, tr. Howard and Paret

---

The first, the supreme, the most far-reaching act of judgement that the statesman and commander have to make is to establish . . . the kind of war on which they are embarking; neither mistaking it for, nor trying to turn it into, something that is alien to its nature. This is the first of all strategic questions and the most comprehensive.

Major General Carl von Clausewitz, *On War*, i, 1832, tr. Howard and Paret

---

We argue that a commander-in-chief must also be a statesman, but he must not cease to be a general. On the one hand, he is aware of the entire political situation; on the other, he knows exactly how much he can achieve with the means at his disposal.

Major General Carl von Clausewitz, *On War*, i, 1832, tr. Howard and Paret

The degree of force that must be used against the enemy depends on the scale of political demands on either side. These demands so far as they are known, would show what efforts each must make; but they seldom are fully known – which may be one reason why both sides do not exert themselves to the same degree.

Major General Carl von Clausewitz, *On War*, viii, 1832, tr. Howard and Paret

---

The probable character and general shape of any war should mainly be assessed in the light of political factors and conditions.

Major General Carl von Clausewitz, *On War*, viii, 1832, tr. Howard and Paret

---

The only question . . . is whether . . . the political point of view should give way to the purely military (if a purely military point of view is conceivable at all) . . . Subordinating the political point of view to the military would be absurd, for it is policy that creates war. Policy is the guiding intelligence and war only the instrument, not vice versa. No other policy exists, then, than to subordinate the military point of view to the political.

Major General Carl von Clausewitz, *On War*, viii, 1832, tr. Howard and Paret

---

Policy is the intelligent faculty, war only the instrument, not the reverse. The subordination of the military view to the political is, therefore, the only thing possible.

Major General Carl von Clausewitz, *On War*, 1832

It is a senseless proceeding to consult the soldiers concerning plans for war in such a way as to permit them to pass purely military judgments on what the ministers have to do; and even more senseless is the demand of theoreticians that the accumulated war material should simply be handed over to the field commander so that he can draw up a purely military plan for the war or campaign.

Major General Carl von Clausewitz, *Kreig und Kriefuehrung*, 1857

Politics uses war for the attainment of its ends; it operates decisively at the beginning and at the end, of course in such manner that if refrains from increasing its demand during the war's duration or from being satisfied with an inadequate success . . . Strategy can only direct its efforts toward the highest goal which the means available make attainable. In this way, it aids politics best, working only for its objectives, but in its operations independent of it . . . There is uncertainty in war, but the aims of policy remain. Policy must go hand in hand with strategy.

Field Marshal Helmuth Graf von Moltke, *Ueber Strategie*, 1871

That the soldier is but the servant of the statesman, as war is but the instrument of diplomacy, no educated soldier will deny. Politics must always exercise an extreme influence on stragegy; but it cannot be gainsaid that interference with the commanders in the field is fraught with the gravest danger.

Colonel George F. R. Henderson, *Stonewall Jackson*, 1898

From Jomini also I imbibed a fixed disbelief in the thoughtlessly accepted maxim that the statesman and general occupy unrelated fields. For this misconception I substituted a tenet of my own, that war is simply a violent political movement.

Rear Admiral Alfred Thayer Mahan, *From Sail to Steam*, 1907

. . . politics is the reason, and war is only the tool, not the other way round. Consequently, it remains only to subordinate the military point of view to the political.

V. I. Lenin (1870–1924), *On War*

History causes the military problem to become the essence of the political problem.

V. I. Lenin, 1921, speech

Policy is a thought process, while war is merely a tool, and not the reverse.

V. I. Lenin (1870–1924), quoted in Gorshkov, *Sea Power of the State,* 1979

The state must determine the nature of overall, and particularly military, policy in advance, outline accordingly the possible objectives of its military goals, work out and establish a definite plan of statewide action which allows for future conflicts and makes timely provision for their succession with appropriate utilization of national energy.

Mikhail V. Frunze (1885–1925)

. . . War is not only a means of politics and is not limited only to combat operations. The Actions of the armed forces are accompanied by the organized and combined pressure and strikes on all fronts of the struggle (economic, political, etc.).

Marshal of the Soviet Union Mikhail N. Tukachesvskiy (1893–1937), quoted in Gareyev, *Frunze, Military Theorist,* 1985

War is not the continuation of policy. It is the breakdown of policy.

Colonel General Hans von Seekt, *Thoughts of a Soldier,* 1930

In the theatre of peacetime, it is the statesman who plays the chief role. Whether the masses greet him with applause or boos, it is for him first of all that they have eyes and ears. Then suddenly war calls another actor from the wings, pushes him to the middle of the stage, and trains the limelight on him: the military chief appears. A drama is about to begin which will be played by statesman and soldier in concert. No matter how great the crowd of extras, how noisy the audience, it is on these two performers that attention will be centered. So closely interwoven is their dialogue that nothing said by either has any relevance, point or effect except with reference to the other. If one of them misses his cue, then disaster overwhelms them both. However widely in fact the work of the civil government differs from that of the High Command, no one would seriously question the interdependence of the two authorities. What policy can hope to succeed if the country's arms are brought low? Of what use is strategical planning if the means of carrying it our are not forthcoming?

Charles de Gaulle, *The Edge of the Sword,* 1932

The national strategy of any war, that is, the selection of national objectives and the determination of the general means and methods to be applied in obtaining them, as well as development of the broad policies applicable to the prosecution of the war, are decisions that must be made by the head of state, acting in conformity with the expressed will of the Government . . .

General of the Army Douglas MacArthur, *Military Situation in the Far East,* presented to the 82nd Congress, 1932

There are some militarists who say: 'We are not interested in politics but only in the profession of arms.' It is vital that these simple-minded militarists be made to realize the relationship that exists between politics and military affairs. Military action is a method used to attain a political goal. While military affairs and political affairs are not identical, it is impossible to isolate one from the other.

Mao Tse-tung, *On Guerrilla Warfare,* 1937

. . . war has its own particular characteristics and in this sense it cannot be equated with politics in general . . . When politics develops to a certain stage beyond which it cannot proceed by the usual means, war breaks out to sweep the obstacles from the way . . . When the obstacle is removed and our political aim attained, the war will stop. But if the obstacle is not completely swept away, they will have to continue till the aim is fully accomplished . . . It can therefore be said that politics is war without bloodshed while war is politics with bloodshed.

Mao Tse-tung, *On Protracted War,* 1938.

A military man in the political world is a complete innocent. A military man turned politician never achieves anything worthwhile. The more confidence he has, the less he can do.

Admiral Yamamato Isoroku, 1938, quoted in Agawa, *The Reluctant Admiral*, 1979

When policy is too stiff, and its war-aims too absolute, strategy forfeits the power to keep its plans flexible, to adopt a new line when one is blocked, and to retrieve advantage from the mishaps inevitable in war. An uncompromising attitude hinders the essential adaptability of strategy to events as well as to conditions. (July 1939)

Captain Sir Basil Liddell Hart, *Thoughts on War*, 1944

All the reproaches that have been levelled against the leaders of the armed forces by their countrymen and by the international courts have failed to take into consideration one very simple fact: that policy is not laid down by soldiers, but by politicians. This has always been the case and is so today. When war starts, the soldiers can only act according to the political and military situation as it then exists. Unfortunately it is not the habit of politicians to appear in the conspicuous places when the bullets begin to fly. They prefer to remain in some safe retreat and to let the soldiers carry out 'the continuation of politics by other means'.

Colonel General Heinz Guderian, *Panzer Leader*, 1953

History shows that gaining military victory is not in itself equivalent to gaining the object of policy. But as most of the thinking about war has been done by men of the military profession there has been a very natural tendency to lose sight of the basic national object, and identify it with the military aim. In consequence, whenever war has broken out, policy has too often been governed by the military aim – and this has been regarded as an end in itself, instead of as merely a means to the end.

Captain Sir Basil Liddell Hart, *Strategy*, 1954

The soldier always knows that everything he does . . . will be scrutinized by two classes of critics – by the government that employs him and by the enemies of that Government.

Field Marshal Viscount Slim, *Unofficial History*, 1959

Politics, from an elevation of military and political factors, selects the most propitious moment to start a war, taking into account all the strategic considerations.

Marshal of the Soviet Union V. D. Sokolovskiy, *Soviet Military Strategy*, 1961

WAR: Protracted War

No, more, Polydamus! Your pleading repels me now.
Aren't you sick of being caged inside those walls?
Time was when the world would talk of Priam's Troy
as the city rich in gold and rich in bronze – but now
our houses are stripped of all their sumptuous treasure
troves, sold off and shipped to Phrygia, lovely Maeonia,
once great Zeus grew angry . . .

Hector in the tenth year of the Trojan War, Homer, *The Iliad*, xviii, tr. Fagles

---

For there has never been a protracted war from which a country has benefited.

Sun Tzu, *The Art of War*, c. 500 BC, tr. Griffith

The German people, both at home and at the front, have suffered and endured inconceivable hardships in the four long years of war. The war has undermined and disintegrated patriotic feeling and the whole national *morale*.

Poisonous weeds grew in this soil. All German sentiment, all patriotism, died in many breasts. Self came first. War profiteers of every kind, not excluding the political variety, who took advantage of the country's danger and the Government's weakness to snatch political and personal advantages, became more and more numerous. Our resolution suffered untold harm. We lost confidence in ourselves.

General Erich Ludendorff, *My War Memories, 1914–1918*, 1919

But once war is forced upon us, there is no other alternative than to apply every available means to bring it to swift end. War's very object is victory, not prolonged indecision.

General of the Army Douglas MacArthur, 19 April 1951, Address to Congress

---

The experience of history brings ample evidence that the downfall of civilized states tends to come not from the direct assaults of foes but from internal decay, combined with the consequences of exhaustion in war. A state of suspense is trying – it has often led nations as well as individuals to commit suicide because they were unable to bear it. But suspense is better than to reach exhaustion in pursuit of the mirage of victory. Moreover, a truce to actual hostilities enables a recovery and development of strength, while the need for vigilance helps keep a nation on 'its toes'.

Captain Sir Basil Liddell Hart, *Strategy*, 1954

WAR: Results of War

So the voice of the king rang out in tears,
the citizens wailed in answer, and noble Hecuba
led the wives of Troy in a throbbing chant of sorrow:
'O my child – my desolation! How can I go on living?
What agonies must I suffer now, now *you* are dead and gone?
You were my pride throughout the city night and day –
a blessing to us all, the men and women of Troy:
throughout the city they saluted you like a god.
You, you were the greatest glory while you live –
now death and fate have seized you, dragged you down!'

Homer, *The Iliad*, xxii, *c.* 800 BC, tr. Fagles

The end and perfection of our victories is to avoid the vices and infirmities of those whom we subdue.

Alexander the Great (356–323 BC), quoted in Plutarch (after AD 119), *The Lives of the Noble Grecians and Romans*

Weapons are instruments of ill omen, war is immoral.

Really they are only to be resorted to when there is no other choice. It is not right to pursue aggressive warfare because one's country is large and prosperous, for this ultimately ends in defeat and destruction. Then it is too late to have regrets. Military action is like a fire – if not stopped it will burn itself out. Military expansion and adventurism soon lead to disaster.

The rule is 'Even if a country is large, if it is militaristic it will eventually perish.'

Liu Ji (AD 1310–1375), *Lessons of War*

I hope to God that I have fought my last battle. It is a bad thing to be always fighting. While in the thick of it I am too much occupied to feel anything; but it is wretched just after. It is quite impossible to think of glory. Both mind and feelings are exhausted. I am wretched even at the moment of victory, and I always say that, next to a battle lost, the greatest misery is a battle gained. Not only do you lose those dear friends with whom you have been living, but you are forced to leave the wounded behind you. To be sure, one tries to do the best for them, but how little that is! At such moments every feeling in our breast is deadened. I am now just begining to regain my natural spirits, but I never wish for any more fighting.

The Duke of Wellington, 1815, to Lady Shelley after Waterloo

. . . It is not enough that our armies are victorious, that our enemies vanish from our territory, nor that the whole world recognizes our independence; we need something more: to be free under the auspices of liberal laws, emanating from the most sacred spring, which is the will of the people.

General Simon Bolivar, 1818

It is painful enough to discover with what unconcern they speak of war and threaten it. I have seen enough of it to make me look upon it as the sum of all evils.

Lieutenant General Thomas 'Stonewall' Jackson, April 1861, letter

---

What a cruel thing is war: to separate and destroy families and friends, and mar the purest joys and happiness God has granted us in this world; to fill our hearts with hatred instead of love for our neighbors, and to devastate the fair face of this beautiful world.

General Robert E. Lee, 25 December 1862, letter to his wife

I begin to regard the death and mangling of a couple thousand men as a small affair, a kind of morning dash – and it may well be that we become so hardened.

General of the Army William T. Sherman, July 1864, letter to his wife

---

The legitimate object of war is a more perfect peace.

General of the Army William T. Sherman, 20 July 1865

Eternal peace is a dream, and not even a good dream, for war is part of God's world ordinance. In war, the noblest virtues flourish that otherwise would slumber and decay – courage and renunciation, the sense of duty and sacrifice, even to the giving of one's life. The experience of war stays with a man, and steels him all his life.

Field Marshal Helmuth Graf von Moltke (1800–1891)

A war, even the most victorious, is a national misfortune.

Field Marshal Helmuth Graf von Moltke, 1880, letter

War loses a great deal of its romance after a soldier has seen his first battle.

Colonel John S. Mosby, *War Reminiscences*, 1887

War, with its many acknowledged sufferings, is above all harmful when its cuts a nation off from others and throws it back upon itself.

Rear Admiral Alfred Thayer Mahan, *The Influence of Sea Power Upon History*, 1890

I am not worried about the war; it will be difficult but we shall win it; it is after the war that worries me. Mark you, it will take years and years of patience, courage, and faith.

Field Marshal Jan Christian Smuts (1870–1950)

War stirs in men's hearts the mud of their worst instincts. It puts a premium on violence, nourishes hatred, and gives free rein to cupidity. It crushes the weak, exhalts the unworthy, and bolsters tyranny. Because of its blind fury many of the noblest schemes have come to nothing and the most generous instincts have more than once been checked. Time and time again it has destroyed all ordered living, devastated hope, and put the prophets to death. But, though Lucifer has used it for his purposes so, sometimes, has the Archangel. With what virtues has it not enriched the moral capital of mankind! Because of it, courage, devotion and nobility have scaled the peaks. It has conferred greatness of spirit on the poor, brought pardon to the guilty, revealed the possibilities of self-sacrifice to the commonplace, restored honour to the rogue, and given dignity to the slave. It has carried ideas in the baggage wagons of its armies, and reforms in the knapsacks of its soldiers. It has blazed a trail for religion and spread across the world influences which have brought renewal to mankind, consoled it, and made it better. Had not innumerable soldiers shed their blood there would have been no Hellenism, no Roman civilization, no Christianity, no Rights of Man, and no modern developments.

Charles de Gaulle, *The Edge of the Sword*, 1932

Mussolini said in the early 1930s: 'War alone brings all human energies to their highest tension, and sets a seal of nobility on the people who have the virtue to face it.' This is rubbish, and dangerous rubbish at that. War does not ennoble. Kant's view that war had made more bad people than it has destroyed is probably nearer the mark. But the interesting thing is that although war almost certainly does not ennoble, the preparation of men to fight in it almost certainly can and very often does.

General Sir John Hackett, *The Profession of Arms*, 1863

WEAPONS

Weapons are ominous tools to be used only when there is no alternative.

Sun Tzu, *The Art of War*, c. 500 BC, tr. Griffith

The sword is the soul of the warrior. If any forget or lose it he will not be excused.

Tokugawa Ieyasu (1543–1616), quoted in Sadler, *The Maker of Modern Japan: The Life of Tokugawa Ieyasu*

Five things there are from which the soldier must never be separated: his gun, his cartridges, his field pack, his rations for at least four days, and his digging tool.

Napoleon, dictation at St. Helena, quoted in ed., Herold, *The Mind of Napoleon*, 1955

The means of destruction are approaching perfection with frightful rapidity.

Lieutenant General Antoine-Henri Baron de Jomini, *Summary of the Art of War*, 1838

New weapons would seem to be regarded merely as an additional tap through which the bath of blood can be filled all the sooner.

Captain Sir Basil Liddell Hart, *Paris, Or the Future of War*, 1925

The moral and material are interdependent. Weapons without courage are ineffective, but so are the bravest troops without sufficient weapons to protect them and their morale. Courage soon oozes when soldiers lose confidence in their weapons. (December 1929)

Captain Sir Basil Liddell Hart, *Thoughts on War*, 1944

No new weapons can be introduced without changing conditions, and every change in condition will demand a modification in the application of the principles of war.

Major General J. F. C. Fuller, *Armoured Warfare*, 1943

There are two universal and important weapons of the soldier which are often overlooked – the boot and the spade. Speed and length of marching has won many victories; the spade saved many defeats and gained time for victory.

Field Marshal Earl Wavell, *The Good Soldier*, 1945

A careful study of past military history and particularly of the 'little picture' of our own infantry operations in the past war leads to the conclusion that weapons when correctly handed in battle seldom fail to gain success. There is no other touchstone to tactical success, and it is a highly proper doctrine which seeks to ingrain in the infantry soldier a confidence that superior use of superior weapons is the surest protection.

Brigadier General S. L. A. Marshall, *Men Against Fire*, 1947

WEATHER

Climate is what you expect but weather is what you get.

Anonymous

It is always necessary to shape operation plans . . . on estimates of the weather, and as this is always changing, one cannot imitate in one season what has turned out well in another.

Frederick the Great, *Instructions to His Generals*, 1747

Complaints have been brought to my attention that the infantry have got their feet wet. That is the fault of the weather. The march was made in the service of the most mighty monarch. Only women, dandies and lazybones need good weather.

Field Marshal Prince Aleksandr V. Suvorov, 1799, to the commander of the Austrian allied force who complained of having to march in bad weather, quoted in Longworth, *The Science of Victory*, 1966

I cannot command winds and weather.

Admiral Lord Nelson, April 1796

. . . the Russian land which he dreamed of enslaving will be strewn with his bones. We will pursue tirelessly. Let winter, blizzards, and the cold come. Will you fear them, children of the north? Your iron breast fears neither the severity of weather nor the malice of enemies. It is the reliable wall of the homeland, against which all collapses.

Field Marshal Prince Mikhail I. Kutuzov, 1812, exhorting his troops during Napoleon's invasion of Russia, *Sbornik dokumentov* (*A Collection of Documents*), 1955

Weather is not only to a great extent a controller of the condition of ground, but also of movement. It is scarcely necessary to point out the influence of heat and cold on the human body, or the effect of rain, fog, and frost on tactical and administrative mobility; but it is necessary to appreciate the moral effect of weather and climate, for in the past stupendous mistakes have resulted through deficiency in this appreciation.

Unless the headquarters staff have intimate experience of the conditions surrounding the fighters, two types of battle are likely to be waged – the first between the brains of the army and the enemy, in which case this action will be rendered impotent on account of the muscles being unable to execute the commands of the brains; and the second between the muscles and the enemy, which battle will be disorganized, not so much through the enemy's opposition as through the receipt of orders which are impossible to carry out.

Major General J. F. C. Fuller, *The Foundation of the Science of War*, 1926

The Admiral cannot take up a position that only in ideal conditions of tide and moon can the operation be begun. It has got to be begun as soon as possible, as long as conditions are practicable, even though they are not the best. People have to fight in war on all sorts of days, and under all sorts of conditions.

Sir Winston S. Churchill, note to General Ismay regarding the Dakar operation, 19 August 1940

This is sheer torture for the troops, and for our cause it is a tragedy, for the enemy is gaining time, and in spite of all our plans we are being carried deeper into winter. It really makes me sad. The best of intentions are wrecked by the weather. The unique opportunity to launch a really great offensive recedes further and further, and I doubt if it will ever recur. God alone knows how things will turn out. One must just hope and keep one's spirits up, but at the moment it is a great test.

Colonel General Heinz Guderian, November 1941, quoted in von Mellenthin, *German Generals of World War II*, 1977

WHEN TO FIGHT

To refrain from intercepting an enemy whose banners are in perfect order, to refrain from attacking an army drawn up in calm and confident array – this is the art of studying circumstances.

Sun Tzu, *The Art of War*, c. 500 BC, tr. Griffith

Now there are five circumstances in which victory may be predicted:

o He who knows when he can fight and when he cannot will be victorious.

o He who understands to use both large and small forces will be victorious.

o He whose ranks are united in purpose will be victorious.

o He who is prudent and lies in wait for an enemy who is not will be victorious.

o He whose generals are able and not interfered with by the sovereign will be victorious.

It is in these five matters that the way to victory is known.

Therefore I say: 'Know the enemy and know yourself'; in a hundred battles you will never be in peril.

When you are ignorant of the enemy but know yourself, your chances of winning or losing are equal.

If ignorant both to your enemy and of yourself, you are certain in every battle to be in peril.

Sun Tzu, *The Art of War*, c. 500 BC, tr. Griffith

If I know that my troops are capable of striking the enemy, but do not know that he is invulnerable to attack, my chance of victory is but half.

If I know that the enemy is vulnerable to attack, but do not know that my troops are incapable of striking him, my chance of victory is but half.

If I know that the enemy can be attacked and that my troops are capable of attacking him, but do not realize that because of the conformation of the ground I should not attack, my chance of victory is but half.

Therefore when those experienced in war move they make no mistakes; when they act, their resources are limitless.

And therefore I say: 'Know the enemy, know yourself'; your victory will never be endangered. Know the ground, know the weather; your victory will then be total.

Sun Tzu, *The Art of War*, c. 500 BC, tr. Griffith

Warfare is like hunting. Wild animals are taken by scouting, by nets, by lying in wait, by stalking, by circling around, and by other such strategems rather than by sheer force. In waging war we should proceed in the same way, whether the enemy be many or few. To try simply to overpower the enemy in the open, hand to hand and face to face, even though you might appear to win, is an enterprise which is very risky and can result in serious harm. Apart from extreme emergency, it is ridiculous to try to gain victory which is so costly and brings only empty glory.

The Emperor Maurice, *The Strategikon*, c. AD 600

Good officers never engaged in general actions unless induced by opportunity or obliged by necessity.

The Emperor Maurice, *The Strategikon*, c. AD 600

I think I also have something left to say relating to the precautions a general should take before he leads his army to battle; in the first place, it is my opinion that he should never engage unless he has an advantage over the enemy, or he is compelled to act. Now the advantage may arise from the terrain or from the order, superiority, or bravery of his army; he may be compelled to engage by a conviction that if he does not, he must inevitably be ruined.

Niccolo Machiavelli, *The Art of War*, 1521

The different circumstances under which a battle should be avoided or declined are, when there is greater danger to be apprehended from defeat than advantage to be derived from victory; when you are very inferior to your enemy in numbers, and are expecting reinforcements; above all, when the enemy is advantageously posted, or when he is contributing to his own ruin by some inherent defect in positions or by the errors and divisions of his generals.

Field Marshal Count Raimond Montecuccoli (1609–1680)

Decline the attack altogether unless you make it with advantage.

Field Marshal Maurice Comte de Saxe, *My Reveries*, 1732

If . . . you are inferior in numbers do not despair of winning, but do not expect any other success than that gained by your skill. It is necessary to seek mountainous country and use artifices, so that if you were to be forced to battle, the enemy would not be able to face you with a front superior to your own, and so that you may be able definitely to protect your flanks.

Frederick the Great, *Instructions for His Generals*, 1747

When you determine to risk a battle, reserve to yourself every possible chance of success, more particularly if you have to deal with an adversary of superior talent, for if you are beaten, even in the midst of your magazines and your communications, woe to the vanquished!

Napoleon, *The Military Maxims of Napoleon*, 1831, tr. D'Aguilar

---

If a solution cannot be reached without battle, this imposes on you the strategic aim to force battle at the time, and under the conditions, most favourable tactically to yourself.

Rear Admiral Alfred Thayer Mahan (1840–1914)

WILL

In battle, two moral forces, even more than two material forces, are in conflict. The stronger conquers. The victor has often lost . . . more men than the vanquished . . . With equal or even inferior power of destruction, he will win who is determined to advance.

Colonel Charles Ardnant du Picq, *Battle Studies*, 1880

The will to conquer: such is the first condition of victory, consequently the first duty of every soldier; and it is also the supreme resolution with which the commander must fill the soul of his subordinates.

Marshal of France Ferdinand Foch, *Principles of War*, 1913

I have been awarded the Iron Cross for my modest share in the battle of Tannenberg. I had never thought that this finest of all military decorations would won by sitting at the end of a telephone line. However, I realize now that there must be someone there who keeps his nerve, and by brute determination and will to victory overcomes difficulties, panics and suchlike nonsense.

General Max Hoffmann, 9 September 1914, *War Diaries and other Papers*, Vol I, 1929

But in war it often happens that the impossible is in the end made possible by the might of one's utmost efforts and a steadfast will.

General Aleksei A. Brusilov, September 1914, *A Soldier's Notebook*, 1931

A battle won is a battle in which one will not confess himself beaten.

Marshal of France Ferdinand Foch, *Precepts and Judgments*, 1919

In war the chief incalculable is the human will.

Captain Sir Basil Liddell Hart, *Encyclopaedia Britannica*, 1929

Hindenburg was right when he said that the man who wins a war is the one whose nerves are the strongest. Our nerves proved particularly feeble, because we had so long been forced by the inadequacy of our technical equipment to make up the deficit by an excessive wastage of life. One cannot fight successfully with bare hands against an enemy provided with all the resources of modern warfare and inspired by patriotism.

General Aleksei A. Brusilov, *A Soldier's Notebook*, 1931

. . . in any war victory is determined in the final analysis by the state of mind of the masses which shed their blood on the field of battle.

Mikhail V. Frunze (1885–1925)

It is not simply the weapons one has in one's arsenal that give one flexibility, but the willingness and ability to use them.

Mao Tse-tung (1893–1976)

. . . If we clear the air of the fog of catchwords which surround the conduct of war, and grasp that in the human will lies the source and mainspring of all conflict, as of all other activities of man's life, it becomes clear that our object in war can only be attained by the subjugation of the opposing *will*. All *acts*, such as defeat in the field, propaganda, blockade, diplomacy, or attack on the centres of government and population, are seen to be but means to that end . . .

Captain Sir Basil Liddell Hart, *Thoughts on War*, 1944

The will does not operate in a vacuum. It cannot be imposed successfully if it runs counter to reason. Things are not done in war primarily because a man wills it; they are done because they are do-able. The limits for the commander in battle are defined by the general circumstances. What he asks of his men must be consistent with the possibilities of the situation.

Brigadier General S. L. A. Marshall, *Men Against Fire*, 1947

What can be successfully willed must first be clearly seen and understood. If amid the confusions of battle the commander sees what is required by the situation, if amid the miscarriage of arrangements and the assailing doubts of other men he measures the means of doing it, and if he then gives his order and holds his men to their duty, this is the ultimate triumph of the will on the field of battle. To reflect on this thought is to note that he exercises his will far less upon his men than upon himself.

Brigadier General S. L. A. Marshall, *Men Against Fire*, 1946

For power of will comes from the exercise of it. In the military commander strength of will does not automatically flourish apace with the growth of knowledge in the material things of war. But unless it does so, experience is had to no broad purpose and knowledge does not become wisdom. Equally, if the will power of the commander is in excess of his knowledge, it cannot be exercised for the good of the commanded. Thus, true strength of will in the commander develops from his study of human nature, for it is in the measure that he acquires knowledge of how other men think that he perfects himself in the control of their thoughts and acts.

Brigadier General S. L. A. Marshall, *Men Against Fire*, 1947

It is fatal to enter any war without the will to win it.

General of the Army Douglas MacArthur, 7 July 1952, Address to the Republican National Convention

---

. . . Your job as an officer is to make decisions and to see them carried out; to force them through against the opposition not only of the enemy, that is fair enough, but against that of your own men. Against colleagues who want it done another way, of allies – there is only one thing worse than having allies and that is not to have them – and of all opposition of man and nature that will bar your way. You cannot be a leader at all without this strength of will, this determination.

Field Marshal Viscount Slim, *Courage and Other Broadcasts*, 1957

WITHDRAWAL

. . . If we happen to be too weak for attack, then we strive to detach ourselves from the embraces of the enemy in order later to gather ourselves into a fist and to strike at the enemy's most vulnerable spot.

Leon Trotsky, 1921, *Military Writings of Leon Trotsky*, 1969

The withdrawal should be thought of as an offensive instrument, and exercises be framed to teach how the enemy can be lured into a trap, closed by counterstroke or devastating circle of fire.

Sir Basil Liddell Hart, *Thoughts on War*, 1944

Of all operations of war, a withdrawal under heavy enemy pressure is probably the most difficult and perilous. Indeed it is recorded of the great Moltke, that when he was praised for his generalship in the Franco–Prussian War, and was told by an admirer that his reputation would rank with such great captains as Napoleon, Frederick, and Turenne, he answered, 'No, for I have never conducted a retreat'.

Major General F. W. von Mellenthin, *Panzer Battles*, 1971

WIVES AND MOTHERS

She . . . ran out of the house like a raving woman
with pulsing heart, and her two handmaidens went along with her.
But when she came to the bastion and where the men were gathered
she stopped, staring, on the wall; and she saw him
being dragged in front of the city, and the running horses
dragged him at random toward the hollow ships of the Achaeans.
The darkness of night misted over the eyes of Andromache.
She fell backward, and gasped the life from her, and far off
threw from her head the shining gear that ordered her head-dress,
the diadem and the cap, and the holding-band woven together,
and the circlet, which Aphrodite the golden once had given her
on that day when Hektor of the shining helmet led her forth
from the house of Eetion, and gave numberless gifts to win her.

Homer, *The Iliad*, c. 800 BC, tr. Lattimore

A woman who hampers a man at the beginning of his career is a hateful abomination, and he always thinks so sooner or later. So you had best not consider me at all in making your decision [to stay in the Service]. The family might not thank me for telling you this but I think it . . . A girl might just ruin a man's life by upsetting it at the beginning. You can decide better if you consider your self as one instead of as two. You must decide alone and then I will go with you *any* where.

The future Mrs. George S. Patton, Jr., Beatrice Ayer, quoted by Patton, Jr. in a letter to his parents, 17 January 1909, *The Patton Papers*, Vol. I, 1972

Glory of Women

You love us when we're heroes, home on leave,
Or wounded in a mentionable place.
you worship decorations; you believe
That chivalry redeems the war's disgrace.
you make us shells. You listen with delight,
By tales of dirt and danger fondly thrilled.
You crown our distant ardours while we fight,
And mourn our laurelled memories when we're killed.
You can't believe that British troops 'retire'
When hell's last horror breaks them, and they run,
Trampling the terrible corpses – blind with blood.
O German mother dreaming by the fire,
While you are knitting socks to send your son
His face trodden deeper in the mud.

Siegfried Sassoon (1886–1967)

. . . Among women, too, it is the young of whom most is demanded. The lot of mothers who lose their sons is hard, but doubly hard is that of the young women who lose the chance of ever becoming wives and mothers. (February 1939)

Captain Sir Basil Liddell Hart, *Thoughts on War*, 1944

July 14, 1861
Camp Clark, Washington

My very dear Sarah:

The indications are very strong that we shall move in a few days – perhaps tomorrow, and lest I should not be able to write again, I feel impelled to write a few lines that may fall under your eye when I shall be no more . . .

I have no misgivings about, or lack of confidence in the cause in which I am engaged, and my courage does not halt or falter. I know how American Civilization now leans on the triumph of the Government, and how great a debt we owe to those who went before us through the blood and suffering of the Revolution. And I am willing – perfectly willing – to lay down all my joys in this life, to help maintain this Government, and to pay that debt . . .

Sarah my love for you is deathless, it seems to bind me with mighty cables that nothing but Omnipotence could break; and yet my love of Country comes over me like a strong wind and bears me unresistably with all these chains to the battle field.

The memories of the blissful moments I have enjoyed with you come crowding over me, and I feel most deeply grateful to God and to you that I have enjoyed them so long. And how hard it is for me to give them up and burn to ashes the hopes of future years, when, God willing, we might still have lived and loved together, and see our boys grown up to honorable manhood, around us. If I do not return, my dear Sarah, never forget how much I loved you, nor when my last breath escapes me on the battle field, it will whisper your name. Forgive my many faults, and the many pains I have caused you. How thoughtless, how foolish I have often times been!

But, O Sarah! if the dead can come back to this earth and flit unseen around those they love, I shall always be with you; in the brightest day and the darkest night . . . *always, always*; and when the soft breeze fans your cheek, it shall be my breath, or the cool air your throbbing temple, it shall be my spirit passing by. Sarah do not mourn me dead; think I am gone and wait for me, for we shall meet again . . .

Major Sullivan Ballou, 2nd Rhode Island Infantry, wrote this letter to wife in Smithfield a week before his death at the battle of First Bull Run, quoted in Burns, *The Civil War*, 1990.

Wait for me

Wait for me, and I'll return.
Only wait very hard . . .
Wait, when you are filled with sorrow
As you watch the yellow rain,
Wait, when the wind sweeps the snowdrifts,
Wait, in the sweltering heat,
Wait, when others have stopped waiting,
Forgetting their yesterdays.
Wait, even when from afar no letters come to you,
Wait, even when others are tired of waiting,
Wait, even when my mother and son think I am no more,
And when friends sit around the fire drinking to my memory,
Wait, and do not hurry to drink to my memory too;

Wait for I'll return,
Defying every death,
And let those who do not wait
Say I was lucky.
They will never understand
That in the midst of death,
You with your waiting saved me.
Only you and I will know how I survived, –
It's because you waited,
As no one else did.

Konstantin M. Simonov, *Wait for Me*, c. 1943 – the most famous Russian poem of World War II.

WOUNDED

Alexander had been hurt by a sword-thrust in the thigh, but his did not prevent him from visiting the wounded on the day after the battle, when he also gave splendid funeral to the dead in the presence of the whole army paraded in full war equipment.

Flavius Arrianus Xenophon (Arrian), of Alexander the Great after the Battle of Issus, 333 BC, *The Campaigns of Alexander*, c. AD 150

After the battle the general should give prompt attention to the wounded and see to burying the dead. Not only is this a religious duty, but it greatly helps the morale of the living.

The Emperor Maurice, *The Strategikon*, c. AD 600

It is right to be very concerned about the wounded. If we neglect them, we will find that the rest of the troops will deliberately not fight well, and our remissness will cause us to lose some who could have been saved.

The Emperor Maurice, *The Strategikon*, c. AD 600

On this day the Turks made four assaults . . . They made all these assaults hoping to wear out our small force . . . During these engagements, many convalescents, although not yet fully recovered, used to come to the works and helped as best they could, because, like men of spirit, they preferred to die fighting rather than be cruelly butchered in their quarters if it were our misfortune that the Turks should win.

Francesco Balbi de Correggio, 1565, at the Siege of Malta

Think also of the poor wounded of both armies. Especially have paternal care for your own and do not be inhuman to those of the enemy.

Frederick the Great, *Instructions to His Generals*, 1747

There is nothing which gains an officer the love of his soldiers more than his care of them under the distress of sickness; it is then he has the power of exerting his humanity in providing them every comfortable necessary, and making their situation as agreeable as possible.

Major General Friedrich Baron von Steuben, *Revolutionary War Drill Manual*, 1794

Pay every attention to the sick and wounded. Sacrifice your baggage, everything for them. Let the wagons be devoted to their use, and if necessary your own saddles . . .

Napoleon (1769–1821), quoted in Chandler, *The Campaigns of Napoleon*, 1966

My wounded are behind me, and I will never pass them alive.

Major General Zachary Taylor, 22 February 1847, at the suggestion of retreat at Buena Vista

---

It is more fruitful to wound than to kill. While the dead man lies still, counting only one man less, the wounded man is a progressive drain upon his side. Comrades are often called upon to bandage him, sometimes even to accompany him back; stretcher-bearers and ambulance drivers to carry him back; doctors and orderlies to tend him in hospital. And on his passage thither the sight of him tends to spread depression among the beholders, acting on morale like the drops of cold water which imperceptibly wear away the stone. (April 1930)

Captain Sir Basil Liddell Hart, *Thoughts on War*, 1944

Men, all I can say is, if I had been a better general, most of you would not be here.

General George S. Patton, Jr., 1945, to wounded soldiers at Walter Reed Army Hospital in Washington, DC

I feel like a fugitive from th' law of averages.

William 'Bill' Mauldin, comment from one of his G.I. veterans in a cartoon, *Up Front*, 1944

During battle, it is very important to visit frequently hospitals containing newly wounded men. Before starting such an inspection, the officer in charge of the hospital should inform the inspecting general which wards contain men whose conduct does not merit compliments.

General George S. Patton, Jr., *War As I Knew It*, 1947

I realized vividly now that the real horrors of war were to be seen in hospitals, not on the battlefield.

Lieutenant General Sir John Glubb, *Into Battle: A Soldier's Diary of the Great War*, 1978

WRITING MILITARY HISTORY

I shall be content if those shall pronounce my History useful who desire to have a view of events as they did really happen, and as are very likely, in accordance with human nature, to repeat themselves at some future time – if not exactly the same, yet very similar.

Thucydides, *The Peloponnesian War*, c. 404 BC, tr. Jowett

Few persons will understand me, but I write for the connoisseurs, trusting that they will not be offended by the confidence of my opinions.

Field Marshal Maurice Comte de Saxe, *My Reveries*, 1732

Experts should not be offended by the assurance with which I deliver my opinions. They should correct them; that is the fruit I expect from my work.

Field Marshal Maurice Comte de Saxe, *My Reveries*, 1732

He could not tell the whole truth without hurting the feelings of many worthy men, and without doing mischief. Expatiating on the subject, he related many anecdotes illustrating this observation, showing errors committed by generals and others – especially at Waterloo – errors so materially affecting his operations that he could not do justice to himself if he suppressed them, and yet by giving them publicity he would ungraciously affect the favour of many worthy men, who only fault was dullness.

General Sir William Napier, relating the Duke of Wellington's reluctance to write a history of his campaigns, quoted in Oman, *Wellington's Army 1809–1814*, 1913

My ambition was to write a book that would not be forgotten in two or three years, and which anyone interested in the subject would certainly take up more than once.

Major General Carl von Clausewitz *On War*, 1832, tr. Howard and Paret

---

If a few prejudiced men, after reading this book and carefully studying the detailed and correct history of the campaigns of the great masters of the art of war, still contend that it has neither principles nor rules, I can only pity them and reply, in the famous words of Frederick, 'A mule which had made twenty campaigns under Prince Eugene would not be a better tactician than at the beginning'.

Lieutenant General Antoine-Henri Baron de Jomini, *Summary of the Art of War*, 1838

I want to tell the story *professionally* & to comment freely on every professional feature as one w[oul]d comment on moves of chess.

Brigadier General Edward Porter Alexander, letter, 1901, referring to his writing of *Military Memoirs of a Confederate: A Critical Narrative*, 1907

He who seeks, finds, if he does not lose heart; and to me, continuously seeking, came from within the suggestion that control of the sea was an historic factor which had never been systematically appreciated and expounded. Once formulated consciously, this thought became the nucleus of all my writing for twenty years to come . . .

Rear Admiral Alfred Thayer Mahan, *From Sail to Steam,* 1907

The military historian who is instructed in the principles of the art of war finds, as it were imposed upon him, the necessity of so constructing his narrative as to present a substantial unity in effect. Such familiar phrase as the 'key of the situation', the decisive point for which he has been taught to look, upon the tenure of which depends more or less the fortune of war, sustains continually before his mind the idea, to which his treatment must correspond, of a central feature round which all else groups; not only subordinate but contributive. Here is no vague collocation of words, but the concrete, pithy expression of a trained habit of mind which dominates writing necessarily, even though unconsciously to the writer.

Rear Admiral Alfred Thayer Mahan, *Naval Administration and Warfare,* 1908

It is not the colouring, but in the grouping, that the true excellence of the military historian is found; just as the battle is won, not by the picturesqueness of the scene, but by the disposition of the forces. Both the logical faculty and the imagination contribute to his success, but the former much exceeds the latter in effect. A campaign, or a battle, skilfully designed, is a work of art, and duly to describe it requires something of the appreciation and combinative faculty of an artist; but where there is no appeal over the imagination, to the intellect, impressions are apt to lack distinctness. While there is a certain exaltation in sharing, through vivid narrative, the emotions of those who have borne a part in some deed of conspicuous daring, the fascination does not equal that wrought upon the mind as it traces the sequence by which successive occurrences are seen to issue in their necessary results, or causes apparently remote converge upon a common end.

Rear Admiral Alfred Thayer Mahan, *Naval Administration and Warfare,* 1908

. . . in war time no soldier is free to say what he thinks; after a war no one cares what a soldier thinks.

General Sir Ian Hamilton, *Ian Hamilton's Dispatches From the Dardenelles,* 1917

If truth is many-sided, mendacity is many-tongued.

History cannot proceed by silences. The chronicler of ill-recorded times has none the less to tell his tale. If facts are lacking, rumour must serve. Failing affidavits, he must build with gossip.

Sir Winston S. Churchill, *Marlborough,* 1933

After twenty years of such work pure documentary history seems to me akin to mythology. For those who still put their faith in it, here is a short story with a moral. When the British front was broken in March, 1918, and French reinforcements came to help in filling the gap, an eminent French general arrived at a certain army corps headquarters, and there majestically dictated orders giving the line on which his troops would stand that night and start their counter-attack in the morning. After reading it, with some perplexity, the British corps commander exclaimed, 'But that line is behind the German front; you lost it yesterday.' To which he received the reply, made with a knowing smile: 'C'est pour l'histoire.'

Captain Sir Basil Liddell Hart, *Through the Fog of War,* 1938

Deep is the gulf between history and historic truth, and perhaps never more so than in books dealing with military history. If one reason is that they are usually written by soldiers untrained as historians, and another that there is frequently some personal link, whether of acquaintance or tradition, between author and subject, a deeper reason lies in a habit of mind. For the soldier, 'My country, right or wrong', must be the watchword. And this essential loyalty, whether it be to a country, to a regiment or to comrades, is so ingrained in him that when he passes from action to reflection it is difficult for him to acquire instead the historian's single-minded loyalty to truth.

Captain Sir Basil Liddell Hart, *Through the Fog of War,* 1938

The military history of a war written shortly after the end of that war is generally unsatisfactory because the silence demanded by considerations of security outweigh the benefit to be derived from publication – even if the scope is limited to recording the lessons learned. If military history is to have any real scholarly value from which something may be learned, it must be comprehensive, exhaustive, complete and objective. Writing that is distorted by bias or subjectivity, or is unable – for security reasons – to give all the facts, will not be of lasting value nor will it make a significant contribution to military science, though it may arouse ephemeral interest.

General Yigal Allon, *The Making of Israel's Army,* 1960

. . . most battles are more like a schoolyard in a rough neighborhood at recess time than a clash between football giants in the Rose Bowl. They are messy, inorganic, and uncoordinated. It is only much later, after the clerks have tidied up their reports and the generals have published their memoirs, that the historian with his orderly mind professes to discern an understandable pattern in what was essentially catch-as-catch-can, if not chaotic, at the time.

Brigadier General S. L. A. Marshall, *Ambush,* 1969

. . . my official duties required me to get a full and final accounting of what had happened on that field.

It was done by assembling the survivors of every unit that had fought, interviewing them as a group, and recording their experiences personal and in common from the beginning of movement till the end of fighting.

The method of reconstructing what develops in combat, relating cause to effect, and eliminating the fog, is my own . . . It works because it is simple and because what one man remembers will stir recall in another. The one inviolable rule, if such group interviewing is to get valid results, is that the question-and-answer routine must be in sequence step-by-step from first to last.

Brigadier General S. L. A. Marshall, *The Fields of Bamboo,* 1971

YOUTH VERSUS AGE

Neither too young nor too old; since the young man does not inspire confidence, the old man is feeble, and neither is free from danger, the younger man lest he err through reckless daring, the older lest he neglect something through physical weakness. The ideal lies between the two, for physical vigour is found in the man who has not yet grown old, and discretion in the man who is not too young. Those who value physical strength without discretion, or discretion without physical strength, have failed to accomplish anything. For a weak mind can contribute no valuable ideas, nor can strength unsupported bring to completion any activity.

Onasander, *The General,* c. AD 58

I consider it a great advantage to obtain command young, having observed as a general thing that persons who come into authority late in life shrink from responsibility, and often break down under its weight.

Admiral David G. Farragut, 1819, journal entry

. . . boldness grows less common in the higher ranks . . . Nearly every general known to us from history as mediocre, even vacillating, was noted for dash and determination as a junior officer.

Major General Carl von Clausewitz, *On War,* 1832

. . . I told him that our real difficulty was not man-power, but brain-power, and that so long as our Generals would think in terms of 1870, there could be no progress.

'1870?' he queried after a long pause, '1870?' – they must be very old men.'

'Very old,' I answered, 'from their knowledge of warfare, most of them might have fought at the battle of Hastings.

Major General J. F. C. Fuller, 1918, to Sir Henry Wilson

Long training tends to make a man ever more expert in execution, but that skill may be gained at the expense of fertility of ideas, originality of conception and elasticity of views. It is too hopeful to expect generals whose minds have become set to make effective use of new weapons or tactics. Youth is surprise and surprise is war. (March 1923)

Captain Sir Basil Liddell Hart, *Thoughts on War*, 1944

. . . the question of age v. youth. The advocates of the claims of age and experience cannot pretend that intellect is any more the prerogative of age than of youth. They may, however, urge that age has had greater opportunity for study. But rare is the senior officer who could honestly claim to have devoted his lifetime to intensive study. More often the hours spent in study steadily decline with the gradual extension of responsibilities. Few, indeed, have been able to devote even a year or two during their career to intensive study. In the case of the majority, study has been desultory. Thus where, as so often happens with genius, the intellect ripens early, the young student may have spent more hours in reading and reflection by the time he is twenty-five or thirty than most men spend in their whole lifetimes. (March 1923)

Captain Sir Basil Liddell Hart, *Thoughts on War*, 1944

Few youths of spirit are content at eighteen with comforts or even caresses. They seek physical fitness, movement and the comradeship of their equals under hard conditions. They seek distinction, not favour, and exult in their manly independence.

Sir Winston S. Churchill, *Marlborough*, 1933

In war it is almost impossible to exaggerate the evil effects of age upon generalship, and through generalship, on the spirit of the army. In peace time it may be otherwise, but in war time the physical, intellectual and moral stresses and strains which are at once set up immediately discover the weak links in a general's harness. First, war is obviously a young man's occupation; secondly, the older a man grows the more cautious he becomes, and thirdly, the more fixed become his ideas. Age may endow a man with experience, but in peace time there can be no moral experience of war, and little physical experience. Nothing is more dangerous in war than to rely upon peace training; for in modern times when war is declared, training has always been proved out of date. Consequently, the more elastic a man's mind is, that is the more it is able to receive and digest new impressions and experiences, the more common sense will be the actions resulting. Youth, in every way, is not only more elastic than old age, but less cautious and far more energetic. In a moment youth will vault into the saddle of a situation, whilst old age is always looking around for someone to give it a leg up.

Physically an old man is unable to share with his men the rough and tumble of war; instinctively he shuns discomfort, he fears sleeping under dripping hedges, dining off a biscuit, or partaking of a star-lit breakfast, not because he is a coward, but because for so many years he has slept between well-aired sheets, dined off a well-laid table and breakfasted at 9 o'clock, that he instinctively feels that if these things are changed he will not be himself, and he is right for he will be an *uncomfortable* old man.

Major General J. F. C. Fuller, *Generalship: Its Diseases and Their Cure*, 1933

. . . it is sad to remember that, when anyone has fairly mastered the art of command, the necessity for that art usually expires – either through the termination of the war or through the advanced age of the commander.

General George S. Patton, Jr., *War As I Knew It*, 1947

I am doubtful whether the fact that a man has gained the Victoria Cross for bravery as a young officer fits him to command an army twenty or thirty years later. I have noticed more than one serious misfortune which arose from such assumptions . . .

Sir Winston S. Churchill, on Sir Redvers Buller, *My Early Life,* 1930

Next comes the vexed question of age. One of the ancient Roman poets has pointed out the scandal of old men at war and old men in love. But at exactly what age a general ceases to be dangerous to the enemy and a Don Juan to the other sex is not easy to determine. It is impossible really to give exact values to the fire and boldness of youth as against the judgment and experience of riper years; if the mature mind still has the capacity to conceive and to absorb new ideas, to withstand unexpected shocks, and to put into execution bold and unorthodox designs, its superior knowledge and judgment will give the advantage over youth. At the same time there is no doubt that a good young general will usually beat a good old one . . .

Field Marshal Earl Wavell, *Soldiers and Soldiering,* 1953

Select Bibliography

Alexander, Edward Porter. *Fighting for the Confederacy: The Personal Recollections of General Edward Porter.* Ed., Gary W. Gallagher. Chapel Hill: University of North Carolina Press, 1989.

Alexander, Harold. *The Alexander Memoirs 1940–1945.* Ed., John North. New York: McGraw–Hill Book Co., 1961.

Allon, Yigal. *The Making of Israel's Army.* New York: The Universe Press, 1970.

Ardnant du Picq, Charles. *Battle Studies.* Harrisburg: Stackpole Books, 1957.

Arrian (Flavius Arrianus Xenophon). *The Campaigns of Alexander the Great.* London: Penguin Books, 1987.

Baker, Thomas M. *The Military Intellectual and Battle: Raimondo Montecuccoli and the Thirty Years War.* Albany: State University of New York Press, 1975.

Beaufre, Andre. *An Introduction to Strategy.* New York: Praeger, 1965.

Bernhardi, Friedrich von. *On War of Today.* London: Hugh Rees, Ltd., 1913.

— *The War of the Future.* New York: D. Appleton & Co., 1921.

de Bourcet, Jean. *The Defence of the Piedmont.* London: Oxford University Press, 1927.

Bradley, Omar N. *A Soldier's Story.* New York: Henry Holt and Co., 1951,

Breasted, James H. *Ancient Records of Egypt.* 4 Vols. Chicago: The University of Chicago Press, 1906.

Brusilov, Aleksei A. *A Soldier's Notebook 1914–1918.* Westport: Greenwood Press, Publishers, 1931.

Caesar, Julius. *The Gallic War and Other Writings of Julius Caesar.* Trans., Moses Hadas. New York: Random House, 1957.

Carver, Michael. *War Since 1945.* London: Weidenfeld and Nicolson, 1980.

Chamberlain, Joshua Lawrence. *The Passing of the Armies: An Account of the Final Campaign of the Army of the Potomac.* New York: G. P. Putnam's Sons, 1915.

Churchill, Winston S. *Blood, Sweat, and Tears.* New York: G. P. Putnam's Sons, 1941.

— *Great Contemporaries.* New York: G. P. Putnam's Sons, 1937.

— *Marlborough: His Life and Times.* 4 Vols. London: Harrap, 1933–38.

— *Maxims and Relections.* Ed., Colin Coote. Boston: Houghton Mifflin Co., 1949.

— *My Early Life.* 1933.

— *The River War: An Historical Account of the Reconquest of the Sudan.* 2 Vols. London: Longmans, Green, and Co., 1899.

— *The Second World War.* 6 Vols. Boston: Houghton Mifflin, 1948–53.

— *The World Crisis.* New York: Scribner's Sons, 1923.

Clausewitz, Carl von. *The Campaign of 1812 in Russia.* Westport: Greenwood Press, Publishers, 1977.

— *On War.* Eds./Trans. Michael Howard and Peter Paret. Princeton: Princeton University Press, 1976.

— *Principles of War.* Harrisburg: Stackpole Books, 1942.

Collins, Arthur S. *Common Sense Training.* San Rafael: Presidio Press, 1978.

Cromwell, Oliver. *Writings and Speeches of Oliver Cromwell.* Cambridge: Harvard University Press, 1947.

Davis, Burke. *Marine! The Life of Chesty Puller.* Boston: Little Brown and Co., Inc., 1962.

Dayan, Moshe. *Moshe Dayan: The Story of My Life.* London: Wiedenfeld & Nicholson, 1976.

De Gaulle, Charles. *The Army of the Future.* New York: Lippincott, 1941.

— *The Edge of the Sword.* Westport: Greenwood Press, 1960 (originally published in France in 1932).

— *War Memoirs*. 5 Vols. New York: Simon and Schuster, 1955–61.

Dönitz, Karl. *Memoirs: Ten Years and Twenty Days*. Cleveland: The World Publishing Co., 1959.

Duffy, Christopher. *The Military Life of Frederick the Great*. New York: Atheneum, 1986.

Dupuy, Trevor N. *A Genius for War: The German Army and the General Staff*. Englewood Cliffs, Prentice Hall, 1977.

— *Understanding War: History and Theory of Combat*. New York: Paragon House Publishers, 1987.

Eisenhower, Dwight D. *At Ease: Stories I Tell My Friends*. New York: Doubleday & Co., 1967.

— *Crusade in Europe*. Garden City: Doubleday & Co., 1948.

— *The Papers of Dwight David Eisenhower: The War Years*. Eds., Alfred D. Chandler, Jr. and Stephen E. Ambrose. Baltimore: John Hopkins, 1970.

Erfurt, Waldemar. *Surprise*. Harrisburg: Military Services Publishing Co., 1957.

Fisher, John A. *Memories*. London: Hodder and Stoughton, 1919.

Fitton, Robert A. *Leader: Quotations from the Military Tradition*. Boulder: Westview Press, 1990.

Foch, Ferdinand. *The Memoirs of Marshal Foch*. London: William Heinemann, Ltd., 1931.

— *Precepts and Judgements*. London: Chapman and Hall, 1919.

— *The Principles of War*. New York: H. K. Fly Co., 1918 (originally published in France in 1913).

Foertsch, Hermann. *The Art of Modern Warfare*. New York: Oskar Piest, 1940.

Fraser, William. *Words on Wellington*. London: George Routledge & Sons, Ltd., n.d.

Frederick II the Great. *Frederick the Great on the Art of War*. Ed., Jay Luvaas. New York: The Free Press, 1966.

— *Instructions to His Generals*, 1747; in Ed., T. R. Phillips, *Roots of Strategy*. Book 1. Harrisburg: Stackpole Books, 1985.

Freytag-Loringhoven, Hugo von. *Generalship in the Great War*. U.S. Army War College, 1934.

— *The Power of Personality in War*; in *Roots of Strategy*, Book 3. Harrisburg: Stackpole Books, 1991.

Fuller J. F. C. *The Conduct of War 1789–1961*. London: Eyre & Spottiswoode, 1961.

— *The Foundations of the Science of War*. London: Hutchinson & Co., 1926.

— *Generalship: Its Diseases and Their Cure*. London: Faber and Faber, 1933.

— *The Generalship of Alexander the Great*. London: Eyre & Spottiswoode, 1958.

— *Grant and Lee: A Study in Personality and Generalship*. London: Eyre and Spottiswoode, 1933.

— *Lectures on F. S. R. II*. London: Sifton Praed & Co., 1931.

— *Lectures on F. S. R. III*. London: Sifton Praed & Co., 1932.

— *Memoirs of an Unconventional Soldier*. Ivor Nicholson and Watson, 1936.

— *A Military History of the Western World*. 3 Vols. New York: Minerva Press, 1967.

— *The Reformation of War*. London: Hutchinson & Co., 1923.

Gabriel, Richard A. and Paul L. Savage. *Crisis in Command: Mismanagement in the Army*. New York: Hill and Wang, 1978.

Gabrieli, Francesco. *Arab Historians of the Crusades*. Berkeley: University of California Press, 1984.

Gareyev, Makhmut. *M. V. Frunze: Military Theorist*. Moscow: Voyenizdat, 1985.

Gavin, James M. *On to Berlin*. New York: Viking, 1978.

Giap, Vo Nguyen. *People's War – People's Army*. Hanoi: Foreign Languages Publishing House, 1961.

Glubb, John Bagot. *A Soldier With the Arabs*. New York: Harper and Brothers, 1957.

Goltz, Colmar von der. *The Conduct of War*. London: Kegan Paul, Trench, Trubner & Co., Ltd., 1908.

— *The Nation in Arms*. London: William Heinemann, Ltd., 1887.

Gordon, John B. *Reminiscences of the Civil War*. New York: Scribner's, 1903.

Grant, Ulysses. S. *Personal Memoirs of U.S. Grant*. New York: Literary Classics of the

United States, 1990.

Guderian, Heinz. *Panzer Leader*. London: Michael Joseph, 1953.

Guevera, Che. *Guerrilla Warfare*. Lincoln: Michael Joseph, 1953.

de Guingand, Francis. *Operation Victory*. New York: Scribner's, 1947.

Hackett, John Winthrop. *The Profession of Arms*. London: Sidgwick and Jackson, Ltd, 1983.

Hamilton, Ian. *Gallipoli Diary*. 2 Vols. New York: George H. Doran Co., 1920.

— *The Soul and Body of an Army*. London: Edward Arnold & Co., 1921.

Harris, Arthur. *Bomber Offensive*. London: Greenhill Books, 1990.

Heinl, Robert Debs, Jr., *Dictionary of Military and Naval Quotations*. Annapolis: U.S. Naval Institute, 1966.

Henderson, George F. R. *Stonewall Jackson and the American Civil War*. Da Capo Press, 1988 (originally published in Great Britain in 1898).

Hindenburg, Paul von. *Out of My Life*. London: Cassell and Co. Ltd, 1920.

Hoffman, Max. *War Diaries and Other Papers*. 2 Vols. London: Martin Secker, 1929.

Holmes, Oliver Wendell. *Touched With Fire*. Cambridge: Harvard university Press, 1948.

Homer. *The Iliad*. Tr., Robert Fagles. New York: Viking Penguin, 1990.

Horrocks, Brian. *Corps Commander*. New York: Scribner's Sons, 1977.

— *Escape to Action*. New York: 1960.

Jackson, Robert. *A Systematic View of the Formation, Discipline, and Organization of Armies*. London: Parker, Furnivall, and Parker, 1845 (Originally published in 1804).

Jang Chu. *The Precepts of Ssu Ma Jang Chu*. A. L. Sadler. *Three Military Classics of China*. Sydney: Australian Medical Publishing Co., Ltd., 1944.

Jomini, Antoine-Henri. *Summary of the Art of War*. Harrisburg: Stackpole Books, 1965.

Kai Ka'us Ibn Iskandar. *A Mirror for Princes* *(1082 AD).New York: E. P. Dutton & Co., Inc., 1951.*

Kesselring, Albert. The Memoirs of Field Marshal Kesselring. William Kimber, 1953.

King, Ernest J. and Walter Muir Whitehill. *Fleet Admiral King: A Naval Record*. New York: W. W. Norton & Co., Inc., 1952.

Kinnamos, John. *Deeds of John and Manuel Comnenus*. New York: Columbia University Press, 1976.

Kolokotrones, Theodoros. *Theodoros Kolokotrones: The Old Man of the Morea*. Brookline: Hellenic College Press, 1983.

Lawrence, T. E., *The Seven Pillars of Wisom*. New York: Doubleday, 1935.

— *T. E. Lawrence to His Biographers, Robert Graves and Liddell Hart*. Garden City: Doubleday and Co., 1963.

Lee, Robert E. *The Wartime Papers of Robert E. Lee*. Ed., Clifford Dowdey. New York: Da Capo Press, 1987.

Leites, Nathan. *Soviet Style in War*. New York: Crane Russak, 1982.

LeMay, Curtis E. *Mission With LeMay*. Garden City: Doubleday & Co., 1965.

Lettow Vorbeck, Paul von. *East African Campaigns*. New York: Robert Speller & Sons Publishers, Inc., 1957.

Liddell Hart, Basil H. *Colonel Lawrence: The Man Behind the Legend*. New York: Dodd, Mead & Co., 1934.

— *Foch: The Man of Orleans*. London: Eyre & Spottiswoode, 1931.

— *The Ghost of Napoleon*. London: Faber & Faber, 1933.

— *Great Captains Unveiled*. London: Blackwood, 1927.

— *A Greater Than Napoleon: Scipio Africanus*. London: Blackwood, 1926.

— *The Other Side of the Hill*, London: Cassell, 1951.

— *The Real War 1914–1918*. London: Faber & Faber, 1930.

— *Reputations: Ten Years After*. London: John Murray, 1928.

— *Strategy: The Indirect Approach*. New York: Frederick A. Praeger, Publisher, 1954.

— *The Sword and the Pen*. New York: Thomas Y. Crowell Co., 1976.

— *Thoughts on War*. London: Faber & Faber, Ltd., 1944.

— *Through the Fog of War*. London: Faber & Faber, 1938.

Liu Ji. *Lessons of War.* Ed., tr., Thomas Cleary. *Mastering the Art of War.* Boston: Shambhala, 1989.

Longstreet, James. *From Manassas to Appomattox.* Philadelphia: Lippencott, 1896.

Longworth, Phillip. *The Art of Victory: The Life and Achievements of Field Marshal Suvorov 1729–1800.* New York: Holt, Rinehart and Winston, 1966.

Luchenbill, Daniel D. *Ancient Records of Assyria and Babylonia.* 2 Vols. Greenwood Press, Publishers, 1926.

Ludendorff, Erich von. *My War Memories 1914–1918.* 2 Vols. London: Hutchinson & Co., 1919.

MacArthur, Douglas. *Reminiscences.* New York: McGraw-Hill Book Co., 1964.

Machiavelli, Niccolo. *The Art of War.* Indianapolis. The Robbs-Merill Co., Inc., 1965.

Madariaga, Salvador de. *Hernan Cortes: Conqueror of Mexico.* Coral Gables: University of Miama Press, 1942.

Mahan, Alfred. *The Influence of Sea Power Upon History 1660–1783.* Boston: Little, Brown & Co., 1890.

— *Life of Nelson: The Embodiment of the Sea Power of Great Britain.* 2 Vols. Boston: Little, Brown & Co., 1897.

— *Naval Administration and Warfare.* Boston: Little, Brown & Co., 1908.

— *Naval Strategy.* Boston: Little, Brown & Co., 1911.

— *From Sail to Steam: Recollections of a Naval Life.* New York: Harper & Brothers, 1907.

Mahan, Dennis Hart. *Advanced-Guard, Out-Post, and Detachment Service of Troops, with the Essential Principles of Strategy and Grand Tactics.* New York: John Wiley, 1863.

Majatovich, Chedomil. *Constantine: The Last Emperor of the Greeks.* London: Sampson Low, Marston & Co., 1982.

Mannerheim, Gustav. *The Memoirs of Marshal Mannerheim.* New York: E. P. Dutton & Co., Inc., 1953.

Manstein, Erich von. *Lost Victories.* London: Metheun & Co., Ltd., 1958.

Mao Tse-tung. *Selected Military Writings of Mao Tse-tung.* Peking: Foreign Languages Press, 1967.

Marshall, George C. *Memoirs of My Service in the World War 1917–1918.* Boston: Houghton Mifflin Co., 1976.

— *Selected Speeches and Statements of General of the Army George C. Marshall.* Washington, DC: The Infantry Journal, 1945.

Marshall, S. L. A. *The Armed Forces Officer.* Washington: U.S. Government Printing Office, 1951.

— *Bringing Up the Rear.* San Rafael: *Presidio Press,* 1979.

— *Men Against Fire.* New York: William Morrow & Co., 1947.

— *The Soldier's Load and the Mobility of a Nation.* Quantico: The Marine Corps Association, 1980.

Mauldin, William. *Up Front.* New York: W. W. Norton & Co., 1968.

Maurice. *The Strategikon.* Tr., George T. Dennis. Philadelphia: University of Pennsylvania Press, 1984.

Meinertzhagen, Richard. *Army Diary 1899–1926.* Edinburgh: Oliver & Boyd, 1957.

Mellenthin, F. W von. *Panzer Battles.* Norman: Oklahoma University Press, 1971.

Moltke, Helmuth von. *Strategy: Its Theory and Application: The Wars for German Unification 1866–1871.* Westport: Greenwood Press, 1971.

Monash, John. *The Australian Victories in France in 1918.* London: Hutchinson, n.d.

Montgomery, Bernard Law. *The Memoirs of Field Marshal Montgomery.* Cleveland: The World Publishing Co., 1958.

Moore, John. *The Diary of Sir John Moore.* Ed., J. F. Maurice. 2 Vols. London: Edward Arnold, 1904.

Morison, Samuel Eliot. *John Paul Jones: A Sailor's Biography.* Boston: Little, Brown & Co., 1959.

Mosby, John Singleton. *Mosby's War Reminiscences.* New York: Pageant BooK Co., 1958.

Napoleon. *The Military Maxims of Napoleon.* Ed., David Chandler. Tr. George C. D'Aguilar. London: Greenhill Books, 1987.

— *The Mind of Napoleon: A Selection from his Written and Spoken Words.* Ed., Christopher J. Herold. New York: Columbia University Press, 1955.

— *Napoleon's Memoirs.* Ed., Somerset de Chair. London: Golden Cockerel Press, 1945.

Newman, Aubrey S. *Follow Me: The Human Element in Leadership.* Novato: Presidio Press, 1981.

Ney, Michel. *The Memoirs of Marshal Ney.* 2 Vols. London: Bull & Churton, 1834.

Ogasawara, Nagayo. *Life of Admiral Togo.* Tokyo: Seit Shorin Press, 1934.

Onasander. *The General.* Loeb Classical Library No. 48. Cambridge: Harvard University Press, 1937.

Ouranos, Nikephorus. *Campaign Organization and Tactics.* Ed., George T. Dennis. *Three Byzantine Military Treatises.* Washington, DC: Dumbarton Oaks, 1985.

Parkinson, Roger. *The Fox of the North: The Life of Kutuzov, General of War and Peace.* New York: David McKay, 1976.

Patton, George S., Jr. *The Patton Papers.* 2 Vols. Boston: Houghton Mifflin Co., 1972–74.

— *War As I Knew It.* Boston: Houghton Mifflin Co., 1947.

Pershing, John J. *My Experiences in the World War.* 2 Vols. New York: Stokes, 1931.

Pfiffer, Robert H. *State Letters of Assyria.* New Haven: American Oriental Society, 1935.

Phocas, Nikephoros. *Skirmishing.* Ed., George T. Dennis. *Three Byzantine Military Treatises.* Washington, DC: Dumbarton Oaks, 1985.

Pilsudski, Joseph. *The Memoirs of a Polish Revolutionary and Soldier.* London: Faber & Faber Ltd., 1931.

Polybius. *The Rise of the Roman Empire.* London: Penguin Books, 1987.

Procopius. *A History of the Wars.* 7 Vols. Loeb Classical Library. Cambridge: Harvard University Press, 1929.

Ralston, David B. *Soldiers and States: Civil–Military Relations in Modern Europe.* Boston: D. C. Heath & Co., 1966

Ridgway, Mathew B. *The Korean War.* Garden City: Doubleday & Co., 1967.

— *Soldier: The Memoirs of Mathew B. Ridgway.* New York: Harper, 1956.

Robertson, William. *From Private to Field-Marshal.* Boston: Houghton Mifflin Co., 1921.

Rojas, Ricardo. *San Martin: Knight of the Andes.* New York: Cooper Square Publishers, 1967.

Rommel, Erwin. *Infantry Attacks.* Washing ton, DC: *Combat Forces Press,* 1956.

— *The Rommel Papers.* Ed., B. H. Liddell Hart. New York: Harcourt Brace & Co., 1953.

Royal, Trevor. *Dictionary of Military Quotations.* London: Routledge, 1990.

Sadler, A. L. *The Maker of Modern Japan: The Life of Tokugawa Ieyasu.* London: George Allen & Unwin Ltd., n.d.

Saxe, Maurice de. *My Reveries Upon the Art of War* (1732). Ed., T. R. Phillips. *Roots of Strategy.* Book 1. Harrisburg: Stackpole Books, 1985.

Schlieffen, Alfred von. *Cannae.* Fort Leavenworth: U.S. Army Commander and General Staff College Press, 1931.

Seekt, Hans von. *Thoughts of a Soldier.* London: Ernest Benn, Ltd., 1930.

Sharon, Ariel. *Warrior: An Autobiography.* New York: Simon and Schuster, 1989.

El Shazly, Saad. *The Crossing of the Suez.* San Francisco: American Mideast Research, 1980.

Sheridan, Philip H. *Personal Memoirs of Phillip Henry Sheridan.* 2 Vols. Charles L. Webster & Co., 1888.

Sherman, William T. *Memoirs of General W. T. Sherman.* New York: Literary Classics of the United States, 1990.

Simpkin, Richard. *Deep Battle: The Brainchild of Marshal Tukhachevskii.* London: Brasey's Defense Publishers, 1987.

Slim, William. *Courage and Other Broadcasts.* London: Cassell & Co., 1957.

— *Defeat Into Victory.* New York: David McKay Co., Inc., 1963.

Smythe, John. *Certain Discourses Military.* Ithica: Cornell University Press, 1966.

Sokolovskiy, V. D. *Soviet Military Strategy.* New York: Crane, Russak & Co., 1975.

Steuben, Frederick William von. *Revolutionary Drill Manual* (1794). New York: Dover Publications, Inc., 1985.

Stockdale, James B. *A Vietnam Experience: Ten Years of Reflections.* Berkeley: Hoover

Institute Press, 1984.

Summers, Harry G., Jr. *On Strategy: The Vietnam War in Context.* Carlisle Barracks: U.S. Army War College, 1981.

Sun Tzu. *The Art of War.* Ed., Samuel Griffith. New York: Oxford University Press, 1969.

Tedder, Arthur W. *With Prejudice: The War Memoirs of Marshal of the Royal Air Force Lord Tedder, G.C.B.* Boston: Little, Brown & Co., 1966.

Thompson, John W. *Fix Bayonets!* New York: Scribner's, 1926.

Thucydides. *The Peloponnesian War.* Tr. Benjamin Jowett. New York: Washington Square, 1970.

Trotsky, Leon. *History of the Russian Revolution.* 3 Vols. New York: Simon & Schuster, 1932.

— *Military Writings.* New York: Merit Publishers, 1969.

Upton, Emory. *The Military Policy of the United States Since 1775.* New York:

Vegetius (Flavius Rennatus). *The Military Institutions of the Romans* (c. AD 378). Harrisburg: Stackpole Books, 1965.

Washington, George. *Maxims of George Washington.* Mount Vernon: The Mount Vernon Ladies Association, 1989.

— *The Writings of George Washington.* 39 Vols. Ed., John C. Fitzgerald. Washington, DC: 1931–44.

Wavell, Archibald P. *Allenby: A Study in Greatness.* 2 Vols. New York: Oxford University Press, 1941.

— *Generals and Generalship.* London: 1941.

— *Soldiers and Soldiering.* London: Jonathan Cape, 1953.

Webb, Henry J. *Elizabethan Military Science: The Books and Practice.* Madison: the University of Wisconsin Press, 1965.

Webb, James H. *Fields of Fire.* New York: Bantam, 1977.

Weygand, Max. *Turenne: Marshal of France.* Boston: Houghton Mifflin Co., 1930.

De Wet, Christiaan R. *A Three Years War.* London: Archibald Constable & Co., Ltd., 1902.

Williams-Ellis, Clough and A. Williams-Ellis. *The Tank Corps.* London: Country Life, 1919.

Wilson, William S., Ed. *The Ideals of the Samurai: Writings of Japanese Warriors.* Burbank: Ohara Publications, Inc., 1982.

Wolseley, Garnet J. *The Soldier's Pocket-book for Field Service.* London: 1869.

— *The Story of a Soldier's Life.* Westminster: Archibald Constable & Co., Ltd., 1903.

Wood, Leonard. *The Military Obligation of Citizenship.* Princeton: Princeton University Press, 1915.

Wu Chi. *Wu Chi on the Art of War.* A. L. Sadler. *Three Military Classics on China.* Sydney: Australian Medical Publishing Co., Ltd, 1944.

Xenophon. *A History of My Times.* Tr. Rex Warner. Hammondsworth: Penguin Books, 1966.

— *The Persian Expedition.* Tr. Rex Warner. Baltimore: Penguin Books, 1965.

Young, Desmond. *The Desert Fox.* New York: Harper & Row, 1950.

Zhuge Liang. *The Way of the General.* Ed., Tr., Thomas Cleary. *Mastering the Art of War.* Boston: Shambhala, 1989.

Zhukov, Georgi K. *Reminiscences and Reflections.* 2 Vols. Moscow: Progress Publishers, 1985.

Biographical Index

NFI: No further information available.

Abrams, Creighton (1914–1974); General, US Army; one of the finest US armour commanders in World War II; succeeded Westmoreland in command of US forces in Vietnam and vastly improved performance in combat by fighting the enemy in his own environment; as Chief of Staff, he began the reform process that eventually led to the Army's professional renaissance. *Quoted*: 131, 179, 354, 367, 397, 405.

Achilles (13th century BC); Prince of the Myrmidons; greatest hero and most deadly warrior of the Trojan War, as described by Homer in *The Iliad*, the Greek name of which is *The Anger of Achilles*; as indicated by this title, the plot revolves around his willingness to sacrifice everything for his honour and glory sake; this thirst for glory made Achilles the boyhood hero and model of Alexander the Great. *Quoted*: 192, 376.

Agamemnon (13th Century BC); King of Golden Mycenae and High King of the Greeks (Anax Andron) in the expedition against Troy, as described by Homer in *The Iliad*; his arrogance provoked the great anger of Achilles that nearly resulted in the destruction of the Greeks before Troy; murdered by his wife and her lover upon his return to Mycenae after the fall of Troy. *Quoted*: 373.

Ahmose, Son of Abana (*c.* 1570–*c.* 1512 BC); Egyptian general who served with distinction in ten campaigns under three successive pharaohs: Ahmose I, Ahmenhotep I, and Thutmose I of the 18th Dynasty; his tomb inscription is the only contemporary account of the expulsion of the Hyksos from Egypt. *Quoted*: 27, 343, 363, 375, 452.

Ajax 'The Lesser' (13th Century BC); one of the great Greek heroes of the Trojan War, as described by Homer in *The Iliad*; known as the 'Fleet-Footed' and killed himself when the armour of Achilles was awarded to Odysseus rather than himself. *Quoted*: 380.

Albert, Archduke of Austria (1817–1895); son of Archduke Charles; fought in the Italian Campaigns (1848–49) and in war with Prussia (1866); commanded the Italian front and won the Battle of Custoza; after the war served as commander-in-chief of the Austrian armed forces and worked for their modernization. *Quoted*: 67, 211.

Alexander III, 'The Great', (356–323 BC); King of Macedonia and conqueror of the Persian Empire and master tactician, strategist, logistician, and leader of men; known by his epithet, *Anikitos* – Invincible. *Quoted*: 148, 192, 193, 368, 472.

Alexander, Edward Porter (1835–1910); Brigadier General, Confederate States Army, commander of the artillery of Longstreet's 1st Corps. *Quoted*: 403, 419, 482.

Alexander, Harold Rupert (1891–1961); 1st Earl Alexander of Tunis, Field Marshal, British Army; commanded Allies in Mediterranean and Italian Theatres in World War II. *Quoted*: 259, 375, 463.

Alfred 'The Great' (AD 849–899); King of Wessex; saviour of England from the Danes in a succession of wars (871–896) and finally recognized as sovereign of the Angles and Saxons (896); founded first English navy, encouraged learning and noted as a lawgiver. *Quoted*: 463.

Allenby, Edmund Henry (1861–1936); Viscount of Meggido and Felixstowe; Field Marshal, British Army; commander of Egyptian Expeditionary Force in World War I; drove Turks out of Palestine and Syria; first Christian conqueror of Jerusalem (1917) since the 13th century; broke Turkish power at the Battle of Meggido

(1918). *Quoted*: 122, 182, 243, 350, 412.

Allon, Yigal (1918–); General; Israeli Defence Force (IDF); one of the founders of the Palmach before 1948; commanded southern front during Israel's War of Independence; one of most experienced and influential officers in the IDF during its creative formative years. *Quoted*: 45, 91, 115, 116, 241, 256, 301, 327, 484.

Alvarez de Toledo, Fernando (1507–1528); the Duke of Alva; Spanish soldier; one of Spain's greatest generals, chiefly known for his bloody attempts to pacify the Netherlands (1567–73); conducted a masterful conquest of Portugal in 1581. *Quoted*: 175, 368.

Amenhotep II (reigned 1450–25 BC); Pharaoh of Egypt; put down revolt in Syria and kept Nubia under firm control. *Quoted*: 35.

Amenhotep III (reigned 1417–1379 BC) Pharaoh of Egypt; campaigned successfully in Nubia and maintained Egyptian supremacy in Asia; his reign was one of magnificence and prosperity; father of Akenaton. *Quoted*: 80.

Amenhotep, Son of Hapi (*c.* 1460 BC–*c.* 1380 BC); high Egyptian official under Pharaoh Ammenhotep III, responsible for all royal works and for military mobilization; apparently combined duties of grand vizier and minister of defence; known for his wisdom. *Quoted*: 66.

Antigonus II Gonatus (*c.* 319 BC–*c.* 239 BC), King of Macedonia and the son of Demetrius I Poliorcetes. *Quoted*: 368.

Antoninus Marcus Aurelius (AD 121–180); Roman Emperor and stoic philosopher; last of the 'Five Good Emperors'; ruled during a period of upheaval in which a great plague and German invasions shook the Empire; defeated the Germans and restored the Danubian frontier; wrote *Meditations*. *Quoted*: 27, 63, 229, 357.

Antony, Mark (Marcus Antoninus) (81–30 BC); Roman triumvir and general; served under Caesar in the conquest of Gaul and rose under his patronage to govern Italy; commanded Caesar's left wing at Pharsallus (48 BC); rallied forces against Caesar's assassins by his powers of oratory; formed with Octavian and Lepidus the second triumvirate to rule Rome (43 BC) and defeated assassins at Philippi (42 BC); took the East as his share of empire to rule and fell in love with Cleopatra, Queen of Egypt; defeated by Parthians (34 BC) and alienated Roman support by attention to Cleopatra; rivalry with Octavian provoked war in which he and Cleopatra were defeated at great naval battle at Actium (31 BC); he fled to Egypt and was deserted by army, whereupon he committed suicide. *Quoted*: 147.

Archidamus II, King of Sparta (*d.* 427 BC); attempted to prevent war between Athens and Sparta and later commanded Spartan forces at the beginning of the Peloponnesian War. *Quoted*: 137, 390.

Archidamus III, King of Sparta (*d.* 338 BC); commanded forces in the Peloponnesian War; defended Sparta against Epaminondas (363); supported Phocians in the Sacred War (355–346); and commanded a mercenary army in Italy protecting Tarentum (338). *Quoted*: 157, 459.

Ardnant du Picq, Charles (1821–1870); Colonel, French Imperial Army; analyst and writer on the dynamics of men in combat; killed in the Franco–Prussian War, his posthumus work, *Battle Studies* (1880) emphasized the moral forces involved in combat and had great influence on the French Army. *Quoted*: 26, 43, 44, 54, 58, 81, 139, 163, 199, 204, 224, 244, 248, 266, 271, 308, 317, 369, 421, 428, 454, 477.

Aristides 'The Just' (*c.* 530–*c.* 468 BC); Athenian stateman and general; commanded a contingent of the Athenian host at Marathon (490 BC); supported creation of Athenian military instead of naval power, in opposition to Themistocles; ostracized (382 BC); returned to serve as *strategos*, loyally supporting Themistocles in Salamis campaign (480–79 BC); commanded Athenians at Platea (479 BC); arranged alliance between Athens and Delian League. *Quoted*: 202.

Arrianus Xenophon, Flavius (Arrian) (*c.* AD 95–*c.* AD 175); Roman governor of Cappodocia and general; known best for his excellent *The Campaigns of Alexander* (*c.* AD 150). *Quoted*: 269, 275, 280, 397.

killed in battle outside Amphipolis. *Quoted*: 172, 362, 395.

Brown, C. R. (*b.* 1899); Vice Admiral, US Navy; a naval aviator since 1924, he commanded in World War II in a number of shore and ship-based aviation units – commanded USS *Hornet* (1945–6); rose to command US 6th Fleet (1957–8) to be CINC, Allied Forces southern Europe; author of *Principles of War* (1949). *Quoted*: 288, 340.

Brusilov, Aleksei Alekseyevich (1853–1926); General, Russian Imperial Army; most successful Russian general of World War I; his 'Brusilov Offensive' of 1916 shattered the Austrian Army in the only true rupture of the trench system in the war but failed to attain decisive results because of lack of support; later served Red Army in Civil War as an adviser; wrote *A Soldier's Notebook*, 1931. *Quoted*: 36, 108, 130, 158, 290, 320, 354, 360, 374, 477, 478.

Bugeaud de la Piconnerie, Thomas-Robert (1784–1849); Duc d'Isly; Marshal of France; after service under Napoleon, he supported the restoration of the Bourbons and later served in North Africa (1836–47), developing effective tactics against the Arabs; created duke for his victory at the Battle of Isly (1844). *Quoted*: 83.

Bullard, Robert Lee (1861–1947); Lieutenant General, US Army; served in the Cuban campaign and the Philippine Insurrection; commanded 1st Infantry Division in France in 1918 and made first US Army divisional assault in World War I; commanded III Corps and then US 2nd Army; after retirement he was a prolific writer on military subjects. *Quoted*: 250, 333.

Burgoyne, John (1722–1792); Major General, British Army; served in Seven Years War, and in the American Revolution commanded the British invasion of the Colonies from Canada in 1777 during which his army was defeated and surrendered at the Battle of Saratoga. *Quoted*: 332, 460.

Burke, Arleigh 'Thirty-One Knot Burke' (*b.* 1901), Admiral, US Navy; served in World War II and Korea; in 1943 he earned distinction by sinking at high speed a large number of Japanese vessels because of his gunnery; became Chief of Naval Operations and served longer than any of his predecessors (1955–61). *Quoted*: 22, 23.

Burne, Alfred H.; Lieutenant Colonel, British Army; author of *The Art of War on Land*, 1944. *Quoted*: 244.

Butler, Sir William (1838–1910); Lieutenant General, British Army. *Quoted*: 333.

Caesar, Gaius Julius (100–44 BC); Roman general and statesman – one of the great captains of military history; established his military reputation as conqueror of Gaul; member of the First Triumvirate with Crassus and Pompey; defeated Pompey at the Battle of Pharsalus (48 BC) during the Civil War; ruled as dictator and enacted many reforms until his murder; wrote *The Gallic War* (*c.* 45 BC) and *The Civil War*, models of lucid, concise prose and military history. *Quoted*: 60, 83, 161, 184, 317, 367, 377, 458.

Cambronne, Pierre-Jacques-Etienne (1770–1842); Vicomte; General, French Army; served with distinction in campaigns of 1812, 1813, and 1914; accompanied Napoleon to Elba; commanded a division of the Imperial 'Old' Guard at Waterloo in 1815. *Quoted*: 418.

Campbell, Sir Alan (NFI); Lieutenant General, British Army; served in the Peninsular campaign. *Quoted*: 329.

Carnot, Lazare-Nicolas-Marguerite, 'The Organizer of Victory' (1753–1823); General, French Army; entered the army as an engineer but became deputy to the National Convention (1792) and member of the Committee for Public Safety (1793); in charge of the mobilization, creation and direction of the armies of republican France that defied the royal armies of Europe. *Quoted*: 72.

Carver, Michael (1915–); Baron; Field Marshal, British Army; Chief of the General Staff (1971–73) and Chief of the Defence Staff (1973–76); distinguished military writer. *Quoted*: 343.

Catiline (Lucius Sergius Catalina) (*c.* 108–62 BC); Roman politician; defeated in election for consul (63, 64 BC) and plotted to murder the consuls and plunder Rome;

exposed by Cicero, he fled to Etruria and was slain (62 BC). *Quoted*: 112.

Cervantes, Miguel de (1547–1616); Spanish soldier and novelist; his left hand was maimed at the Battle of Lepanto in 1571; continued to serve against the Turks; was captured and imprisoned in Tunis and finally escaped; wrote the masterpiece *Don Quixote*, the first of the modern novels. *Quoted*: 30, 183, 331, 451, 459.

Chabrias (*fl.* 390–356 BC); Athenian General; skilled commander continuously employed in this period by either Athens or foreign powers such as Egypt; his victory off Naxos in 376 and other operations did much to aid the rise of the Second Athenian Empire; killed in battle. *Quoted*: 234.

Chamberlain, Joshua Lawrence (1828–1914); Major General, US Army; appointed colonel of the 20th Maine Volunteers (1863) and commanded regiment throughout the Civil War; wounded six times in the war, he played the pivotal role at Gettysburg (1863) by holding Little Round Top, for which he was later awarded the Medal of Honor (1888); repeatedly promoted for distinction by General Grant, he was given the honour of accepting the surrender of the Army of Northern Virginia; in an act of great chivalry, he brought the Army of the Potomac to present arms as the Confederates marched past. *Quoted*: 203.

Chang Yu; (Sung Dynasty c. AD 1000); Chinese philosopher, historian, and commentator to Sun Tzu, *The Art of War*. *Quoted*: 348, 398.

Charles XII (1682–1718); King of Sweden; brilliant soldier and one of the 'Great Captains' of military history; a lover of war and soldiering, he formed his reputation at as young an age as Alexander by defeating a coalition of Russia, Denmark, Poland and Saxony; however his constant campaigning exhausted Sweden and his defeat at Poltava (1709) destroyed much of the Swedish Army; he was killed at the Siege of Frederikstad in Norway. *Quoted*: 226, 462.

Charles, Archduke of Austria (1771–1847); Austrian General; distinguished himself in wars of the French Revolution and Napoleon; instituted sweeping army reforms (1805); inflicted the first defeat on Napoleon at the Battle of Aspern–Essling (1809) and yielded a Pyrrhic victory to him at the following Battle of Wagram (1809); author of works on strategy. *Quoted*: 98, 248, 411.

Chia Lin; (Tang Dynasty AD 618–905) Chinese commentator to Sun Tzu, *The Art of War*. *Quoted*: 75.

Chiang Kai-shek (1887–1975); Chinese general and politician; trained and served with Japanese Army (1910–11); helped establish Chinese Republic with Sun Yat-sen; developed and led Kuomintang Army to become President of Nationalist China; waged war on the Communists; leader in defence of China against Japanese; lost post-war civil war against Communists; retreated to Taiwan with his followers to maintain Government. *Quoted*: 255, 375.

Chuikov, Vasily Ivanovich (1900–1982); Marshal of the Soviet Union; Soviet adviser to Chiang Kai-shek; led defence of Stalingrad (1942); commanded élite 8th Guards Army in the capture of Berlin (1945); accepted surrender of Germany for the Soviet Union. *Quoted*: 20, 110, 124.

Churchill, Sir Winston Leonard Spencer (1874–1965); British statesman, historian, journalist, and soldier; one of the greatest men of the twentieth century – a true Renaissance man of talent and vision in government, foreign policy, and war; as Prime Minister, he rallied the free world in its war against the Axis (1940–45). *Quoted*: 43, 44, 49, 59, 61, 79, 80, 83, 88, 91, 108, 113, 114, 119, 122, 140, 147, 151, 159, 178, 200, 203, 223, 239, 244, 245, 250, 264, 267, 272, 280, 284, 285, 290, 311, 330, 334, 344, 350, 351, 361, 367, 370, 383, 384, 388, 392, 396, 401, 412, 426, 429, 434, 445, 447, 455, 457, 465, 466, 468, 475, 483, 485, 486.

Chu Teh (1886–1976); Chinese general; one of the original founders of the Peoples' Liberation Army, commanded in war against Japan and in Chinese Civil War. *Quoted*: 261.

El Cid Campeador (El Cid – Rodgrigo Diaz de Vivar); Castillian soldier and Spanish national hero subject of legends and epics

Bahrain (1976–7); Deputy Chief of Naval Operations (1977–80); commander-in-chief, Allied Forces South Europe (1980–3); commander-in-chief, Pacific (1983–5); Chairman of the Joint Chiefs of Staff (1985–9). *Quoted*: 302.

Cunningham, Andrew Browne (1883–1963); 1st Viscount Cunningham; Admiral, Royal Navy; commander-in-chief of British naval forces in the Mediterreanean during World War II through 1943; served as First Sea Lord and Chief of Naval Staff (1943–46). *Quoted*: 37, 437.

Cyrus the Younger (*c.* 423–401 BC); Persian satrap and usurper; as satrap of western Asia Minor, he gathered a large army including Xenophon and his 10,000 Greeks to overthrow the Great King but was slain in battle at Cunaxa. *Quoted*: 442.

Dahlgren, John Adolphus Bernard (1809–1970); Rear Admiral, US Navy; promoted the establishment of a fully-fledged ordnance department for the Navy; developed a number of new guns including the famous Dalhgren Gun; appointed Chief of the Bureau of Ordnance (1862) at the insistence of President Lincoln; promoted to rear admiral (1863), he commanded the South Atlantic Blockading Squadron, attempted the impossible reduction of the defences of Charleston, and took part in the capture of Savannah (1864); after the war, he commanded the South Pacific Squadron, and served again as chief of the Bureau of Ordnance and commandant of the Washington Navy Yard. *Quoted*: 399.

Daidoji Yuan; Japanese samurai, Confucian scholar, expert in military affairs, and author of *The Code of the Samurai*; wrote during the early 17th century. *Quoted*: 69.

Dayan, Moshe (1915–1981); Israeli soldier and politician; founded Haganah militia (1939); commanded Jerusalem front in War (1948); Chief of Staff of Israeli Defence Force (1953–8); credited with success of the Sinai Campaign (1956); planned and conducted Six Day War (1967); Minister of Defence (1967, 1969–71); Foreign Minister (1977–79). *Quoted*: 120, 169, 181, 221, 302.

Decatur, Stephen (1799–1820); Admiral, US Navy; daring and skilled commander during the war with the Barbary Pirates and the War of 1812. *Quoted*: 311.

Defoe, Daniel (1659–1731); English journalist and novelist; served as government writer and secret agent; best known for his novels *Robinson Crusoe, Moll Flanders, Journal of the Plague Year,* and *Memoirs of a Cavalier*. *Quoted*: 37.

De Gaulle, General Charles-Andres-Marie-Joseph (1890–1970); French soldier and statesman; one of the earliest advocates of armoured warfare; as brigadier general in 1940 organized the Free French forces that kept alive French sovereignty and national spirit and played an important role in the liberation of France; headed two post-war governments, freed French colonies, and restored France to its former rank in world affairs. *Quoted*: 18, 30, 64, 72, 88, 90, 123, 142, 146, 183, 191, 195, 196, 198, 208, 221, 225, 226, 228, 229, 236, 261, 306, 315, 324, 334, 342, 344, 379, 386, 397, 425, 427, 446, 458, 467, 470, 474.

Demosthenes (384–322 BC); Athenian orator and politician; regarded in antiquity as the finest of the Athenian orators; as a politician consistently opposed Macedonian expansion and proposed the alliance with Thebes that led to their crushing defeat at Chaeronea in 338. *Quoted*: 72, 84, 261.

Dietrich, Josef 'Sepp' (1892–1966); Colonel General, Waffen SS; veteran of the German World War I tank corps, he became an ardent follower of Hitler and commanded his bodyguard which grew into the Waffen SS and commanded its leading combat unit in its expansion from battalion to army; he was noted for his aggressiveness and the loyalty of his men and earned his reputation for ruthlessness in his recapture of Kharkov (1943) and the Ardennes (1944). *Quoted*: 326.

Dill, Sir John Greer (1881–1944); General, British Army; served in the Boer War and World War I; served as director of military operations and intelligence (1935–6); commanded I Corps in France (1940); served as Chief of the Imperial General Staff (1940–1) and as the chief British representative on the Anglo-American board of strategy (1941–4). *Quoted*: 363.

in First World War (1917–18); helped plan the first great tank assault in history at Cambrai (1917) and established the perfect role of the tank commander – by leading his tanks from the front, possibly the only general officer to do so in all of the Great War. *Quoted*: 163, 271, 333, 429.

Emerson, Ralph Waldo (1803–82); American essayist and poet; emerged as the leader of a period of literary brilliance in New England (*c.* 1835–65) and as an anti-slavery activist. *Quoted*: 351.

Erfurth, Waldemar; General, German Army; served in World Wars I and II, writing a major work on the element of surprise in war; in World War II, he headed the historical section of the General Staff, supervised German Military publications and later headed German military commission to Finland in 1943. *Quoted*: 54, 417.

Esarhaddon (*d.* 669 BC); King of Assyria; rebuilt Babylon destroyed by his father, Sennacherib; campaigned successfully against Chaldeans and Medes and conquered Egypt (671 BC). *Quoted*: 23.

Eugene (François-Eugene-Maurice de Savoie-Carignan) (1663–1736); Prince of Savoy; Field Marshal in Austrian Service; originally he was the French Comte de Soissons, but lack of advancement caused him to renounce his country and seek fame in Austrian Service in which he became possibly the most famous soldier of his age in wars against the French and Turks. *Quoted*: 226, 241, 270.

Euripides (*c.* 484–*c.* 406 BC); greatest of the Athenian tragedians; his plays showed a unique understanding of the consequences of war. *Quoted*: 106, 112, 117, 119, 120, 227, 315, 316, 394, 450, 453.

Ewell, Richard Stoddert (1817–72); Lieutenant General, Confederate States Army; commissioned in the US Army, he served in the Mexican War and against the Apaches in New Mexico; served with Stonewall Jackson in the Shenandoah Valley; lost a leg at Second Bull Run; returning to duty, he commanded Jackson's II Corps at Gettysburg where he failed to crush the Union right flank; fought General Grant to a standstill at Wilderness (1864) and held the 'Bloody Angle' at Spotsylvania a few days later; captured with his corps just before the surrender at Appomattox (1865). *Quoted*: 210.

Farragut, David Glasgow (1801–1870); Admiral, US Navy; commanded the squadron that broke the Confederate hold on the mouth of the Mississippi River during the Civil War and seized New Orleans; defeated Confederate squadron in the Battle of Mobile Bay (1864); greatest Union naval officer of the Civil War. *Quoted*: 127, 176, 204, 284, 484.

Fernandez de Cordoba, Gonzalo (1453–1515); Spanish soldier known as el Gran Capitan, 'The Great Captain', – the origin of the term; distinguished himself in the wars against the Moors; won repeated and brilliantly fought victories against the French in Italy and the Turks in Greece.

Fernando I (1751–1825); King of the Two Sicilies; a weak, inept, cruel and reactionary ruler; joined Anglo-Austrian coalition against France (1793) and was driven out of his Kingdom of Naples by the French; he ruled Sicily until restored to Naples in 1815 when he joined the kingdoms; his cruelty sparked a revolution which forced him to grant a constitution; he overthrew the constitutional government and began a period of repression and brutality. *Quoted*: 448.

Ferry, Abel; Lieutenant, French Army; a deputy of the French parliament, was mobilized as a reservist and in 1916 wrote an influential pamphlet on the effects of modern weapons on massed infantry attacks, decrying the slaughter of French troops. *Quoted*: 267.

Fisher, John Arbuthnot (1841–1920); 1st Baron of Kilverstone; Admiral, Royal Navy; largely responsible for preparing the Royal Navy for World War I; disapproved of the Gallipoli Campaign and resigned in protest. *Quoted*: 292, 343, 353, 388, 416.

Foch, Ferdinand (1851–1929); Marshal of France; as instructor of strategy in the Ecole de Guerre, heavily influenced the French Army to adopt the offensive tactics that failed so badly in 1914; served initially as a corps commander and played an important part in the French victory in the First Battle of the Marne (1914); eventually became supreme commander of Allied forces in 1918 and orchestrated the Allied victory, having modified his theories to match realities; wrote *Principles of War* (1913) and *Precepts and Judgments* (1919). *Quoted*: 18, 21, 25, 36, 43, 48, 78, 79, 86, 92, 93, 128, 136, 140, 146, 153, 162, 175, 176, 192, 200, 216, 246, 251, 258, 267, 283, 285, 287, 290, 292, 295, 323, 356, 361, 369, 386, 416, 419, 431, 440, 447, 454, 477.

Foertsch, Herrmann; German General Staff Colonel; military writer who wrote the influencial *Modern Warfare* (1940). *Quoted*: 404, 406, 424, 447.

Folard, Chevalier Jean-Charles de (1669–1752); served with the French Army in War of Spanish Succession; with Knights of Malta against the Turks (1713–14); and with Swedes under Charles XII (from 1714); wrote *Nouvelles découvertes sur la guerre* (1724) which advocated infantry columns in the attack rather than linear formations. *Quoted*: 110, 464.

Forrest, Nathan Bedford (1821–1877); Lieutenant General, Confederate States Army; entirely unschooled in the military arts, he won the admiration of leaders of both sides of the Civil War for his instinctive grasp of strategy and tactics, his courage, daring, and concern for his men; known as 'That Devil Forrest' by Union forces, he was possibly one of the greatest soldiers of the war. *Quoted*: 99, 174, 293.

Fortescue, Sir John; British military historian; the most eminent and perceptive of the Victorian military historians. *Quoted*: 83.

Franco, Francisco (1892–1975); Spanish soldier and ruler; led the anti-Communist forces against the Communist dominated Loyalists during the Spanish Civil War; ruled Spain til his death; although a Fascist and supporter of the Axis, he remained neutral during World War II. *Quoted*: 215.

Fraser, Sir David (1920–); General, British Army; served in World War II as a regimental officer and went on thereafter to command at battalion through division levels; served as Director of Plans, assistant Chief of Defence Staff, Vice-Chief of the General Staff, British Military Representative to NATO, and finally as Commandant of the Royal College of Defence Studies in London. *Quoted*: 24, 161.

Frederick II 'The Great' (1712–1786); King of Prussia; warlike and aggressive Soldier-King considered to be one of the 'Great Captains' of history; engaged in a number of wars to enlarge Prussia; his model of the Prussian Army was largely imitated throughout Europe; wrote *Military Instructions to His Generals*, 1747. *Quoted*: 19, 22, 26, 31, 35, 41, 46, 47, 58, 60, 61, 62, 63, 74, 75, 78, 97, 98, 112, 128, 129, 132, 135, 138, 152, 157, 161, 162, 167, 175, 177, 180, 183, 185, 197, 207, 218, 227, 239, 243, 253, 254, 257, 260, 263, 270, 275, 280, 283, 286, 290, 291, 292, 294, 296, 312, 327, 337, 346, 348, 349, 354, 356, 357, 362, 373, 379, 380, 389, 391, 405, 414, 432, 439, 440, 446, 464, 467, 475, 477, 481.

Frederick of Prussia, Prince (NFI). *Quoted*: 219.

Frederick William I (1688–1740); King of Prussia; spent most his reign improving and strengthening economically his kingdom, particularly his army; he left his son, Frederick the Great, a strongly established state. *Quoted*: 332.

French, Sir John Denton (1852–1925); 1st Earl of Ypres; Field Marshal, British Army; served with distinction in the Nile Expedition (1984–5) and Boer War (1899–1901); served as Chief of the Imperial General Staff (1912–4); commanded British Expeditionary Force in World War I (1914–5); although the BEF performed well, his leadership and understanding of modern war were inadequate and resulted in heavy casualties. *Quoted*: 369.

Freytag-Loringhoven, Hugo von (*d.* 1921); Baron, Major General, German Imperial Army; military writer of renown; served as Deputy Chief of Staff in World War I; wrote *The Power of Personality in War* (1911), *Generalship in the World War* (1920). *Quoted*: 111, 188, 191, 209, 220, 235, 356, 380, 466.

Frunze, Mikhail Vasilyevich (1885–1925);

one of the founders of the Red Army; major commander in Russian Civil War against the Whites; developed unified military doctrine directly opposed to Trotsky. *Quoted*: 103, 146, 178, 205, 218, 240, 264, 299, 374, 423, 470, 478.

Fuller, John Charles Frederick (1878–1966); Major General, British Army; father of modern armoured warfare; his ideas on use of tanks decisively influenced Germans and Soviets; also an advocate of military reform and a rigorous, intellectually-vigorous professionalism in the British Army; prolific writer. *Quoted*: 20, 28, 29, 30, 34, 36, 38, 52, 56, 64, 71, 95, 99, 104, 106, 117, 130, 132, 145, 146, 149, 153, 155, 162, 168, 174, 175, 189, 191, 195, 197, 200, 201, 202, 211, 213, 220, 223, 224, 225, 247, 248, 264, 281, 287, 288, 291, 306, 308, 318, 324, 339, 341, 347, 361, 365, 370, 384, 391, 401, 403, 407, 416, 417, 421, 422, 425, 448, 457, 463, 474, 475, 485.

Gabriel, Richard Alan (1942–); Lieutenant Colonel, US Army Reserve; military writer of the Post-Vietnam era specializing in military reform issues touching on ethics, cohesion, and leadership; wrote *Crisis in Command* (1978). *Quoted*: 230, 237, 298, 408.

Gallery, Daniel Vincent (1901–77); Rear Admiral, United States Navy; in 1944 while in command of an aircraft carrier and five destroyers, he captured *U-505*, the first capture of an enemy warship on the high seas by a US warship since 1815; served as assistant chief of naval operations (1946–51), followed by command of a carrier division (1951–2), the Atlantic Fleet anti-submarine force (1952), and finally by a number of shore commands; authored *Clear the Decks* (1951), the story of the capture of *U505*. *Quoted*: 274.

Gallili, Yisrael (General, Israeli Defence Force). *Quoted*: 143.

Galway (Henri Massue) (1648–1720); 1st Earl of Galway; *aide-de camp* to Turenne; commanded regiment of French Huguenots in service of William III; served as commander-in-chief in Ireland and later of English forces in Portugal (1706–7) in War of Spanish Succession. *Quoted*: 160.

Garibaldi, Giuseppi (1807–1882); Italian military and national leader; served in armies of Sardinia and the Roman Republic; led 1,000-man expedition known as the 'Redshirts' to Sicily and overthrew Kingdom of Two Sicilies which united with Sardinia and proclaimed Italian Kingdom. *Quoted*: 382.

Gavin, James Maurice (1907–); Lieutenant General, US Army; commanded 82nd Airborne Division in World War II as youngest US division commander; he was a vociferous post-war critic of the downgrading of US ground forces and later of the Vietnam War. *Quoted*: 120, 468.

Gehlen, Reinhard; General, German Army; Chief of Foreign Armies East (1943–45) military intelligence for the Eastern Front; after the war cooperated with the Allies and eventually created West Germany's intelligence service. *Quoted*: 78, 218, 259.

Genghis Khan (c. AD 1162–AD 1227); Mongol conqueror; united Mongol and other steppe tribes, created one of the most efficient, mobile, and deadly military systems in history; conquered northern China, Khwarezm Empire, Iran, and part of Russia. *Quoted*: 316, 453.

Getman, Andrey Lavrent'yevich (1903–); Army General, Soviet Army; in World War II, commanded successively as commander of a tank division, corps, and as deputy commander of 1st Guards Tank Army; after the war served as commander of an army and a military district. *Quoted*: 289.

Giap, Vo Nguyen (1910–); General, Vietnamese Army; founder and leader of the Vietnamese military arm of the Communist Party which became North Vietnamese Army; commanded the Vietnamese forces which defeated the French (1954); forced the United States into a protracted war that led to its strategic defeat; destroyed the South Vietnamese Army and finally conquered the South (1975). *Quoted*: 74, 100, 145, 214, 376, 461, 462.

Gibbon, Edward (1737–1794); temporary captain in the South Hampshire militia during the Seven Years War; consummate historian; wrote *Autobiography* and *The Rise and Fall of the Roman Empire. Quoted*: 262.

Glover, Sir James Malcolm (1929–); General, British Army; served in as Brigade Major, 48 Gurkha Infantry Brigade (1960–2); commanded 3rd Battalion Royal Green Jackets (1970–1), 19 Airportable Brigade (1974–5), Land Forces, Northern Ireland (1979–80), and UK Land Forces (1985–7); served as Deputy Chief of Defence Staff (Intelligence) (1981–3) and as Vice Chief of General Staff and Member of the Army Board (1983–5). *Quoted*: 65.

Glubb, John Bagot (1897–1986); British Soldier; veteran of World War I, he later served in Iraq where he gained a reputation as friend of the Arabs; organized and commanded Arab Legion (1939–56) in Palestine which later became the Jordanian Army; known as 'Glubb Pasha' he combined respect for Arab traditions with introduction of European techniques and methods; his legion gained legendary reputation for its efficiency and was only Arab army to inflict defeats on the Israelis in 1948. *Quoted*: 217, 482.

Gneisenau, August Wilhelm von (1760–1831); Field Marshal, Prussian Army; prominent reformer of the Prussian Army 1807–09; served as Blücher's chief of staff 1813–1816; played critical role in redirecting the Prussian Army after its defeat at Ligny to victory at Waterloo. *Quoted*: 36, 286, 314, 405.

Goltz, Wilhelm Leopold Colmar von der (1843–1916); Field Marshal, German Imperial Army; also called 'Goltz Pasha'; recognized the Turkish Army (1883–96); military governor of Belgium; commanded Turkish First Army in Mesopotamia that invested and captured a British army at Kut (1916); also known as a military theorist and writer, *The Nation in Arms* (1883) and *The Conduct of War* (1908). *Quoted*; 58, 103, 114, 128, 155, 171, 223, 423, 456, 465.

Gonzalo de Cordoba 'El Gran Capitan' (Gonzalo Fernandez de Cordoba) (1453–1515); distinguished in wars against the Moors and one of two commissioners to receive surrender of Granada (1492); defeated the French in Italy (1495–7), captured Cephelonia from the Turks (1500); defeated French again in Italy (1502–3); the first of the 'Great Captains'. *Quoted*: 278.

Gordon, John Brown (1832–1904); Lieutenant General, Confederate States Army; served with the Army of Northern Virginia and commanded a wing of the Army by the end of the Civil War; dynamic young officer, lieutenant general at age of thirty-three; wrote *Reminiscences of the Civil War* (1904). *Quoted*: 46, 306.

Gorshkov, Sergei Georgyevich (1910–1988); Admiral of the Fleet, Soviet Navy; distinguished himself in World War II in actions with the Black Sea Fleet; made chief of the Soviet Navy in the 1950s, a post he retained for thirty years; in that time he advocated a strong blue-water navy and oversaw its construction to rival the US Navy. *Quoted*: 65, 345, 362, 389.

Grandmaison de; Colonel, French Army; Foch's prize pupil at the French Staff College (c. 1908–13) and chief proponent of his offensive strategy that led the French Army to appalling losses in 1914–15. *Quoted*: 43.

Grant, Ulysses S. (1822–1885); General of the Army, US Army, and 18th President of the United States; served in the Mexican War but left the Army for seemingly oblivion; rejoined the Army in the Civil War and gave the Union its first serious victory when he captured Fort Donelson in 1862; rose through a series of victories in the West to command of all the Union armies and eventually forced Lee's surrender in 1865; Grant was the first great commander to understand the nature of modern war and apply it successfully; his *Memoirs of U.S. Grant* (1885) are a model of American prose and lucid military exposition. *Quoted*: 26, 32, 37, 73, 94, 103, 107, 113, 132, 136, 149, 158, 220, 254, 266, 270, 304, 309, 316, 343, 352, 359, 362, 375, 385, 387, 392, 418, 450, 452, 468.

Graves, Robert Ranke (1895–1985); British poet, novelist, and critic; served in First World War and later was a close friend of T. E. Lawrence and Liddell Hart; wrote romantic poetry in traditional metres and

Allied naval forces in South Pacific (1942–4); defeated Japanese in battle for the Solomons; assumed command of US Third Fleet (1944–5) and defeated the Japanese in the battle of Leyte Gulf (1944); promoted Fleet Admiral (1945). *Quoted*: 108.

Hamilton, Sir Ian Montieth (1853–1947); General, British Army; distinguished colonial general and military writer; commanded Gallipoli Campaign (1915) and was relieved of command as a scapegoat; wrote *Soul and Body of an Army* (1920) and *Gallipoli Diary* (1921). *Quoted*: 58, 59, 114, 117, 170, 191, 209, 220, 223, 267, 297, 306, 308, 310, 311, 315, 321, 347, 359, 366, 386, 389, 396, 426, 454, 456, 457, 483.

Hammerstein Equord, Kurt von; General, German Army (NFI). *Quoted*: 297.

Hannibal (274–183 BC); Carthaginian General; one of the most daring 'Great Captains', in history, he was a sworn enemy of Rome; crossed the Alps to take the Second Punic War to Rome; won a number of crushing victories of which Cannae (216 BC) was the greatest; ravaged Italy but could not break Rome; recalled to Carthage to deal with a Roman invasion and was defeated at Zama (202 BC) by Scipio Africanus; headed Carthaginian government until forced into exile by Rome; served Antiochus II of Syria against the Romans until the defeat at Magnesia (190 BC); hounded by the Romans, he committed suicide rather than fall into their hands. *Quoted*: 80, 135, 148.

Harbord, James (1867–1947), Lieutenant General, US Army; served in the Cuban campaign and Philippine Insurrection; in World War I, served as Chief of Staff of AEF, commanded 4th Brigade of 2nd Infantry Division at Belleau Wood, later commanded 2nd Division, followed by command of Service of Supply which he performed with great distinction; became Chief of Staff of the Army in 1922. *Quoted*: 40.

Harmon, Ernest Nason (1894–); Major General, US Army; as commander successively of the 1st and 2nd Armored Divisions, he played a crucial role in the Battles of Cassino, Anzio, and the Bulge; commanded US Constabulary force in Germany during the Occupation; wrote *Combat Commander* (1970). *Quoted*: 274.

Harris, Sir Arthur (1892–1984); Air Marshal, Royal Air Force; commanded Bomber Command in World War II and was responsible for the British strategic bombing of Germany which led to the large-scale destruction of German cities; wrote *Bomber Offensive* (1947). *Quoted*: 59.

Harrison, Henry (16th Century British soldier). *Quoted*: 394.

Hattusili I (1650–1620 BC); King of Hittite Old Kingdom; moved capital to Hattusas and expanded empire into northern Syria. *Quoted*: 116.

Hawkwood, Sir John de (*d.* 1394); English mercenary; distinguished himself at Crécy and Poitiers and became a famous English condottiore in 14th-century Italy, serving Florence; known in Italian as 'Acuto'. *Quoted*: 207.

Hector (13th century BC); son of King Priam, the greatest of the heroes of Troy, and its war leader in the Trojan War, as described by Homer in *The Iliad*; Hector has been described as the greatest gentleman of all the heroes, a champion and shepherd of his people from a sense of duty rather than lust for glory as was the case with Achilles. *Quoted*: 60, 150, 173, 471.

Hemingway, Ernest Miller (1899–1961); American writer and war correspondent; served in First World War as a volunteer ambulance driver in Italy; served in Second World War as a war correspondent; won Nobel Prize for literature (1954); best known war novels were *A Farewell to Arms* (1929) and *For Whom the Bell Tolls* (1940). *Quoted*: 127.

Henderson, George Francis (1852–1903); Colonel, British Army; saw active service in India and Egypt; taught military history at Sandhurst and the Staff College; served as intelligence officer to British commander-in-chief during the Boer War; his particular interest was the American Civil War; his biography of Stonewall Jackson (1898) became a military classic; as an instructor and historian, he deeply influenced many future officers. *Quoted*: 205, 228, 339, 469.

Henry IV (1553–1610), King of France; often called 'Henry of Navarre'; founder of the Bourbon dynasty in France; successful Protestant general during the French religious civil wars; embraced Catholocism to win the French throne. *Quoted*: 73, 147, 289, 378.

Herodotus (*c.* 480–420 BC); Greek historian known as the 'Father of History' for his systematic treatment and mastery of style; travelled about the known world during the course of his research for his histories; best known for his history of the Graeco-Persian Wars (500–479 BC). *Quoted*: 213, 282.

Hill, Daniel Harvey (1821–1889); Lieutenant General, Confederate States Army; a suberb subordinate of both Generals Jackson and Longstreet, he served in the Seven Days Battle, Antietam, Fredericksburg, and commanded a corps at Chickamauga in 1863; one of Lee's favorite generals for whom he called while in delirium on his deathbed. *Quoted*: 151, 262, 346, 416.

Hindenburg, Paul von (1847–1934); Field Marshal, Imperial German Army; victor of Battle of Tannenberg, 1914, and from 1916 to 1918 iron-nerved Chief of the German General Staff; wrote *Out of My Life*, 1920. *Quoted*: 59, 67, 77, 96, 187, 188, 195, 242, 318, 348, 351, 394.

Hippocrates of Cos (*c.* 460–380 BC); Greek physician, the Father of Modern Medicine. *Quoted*: 249.

Hoffmann, Max; (*d.* 1929); Major General, German Imperial Army; brilliant staff officer responsible for the German riposte which destroyed two Russian armies in the Battle of Tannenberg (1914); served through most of World War I on the staff of the German armies on the Eastern Front; wrote *War Diaries and Other Papers* (1929). *Quoted*: 21, 127, 158, 200, 211, 477.

Holmes, Oliver Wendell (1841–1935); Union captain during the Civil War and later Associate Justice of the United States Supreme Court. *Quoted*: 85, 113, 182, 197, 364, 377, 462.

Homer (between 1050 BC and 850 BC); Ionian Greek epic poet and bard who composed the *Iliad* and the *Odyssey* and/or combined and polished various epic poems handed down in the bardic tradition from Mycenean times. *Quoted*: 49, 68, 80, 105, 112, 119, 126, 150, 154, 166, 173, 192, 262, 310, 326, 368, 373, 376, 380, 425, 448, 456, 458, 462, 471, 472, 479.

Hopton, Sir Ralph (1596–1652); 1st Baron Hopton; English general; originally of Puritan leanings, he served as a Royalist lieutenant general and won victories at Stratton and Landsdowne; created baron (1643); commanded Royalist army and forced to surrender (1646); left England with Prince Charles (1648). *Quoted*: 185.

Horrocks, Sir Bryan Gwynne (1895–1985); Lieutenant General, British Army; served in First World War as a platoon commander and in Second World War successively as 44th (HC) Division, 9th Armoured Division, XIII Corps, X Corps in Egypt and Africa, IX in Tunis, and XXX Corps in Europe; known as one of Field Marshal Montgomery's favourite and most trusted corps commanders; commanded Western Command (1946–8), British Army of the Rhine (1948–9). *Quoted*: 46, 182, 302.

Hottell, John Alexander (Major, US Army). *Quoted*: 260.

Ho Yen Hsi; Chinese commentator on Sun Tzu's *The Art of War*; no other information. *Quoted*: 121, 219, 331, 449.

Jackson, Andrew 'Old Hickory' (1867–45); Major General, United States Army, and later seventh President of the United States; commissioned major general of the Tennessee Militia (1802) and defeated Creeks at the battle of Horseshoe Bend (1814); commissioned major general, US Army (1814); captured Pensacola (1814) and defeated British army at the battle of New Orleans (1815); put down Seminole uprising (1818); served as president (1829–33) in which role he gave great encouragement to the westward expansion of the United States. *Quoted*: 72, 114, 151, 371.

Jackson, Robert (1750–1827); British military doctor; served in British Army during the American Revolution and later as the inspector-general of British hospitals during the Peninsular War; wrote a highly perceptive book on the art of war and leadership, *A View of the Formation, Discipline, and Economy of Armies*, 1804. *Quoted*: 101, 172, 190, 204, 222, 224, 289.

Jackson, Thomas (1824–1863); Lieutenant General, Confederate States Army; one of the great American military eccentrics and great battlefield fighting generals, he first established his reputation in 1862, clearing Union forces which heavily outnumbered him from the Shenandoah Valley in the Valley Campaign, considered a model of daring, mobility, and concentration; later commanded the 2nd Corps of the Army of Northern Virginia forming an invincible relationship with Lee that saved the Army at Antietam and brought victory at Fredericksburg and Chancellorsville in 1863 where he was accidently killed by his own men. *Quoted*: 20, 50, 85, 99, 120, 121, 136, 151, 170, 174, 203, 246, 283, 293, 329, 330, 335, 350, 391, 402, 411, 416, 473.

Jervis, John (1735–1823); Earl of St. Vincent; Admiral, Royal Navy; created earl for his victory over the French at Cape St. Vincent (1797); averted mutiny in the Channel Fleet (1798); commanded Channel Fleet (1799–1801 and 1806–07); as First Lord of the Admiralty (1801–04) eliminated corruption in naval dockyards and reformed naval administration. *Quoted*: 107, 372, 388, 411, 415.

Joan of Arc (1412–1431); Catholic Saint and French national heroine, also called 'The Maid of Orleans'; simple peasant maid who convinced the Dauphin, later Charles VII, that God had sent her to liberate France from the English invaders; aroused such enthusiasm throughout France and achieved a number of victories that changed the trend of the war which finally led to the defeat of the English; captured by the Burgundians who sold her to the English who burnt her at the stake as a witch; canonized in 1920. *Quoted*: 166.

Joffre, Joseph-Jacques-Césaire (1852–1931); Marshal of France; commander of the French Army at the beginning of World War I, (1914–16); saved France in the Battle of the Marne in 1914. *Quoted*: 372.

de Jomini, Antoine-Henri (1779–1869); Baron, Swiss soldier; Lieutenant General in Russian service; served as staff officer for Napoleon but changed sides to serve Allies; influential military writer in 19th century; wrote *Summary of the Art of War*, 1838. *Quoted*: 32, 33, 39, 42, 43, 46, 58, 66, 69, 75, 83, 84, 92, 99, 100, 107, 110, 114, 163, 167, 186, 188, 199, 248, 254, 258, 260, 262, 270, 278, 294, 297, 310, 333, 338, 343, 348, 350, 374, 389, 402, 421, 422, 433, 456, 474, 482.

Jones, John Paul (1747–1792); Admiral, US Navy; upon commission in the new US Navy in 1775, he was the first to raise the US Grand Union Flag at sea; his daring exploits in the American Revolution, especially the engagement between the USS *Bonhomme Richard* and HMS *Serapis*, helped win international recognition and gave birth to a national legend; regarded as the spiritual founder of the US Navy; later served Catherine the Great against the Turks. *Quoted*: 49, 118, 203, 253, 378, 387, 411, 415.

Josephus, Flavius (originally Joseph ben Mathias (*c.* 37–*c.* 100); Jewish historian and general; participated in Jewish Revolt (66) against Rome and held Jopata 47 days against Roman siege by Vespasian and then surrendered (67); won Vespasian's favour and romanized his name; accompanied him to Rome after fall of Jerusalem (70); wrote *A History of the Jewish War*, *Antiquities of the Jews* and an *Autobiography*. *Quoted*: 440.

Joubert, Barthelemy-Catherine (1769–99); general, French Army; served with distinction in Italy (1793–9); as chief of the Army of Italy (1798–9) and then under General Moreau (1799); killed at Novi. *Quoted*: 448.

Kai Ka'us Ibn Iskander (11th century); Turkic prince and warrior who wrote *A Mirror for Princes*, *c.* 1082 . *Quoted*: 19, 116, 166.

Katukov, Mikhail Yefimovich (1900–1976); Marshal of Tank Troops, Soviet Army; joined the Red Army in 1919 and served through Civil War; in World War II, served at brigade through tank army levels, finally commanding 1st Guards Tank

ies at 2nd Bull Run (1862), Fredericksburg, and Chancellorsville (1863); he lost the Battle of Gettysburg (1863) which became known as the high tide of the Confederacy; and continued to fight a stubborn defensive campaign against U.S. Grant until overwhelmed and forced to surrender at Appomattox (1865). *Quoted*: 17, 36, 70, 75, 85, 102, 130, 140, 151, 193, 261, 284, 352, 362, 372, 378, 419, 442, 462, 473.

Lefebrve, Pierre-François (1755–1820); Duc de Danzig; Marshal of France; rose from the ranks during the French Revolution to be one of Napoleon's marshals; captured Danzig (1807); commanded Imperial Guard (1812–14). *Quoted*: 74.

Lejeune, John Archer (1867–1942); Major General, US Marine Corps; commanded Marine 4th Brigade and later 2nd Infantry Division in France in 1918; served as Commandant of the Marine Corps 1920–9 and raised professional standards of Corps and introduced tactical doctrine and techniques of amphibious warfare. *Quoted*: 449.

Leland, Edwin Sterling (1936–); Major General, US Army; commander of the innovative US Army National Training Center (NTC) (1984–6) and later commander of the 1st Armored Division. *Quoted*: 224.

LeMay, Curtis Emerson (1906–1990); General, US Air Force; brilliant commander of strategic air force elements in World War II; conducted strategic bombing campaign against Japan; directed the Berlin Airlift in 1948; later commanded the Strategic Air Command (SAC)(1948–1957) and later Chief of Staff of the Air Force (1961–56); wrote *Mission with LeMay* (1965). *Quoted*: 53, 57, 88, 106, 124, 172, 177, 206, 315, 353.

Lenin, Vladimir Ilych (Vladimir Ilych Ulyanov) (1870–1924); Russian Communist revolutionary; with the help of Leon Trotsky, overthrew the weakened tsarist government in Russia and firmly fixed Soviet power in place for the next 74 years; master conspirator and organizer, he moulded the Bolsheviks into an effective and dynamic cadre able to seize power and hold it in the vicious Russian Civil War that followed; he instituted the most oppressive and destructive features of the regime: secret police, bureaucracy, concentration camps, and mass killings. *Quoted*: 99, 128, 176, 248, 339, 433, 451, 469, 470.

Leopold I 'The Old Dessauer' (1676–1747); Prince of Anhalt-Dessau; Prussian Field Marshal; commanded Prussian contingent during the War of the Spanish Succession, distinguishing himself at Hochstadt (1703), Blenheim (1704), Cassano (1705), Turin (1706), Tournai and Malplaquet (1709); defeated Charles XII at Strakund and Rügen (1715); moulded Prussian Army (1715–40) into a disciplined and efficient force that made possible the victories of Frederick the Great; introduced the musket ramrod and modern bayonet; retired after defeating the Austrians at Kesseldorf (1745). *Quoted*: 329.

Letterman, Jonathan (1824–72); Major, US Army; military doctor who saw much service in the Indian campaigns before the Civil War; appointed medical director of the Army of the Potomac, he completely reorganized the medical service by introducing a system of forward first-aid stations at the regimental level, mobile field hospitals behind divisions, and base hospitals, all linked by an efficient ambulance corps, and supported by an effective medical supply system; his organization was adopted by all other Union armies. *Quoted*: 249.

Lettow-Vorbeck, Paul von (1870–1964); Major General, Imperial German Army; commanded German forces in East Africa in one of the greatest economy of force operations in history; with fewer than 20,000 men, he drew one-third of a million Allied troops into the theater; known for chivalrous conduct and daring; wrote *East African Campaigns* (1957). *Quoted*: 153, 219, 221, 241, 365, 419.

Li Ch'uan; Tang Dynasty writer on military subjects and commentator on Sun Tzu's *The Art of War*. *Quoted*: 206, 299, 384.

Liddell Hart, Basil Henry (1895–1970); English military historian and strategist; experience in World War I led him to

death of Marshal Turenne (1675). *Quoted*: 73, 101, 441, 449.

Luck, Hans von (1911–); Colonel, German Army; protégé of Field Marshal Rommel, served as his reconnaissance officer both in France and North Africa; later rose to command 21st Panzer Division; wrote *Panzer Commander* (1990). *Quoted*: 71.

Ludendorff, Erich (1865–1937), General, Imperial German Army; brilliant *alter ego* to von Hindenburg in First World War as his Chief of Staff first on the Eastern Front (1914–16) then as Quartermaster General (1916–18); described as the brains of the team, he became the virtual dictator of Germany in the last half of the war; attempted to overthrow Weimar Republic in league with Hitler in 1923 *putsch*; wrote My *War Memories 1914–1918* (1919). *Quoted*: 26, 56, 76, 85, 140, 178, 186, 188, 208, 215, 220, 241, 247, 255, 268, 295, 323, 346, 360, 403, 425, 472.

Lyautey, Louis-Hubert (1854–1934); Marshal of France; colonial administrator and soldier; conquered Madagascar and pacified Morocco; Minister of War (1916–1917). *Quoted*: 454.

MacArthur, Douglas (1880–1964); General of the Army, US Army; one of the greatest of American commanders; Chief of Staff of the Army (1930–35); commanded victorious island-hopping advance of US and Allied Forces in the Southwest Pacific against Japan in World War II; as Supreme Commander of Allied forces in the Pacific, he accepted the surrender of Japan; ruled Japan as virtual proconsul; his reforms set Japan on the road to democracy and prosperity; as commander of UN Forces in Korea, his landing at Inchon (1950) resulted in destruction of most of the North Korean Army; he suffered a major reverse with the surprise entry of Chinese forces into the war; he was relieved by President Truman over refusal to accept the civilian authority in the conduct of the Korean War. *Quoted*: 35, 55, 70, 71, 87, 95, 109, 124, 126, 136, 162, 201, 209, 223, 227, 243, 268, 271, 297, 334, 340, 365, 372, 383, 400, 401, 435, 442, 455, 470, 472, 478.

McAuliffe, Anthony Clement (1898–1975); General, US Army; served with 101st Airborne Division in Normandy and Netherlands; as temporary division commander, he held the town of Bastogne in the path of the German offensive during the Battle of the Bulge, inspiring his men to a last-ditch stand. *Quoted*: 444.

McClellan, George Brinton (1826–85); Major General, US Army; served with distinction in the Mexican War and was an observer in the Crimean War, after which he left the service; commissioned major general (1861), he organized and trained the Army of the Potomac as an efficient fighting force; a brilliant organizer, he displayed little of the aggressiveness and daring of a great commander and consistently overestimated Confederate strength both in the Peninsular campaign and the Battle of Antietam (1862), throwing away great advantages. *Quoted*: 29.

McDowell, Irvin (1818–1885); Major General, US Army; served in the Mexican War; commanded the first Union Army in action in the Civil War and was defeated at First Bull Run, 1861. *Quoted*: 335.

Machiavelli, Niccolo (1469–1527), Italian political philosopher and military theorist; noted for his cold-blooded advocacy of power politics; wrote *The Prince*, 1513; *Discourses*, 1517; and *The Art of War*, 1521. *Quoted*: 23, 38, 49, 58, 62, 73, 84, 98, 121, 125, 132, 138, 147, 154, 183, 215, 222, 253, 256, 278, 279, 283, 304, 308, 312, 331, 348, 364, 368, 389, 390, 395, 402, 411, 431, 437, 446, 449, 456, 459, 463, 476.

McNair, Lesley James (1883–1944); General, US Army; participated in the Mexican Punitive Expedition (1916) and served in France on the staff of the American Expeditionary Force (AEF) (1917–19); in Second World War, he commanded the Army Ground Forces, responsible for mobilization, training and organization of the entire Army; he pioneered the simulation of battle conditions in training and insisted upon high levels of physical and mental preparation; designated to command the diversionary and imaginary 1st Army Group during the Normandy invasion, he was killed on a tour of the front by an American bomb that fell short. *Quoted*: 244, 291, 440, 442.

Mahan, Alfred Thayer (1840–1914), Rear Admiral, United States Navy; son of Dennis Hart Mahan; one of the first commandants of the US Naval War College and one of the great military theorists of the modern era; his book *The Influence of Sea Power on History* (1890) revolutionized naval strategy throughout the world. *Quoted*: 22, 52, 56, 62, 76, 79, 94, 98, 104, 127, 135, 148, 160, 181, 193, 200, 205, 222, 224, 244, 246, 249, 264, 284, 287, 293, 295, 311, 339, 342, 350, 372, 385, 388, 399, 406, 409, 412, 423, 431, 437, 456, 465, 469, 473, 477, 483.

Mahan, Dennis Hart (1802–1871); engineer and educator; dean of the faculty at West Point, he set an indelible professional seal of excellence on the institution through his rigorous instruction and broad sense of the military requirements of the Academy; nearly all the great Civil War generals were his students. *Quoted*: 17, 199, 222, 286, 338, 399, 421.

Makarov, Stepan Osipovich (1849–1904); Admiral, Russian Imperial Navy; hero of the Russo–Turkish War (1878); distinguished and professional officer; briefly commanded Russian naval forces at Port Arthur (1904) until killed by mine that sank his flagship. *Quoted*: 24, 134, 392.

Malone, Dandridge M. (Colonel, US Army). *Quoted*: 442.

Manchester, Edward Montagu (1602–71); 2nd Earl of Manchester; one of the twelve peers who petitioned King Charles I to summon the Long Parliament; major general of the eastern counties, he was in nominal command at the battle of Marston Moor (1644); accused by Cromwell of incompetence, he resigned (1645); opposed trial of the king and creation of the Commonwealth and eventually supported the restoration of the Stuarts. *Quoted*: 458.

Mannerheim, Carl Gustav Emil von (1867–1951); Finnish statesman and field marshal; originally served in Russian Army (1887–1917) in which he won distinction in Russo-Japanese War (1904–5), rising to the rank of lieutenant general; after the 1917 October Revolution, he organized the White Guards in Finland (1918) and suppressed the Bolsheviks; planned and built the Mannerheim Line of defences against which the Red Army battered itself in the Winter War (1940–1); commanded the Finnish Army in its tenacious defence and later in the recapture of lost Finnish territory in Second World War; as president of Finland (1944–6), he extracted Finland from the war without it succumbing to a communist takeover. *Quoted*: 82, 156, 334, 348.

Manstein, Fritz Erich von (1887–1973); Field Marshal, German Army; conqueror of the Crimea in 1942; saved German position on the Eastern Front after Stalingrad in brilliant Third Battle of Kharkov in 1943; wrote *Lost Victories*, 1957. *Quoted*: 91, 285.

Manteuffel, Hasso von (1897–1978); General der Panzertrupen, German Army; one of the youngest and most able of the German Panzer commanders; commanded Panzergrenadier Division *Grossdeutschland* and later Fifth Panzer Army in the Battle of the Bulge (1945). *Quoted*: 152, 319.

Mao Tse-tung (1893–1976); Chinese soldier and statesman; one of the founders of the Chinese Communist Party, he eventually became its leader and foremost theoretician; established the Peoples Liberation Army (PLA) and led it in civil war against the Nationalists, then played a major role in the war against the Japanese invaders of China, and finally destroyed the Nationalist armies and government in the Civil War (1949); led China into the Korean War against the United Nations; domestic reforms conducted with great bloodshed, and his attempts to decentralize government, reduce the power of the bureaucracy, and reinvigorate revolutionary principles brought China repeated catastrophes and hardships. *Quoted*: 17, 37, 77, 90, 100, 121, 142, 179, 187, 205, 218, 265, 287, 300, 309, 324, 334, 357, 376, 424, 460, 461, 466, 470, 478.

Marius, Gaius (*c*. 157–86 BC); Roman general and politican; victor over Jugurtha in 106;

he reorganized Roman Army and dealt utter defeat to the invading Cimbri and Teutones (101/102); engaged in civil war with Sulla. *Quoted*: 174.

Marlborough (John Churchill) (1650–1722); Duke of Marlborough; English soldier and statesman; one of the 'Great Captains', the military genius that commanded the coalition armies in the wars against French expansionism under Louis XIV. *Quoted*: 167, 238, 257, 453.

Marshall, George Catlett (1880–1959); General of the Army, US Army; as the junior general officer in the Army he was promoted to Chief of Staff by President Roosevelt; as such he oversaw the expansion and strategy of the U.S. Army in World War II; called 'The Organizer of Victory'. *Quoted*: 43, 72, 86, 89, 155, 179, 216, 231, 235, 236, 252, 265, 272, 290, 314, 363, 371, 382, 396, 404, 407.

Marshall, Samuel Lyman Atwood (1900–1977); Brigadier General, US Army Reserve; military historian and analyst of men in combat whose works, especially *Men Against Fire* (1947) had a profound effect on the training in the Army after World War II; prolific military writer of small unit actions. *Quoted*: 21, 27, 40, 46, 47, 52, 64, 81, 87, 88, 90, 97, 100, 102, 105, 114, 115, 123, 124, 131, 133, 136, 143, 144, 145, 155, 160, 164, 169, 172, 174, 183, 203, 205, 206, 207, 209, 212, 217, 218, 221, 224, 227, 231, 232, 236, 238, 241, 243, 255, 261, 266, 267, 271, 272, 273, 276, 277, 280, 282, 285, 298, 300, 301, 307, 310, 311, 316, 318, 321, 336, 339, 366, 372, 379, 384, 393, 398, 401, 402, 404, 407, 408, 410, 412, 413, 430, 444, 445, 446, 459, 474, 478, 484.

Martel, Giffard Le Quesne (1889–1958); Lieutenant General, British Army; one of the founders of the British Royal Tank Corps in World War I and continuous advocate of mechanized warfare; his writings were more appreciated and applied by the Germans than the British; Commander, Royal Tank Corps (1942). *Quoted*: 40, 230.

Mauldin, William 'Bill' Henry (1921–); American editorial cartoonist and author; as an Army sergeant, he became famous for his irreverent cartoon series of two GIs, 'Willie and Joe', which appeared in the Army overseas newspaper, *Stars and Stripes* and for which he received a Pulitzer Prize (1945). *Quoted*: 47, 216, 309, 482.

Maurice (AD 539–602) (Mauricius Flavius Tiberius); East Roman Emperor and general; defeated Persians and Avars; wrote *The Strategikon*, *c.* AD 600. *Quoted*: 19, 95, 101, 104, 105, 106, 107, 109, 121, 122, 127, 132, 138, 156, 166, 178, 182, 185, 194, 202, 207, 210, 229, 231, 233, 238, 240, 243, 249, 253, 257, 263, 282, 292, 298, 299, 303, 322, 326, 328, 331, 346, 347, 348, 349, 353, 357, 368, 378, 389, 390, 392, 397, 398, 401, 425, 431, 436, 438, 440, 449, 456, 476, 481.

Meinertzhagen, Richard (1878–1967); colonel, British Army and archaeologist, ornithologist, and Zionist; one of the great English military eccentrics; brilliant and ruthlessly effective as the theater intelligence officer during World War I in East Africa (1914–16) and Palestine (1917–18) under Allenby; his deception at Bersheba allowed Allenby to break the Turkish hold on Palestine and later Syria; wrote *Army Diary*, 1960. *Quoted*: 175, 236, 345.

Mellenthin, F. W. von (1904–); Major General, German Army; distinguished General Staff officer who served with Rommel in North Africa, as chief of staff for 48th Panzer Corps, Fourth Panzer Army; and Army Group G, West; wrote *Panzer Battles* (1971). *Quoted*: 274, 436, 479.

Menelaus (13th century BC); King of Sparta and one of the great Greek heroes in the Trojan War for the return of whose stolen wife, the Beautiful Helen, the war was fought – as described by Homer in *The Iliad*. *Quoted*: 112.

Menon, V. K. Krishna (1897–1974); Indian diplomat and Minister of Defence in the early 1960s. *Quoted*: 279.

Menzies, John; author of *Reminiscences of an Old Soldier* (1883) (NFI). *Quoted*: 357.

Merneptah (13th century BC); Pharaoh of Egypt (1236–1223 BC); son of Ramses II; defeated the great invasion of Egypt by a coalition of the Sea Peoples and Libyans. *Quoted*: 346, 453.

the Whiskey Rebellion (1794). *Quoted*: 262.

Morshead, Leslie (1890–?); Lieutenant General, Australian Army; served in World War I and attained battalion command at age 26 and earned great distinction; served as a reservist between the wars; commanded 9th Infantry Division in its epic defence of Tobruk in 1941; later served as corps commander in South-west Pacific theatre; known for his aggressive tactics and leadership. *Quoted*: 218, 327.

Mortimore, Captain H. D. (NFI). *Quoted*: 425.

Mosby, John Singleton (1833–1966); Colonel, Confederate States Army; incomparable commander of Confederate irregular forces in the Shenandoah Valley during the Civil War; tied down large numbers of Union troops in security operations performing a major economy of force operation for General Lee; wrote *Mosby's War Reminiscences* (1887). *Quoted*: 52, 94, 163, 171, 177, 208, 239, 267, 355, 412, 460, 473.

Moser, Otto von (General, German Army). *Quoted*: 39.

Moskalenko, Kirill Semenovich (1902–?); Marshal of the Soviet Union; joined Red Army in 1920 and served in Civil War; after the war served as artillery officer at regimental through corps levels; in World War II, he commanded 38th, 1st Tank, and 1st Guards, and 40th Armies with great distinction, particularly in crossing the Dnieper and liberating Kiev with 38th Army. After the war, served as commander of the Moscow Air Defence Region, Moscow Military District, Strategic Rocket Forces, and as deputy minister of defence. Reputed to be the officer who personally either arrested or shot Chief of NKVD, Beria, in 1953 in alliance with Khrushchev. *Quoted*: 109, 179.

Mountbatten, Louis (1900–1979); 1st Earl Mountbatten; British Admiral and statesman; chief of Combined Operations (Commandos) (1942–43); Supreme Allied Commander, Southeast Asia theater (1943–46); directed reconquest of Burma; as Viceroy of India (1947); he oversaw the transfer of power to India and Pakistan; First Sea Lord (1955–59); Chairman, UK Defence Staff and Chiefs of Staff Committee (1959–65); murdered by Irish Republican Army (IRA) terrorists. *Quoted*: 148, 420.

Nabeshima Naoshige (AD 1538–1618); Japanese general and warlord, assisted rise of Tokugawa Ieyasu; wrote *Lord Nabeshima's Wall Inscriptions*. *Quoted*: 204, 409.

Nakhimov, P. S. (1802–55); Admiral, Imperial Russian Navy; one of the most aggressive and successful admirals in the Imperial Russian Navy; won major victories against the Turks in the Crimean War and was the motivating force in the defense of Sebastopol in which he was killed. *Quoted*: 56, 193, 234, 395.

Napier, Sir William (1785–1860); Lieutenant General, British Army; served in Portugal 1809–11; retired and became Governor of Guernsey; wrote *History of the War in the Peninsula* (1828–40) and *History of the Conquest of Scinde* (1844–6). *Quoted*: 190, 215, 254, 284, 374, 411, 447, 452, 482.

Napoleon I, (1769–1821); Emperor of the French; one of the greatest of the 'Great Captains' in history; rose from obscurity during the French Revolution to become in succession most successful general of the Republic, First Consul, and finally Emperor (1804); clearly a military genius and leader of extraordinary talents who brought all the military advances and reforms of his time together into a lethal military technique that destroyed all the traditional armies of Europe; he was finally defeated by over-extension and exhaustion of his forces due to the disastrous decisions to occupy Spain and invade Russia and by the adoption by his enemies of many of his lessons; finally defeated at Waterloo (1815) and exiled to the Island of St. Helena until his death. *Quoted*: 19, 21, 27, 28, 29, 31, 32, 38, 41, 42, 45, 46, 50, 51, 53, 58, 61, 63, 66, 71, 74, 81, 84, 85, 90, 92, 94, 98, 102, 103, 104, 107, 109, 110, 111, 113, 117, 120, 123, 125, 129, 136, 148, 152, 154, 157, 158, 160, 161, 162, 171, 173, 175, 176, 179, 181, 182, 185, 186, 190, 193, 194, 195, 199, 203, 204, 212, 213, 215, 233, 234, 239, 241, 243,

to his being able to direct and coordinate a development programme that led to the launching (1954) of *Nautilus*, the first nuclear-powered submarine; his singleness of purpose and devotion to nuclear power led to his domination of the design and control of naval nuclear propulsion systems and the training of their crews; his influence with Congress was such that he was kept on active duty long after retirement age. *Quoted*: 264.

Ridgway, Mathew Bunker (1895–); General, US Army; commanded the 82nd Airborne Division and later the XVII Airborne Corps in World War II, and the 8th Army and later the United Nations Command in Korea; later as Chief of Staff of the Army (1953–55) argued with Eisenhower on the need for a viable army in the face of the President's emphasis on nuclear weapons; wrote *Soldier* (1956) and *The Korean War* (1967). *Quoted*: 50, 64, 68, 91, 116, 124, 132, 182, 234, 237, 269, 320, 325, 352, 363, 383, 393, 396, 405, 409, 411, 444.

Robertson, Sir William Robert (1860–1933); Field Marshal, British Army; rose from private to field Marshal and served as Chief of the Imperial General Staff during World War I; wrote *From Private to Field Marshal* (1921). *Quoted*: 77.

Rokossovskiy, Konstantin Konstantinovich (1896–1968); Marshal of the Soviet Union; World War I veteran, joined Red Army in 1918 and fought in Russian Civil War; in World War II served ably as commander successively of seven different fronts; from 1949 to 1956 served as Polish Minister of Defence at Stalin's command, using Polish ancestry as pretext.

Rommel, Erwin Johannes (1891–1944); Field Marshal, German Army; led the spearhead of Guderian's Panzer army in the defeat of France in 1940; known as the 'Desert Fox' for his exploits in desert warfare in North Africa; initially commanded German forces opposing Normandy invasion until wounded; suicide by order of Hitler for implication in assassination plot; wrote *Infantry Attacks* (1937) and *The Rommel Papers* published posthumously (1953). *Quoted*: 21, 25, 40, 49, 100, 133, 160, 168, 177, 212, 224, 240, 250, 256, 263, 301, 307, 319, 340, 379, 386, 398, 400, 428, 430, 437, 440, 442.

Roosevelt, Theodore (1858–1919); twenty-sixth president of the United States; served as Assistant Secretary of the Navy (1897–8); organized the first volunteer cavalry regiment known as the 'Rough Riders' in the Spanish-American War and served with it in combat in Cuba (1898); as president (1901–9), he built the Panama Canal, negotiated the end to the Russo-Japanese War (1904–5) for which he won the Nobel Peace Prize, and announced the Roosevelt Corollary, making the US the defender of the Western Hemisphere (1904); domestic policy emphasized business regulation and conservation. *Quoted*: 64, 108, 206, 314, 333, 452.

Root, Elihu (1845–1937); US Secretary of War 1899–1904 most known for his reforms of the US Army that created the Army War College and established the modern, professional basis of the Army that allowed it to succeed in World War I. *Quoted*: 254, 333.

Rumantsyev, Pyotr Aleksandrovich (1725–1796); General, Russian Imperial Army; distinguished himself in the Seven Years War and the Russo–Turkish War (1768–1774). *Quoted*: 134.

Rundstedt, Gerd von (1875–1953); Field Marshal, German Army; served in various staff positions through corps in World War I; commanded the breakthrough in the West against the British and French in 1940; commanded Army Group South in invasion of Soviet Union; commanded German forces in the West during Allied invasion at Normandy; commanded German forces in counter-offensive of Battle of the Bulge (1944), an offensive of which he did not approve. *Quote*: 189.

Saladin (Arabic – Salah ad-Din) (1137 or 1138–1193); Muslim hero and Sultan; suppressed Fatimid dynasty and was proclaimed as Sultan of Egypt and Syria (1174); united Muslim territories of Syria, Northern Mesopotamia; reconquered much of Palestine from the Crusaders after winning the Battle of Hattin (1187) but

Schulenburg, Johann M. von (17–18th centuries); Saxon field marshal; successfully commanded the major assault of the allied German forces at Malplaquet (1709). *Quoted*: 283.

Schwarzkopf, H. Norman (1932–); General, US Army; as Commander, US Central Command (CENTCOM), he organized the defense of Saudi Arabia (1990) against the threat of an Iraqi invasion and then planned and conducted a masterful campaign employing and integrating coalition forces to rout and destroy the Iraqi Army in Operation 'Desert Storm' (1991). *Quoted*: 28, 56, 133, 159, 244, 276, 336.

Scipio Aemilianus Africanus Numantinus (Publius Cornelius) (184–129 BC) 'Scipio the Younger'; grandson of Scipio the Elder and the destroyer of Carthage. *Quoted*: 66, 167, 211, 246, 371.

Scipio Africanus (Publius Cornelius) (236–184 BC); called 'Scipio the Elder'; one of the Great Captains, he served at Cannae (216 BC), and commanded Roman forces in Spain to defeat Carthaginians and conquer the province; invaded Carthage forcing the return of Hannibal from Italy and defeated him at the Battle of Zama (202 BC). *Quoted*: 23, 107, 148, 279.

Scott, Winfield (1786–1866); Lieutenant General, US Army; hero of the War of 1812, he was given command of the expedition to Vera Cruz during the Mexican War; from there he marched on Mexico City in the footsteps of Cortes, winning a number of hard-fought victories against consistently superior odds, and finally captured the city which ended the war; he was deeply respected by the men who had served under him and who were to command armies in the Civil War. *Quoted*: 75, 281.

Seekt, Hans von (1866–1936); Colonel General, German Army; as chief of staff, 11th Army, he was the architect of German victories at Gorlice-Tannow (1915) and conquests of Serbia (1915) and Roumania (1916–17); chief of staff of Turkish Army (1917–18; head of Reichswehr (1920–26), successfully remodelled German Army after its World War I defeat so that it was ready for expansion under Hitler; supported cooperation with the Soviet Union; adviser to Chinese Nationalist Army (1934–35); wrote *Thoughts of a Soldier* (1930). *Quoted*: 18, 57, 67, 96, 108, 141, 151, 180, 189, 201, 309, 342, 352, 366, 371, 382, 433, 449, 470.

Sennacherib (*d.* 681 BC); King of Assyria; campaigned repeatedly in Palestine, Phoenicia, and Babylon; fought great but inconclusive Battle of Khalule (691 BC) against Elamites; restored Nineveh to great splendour; murdered by one or more of his sons. *Quoted*: 41, 135.

Septimius Severus, Lucius (AD 146–211); Roman Emperor; commander of Roman forces in Pannonia which declared him emperor; defeated rivals in civil war and later was victorious over the Parthians which allowed him to annex Mesopotamia; reorganized the Empire on a military basis. *Quoted*: 312.

Sesostris III (*d.* 1843 BC); Pharaoh of Egypt (1878–1843 BC); conducted campaigns into Nubia and Syria; built pyramid at Dashur; raised Egypt to the state of a great power. *Quoted*: 35, 336.

Sharon, Ariel (1928–); General, Israeli Defence Force; one of the most prominent and effective of Israel's generals; organized the effective counter-terrorism special operations efforts of the IDF and then reorganized the paratroopers into an élite arm; played a decisive and brilliant battlefield command role in the 1956, 1967, and 1973 wars; in 1973 conducted the counterattack across the Suez Canal that reversed the fortunes of war on that front; as Defence Minister, was responsible for Israel's invasion of Lebanon; wrote *Warrior: An Autobiography* (1989). *Quoted*: 39, 131, 225, 326, 330, 428.

Shazly, Saad (1935–) El; Lieutenant General, Egyptian Army; organized and commanded the brilliant surprise attack crossing of the Suez Canal in the 1973 War; relieved by President Sadat as a scapegoat for the failures attributed to Sadat's interference with the campaign. *Quoted*: 161, 182, 217, 234.

Sheridan, Philip H. (1831–1888); General of the Army, US Army; one of Grant's most

Takeda Shingen (AD 1521–1573); Japanese general, one of the best-known during the Warring States Period, famous as a strategist; wrote *The Iwamizudera Monogatari*. *Quoted*: 204.

Tamerlane (AD 1336–1405); Timur Lenk, i.e., Timur the Lame (Tamerlane); Turkic conqueror of Central Asia, Persia, Iraq, Armenia; sacked Dehli, Baghdad, Damascus, and Moscow, and crushed the Ottoman Sultan at the Battle of Angora (1402). *Quoted*: 434.

Taylor, Maxwell Davenport (1901–); General, US Army; Deputy Commander of the 82nd Airborne Division and later Commander of the 101st Airborne Division in World War II; commanded 8th Army in Korea in last few months of bitter fighting (1953); Chief of Staff of the Army (1955–58). *Quoted*: 82, 165, 225, 276, 298, 352.

Taylor, Zachary (1784–1850); Major General, US Army, and 12th President of the United States; known as 'Old rough and Ready', saw considerable service as an Indian fighter; achieved fame for a number of victories over Mexican armies in northern Mexico especially Palo Alto, Resaca de la Palma, (1845) and Monterrey and Buena Vista (1846); as a leader and commander he had a powerful influence on the young officers who were later to command armies in the Civil War. *Quoted*: 418, 481.

Tedder, Arthur William (1890–1967); 1st Baron Tedder; Air Chief Marshal, Royal Air Force (RAF); one of the great British air commanders; coordinated air operations in the campaign that drove the Germans out of Africa,; air commander in chief in Mediterranean Theater; and finally deputy commander of Allied Expeditionary Force in Europe under Eisenhower (1943–45); served as Chief of Air Staff (1946–50); wrote *With Prejudice*, 1966. *Quoted*: 25, 250.

Themistocles (*c.* 534 BC–*c.* 460 BC); Athenian general and statesman; as archon began development of Piraeus harbor and convinced the *demos* to spend the city's income of silver on naval power; was the driving force that held the Greek alliance together against the Persian invasion by Xerxes and commanded the Athenian fleet at the Battle of Salamis (480 BC); after the war he continued to urge expansion of the fleet and fortification of Athens but was ostracized (470 BC). *Quoted*: 134, 387, 411.

Thompson, John W. (1893–1944); Colonel, US Marine Corps; joined the Marines in 1917 and was highly decorated; after the war rose to command battalion in the Fleet Marine Force and wrote and illustrated a number of highly successful stories later compiled into *Fix Bayonets! Quoted*: 164, 193, 360, 453.

Thucydides (*d.* after 401 BC); Athenian admiral and historian; served Athens in the Peloponnesian War but was exiled for failure in the unsuccessful expedition to Amphipolis (424 BC); wrote *The History of the Peloponnesian War* (*c.* 404 BC), is considered to be the first critical historian and the foremost historian of antiquity. *Quoted*: 17, 112, 137, 210, 279, 313, 321, 326, 368, 383, 392, 414, 441, 443, 456, 459, 463, 482.

Thutmose III (*d.* 1450 BC); Pharaoh of Egypt; one of the greatest of Egyptian kings, he brought Egypt to the zenith of its power; extended hegemony into Syria with the victory at the great battle of Megiddo (1479 BC) and ranged in seventeen campaigns as far north as the Euphrates; engaged in massive building efforts in Egypt. *Quoted*: 240.

Togo Heihachiro, (1846–1934); Marquis; Admiral, Imperial Japanese Navy; commander of the Japanese Fleet at the Battle of Tsushima (1905) in which he destroyed the Russian Baltic Fleet, the victory which essentially ended the Russo–Japanese War. *Quoted*: 36, 78, 151, 168, 182, 255, 259, 333, 354, 409, 441.

Tokugawa Ieyasu (1543–1616); Shogun; unified Japan in the Battle of Sekigahara (1600) by defeating rival daiymos and appointed Shogun (1603) by the Emperor thus assuming complete civil and military control of Japan and establishing Tokugawa Shogunate. *Quoted*: 181, 233, 241, 474.

U.S. Marine Corps, and writer; decorated for valor in Vietnam, he suffered wounds that left him disabled; active in veterans' affairs and served as Secretary of the Navy under Ronald Reagan; wrote *Fields of Fire* (1978). *Quoted*: 97.

Wellesley, Arthur (1769–1852); 1st Duke of Wellington; British General and Statesman; made his reputation by victories in India (1803); took command of British forces in the Iberian Peninsula after the death of Sir John Moore (1808–1813); conducted a brilliant series of campaigns across Portugal, Spain, and Southern France almost always at great odds and emerged consistently the victor against Napoleon's best marshals; commanded Allied army at the Battle of Waterloo in 1815 in which he delivered a final crushing blow to Napoleon himself. *Quoted*: 21, 29, 46, 52, 74, 94, 96, 133, 139, 140, 151, 170, 173, 175, 178, 180, 208, 229, 231, 239, 257, 299, 317, 330, 332, 361, 367, 368, 369, 374, 412, 434, 448, 458, 468, 472.

Westmoreland, William Childs (1914–); General, US Army; commander of US forces in Vietnam (1964–68); employed attrition strategies in attempt to destroy Viet Cong main force units and regular North Vietnamese Army forces; suffered major political defeat by surprise communist Tet Offensive despite inflicting an equally major military defeat on the enemy; served as Chief of Staff of the Army (1968–72). *Quoted*: 336.

de Wet, Christiaan Rudolf (1854–1922); Boer soldier and politician; legendary commander of Boer guerrilla forces in the Orange Free State in the Boer War (1899–1902); organized Afrikaaner rebellion in 1914 but was defeated by General Louis Botha. *Quoted*: 460.

William I of Orange (1533–1584); Stadholder; Dutch soldier and statesman; known as William the Silent, he led the Dutch provinces in the 'War of Liberation' against rule of Spain; assassinated. *Quoted*: 133.

William III of Orange (1650–1702); Stadholder and later King of England; Dutch soldier and statesman; defeated French attack on Netherlands (1674); offered English crown through marriage to daughter of James II; defeated James at Battle of the Boyne in Ireland (1690); joined the grand alliances against France (1689–97 and 1701); died of fall from a horse. *Quoted*: 147.

Williams-Ellis, Clough; Major, British Army; member of the Tank Corps in World War I and author of *The Tank Corps* (1919). *Quoted*: 248, 426.

Wilson, Louis H. (1920–); General, United States Marine Corps; won the Medal of Honor on Guam (1944); served in successive staff assignments until he commanded I Marine Amphibious Force (1970–1) and then Fleet Marine Force (1972–5); appointed as Commandant, Marine Corps (1975). *Quoted*: 207.

Winder, William Henry (1775–1824); Brigadier General, United States Army; commanded the US force of mostly militia that tried to oppose the British advance on Washington in the battle of Bladensburg (1814); the militia's lack of steadiness and interference with his command on the battlefield by the Secretary of War, Secretary of State and President Madison were the chief cause of the defeat. *Quoted*: 130.

Wingate, Orde (1903–1944); Major General, British Army; classic British military eccentric; a Zionist, deeply religious, and an inspiring leader; founded and commanded Chindit long-range penetration units in Burma; he was a visionary and innovative soldier. *Quoted*: 329, 417.

Wolfe, Sir James (1727–59); Major General, British Army; served in Flanders and Germany against the Young Pretender (1742–7); fought at Dettingen, Falkirk, and Culloden Moor; played a brilliant role in the siege of Louisbourg (1758); commanded the expedition that seized Quebec by scaling the heights to the Plains of Abraham (1759) where he defeated General Montcalm and himself was slain; known as a military genius for the brilliance of his tactics, his boldness and his humane and inspirational leadership; when others complained to George III that he was mad, the king exclaimed, 'Mad is he? I wish he would bite my other gener-

of the late Han Dynasty; considered to be the greatest general of the Christian Era; wrote *The Way of the General. Quoted*: 40, 57, 95, 106, 125, 137, 165, 184, 215, 232, 233, 242, 298, 303, 308, 352, 377.

Zhukov, Georgi Konstantinovich (1896–1974); Marshal of the Soviet Union; rose from Tsarist cavalry sergeant to dominant, greatest Soviet commander in World War II; known for his ruthless drive, he was a highly effective and innovative lieutenant for Stalin. After Stalin's death he attempted to make the army more independent of the Communist Party but was sacked by Khrushchev (1957). *Quoted*: 84, 206, 247, 251, 252, 282, 288, 320, 366, 373, 392, 443.